Biographical Information.

Myron Evans is currently the only scientist on the B
chemical physicist to be appointed in British history, the other two being John Dalton and Michael Faraday. He is currently the Director of the Alpha Foundation's Institute for Advanced Studies (A.I.A.S.) and the author or editor of some seven hundred scientific papers and forty monographs. These are to be collected by the Library of Congress and a list of these publications has been collected by the Nils Bohr Library of the American Institute of Physics. He was educated at the then University College of Wales Aberystwyth and is sometime Junior Research Fellow of Wolfson College Oxford. Numerous honours include the Harrison Memorial Prize and Meldola Medal of the Royal Society of Chemistry. He has made scientific inferences and discoveries, including the explanation of the far infra red spectra of materials; the first computer simulation of the far infra red; the discovery of the gamma or boson peak of the far infra red; the development of non-equilibrium and non-linear molecular dynamics simulation methods at Aberystwyth, IBM and elsewhere; the development of fundamental methods in statistical mechanics; the application of computer simulation to non-linear optics; the inference of the fundamental Evans spin field B(3) of electromagnetism at Cornell in 1991; and the development of Einstein Cartan Evans unified field theory as described for the first time in this book.

GENERALLY COVARIANT UNIFIED FIELD THEORY

THE GEOMETRIZATION OF PHYSICS

Myron W. Evans

Published 2005 by arima publishing

www.arimapublishing.com

ISBN 1-84549-054-1

© Myron W. Evans 2005

All rights reserved

This book is copyright. Subject to statutory exception and to provisions of relevant collective licensing agreements, no part of this publication may be reproduced, stored in a retrieval system, or transmitted in any form or by any means, without the prior written permission of the author.

Printed and bound in the United Kingdom

This book is sold subject to the conditions that it shall not, by way of trade or otherwise, be lent, re-sold, hired out, or otherwise circulated without the publisher's prior consent in any form of binding or cover other than that which it is published and without a similar condition including this condition being imposed on the subsequent purchaser.

arima publishing
ASK House, Northgate Avenue
Bury St Edmunds, Suffolk IP32 6BB
t: (+44) 01284 700321

www.arimapublishing.com

I dedicate this book to
"Objective Natural Philosophy"

Mai

Harddaws teg a'm anrhegai,
Hylaw ŵr mawr hael yw'r Mai.
Anfones ym iawn fwnai,
Glas defyll glan mwyngyll Mai.
Ffloringod brig n'm digiai,
Fflŵr dy lis gyfoeth mis Mai.
Diongl rhag brad y'm cadwai,
Dan esgyll dail mentyll Mai.
Llawn wyf o ddig na thrigai
(Beth yw i mi?) Byth y Mai.
–Dafydd ap Gwilym (14th century)

* * *

Dancing traceries repay
The generous Lord of May.
Sent me warmly to portray
With hazel bush the green of May.
High florins that wan't betray
Fleur de lis of treasured May.
Groves that keep treason away
Cloak me too in leafy May.
Angered that time won't delay,
I dread the leaving of May.
–Rough translation by Myron Evans

* * *

In my craft or sullen art
Exercised in the still night
When only the moon rages
....
Who pay no praise or wages
Nor heed my craft or art.
–Dylan Thomas

* * *

There are two hungers, hunger for bread
And hunger of the uncouth soul
For the light's grace.......
....
From which to draw, drop after drop,
The terrible poetry of his kind.
–R.S. Thomas
The Dark Well

* * *

Y Glowr
Poer y Ilwch o'r pair llachar - a baban
Baw a'i boen aflafar
Ar gymal grib ac alar
Rhed y cwec ar hyd y cwar.
–Myron Evans

* * *

On Peering into the Entrance of a Drift Mine, Nixon Colliery, Mid Nineteen Fifties
I am the Lord of the Flies, this my cave,
You will be the vomit that they feed on
For three hourly pennies per killing day,
The dirty putrescence of a Friday
Shall eat your wages like a methane storm
In the black black garden of the Empire.
Don't think, boy, that you can escape me,
My blank eyes are like the seam before you,
Useless for seeing, the day wasn't here,
The flies gather round me in galleries,
Driven by the smell of death, they firefly;
Briefly they will live, suddenly they die.
Out of the way, boy, there's a tram coming,
Didn't you hear just now the Sirens sing?
–Myron Evans

Preface

The Alpha Institute for Advanced Study (AIAS) is now one of the leading theoretical physics institutes in the world, in the sense that it has advanced beyond the standard model in several ways (*www.aias.us*). It has succeeded in developing a generally covariant (i.e., objective and causal) unified field theory. Since this has achieved Einstein's own aim (1925 to 1955), it is timely that this book should appear in the year (2005) that marks the centenary of special relativity (1905) and the ninetieth anniversary of general relativity (1915). The World Year of Physics (2005) provides the opportunity to make this achievement widely known to the general public through the auspices of the United Nations. There can be no reasonable doubt that the AIAS work of the past two years has been accepted worldwide among physicists and other scientists and has become mainstream physics. There has been sustained high quality interest, and in this year(2004) there have been about 1.6 million page views of *www.aias.us* at the time of writing (31 Dec 2004). This is equivalent to the level of visiting enjoyed by the average complete University (all departments combined). From November 2003 to the present, the online interest has increased several fold; and, in view of the evident importance of AIAS work to recyclable energy and other advanced technologies, we have received several visits from the secretariat of the French President, the U.S. Department of State, the German Bundestag, and senior branches of the British Government, the Swiss Federal Presidential staff and the Canadian Federal Parliament. Additionally, there have been numerous visits at ministerial level from several major industrialised countries; from all the major research universities worldwide; from all the major corporations, institutes, and organizations; and from communities and individuals in about 80 countries.

There are twenty five key papers available on the new theory at the time of writing, and these have been peer-reviewed about fifty times. The Evans field theory is self-checking, it is differential geometry. This book summarizes the theory to date in one convenient volume.

I extend my deep thanks and sincere appreciation to all the 50 or so AIAS Fellows and Emeriti for their untiring help and input during years which were made very difficult for me personally. I also extend my congratulations to the AIAS staff on the impact which their work is making internationally. The web statistics feedback software currently available leaves no reasonable doubt as to the importance and worldwide impact of AIAS work. Surely we have earned a high profile in the World Year of Physics, since we have reached the goal in pursuit of which Einstein worked unceasingly for thirty years: the unification of general relativity with quantum mechanics and the unification of all physics in terms of objective and causal general relativity. The latter is a brilliant gemstone of human thought. There can be no reasonable doubt that AIAS is one of the prime inheritors and one of the prime proponents of Einstein's awesome theory of general relativity.

The Ted Annis Foundation, Craddock Inc., Applied Science Associates and several generous individuals are warmly thanked for funding this work. The scholar Paul Pinter has recently developed a theory of evolution and the causal origin of life based on the Evans field theory. NASA is strongly interested in evaluating the Pinter hypothesis, has reviewed it positively, and has encouraged the submission of a formal proposal.

This book summarizes the development of what is evidently recognized as the first generally covariant unified field theory. The first two chapters are notes on geometry, intended to gradually introduce the concept of the antisymmetric metric tensor, and start on a simple level in three-dimensional Euclidean space. The notes introduce the reader to the idea that the electromagnetic field is spinning spacetime, and not an entity superimposed on flat spacetime. The following chapters gradually develop the unified field theory and apply it systematically to key points in contemporary physics. A unified field theory of gravitation and electromagnetism is developed and based on the concepts of symmetric and antisymmetric metrics. A generally covariant wave equation (the Evans wave equation) is developed from the tetrad postulate of differential geometry, and the equations of the grand unified field theory are recognized as those of the structure relations and Bianchi identities of differential geometry. The Dirac equation is derived from the Evans wave equation, together with all the main wave equations of physics. A series of flow charts, depicting how those equations are inter-related, are given in the First Appendix of this book. The strong and gravitational fields are developed within our unified field theory, which is the first generally covariant theory of the electromagnetic, weak, and strong uclear fields, and therefore of the electro-weak field and radioactivity.

The subsidiary proposition of the Evans wave equation, the Evans lemma, is developed rigorously from the tetrad postulate of differential geometry, and a convenient summary table is given of over one hundred basic concepts and equations. The Evans wave equation is derived from a Lagrangian formulation and least action principle, and the origin of the Planck constant is traced in general relativity. The Hamilton principle of least action in dynamics and the Fermat principle of least time in electrodynamics are put into generally covariant form and unified in the Evans principle of least curvature. The weak-field limit of the Evans wave equation is used to derive a simple electrogravitic equation unifying the Newton and Coulomb inverse square laws. The unified field theory is used to develop physical optics and to provide an explanation for the Sagnac and Aharonov-Bohm effects in general relativity. The Evans phase law is developed from general relativity and shown to reduce to the Berry phase and various geometrical and dynamical phases in physics. A correctly and generally covariant (i.e., objective) description of the electromagnetic phase requires the fundamental Evans spin field, which is therefore a fundamental observable of all optics and spectroscopy. The Evans wave equation is shown to recycle to the Lorentz boost in the appropriate limit of special relativity. The electromagnetic sector of the unified field theory is developed in terms of differential geometry, notably the torsion two-form.

Having developed the theory and tested it in these ways, we introduce some new concepts: the evolution of curvature, an oscillatory universe without singularity, novel and generally covariant force and field equations, a generally covariant Heisenberg equation, and a re-interpretation of the Heisenberg uncertainty principle in terms of differential geometry. The first generally covariant theory of electrodynamics, O(3) electrodynamics, is deduced from our generally covariant unified field theory.

The important question is addressed of how gravitation and electromagnetism interact within Evans spacetime, a spacetime in which both the torsion and curvature tensors coexist in general. It is found that this interaction is governed by the Bianchi identity of differential geometry. It is therefore possible in theory to build counter gravitational devices based on the engineering of gravitation with electromagnetic devices. Conversely it is possible to build devices drawing electromagnetic energy or electromotive force from Evans spacetime. These are two out of several novel technologies which emerge from the unified field theory now known as the Evans field theory.

The first generally covariant theory of quantum electrodynamics is developed without use of the Feynman calculus and without introducing the pathological singularities of the path-integral method. The anomalous magnetic moment of the electron is calculated from the Evans field theory. Finally a generally covariant electro-weak theory is developed, and this shows that gravitation is able to influence radio-activity. Such an inference is well known experimentally, but the standard model is not able to describe it.

The camera-ready form of this book we owe to the patient and meticulous labor of Linda Caravelli, Franklin Amador, Dave Feustel and Jackie Gratrix. The superb job they have done is herewith gratefully acknowledged.

Craigcefnparc, Wales
31 Dec 2004

Myron W. Evans
Alpha Institude for Advanced Study

Contents

Part I Mathematical Prerequisite

Part II Developed Theory

Part I

Mathematical Prerequisite

1

Basic Geometrical Concepts

1.1 Basic Definitions

One of the main aims of this introduction is to show that O(3) electrodynamics is a theory of general relativity in which the electromagnetic field is defined by an antisymmetric metric tensor ("metric" for short), a tensor that can be built up from consideration of a rotating metric vector. In this first section we define the basic concepts needed for this task [1-5].

The metric is developed from the first principles of curvilinear coordinate analysis [6]. We start with basic definitions. First consider the straight line in three-dimensional space,

$$\boldsymbol{r} = x\boldsymbol{i} + y\boldsymbol{j} + z\boldsymbol{k}. \tag{1.1}$$

The unit vectors are defined as

$$\boldsymbol{i} := \frac{\partial \boldsymbol{r}}{\partial x} \bigg/ \left| \frac{\partial \boldsymbol{r}}{\partial x} \right|, \quad \boldsymbol{j} := \frac{\partial \boldsymbol{r}}{\partial y} \bigg/ \left| \frac{\partial \boldsymbol{r}}{\partial y} \right|, \quad \boldsymbol{k} := \frac{\partial \boldsymbol{r}}{\partial z} \bigg/ \left| \frac{\partial \boldsymbol{r}}{\partial z} \right|, \tag{1.2}$$

and the metric vectors as

$$\boldsymbol{q}_x := |\partial \boldsymbol{r}/\partial x|\, \boldsymbol{i} = \boldsymbol{i}, \quad \boldsymbol{q}_y := |\partial \boldsymbol{r}/\partial y|\, \boldsymbol{j} = \boldsymbol{j}, \quad \boldsymbol{q}_z := |\partial \boldsymbol{r}/\partial z|\, \boldsymbol{k} = \boldsymbol{k}. \tag{1.3}$$

The metric element is

$$q_x = |\partial \boldsymbol{r}/\partial x| = 1 = q_y = q_z, \tag{1.4}$$

and the line element is defined as

$$d\boldsymbol{r} = \frac{\partial \boldsymbol{r}}{\partial x} dx + \frac{\partial \boldsymbol{r}}{\partial y} dy + \frac{\partial \boldsymbol{r}}{\partial z} dz = \boldsymbol{q}_x dx + \boldsymbol{q}_y dy + \boldsymbol{q}_z dz. \tag{1.5}$$

The metric tensor is

$$q_{ij} = g_{ji} = \frac{\partial \boldsymbol{r}}{\partial x} \cdot \frac{\partial \boldsymbol{r}}{\partial y}, \text{ etc.} \tag{1.6}$$

If $g_{ij} = 0$ for $i \neq j$ then the coordinate system is *orthogonal.*

If we consider the functional relations that define the complex circular basis [1-5] of three-dimensional space,

$$e^{(1)} = (1/\sqrt{2})(x - iy), \quad e^{(2)} = (1/\sqrt{2})(x + iy), \quad e^{(3)} = z, \tag{1.7}$$

then

$$x = (1/\sqrt{2})\left(e^{(1)} + e^{(2)}\right), \quad y = (i/\sqrt{2})\left(e^{(1)} - e^{(2)}\right), \quad z = e^{(3)} \tag{1.8}$$

are curvilinear coordinate relations in this three-dimensional space. The curve (1.1) can therefore be written as

$$\boldsymbol{r} = (1/\sqrt{2})\left(e^{(1)} + e^{(2)}\right)\boldsymbol{i} + (i/\sqrt{2})\left(e^{(1)} - e^{(2)}\right)\boldsymbol{j} + e^{(3)}\boldsymbol{k}, \tag{1.9}$$

giving the three unit vectors in the complex circular basis as

$$\begin{aligned} \boldsymbol{e}^{(1)} &= \tfrac{\partial \boldsymbol{r}}{\partial e^{(1)}} / \left| \tfrac{\partial \boldsymbol{r}}{\partial e^{(1)}} \right| = (1/\sqrt{2})(\boldsymbol{i} - i\boldsymbol{j}), \\ \boldsymbol{e}^{(2)} &= \tfrac{\partial \boldsymbol{r}}{\partial e^{(2)}} / \left| \tfrac{\partial \boldsymbol{r}}{\partial e^{(2)}} \right| = (1/\sqrt{2})(\boldsymbol{i} + i\boldsymbol{j}), \\ \boldsymbol{e}^{(3)} &= \tfrac{\partial \boldsymbol{r}}{\partial e^{(3)}} / \left| \tfrac{\partial \boldsymbol{r}}{\partial e^{(3)}} \right| = \boldsymbol{k}. \end{aligned} \tag{1.10}$$

In this basis, the line element is

$$d\boldsymbol{r} = \frac{\partial \boldsymbol{r}}{\partial e^{(1)}} de^{(1)} + \frac{\partial \boldsymbol{r}}{\partial e^{(2)}} de^{(2)} + \frac{\partial \boldsymbol{r}}{\partial e^{(3)}} de^{(3)}, \tag{1.11}$$

and the metric vectors are

$$\boldsymbol{q}^{(1)} = \partial \boldsymbol{r}/\partial e^{(1)} = \boldsymbol{e}^{(1)}, \quad \boldsymbol{q}^{(2)} = \partial \boldsymbol{r}/\partial e^{(2)} = \boldsymbol{e}^{(2)}, \quad \boldsymbol{q}^{(3)} = \partial \boldsymbol{r}/\partial e^{(3)} = \boldsymbol{e}^{(3)}. \tag{1.12}$$

These vectors form the O(3) symmetry cyclic relations

$$\boldsymbol{q}^{(1)} \times \boldsymbol{q}^{(2)} = i\boldsymbol{q}^{(3)*}, \quad \boldsymbol{q}^{(2)} \times \boldsymbol{q}^{(3)} = i\boldsymbol{q}^{(1)*}, \quad \boldsymbol{q}^{(3)} \times \boldsymbol{q}^{(1)} = i\boldsymbol{q}^{(2)*}, \tag{1.13}$$

where * denotes complex conjugation [1-5].

Consider the Cartesian unit vector system $(\boldsymbol{i}, \boldsymbol{j}, \boldsymbol{k})$. The metric tensor is formed from

$$\boldsymbol{q}_x = \boldsymbol{i}, \quad \boldsymbol{q}_y = \boldsymbol{j}, \quad \boldsymbol{q}_z = \boldsymbol{k} \tag{1.14}$$

and is given by

$$\begin{aligned} g_{11} &= \boldsymbol{q}_x \cdot \boldsymbol{q}_x = \boldsymbol{i} \cdot \boldsymbol{i} = 1, \\ g_{22} &= \boldsymbol{q}_y \cdot \boldsymbol{q}_y = \boldsymbol{j} \cdot \boldsymbol{j} = 1, \\ g_{33} &= \boldsymbol{q}_z \cdot \boldsymbol{q}_z = \boldsymbol{k} \cdot \boldsymbol{k} = 1, \\ g_{ij} &= \delta_{ij}, \end{aligned} \tag{1.15}$$

This symmetric metric tensor with unit diagonal coefficients represents an orthogonal coordinate system in flat, Euclidean space. It is formed from the dot products of the unit vectors $\boldsymbol{i}, \boldsymbol{j}, \boldsymbol{k}$, because these unit vectors are the same as the metric vectors (Eq. (1.3)).

It is possible to form an anti-symmetric metric tensor by considering the well-known O(3) symmetry of the cross product of unit vectors

$$\boldsymbol{i} \times \boldsymbol{j} = \boldsymbol{k}, \quad \boldsymbol{j} \times \boldsymbol{k} = \boldsymbol{i}, \quad \boldsymbol{k} \times \boldsymbol{i} = \boldsymbol{j}. \tag{1.16}$$

This cyclic O(3) symmetry is also given by the unit Cartesian metric vectors defined in Eq. (1.3):

$$\boldsymbol{q}_x \times \boldsymbol{q}_y = \boldsymbol{q}_z, \quad \boldsymbol{q}_y \times \boldsymbol{q}_z = \boldsymbol{q}_x, \quad \boldsymbol{q}_z \times \boldsymbol{q}_x = \boldsymbol{q}_y, \tag{1.17}$$

and by the well-known O(3) rotation generator matrices [7]

$$J_x := \begin{pmatrix} 0 & 0 & 0 \\ 0 & 0 & -i \\ 0 & i & 0 \end{pmatrix}, \quad J_y := \begin{pmatrix} 0 & 0 & i \\ 0 & 0 & 0 \\ -i & 0 & 0 \end{pmatrix}, \quad J_z := \begin{pmatrix} 0 & -i & 0 \\ i & 0 & 0 \\ 0 & 0 & 0 \end{pmatrix}, \tag{1.18}$$

where

$$[J_i, J_j] = i\epsilon_{ijk} J_k, \quad C_{ijk} = i\epsilon_{ijk}, \tag{1.19}$$

and where ϵ_{ijk} is the Levi-Civita symbol, or fully antisymmetric rank-three unit tensor. For the O(3) group the Levi-Civita symbol also gives the three O(3) group structure constants.

The three antisymmetric metric tensors

$$\begin{gathered} q^{ij}{}_x := iJ_x = \begin{pmatrix} 0 & 0 & 0 \\ 0 & 0 & 1 \\ 0 & -1 & 0 \end{pmatrix}, \quad q^{ij}{}_y := iJ_y = \begin{pmatrix} 0 & 0 & -1 \\ 0 & 0 & 0 \\ 1 & 0 & 0 \end{pmatrix}, \\ q^{ij}{}_z := iJ_z = \begin{pmatrix} 0 & 1 & 0 \\ -1 & 0 & 0 \\ 0 & 0 & 0 \end{pmatrix}, \end{gathered} \tag{1.20}$$

are equivalent to the three metric vectors $\boldsymbol{q}_x, \boldsymbol{q}_y, \boldsymbol{q}_z$ and therefore to the three unit vectors i, j, k and form the O(3) symmetry cyclic relations

$$[q^{ij}{}_x, q^{ij}{}_y] = -q^{ij}{}_z, \quad [q^{ij}{}_y, q^{ij}{}_z] = -q^{ij}{}_x, \quad [q^{ij}{}_z, q^{ij}{}_x] = -q^{ij}{}_y, \tag{1.21}$$

giving an O(3) symmetry basis set for the representation of three dimensional space.

The elements of the antisymmetric metric tensor and its equivalent metric vector are related in contravariant-covariant tensor notation [7] by

$$q_k := -\frac{1}{2}\epsilon_{ijk} q^{ij}, \tag{1.22}$$

thus identifying the metric vector as an axial vector.

In order to build up the antisymmetric tensor corresponding to $\boldsymbol{k} = \boldsymbol{q}_z$, we have

$$|\boldsymbol{q}_z| = q^3 = -q_3 = \frac{1}{2}(\epsilon_{123}\, q^{12} + \epsilon_{213}\, q^{21}). \tag{1.23}$$

The Levi-Civita elements are

$$\epsilon_{123} = -\epsilon_{213} = 1, \tag{1.24}$$

and so we obtain

$$q^{ij}{}_z = -\begin{pmatrix} 0 & -q^{12} & 0 \\ q^{21} & 0 & 0 \\ 0 & 0 & 0 \end{pmatrix} = -\begin{pmatrix} 0 & -1 & 0 \\ 1 & 0 & 0 \\ 0 & 0 & 0 \end{pmatrix}. \tag{1.25}$$

The complete antisymmetric metric tensor corresponding to the sum $\boldsymbol{i}+\boldsymbol{j}+\boldsymbol{k}$ is obtained from

$$q_1 = -\frac{1}{2}\left(\epsilon_{231}\, q^{23} + \epsilon_{321}\, q^{32}\right), \quad q_2 = -\frac{1}{2}\left(\epsilon_{312}\, q^{31} + \epsilon_{132}\, q^{13}\right), \tag{1.26}$$

and is given by

$$q^{ij} = -\begin{pmatrix} 0 & -1 & 1 \\ 1 & 0 & -1 \\ -1 & 1 & 0 \end{pmatrix}. \tag{1.27}$$

Similarly the anti-symmetric metric tensors corresponding to the metric vectors in the complex circular basis are given by

$$q^{ij(1)} = \frac{1}{\sqrt{2}}\begin{pmatrix} 0 & 0 & i \\ 0 & 0 & 1 \\ -i & -1 & 0 \end{pmatrix}, \quad q^{ij(2)} = \frac{1}{\sqrt{2}}\begin{pmatrix} 0 & 0 & -i \\ 0 & 0 & 1 \\ i & -1 & 0 \end{pmatrix}. \tag{1.28}$$

Therefore starting from the displacement vector (1.1) in Euclidean space-we have shown in Eq. (1.3) that the three metric vectors are the three vectors and have demonstrated that there exists a symmetric metric tensor (1.15) and an anti-symmetric metric tensor (1.27). In the complex circular representation the anti-symmetric metric tensor is Eq. (1.28).

1.2 The Metric Tensor In General Relativity

The symmetric metric tensor is well known in the theory of general relativity, in particular the Einstein field equation for gravitation (7) is

$$R_{\mu\nu} - \frac{1}{2}Rq_{\mu\nu} = kT_{\mu\nu}. \tag{1.29}$$

This is an equation in symmetric tensors in curved spacetime, denoted by the subscripts μ and ν. Here $R_{\mu\nu}$ denotes the Ricci tensor, R the scalar curvature, $q_{\mu\nu}$ the symmetric metric tensor, and $T_{\mu\nu}$ the canonical energy-momentum tensor. The constant k is the Einstein gravitational constant. The gravitational field is defined in general relativity by the Einstein field tensor

$$G_{\mu\nu} := R_{\mu\nu} - \frac{1}{2} R q_{\mu\nu}, \tag{1.30}$$

and thus the gravitational field depends directly on the symmetric metric tensor. In flat spacetime the Ricci tensor and scalar curvature disappear, so there is no gravitational field and no energy-momentum. Einstein's theory of general relativity is based on Riemann geometry, in which the curvature of spacetime is defined in terms of the Riemann curvature tensor

$$R^{\kappa}{}_{\lambda\mu\nu} = \partial_{\nu} \Gamma^{\kappa}{}_{\lambda\mu} - \partial_{\mu} \Gamma^{\kappa}{}_{\lambda\nu} + \Gamma^{\rho}{}_{\lambda\mu} \Gamma^{\kappa}{}_{\rho\nu} - \Gamma^{\rho}{}_{\lambda\nu} \Gamma^{\kappa}{}_{\rho\mu} \tag{1.31}$$

through the Christoffel symbols

$$\Gamma^{\rho}{}_{\mu\alpha} = \frac{1}{2} q^{\rho\lambda} \left(\partial_{\mu} q_{\lambda\alpha} + \partial_{\alpha} q_{\mu\lambda} - \partial_{\lambda} q_{\alpha\mu} \right). \tag{1.32}$$

The latter are defined in terms of the elements of the symmetric metric tensor.

The Ricci tensor is defined by the tensor contraction

$$R_{\kappa\rho} = q^{\mu\lambda} R_{\mu\kappa\rho\lambda} \tag{1.33}$$

and the scalar curvature by

$$R = q^{\kappa\rho} R_{\kappa\rho}. \tag{1.34}$$

In these equations summation is used over repeated indices, the Einstein summation convention. From Eq. (1.33) and (1.34)- it can be seen that the Ricci tensor and the scalar curvature are defined in terms of the elements of the metric tensor, so the canonical energy momentum and the gravitational field are also defined in terms of these elements.

Thus everything in gravitation depends on the symmetric metric tensor.

Einstein's 1916 theory of general relativity developed from his 1905 theory of special relativity, so that general relativity is in principle applicable to all theories of physics, including electromagnetic field theory. This is the principle of general relativity. In this introduction we apply the principle of general relativity to the electromagnetic field and unify the latter with the gravitational field. Unification is achieved in two ways: firstly, by developing an equation of electromagnetism that is formally identical with Eq. (1.29) for gravitation; secondly, by applying Clifford algebra [8] to Eq. (1.29), following Sachs. In both unification schemes electromagnetism is developed from the principle of general relativity using Riemann geometry. The electromagnetic field in general relativity is represented by a choice of metric, and the metric is determined with reference to experimental data, in particular the existence of left and right-circular polarization [1-5]. These states of circular polarization in electromagnetic radiation are described by the complex circular basis defined by the metric vectors in the O(3) symmetry cyclical relations of Eq. (1.13) or, alternatively, by the following O(3) symmetry cyclical relations between the unit vectors of the complex circular representation:

$$\begin{aligned} \boldsymbol{e}^{(1)} \times \boldsymbol{e}^{(2)} &= i\boldsymbol{e}^{(3)*}, \\ \boldsymbol{e}^{(2)} \times \boldsymbol{e}^{(3)} &= i\boldsymbol{e}^{(1)*}, \\ \boldsymbol{e}^{(3)} \times \boldsymbol{e}^{(1)} &= i\boldsymbol{e}^{(2)*}. \end{aligned} \tag{1.35}$$

It is also observed experimentally that electromagnetic radiation is a wave phenomenon [1-5]: The electromagnetic field is a propagating, circularly-polarized wave. In general relativity, the electromagnetic wave is therefore a metric that rotates and translates. The wave property of the electromagnetic field is described in the simplest way through the wave equation for a wave propagating along the Z axis:

$$\frac{\partial^2 u}{\partial Z^2} - \frac{1}{c^2}\frac{\partial^2 u}{\partial t^2} = 0, \tag{1.36}$$

where c is the speed of light, a universal constant of general relativity. A simple solution of Eq. (1.36) is

$$u = \exp\left[i(\omega t - \boldsymbol{\kappa} \cdot \boldsymbol{z})\right], \tag{1.37}$$

and we refer to this solution as the electromagnetic phase factor. Here ω is the angular frequency of the wave and κ is its wave-number. The electromagnetic phase:

$$\phi = \omega t - \boldsymbol{\kappa} \cdot \boldsymbol{z} \tag{1.38}$$

is a scalar invariant of special and general relativity [8,9].

1.3 Rotating Metric Vector. The Magnetic Component Of Propagating Electromagnetic Radiation

Consider the metric vector

$$\boldsymbol{q}^{(1)} = \boldsymbol{e}^{(1)} = (1/\sqrt{2})(\boldsymbol{i} - i\boldsymbol{j}) \tag{1.39}$$

and multiply it by the phase factor (1.37). In so doing we form the phase-dependent antisymmetric metric tensor ("metric" for short):

$$q'^{ij(1)} = e^{i\phi} q^{ij(1)} = \frac{1}{\sqrt{2}} \begin{pmatrix} 0 & 0 & ie^{i\phi} \\ 0 & 0 & e^{i\phi} \\ -ie^{i\phi} & -e^{i\phi} & 0 \end{pmatrix}, \tag{1.40}$$

and we consider this metric to be a description of one part of the propagating electromagnetic field, in particular the magnetic part of the field. The fundamental reason for this is that the metric in Eq. (1.40) corresponds to the axial vector [1-5]

$$\boldsymbol{q}'^{(1)} = e^{i\phi}\boldsymbol{q}^{(1)} = (1/\sqrt{2})(\boldsymbol{i} - i\boldsymbol{j})e^{i\phi}, \tag{1.41}$$

and the magnetic field is an axial vector given by

$$\boldsymbol{B}^{(1)} = B^{(0)}\boldsymbol{q}'^{(1)}. \tag{1.42}$$

Therefore, the magnetic component of the propagating electromagnetic field has been built up from the experimental observations that electromagnetic radiation is a wave phenomenon and that electromagnetic radiation is left- and right-circularly polarized. The principle of general relativity has been applied in Eq. (1.41) and the magnetic field made directly proportional to the metric in Eq. (1.42). There are three magnetic field components in the electromagnetic field. In the complex circular basis these are

$$\begin{aligned} \boldsymbol{B}^{(1)} &= B^{(0)}\boldsymbol{q}'^{(1)} = (1/\sqrt{2})B^{(0)}(\boldsymbol{i} - i\boldsymbol{j})e^{i\phi}, \\ \boldsymbol{B}^{(2)} &= B^{(0)}\boldsymbol{q}'^{(2)} = (1/\sqrt{2})B^{(0)}(\boldsymbol{i} + i\boldsymbol{j})e^{-i\phi}, \\ \boldsymbol{B}^{(3)} &= B^{(0)}\boldsymbol{q}^{(3)} = B^{(0)}\boldsymbol{k}, \end{aligned} \tag{1.43}$$

and the three components are interrelated by the O(3) cyclic relation:

$$\begin{aligned} \boldsymbol{B}^{(1)} \times \boldsymbol{B}^{(2)} &= iB^{(0)}\boldsymbol{B}^{(3)*}, \\ \boldsymbol{B}^{(2)} \times \boldsymbol{B}^{(3)} &= iB^{(0)}\boldsymbol{B}^{(1)*}, \\ \boldsymbol{B}^{(3)} \times \boldsymbol{B}^{(1)} &= iB^{(0)}\boldsymbol{B}^{(2)*}. \end{aligned} \tag{1.44}$$

This is the B cyclic theorem of O(3) electrodynamics [1-5]. There are two circularly-polarized and phase dependent complex conjugate components transverse to the direction of propagation:

$$\boldsymbol{B}^{(1)} = B^{(0)}\boldsymbol{q}^{(1)}e^{i\phi}, \quad \boldsymbol{B}^{(2)} = B^{(0)}\boldsymbol{q}^{(2)}e^{-i\phi}, \tag{1.45}$$

and one phase-independent component in the direction of propagation, the $\boldsymbol{B}^{(3)}$ field [1-5] of electromagnetic radiation

$$\boldsymbol{B}^{(3)} = B^{(0)}\boldsymbol{q}^{(3)} \tag{1.46}$$

In general relativity, the existence of these three components follows directly from the existence of the three metric vectors $\boldsymbol{q}^{(1)}, \boldsymbol{q}^{(2)}, \boldsymbol{q}^{(3)}$ in three-dimensional space. O(3) electrodynamics is therefore simply a consequence of the fact that space has an O(3) symmetry.

Consider the physical meaning of the rotating metric in Eq. (1.40) or Eq. (1.41). We can write Eq. (1.41) as

$$\boldsymbol{q}^{(1)'} = S(Z)\boldsymbol{q}^{(1)}. \tag{1.47}$$

The metric vector $q^{(1)'}$ rotates anticlockwise with respect to a fixed frame (Y, X). This rotation is a rotation with time for a given point Z. It could equally be viewed as a clockwise rotation of the frame with respect to the fixed vector $\boldsymbol{q}^{(1)}$. The phase factor

$$S(Z) := \exp[i(\omega t - \kappa Z)] \tag{1.48}$$

generates the rotation, and if we proceed from a point Z to another point Z' the frame (X, Y) has rotated. Therefore S is the rotation generator of the metric vector $q^{(1)}$. In general Yang-Mills gauge field theory [7], the rotation generator analogously rotates the general n-dimensional field ψ:

$$\psi' = S\psi. \tag{1.49}$$

Differentiation of Eq. (1.47) with respect to Z gives

$$\frac{\partial \boldsymbol{q}^{(1)'}}{\partial Z} = S\frac{\partial \boldsymbol{q}^{(1)}}{\partial Z} + \frac{\partial S}{\partial Z}\boldsymbol{q}^{(1)}, \tag{1.50}$$

where

$$\frac{\partial S}{\partial X} = \frac{\partial}{\partial Z}e^{i(\omega t - \kappa Z)} = -i\kappa S. \tag{1.51}$$

It can be seen from a comparison Eq. (1.47) and (1.50) that, under a rotation the derivative $\partial \boldsymbol{q}^{(1)'}/\partial Z$ does not transform in the same way as the vector $\boldsymbol{q}^{(1)'}$ itself. In the language of general gauge field theory, the ordinary derivative $\partial \boldsymbol{q}^{(1)'}/\partial Z$ does not transform covariantly with $\boldsymbol{q}^{(1)'}$. In order for $\partial \boldsymbol{q}^{(1)'}/\partial Z$ to transform covariantly with $\boldsymbol{q}^{(1)'}$, we need to introduce the covariant derivative. In general relativity, the covariant derivative is denoted by the symbol D_ν and is defined by

$$D_\nu A^\mu := \partial_\nu A^\mu + \Gamma^\mu{}_{\lambda\nu} A^\lambda, \tag{1.52}$$

i.e., defined with the Christoffel symbol (1.32).

We can identify the covariant derivative from Eq. (1.50) by writing the latter as

$$\begin{array}{cc} (\partial/\partial Z + i\kappa)\,\boldsymbol{q}^{(1)'} = & S\partial \boldsymbol{q}^{(1)}/\partial Z. \\ \text{(rotating frame)} & \text{(static frame)} \end{array} \tag{1.53}$$

On the left-hand size, the covariant derivative

$$\frac{D\boldsymbol{q}^{(1)'}}{DZ} := (\partial/\partial Z + i\kappa)\,\boldsymbol{q}^{(1)'} \tag{1.54}$$

operates on the vector $\boldsymbol{q}^{(1)'}$ in a rotating frame. On the right-hand side the ordinary derivative $\partial/\partial Z$ operates on the vector $\boldsymbol{q}^{(1)}$ in the fixed frame. The general form of Eq. (1.52) in general gauge field theory is [7]

$$D'_\mu \psi' = SD_\mu \psi, \tag{1.55}$$

where it can be seen, from a comparison of Eq. (1.49) and (1.55), that the covariant derivative of the n-dimensional vector field ψ, denoted $D_\mu \psi$, transforms under rotation in the same way as the field ψ itself. The concept of covariant derivative is central to gauge field theory and to general relativity; and in order to develop electrodynamics as a theory of general relativity in Riemann geometry we need the covariant derivative.

Equation (1.53) can be generalized to spacetime and becomes

$$D_\nu q^\mu := (\partial_\nu - i\kappa_\nu)q^\mu, \tag{1.56}$$

where

$$\kappa_\nu = (\omega/c, -\boldsymbol{k}) \tag{1.57}$$

is the wave four-vector. In order to make a comparison of the covariant derivative (1.56) of electrodynamics with the covariant derivative (1.52) of Riemann geometry, we compare the equation

$$D_\nu q^\mu = (\partial_\nu - i\kappa_\nu)q^\mu, \tag{1.58}$$

and

$$D_\nu q^\mu = \partial_\nu q^\mu + \Gamma^\mu{}_{\lambda\nu}\, q^\lambda. \tag{1.59}$$

The metric four-vector q^μ appearing in Eq. (1.56) is defined by

$$q^\mu := \left(0, \boldsymbol{q}^{(1)'}\right) \tag{1.60}$$

and corresponds to the metric tensor in four dimensions,

$$q'^{\mu\nu(1)} = \frac{1}{\sqrt{2}} \begin{pmatrix} 0 & 0 & 0 & 0 \\ 0 & 0 & 0 & i \\ 0 & 0 & 0 & 1 \\ 0 & -i & -1 & 0 \end{pmatrix} e^{i\phi}. \tag{1.61}$$

Comparison of Eqs. (1.56) and (1.59) leads to

$$-i\kappa_\nu q^\mu = \Gamma^\mu{}_{\lambda\nu}\, q^\lambda; \tag{1.62}$$

and, with $\nu = 3, \mu = 1$, this leads to

$$-i\kappa_3 q^1 = \Gamma^1{}_{\lambda 3}\, q^\lambda = \Gamma^1{}_{13}\, q^1 + \Gamma^1{}_{23}\, q^2. \tag{1.63}$$

From this equation, the Christoffel symbol elements can be identified as

$$\Gamma^1{}_{23} = -\kappa, \quad \Gamma^1{}_{13} = 0. \tag{1.64}$$

The covariant derivative in Eq. (1.56) is the one relevant to anticlockwise motion, and the metric 4-vector (1.60) also corresponds to anticlockwise motion.

In the four dimensions of spacetime, it can be seen that the magnetic fields of O(3) electrodynamics are defined by the antisymmetric, phase-dependent, 4×4 metrics:

$$
\begin{aligned}
B^{\mu\nu(1)} &= B^{(0)} q'^{\mu\nu(1)} \\
&= \frac{B^{(0)}}{\sqrt{2}} e^{i\phi} \begin{pmatrix} 0 & 0 & 0 & 0 \\ 0 & 0 & 0 & i \\ 0 & 0 & 0 & 1 \\ 0 & -i & -1 & 0 \end{pmatrix} = - \begin{pmatrix} 0 & 0 & 0 & 0 \\ 0 & 0 & -B^{(1)}{}_Z & B^{(1)}{}_Y \\ 0 & B^{(1)}{}_Z & 0 & -B^{(1)}{}_X \\ 0 & -B^{(1)}{}_Y & B^{(1)}{}_X & 0 \end{pmatrix}, \\
B^{\mu\nu(2)} &= B^{(0)} q'^{\mu\nu(2)} \\
&= \frac{B^{(0)}}{\sqrt{2}} e^{-i\phi} \begin{pmatrix} 0 & 0 & 0 & 0 \\ 0 & 0 & 0 & -i \\ 0 & 0 & 0 & 1 \\ 0 & i & -1 & 0 \end{pmatrix} = - \begin{pmatrix} 0 & 0 & 0 & 0 \\ 0 & 0 & -B^{(2)}{}_Z & B^{(2)}{}_Y \\ 0 & B^{(2)}{}_Z & 0 & -B^{(2)}{}_X \\ 0 & -B^{(2)}{}_Y & B^{(2)}{}_X & 0 \end{pmatrix}, \\
B^{\mu\nu(3)} &= B^{(0)} q^{\mu\nu(3)} \\
&= B^{(0)} \begin{pmatrix} 0 & 0 & 0 & 0 \\ 0 & 0 & -1 & 0 \\ 0 & 1 & 0 & 0 \\ 0 & 0 & 0 & 0 \end{pmatrix} = - \begin{pmatrix} 0 & 0 & 0 & 0 \\ 0 & 0 & -B^{(3)}{}_Z & B^{(3)}{}_Y \\ 0 & B^{(3)}{}_Z & 0 & -B^{(3)}{}_X \\ 0 & -B^{(3)}{}_Y & B^{(3)}{}_X & 0 \end{pmatrix}.
\end{aligned} \tag{1.65}
$$

1.4 Rotating Metric Vector, The Electric Component Of Propagating Electromagnetic Radiation

The electric field component of propagating electromagnetic radiation is a rotating metric vector perpendicular to the magnetic field through the Faraday law of induction,

$$
\boldsymbol{\nabla} \times \boldsymbol{E} + \frac{\partial \boldsymbol{B}}{\partial t} = 0. \tag{1.66}
$$

The metric that describes the electric component is found as follows. We first construct the rank-two and antisymmetric space unit tensors ϵ^{ij} and ϵ_{ij} from the unit axial vector components:

$$
\begin{aligned}
\epsilon^{ij} &= -\epsilon^{ijk}\epsilon^k = -\epsilon_{ijk}\epsilon^k = \begin{pmatrix} 0 & -\epsilon^3 & \epsilon^2 \\ \epsilon^3 & 0 & -\epsilon^1 \\ -\epsilon^2 & \epsilon^1 & 0 \end{pmatrix}, \\
\epsilon_{ij} &= \epsilon_{ijk}\epsilon_k = \epsilon^{ijk}\epsilon_k = \begin{pmatrix} 0 & \epsilon_3 & -\epsilon_2 \\ -\epsilon_3 & 0 & \epsilon_1 \\ -\epsilon_2 & -\epsilon_1 & 0 \end{pmatrix},
\end{aligned} \tag{1.67}
$$

where

$$
\epsilon^1 = -\epsilon_1 = 1, \quad \epsilon^2 = -\epsilon_2 = 1, \quad \epsilon^3 = -\epsilon_3 = 1. \tag{1.68}
$$

The tensors ϵ^{ij} and ϵ_{ij} are the space components of the rank-two antisymmetric spacetime unit tensors:

$$\epsilon^{\mu\nu} = \begin{pmatrix} 0 & 0 & 0 & 0 \\ 0 & 0 & -\epsilon^3 & \epsilon^2 \\ 0 & \epsilon^2 & 0 & -\epsilon^1 \\ 0 & -\epsilon^2 & \epsilon^1 & 0 \end{pmatrix}, \tag{1.69}$$

$$\epsilon_{\mu\nu} = \begin{pmatrix} 0 & 0 & 0 & 0 \\ 0 & 0 & \epsilon_3 & -\epsilon_2 \\ 0 & -\epsilon_3 & 0 & -\epsilon_1 \\ 0 & \epsilon_2 & -\epsilon_1 & 0 \end{pmatrix}. \tag{1.70}$$

The dual pseudo-tensor $\epsilon_{\rho\sigma}$ is defined as follows:

$$\tilde{\epsilon}^{\mu\nu} = \frac{1}{2}\epsilon^{\mu\nu\rho\sigma}\epsilon_{\rho\sigma} = \begin{pmatrix} 0 & -\epsilon^1 & -\epsilon^2 & -\epsilon^3 \\ \epsilon^1 & 0 & 0 & 0 \\ \epsilon^3 & 0 & 0 & 0 \\ \epsilon^2 & 0 & 0 & 0 \end{pmatrix}. \tag{1.71}$$

The quantities $\epsilon^\mu, \epsilon^{\mu\nu}$, and $\tilde{\epsilon}^{\mu\nu}$ are all metric representations, because they are constructed from the unit axial vector components $\boldsymbol{i}, \boldsymbol{j}, \boldsymbol{k}$. The metric components can be represented as an axial four-vector q^μ, an antisymmetric tensor $q^{\mu\nu}$, or an antisymmetric pseudo-tensor dual to $q_{\rho\sigma}$:

$$q^\mu = (0, q^1, q^2, q^3), \tag{1.72}$$

$$q^{\mu\nu} = \begin{pmatrix} 0 & 0 & 0 & 0 \\ 0 & 0 & -q^3 & q^2 \\ 0 & q^2 & 0 & -q^1 \\ 0 & -q^2 & q^1 & 0 \end{pmatrix}, \tag{1.73}$$

$$\tilde{q}^{\mu\nu} = \begin{pmatrix} 0 & -q^1 & -q^2 & -q^3 \\ q^1 & 0 & 0 & 0 \\ q^2 & 0 & 0 & 0 \\ q^3 & 0 & 0 & 0 \end{pmatrix}; \tag{1.74}$$

and the magnetic field components are defined as being the metric components within a scalar magnitude $B^{(0)}$.

For a circularly polarized electromagnetic field, the magnetic field components are

$$B^{1(1)} = B^{(0)}q^{1(1)} = B^{(0)}{q^{(1)}}_x = \frac{B^{(0)}}{\sqrt{2}}e^{i\phi}, \tag{1.75}$$

$$B^{2(1)} = B^{(0)}q^{2(1)} = B^{(0)}{q^{(1)}}_y = -i\frac{B^{(0)}}{\sqrt{2}}e^{i\phi}, \tag{1.76}$$

$$B^{1(2)} = B^{(0)}q^{1(2)} = B^{(0)}{q^{(2)}}_x = \frac{B^{(0)}}{\sqrt{2}}e^{-i\phi}, \tag{1.77}$$

$$B^{2(2)} = B^{(0)}q^{2(2)} = B^{(0)}{q^{(2)}}_y = i\frac{B^{(0)}}{\sqrt{2}}e^{-i\phi}, \tag{1.78}$$

$$B^{3(3)} = B^{(0)}q^{3(3)} = B^{(0)}{q^{(3)}}_z = B^{(0)}; \tag{1.79}$$

$$\boldsymbol{B}^{(1)} = (B^{(0)}/\sqrt{2})(\boldsymbol{i} - i\boldsymbol{j})e^{i\phi}, \tag{1.80}$$

$$\boldsymbol{B}^{(2)} = (B^{(0)}/\sqrt{2})(\boldsymbol{i} + i\boldsymbol{j})e^{-i\phi}, \tag{1.81}$$

$$\boldsymbol{B}^{(3)} = B^{(0)}\boldsymbol{k}. \tag{1.82}$$

The Faraday law of induction is one of the laws of electrodynamics that can be deduced from general relativity by the following metric transforms:

$$\boldsymbol{q}^{(1)} \to -i\boldsymbol{q}^{(1)}, \quad \boldsymbol{q}^{(2)} \to i\boldsymbol{q}^{(2)}, \quad \boldsymbol{q}^{(3)} \to -i\boldsymbol{q}^{(3)}. \tag{1.83}$$

These transform the magnetic field components $\boldsymbol{B}^{(1)}, \boldsymbol{B}^{(2)}, \boldsymbol{B}^{(3)}$ into the electric field components

$$\begin{aligned} \boldsymbol{E}^{(1)} &= -(E^{(0)}/\sqrt{2})(i\boldsymbol{i} + \boldsymbol{j})e^{i\phi}, &(1.84)\\ \boldsymbol{E}^{(2)} &= (E^{(0)}/\sqrt{2})(i\boldsymbol{i} - \boldsymbol{j})e^{-i\phi}, &(1.85)\\ \boldsymbol{E}^{(3)} &= -iE^{(0)}\boldsymbol{k}, &(1.86) \end{aligned}$$

where

$$E^{(0)} = cB^{(0)} \tag{1.87}$$

in S.I. units [1-5,7]. These electric field components form the O(3) symmetry cyclic relations:

$$\begin{aligned} \boldsymbol{E}^{(1)} \times \boldsymbol{E}^{(2)} &= E^{(0)}\boldsymbol{E}^{(3)*},\\ \boldsymbol{E}^{(2)} \times \boldsymbol{E}^{(3)} &= E^{(0)}\boldsymbol{E}^{(1)*},\\ \boldsymbol{E}^{(3)} \times \boldsymbol{E}^{(1)} &= E^{(0)}\boldsymbol{E}^{(2)*}, \end{aligned} \tag{1.88}$$

and so the electric field components can also be defined in terms of rotation generators of the O(3) group.

The three electric field components of propagating electromagnetic radiation are therefore defined in general relativity as being the components generated from the magnetic field components $\boldsymbol{B}^{(1)}, \boldsymbol{B}^{(2)}, \boldsymbol{B}^{(3)}$ by the metric transformation (1.83). The metric components produced by this transformation also obey an O(3) symmetry cyclic relation:

$$\begin{aligned} (-i\boldsymbol{q}^{(1)}) \times (i\boldsymbol{q}^{(2)}) &= i\boldsymbol{q}^{(3)^*},\\ (i\boldsymbol{q}^{(2)}) \times (-i\boldsymbol{q}^{(3)}) &= i\boldsymbol{q}^{(1)^*},\\ (-i\boldsymbol{q}^{(3)}) \times (i\boldsymbol{q}^{(1)}) &= i\boldsymbol{q}^{(2)^*}; \end{aligned} \tag{1.89}$$

and so the metric transformation (1.83) can be identified as that which produces the O(3) rotation generator J_{ij} from the antisymmetric matrices ϵ_{ij}, i.e., as the transforms

$$
\begin{aligned}
J_x &:= J_{23} = -i\epsilon_{23} = -i \begin{pmatrix} 0 & 0 & 0 \\ 0 & 0 & 1 \\ 0 & -1 & 0 \end{pmatrix}, \\
J_y &:= J_{13} = -i\epsilon_{13} = -i \begin{pmatrix} 0 & 0 & -1 \\ 0 & 0 & 0 \\ 1 & 0 & 0 \end{pmatrix}, \\
J_z &:= J_{12} = -i\epsilon_{12} = -i \begin{pmatrix} 0 & 1 & 0 \\ -1 & 0 & 0 \\ 0 & 0 & 0 \end{pmatrix}.
\end{aligned} \tag{1.90}
$$

The electric and magnetic field components generated by this transformation obey the Faraday induction laws:

$$\nabla \times \boldsymbol{E}^{(1)} + \frac{\partial \boldsymbol{B}^{(1)}}{\partial t} = 0, \tag{1.91}$$

$$\nabla \times \boldsymbol{E}^{(2)} + \frac{\partial \boldsymbol{B}^{(2)}}{\partial t} = 0, \tag{1.92}$$

$$\nabla \times \boldsymbol{E}^{(3)} + \frac{\partial \boldsymbol{B}^{(3)}}{\partial t} = 0. \tag{1.93}$$

The electric field components can be expressed as follows:

$$\tilde{E}^{\mu} = (0, E^1, E^2, E^3), \tag{1.94}$$

$$\tilde{E}^{\mu\nu} = \begin{pmatrix} 0 & 0 & 0 & 0 \\ 0 & 0 & E^3 & -E^2 \\ 0 & -E^3 & 0 & E^1 \\ 0 & E^2 & -E^1 & 0 \end{pmatrix}, \tag{1.95}$$

$$E^{\mu\nu} = \begin{pmatrix} 0 & -E^1 & -E^2 & -E^3 \\ E^1 & 0 & 0 & 0 \\ E^2 & 0 & 0 & 0 \\ E^3 & 0 & 0 & 0 \end{pmatrix}, \tag{1.96}$$

i.e., the electric field components can be used to form an axial four-vector (rank-one tensor) E^{μ}, a rank-two anti-symmetric tensor $E^{\mu\nu}$, and its dual rank-two pseudo-tensor $\tilde{E}^{\mu\nu}$. These quantities are produced in general relativity through the metrics (1.72)–(1.74) multiplied by the factor $-i$:

$$\tilde{E}^{\mu} = -iE^{(0)}q^{\mu}, \tag{1.97}$$

$$\tilde{E}^{\mu\nu} = -iE^{(0)}q^{\mu\nu}, \tag{1.98}$$

$$E^{\mu\nu} = -iE^{(0)}\tilde{q}^{\mu\nu}. \tag{1.99}$$

We know that general relativity is a theory of spacetime in which the derivatives are covariant derivatives. In general relativity therefore, the Faraday law of induction is part of the Jacobi identities

$$D_\mu \tilde{G}^{\mu\nu(i)} = 0, \quad i = 1, 2, 3, \tag{1.100}$$

$$DG^{\mu\nu(i)} + D^\mu G^{\nu\lambda(i)} + D^\nu G^{\lambda\mu(i)} = 0, \quad i = 1, 2, 3, \tag{1.101}$$

where the electromagnetic field tensor is the antisymmetric tensor

$$\begin{aligned} G^{\mu\nu} &= cB^{(0)}(q^{\mu\nu} - i\tilde{q}^{\mu\nu}) \\ &= \begin{pmatrix} 0 & -E^1 & -E^2 & -E^3 \\ E^1 & 0 & -cB^3 & cB^2 \\ E^2 & cB^3 & 0 & -cB^1 \\ E^3 & -cB^2 & cB^1 & 0 \end{pmatrix}, \end{aligned} \tag{1.102}$$

whose dual pseudo-tensor is

$$\begin{aligned} \tilde{G}^{\mu\nu} &= cB^{(0)}(\tilde{q}^{\mu\nu} - iq^{\mu\nu}) \\ &= \begin{pmatrix} 0 & -cB^1 & -cB^2 & -cB^3 \\ cB^1 & 0 & E^3 & -E^2 \\ cB^2 & -E^3 & 0 & E^1 \\ cB^3 & E^2 & -E^1 & 0 \end{pmatrix}. \end{aligned} \tag{1.103}$$

The covariant derivative is defined as

$$D_\mu := \partial_\mu - igA_\mu, \tag{1.104}$$

where A_μ is the electromagnetic potential four tensor

$$A_\mu = (A_0, -A_1, -A_2, -A_3) \tag{1.105}$$

and where

$$g = \kappa / A_0. \tag{1.106}$$

The rank-two electromagnetic four-tensor in general relativity is therefore defined as a sum of metric tensor and metric pseudo tensor:

$$G^{\mu\nu} = cB^{(0)}(q^{\mu\nu} - i\tilde{q}^{\mu\nu}) = E^{(0)}(q^{\mu\nu} - i\tilde{q}^{\mu\nu}). \tag{1.107}$$

Similarly, the dual rank-two electromagnetic pseudo four-tensor is

$$\tilde{G}^{\mu\nu} = cB^{(0)}(\tilde{q}^{\mu\nu} - iq^{\mu\nu}) = E^{(0)}(\tilde{q}^{\mu\nu} - iq^{\mu\nu}). \tag{1.108}$$

For a circularly polarized wave, the magnetic and electric field components of propagating electromagnetic radiation

$$\boldsymbol{E}^{(1)} = -ic\boldsymbol{B}^{(1)}, \quad \boldsymbol{E}^{(2)} = ic\boldsymbol{B}^{(3)}, \quad \boldsymbol{E}^{(3)} = -ic\boldsymbol{B}^{(3)}, \tag{1.109}$$

and the group Jacobi identity (1.100) gives the Faraday law of induction (1.66) if

$$\partial_\mu \tilde{G}^{\mu\nu(i)} = 0, \quad i = 1, 2, 3, \tag{1.110a}$$

$$(A_\mu \tilde{G}^{\mu\nu})^{(i)} = 0, \quad i = 1, 2, 3. \tag{1.110b}$$

In vector notation, these equations are relations between the magnetic and electric components of a propagating, circularly polarized, electromagnetic field. Equation (1.110a) gives Gauss's law and Faraday's induction law of O(3) electrodynamics [1-5]:

$$\partial_\mu \widetilde{G}^{\mu\nu(i)} = 0 \Rightarrow \boldsymbol{\nabla} \cdot \boldsymbol{B}^{(i)} = 0, \quad \boldsymbol{\nabla} \times \boldsymbol{E}^{(i)} + \frac{\partial \boldsymbol{B}^{(i)}}{\partial t} = 0, \quad i = 1, 2, 3. \tag{1.111}$$

Equation (1.110b) gives

$$(A_\mu \widetilde{G}^{\mu\nu})^{(i)} = 0 \Rightarrow$$

$$\boldsymbol{A}^{(1)} \cdot \boldsymbol{B}^{(2)} = \boldsymbol{A}^{(2)} \cdot \boldsymbol{B}^{(1)} = 0, \quad \boldsymbol{A}^{(1)} \times \boldsymbol{E}^{(2)} = \boldsymbol{A}^{(2)} \times \boldsymbol{E}^{(1)} = 0, \text{ etc.,} \tag{1.112}$$

that is,

$$\boldsymbol{A}^{(i)} \cdot \boldsymbol{B}^{(i)} = 0, \quad \boldsymbol{A}^{(i)} \times \boldsymbol{E}^{(i)} = 0, \quad i \neq j.$$

If we apply the metric transform (1.83) to the Gauss and Faraday laws, we obtain the Coulomb and Ampère-Maxwell laws without a source charge-current density:

$$\begin{aligned} &\partial_\mu G^{\mu\nu(i)} = 0 \\ &\Rightarrow \boldsymbol{\nabla} \cdot \boldsymbol{E}^{(i)} = 0, \quad \boldsymbol{\nabla} \times \boldsymbol{B}^{(i)} - \tfrac{1}{c^2} \tfrac{\partial \boldsymbol{E}^{(i)}}{\partial t} = 0, \quad i = 1, 2, 3. \end{aligned} \tag{1.113}$$

The metric transform (1.83) is therefore identified as the well-known dual transform [1-5,7]

$$\boldsymbol{E} \rightarrow -ic\boldsymbol{B}, \quad \boldsymbol{B} \rightarrow (i/c)\boldsymbol{E} \tag{1.114}$$

between electric and magnetic field components. The dual transform (1.114) produces the electromagnetic dual pseudo-tensor from the original tensor:

$$G^{\mu\nu} \rightarrow \tilde{G}^{\mu\nu} \quad \text{from} \quad \boldsymbol{E} \rightarrow -ic\boldsymbol{B}, \tag{1.115}$$

and also results in the following:

$$B^{\mu\nu} = \begin{pmatrix} 0 & 0 & 0 & 0 \\ 0 & 0 & -B^3 & B^2 \\ 0 & B^3 & 0 & -B^1 \\ 0 & -B^2 & B^1 & 0 \end{pmatrix} = B^{(0)} q^{\mu\nu}$$

$$\Rightarrow E^{\mu\nu} = \begin{pmatrix} 0 & -E^1 & -E^2 & -E^3 \\ E^1 & 0 & 0 & 0 \\ E^2 & 0 & 0 & 0 \\ E^3 & 0 & 0 & 0 \end{pmatrix} = -iE^{(0)} \tilde{q}^{\mu\nu}. \tag{1.116}$$

So we can see that the electric tensor $E^{\mu\nu}$ is produced from the magnetic tensor $B^{\mu\nu}$, and vice versa. In general relativity this is the result of the metric transform (1.83), i.e., a result of *geometry*.

The Eqs. (1.100) define the O(3) covariant derivatives of general relativity

$$\begin{aligned} D_\mu \tilde{G}^{\mu\nu(1)} &= \partial_\mu \tilde{G}^{\mu\nu(1)} - ig\epsilon_{(1)(2)(3)} A_\mu^{\ (2)} \tilde{G}^{\mu\nu(3)} = 0, \\ D_\mu \tilde{G}^{\mu\nu(2)} &= \partial_\mu \tilde{G}^{\mu\nu(2)} - ig\epsilon_{(2)(3)(1)} A_\mu^{\ (3)} \tilde{G}^{\mu\nu(1)} = 0, \\ D_\mu \tilde{G}^{\mu\nu(3)} &= \partial_\mu \tilde{G}^{\mu\nu(3)} - ig\epsilon_{(3)(1)(2)} A_\mu^{\ (1)} \tilde{G}^{\mu\nu(2)} = 0. \end{aligned} \tag{1.117}$$

1.5 Summary

The metric vectors defined by the displacement vector in Eq. (1.1) have been shown in Eq. (1.3) to be equal to the Cartesian unit vectors $\boldsymbol{i}, \boldsymbol{j}, \boldsymbol{k}$. In the complex circular basis appropriate for circularly polarized radiation, the metric vectors obey the O(3) symmetry cyclic relations in Eq. (1.13). The metric vectors are written as antisymmetric rank-two tensors using Eq. (1.22), and these tensors are written in the complex circular basis in Eq. (1.28). In Sec. 1.2, the role of the metric tensor is discussed in the context of general relativity and Riemann geometry. The generally covariant metric for the electromagnetic field in general relativity is given in Eq. (1.40), which is based on the empirical observations that electromagnetic radiation is circularly polarized and is a propagating wave phenomenon. Therefore, the metric is built up from observation without any recourse to field equations. The metric is used to define the three magnetic fields $\boldsymbol{B}^{(1)}, \boldsymbol{B}^{(2)}, \boldsymbol{B}^{(3)}$ of O(3) electrodynamics in Eq. (1.43). The covariant derivative relevant to the rotating metric in Eq. (1.40) is defined in Eq. (1.56), and the Christoffel symbol deduced from the covariant derivative in Eq. (1.64). In Sec. (1.4), the electric field components are defined through the dual metric transforming Eqs. (1.83) and (1.90). The dual metric transform generates the three Faraday induction laws of O(3) electrodynamics in Eqs. (1.91)–(1.93). The generally covariant O(3) electromagnetic field equations are defined as the Jacobi identity of the complete antisymmetric field tensor in Eq. (1.100). The generally covariant derivative used in this Jacobi identity is defined in terms of the potential four-vector in Eq. (1.104), and the Jacobi identity is developed as three sets of Maxwell-Heaviside type equations, one for each of the three complex circular indices (1),(2),(3), in Eqs. (1.111) and (1.113). The dual metric transform Eq. (1.83), leads to the Heaviside-Larmor-Rainich dual transform defined in Eq. (1.114). The generally covariant O(3) electromagnetic field equations have therefore been defined in terms of a Jacobi identity of the antisymmetric, phase-dependent, metric and its dual. The metric has been built up from geometrical considerations and from the fact that the electromagnetic field is observed experimentally to be circularly polarized and a phase-dependent propagating wave phenomenon. The generally covariant O(3) field equations are deduced from the geometrical properties of the metric. The O(3) electromagnetic field has been shown to be a property of curved spacetime, as required by the principle of general relativity.

A

The Rank–Three And Rank–Four Totally Antisymmetric Unit Tensor

The rank-three totally antisymmetric unit tensor, the Levi-Civita symbol, is defined as

$$\epsilon^{ijk} = \epsilon_{ijk}. \tag{A.1}$$

The tensor is 1 or -1 according to the following index rules:

$$\epsilon_{123} = \epsilon_{231} = \epsilon_{312} = 1, \quad \epsilon_{132} = \epsilon_{213} = \epsilon_{321} = -1, \tag{A.2}$$

and is therefore antisymmetric in the permutation of any two indices. It is used in tensor analysis to define the cross product of two vectors or to generate an antisymmetric rank-two tensor from an axial vector.

The rank-four totally antisymmetric unit tensor is used to generate the dual pseudo-tensor from a rank-two four-dimensional antisymmetric tensor, and is defined as

$$\epsilon^{\mu\nu\rho\sigma} = -\epsilon_{\mu\nu\rho\sigma}, \quad \epsilon^{0123} = -\epsilon_{0123} = 1. \tag{A.3}$$

It has the following properties:

$$\begin{aligned} &\epsilon^{0123} = \epsilon^{1230} = \epsilon^{2301} = \epsilon^{3012} = 1, \\ &\epsilon^{1023} = \epsilon^{2130} = \epsilon^{3201} = \epsilon^{0312} = -1, \\ &\epsilon^{1032} = \epsilon^{2103} = \epsilon^{3210} = \epsilon^{0321} = 1, \\ &\epsilon^{1302} = \epsilon^{2013} = \epsilon^{3120} = \epsilon^{0231} = -1, \\ &\text{etc.,} \end{aligned} \tag{A.4}$$

and is therefore antisymmetric in the permutation of any two of the four indices.

2

Duality and the Antisymmetric Metric

2.1 Euclidean Spacetime. Special Relativity

In this chapter it is proven that if there exists a symmetric metric in general relativity, then there must exist an antisymmetric metric: The fundamental reason is that one metric is implied by the other through duality in differential geometry [7]. This result of differential geometry leads to the important new result in physics that if the gravitation be described through Riemannian geometry with the well-known Einstein field equation, Eq. (1.29), then there exists an equation of the same type for electrodynamics in general relativity. This equation is derived and explained in this chapter using differential geometry with a antisymmetric metric. Gravitation therefore is a manifestation of curved spacetime with a symmetric metric, and electromagnetism is a manifestation of spinning spacetime with an antisymmetric metric. From differential geometry we can infer that electromagnetism is implied by gravitation in non-Minkowski, four-dimensional spacetime.

This result can be constructed from a consideration of Eq. (1.67), a duality in Euclidean space:

$$\epsilon^{ij} = -\epsilon^{ijk}\epsilon^{k}, \tag{2.1}$$

$$\epsilon_{ij} = \epsilon^{ijk}\epsilon_{k}. \tag{2.2}$$

Consider the metric 3-vector defined in Euclidean space (Chap. 1) as

$$q^{k} = \epsilon^{k} = (1, 1, 1). \tag{2.3}$$

The symmetric metric tensor in Euclidean space can be defined by

$$q^{k} = q^{k\ell(S)}q_{\ell}, \quad q^{k\ell(S)} = -\begin{pmatrix} 1 & 0 & 0 \\ 0 & 1 & 0 \\ 0 & 0 & 1 \end{pmatrix}. \tag{2.4}$$

Define the scalar metric quantity q by

$$q = q^k q_k = q^k \epsilon_k = -3. \tag{2.5}$$

The metric 3-vector q^k is dual to the antisymmetric metric tensor in three space dimensions,

$$q^{ij(A)} = -\epsilon^{ijk} q^k = \begin{pmatrix} 0 & -1 & 1 \\ 1 & 0 & -1 \\ -1 & 1 & 0 \end{pmatrix}; \tag{2.6}$$

and, from Eqs. (2.5) and (2.6) the following duality relation between the symmetric and antisymmetric metric tensor is obtained:

$$q = -\frac{1}{2} q^{ij(A)} \epsilon_{ij} = q^{k\ell(S)} \epsilon_\ell \epsilon_k. \tag{2.7}$$

In three Euclidean-space dimensions it therefore has been shown that the existence of one metric implies that of the other through the fundamental duality relations (1.67). Because ϵ_k is dual to ϵ_{ij} by geometry, then $q^{k\ell(S)}\epsilon_\ell$ must be dual to $q^{ij(A)}$. So the existence of the well-known symmetric metric (2.4) implies the existence of the antisymmetric metric (2.6) by three-dimensional Euclidean geometry.

This simple result is generalized in the chapter using the language of forms and differential geometry, where-upon it becomes clear that the homogeneous field equation of electromagnetism is a Bianchi identity on a closed two-form involving the antisymmetric metric in a non-Minkowski spacetime. The inhomogeneous field equation of electrodynamics can be derived by taking covariant derivatives of both sides of a novel field equation of electrodynamics whose structure is the same as that of the well-known Einstein field equation (1.29) of gravitation.

The duality Eq. (2.7) implies that, if there exists a line element

$$\omega_1 = ds^2 = q^{k\ell(S)} dx_\ell dx_k, \tag{2.8}$$

a zero-form (scalar) of differential geometry, then there exists the zero-form

$$\omega_2 = ds^2 = -\frac{1}{2} q^{ij(A)} dx_i \wedge dx_j, \tag{2.9}$$

because dx_k is dual to the wedge product (7) $dx_i \wedge dx_j$. Therefore, the line element in three-dimensional differential geometry can be expressed as the zero-form ω_1 or the zero-form ω_2. The two-form $q^{ij(A)}$ is a closed two-form whose exterior derivative vanishes identically:

$$d \wedge q^{ij(A)} = 0 = (d \wedge q^{(A)})_{ij}. \tag{2.10}$$

In Euclidean spacetime, the converse of the Poincarélemma of differential geometry implies that:

$$q^{(A)}{}_{ij} = d \wedge q_j \tag{2.11}$$

i.e., that the two-form $q^{(A)}_{ij}$ is the exterior derivative of a one-form q_j. Equation (2.10) is the Bianchi identity applied in flat or Euclidean space.

These results of three-dimensional Euclidean space may be generalized to four-dimensional Minkowski spacetime, in which the symmetric metric tensor is defined by

$$q^\rho = q^{\rho\sigma(S)} q_\sigma. \tag{2.12}$$

The antisymmetric metric tensor then follows from the spacetime generalization of Eq. (2.7):

$$q = -\frac{1}{6} q^{\mu\nu(A)} \epsilon_{\mu\nu} = q^{\rho\sigma(S)} \epsilon_\sigma \epsilon_\rho, \tag{2.13}$$

an equation that shows that if $q^{\rho\sigma(S)}$ is defined from Eq. (2.12) as

$$q^{\rho\sigma(S)} = \begin{pmatrix} 1 & 0 & 0 & 0 \\ 0 & -1 & 0 & 0 \\ 0 & 0 & -1 & 0 \\ 0 & 0 & 0 & -1 \end{pmatrix}, \tag{2.14}$$

then $q^{\mu\nu(A)}$ is defined from Eq. (2.13) as

$$q^{\mu\nu(A)} = \begin{pmatrix} 0 & -1 & -1 & -1 \\ 1 & 0 & -1 & 1 \\ 1 & 1 & 0 & -1 \\ 1 & -1 & 1 & 0 \end{pmatrix}. \tag{2.15}$$

In the language of differential geometry, if there exists the spacetime Minkowski line element, the zero-form

$$\omega_1 = ds^2 = q^{\rho\sigma(S)} dx_\sigma dx_\rho, \tag{2.16}$$

then the line element ds^2 exists as the dual zero-form

$$\omega_2 = ds^2 = -\frac{1}{6} q^{\mu\nu(A)} dx_\mu \wedge dx_\nu, \tag{2.17}$$

because dx_ρ is dual to the wedge product $dx_\mu \wedge dx_\nu$. The Bianchi identity (2.10) is also a Bianchi identity of four-dimensional Minkowski spacetime. The two-form (2.17) is a closed two-form, and so is the exterior derivative of a one-form by the converse of the Poincarélemma.

The symmetric and antisymmetric metrics $q^{\rho\sigma(S)}$ and $q^{\mu\nu(A)}$ are covariants of special relativity, i.e., retain their form under a Lorentz transform. This means that the symmetric metric is diagonal symmetric before and after a Lorentz transform, and the antisymmetric metric is antisymmetric off-diagonal before and after a Lorentz transform. The symmetric and antisymmetric metrics in Euclidean spacetime are constants, so the Christoffel symbols, Riemann tensors, Ricci tensors and scalar curvatures derived from these metrics all disappear. It is well known that in Einstein's general relativistic theory of gravitation, flat spacetime means that gravitation is absent, a result that is implied by the Einstein field equation (1.29). In the received view of electrodynamics, however, the electromagnetic field is an entity superimposed on flat spacetime. In the new view of electrodynamics forged in this book and elsewhere [1-5], the electromagnetic field is a manifestation of spinning spacetime. If the gravitational field is described through the symmetric metric in Eq. (1.29), the Einstein field equation, then the electromagnetic field is described through the antisymmetric metric dual to the symmetric metric using a field equation with the same structure as Eq. (1.29).

2.2 Non-Minkowski Spacetime. General Relativity

By developing the results in Sec. (2.1) in non-Minkowski spacetime, it is possible to develop a theory of general relativity in which a symmetric metric is dual to an antisymmetric metric through Eq. (2.17). The gravitational field is defined by the Einstein field equation (1.29) through the symmetric metric, and the electromagnetic field is dual to this gravitational field through Eq. (2.17) developed for non-Minkowski spacetime. The two fields are therefore related to each other through a duality transformation of differential geometry and are two parts of the same thing. This means that one field can influence another, i.e., gravitation can influence electromagnetism and vice-versa.

In order to extend Eq. (2.9) to a non-Euclidean three-dimensional space, we consider the unit vectors and metric vectors in general curvilinear co-ordinates and extend the analysis to non-Minkowski spacetime. The metric 4-vector in this spacetime is written as an antisymmetric tensor which is used to define a two-form of differential geometry. Gravitation is then defined by an Einstein equation for the symmetric metric and electromagnetism by an Einstein equation for the antisymmetric metric. The homogeneous field equation of electromagnetism is the Bianchi identity analogous to Eq. (2.10) for Euclidean spacetime, and the inhomogeneous field equation of electromagnetism is obtained by taking covariant derivatives on both sides of the Einstein equation for electromagnetism. When the non-Minkowski metric is defined by Eq. (1.61), we recover the homogeneous and inhomogeneous field equations of O(3) electrodynamics. In general curvilinear co-ordinates both the symmetric and antisymmetric metrics are defined in terms of the same set of scale factors, and this result can be used in principle to measure the effect of one field on the other. Electromagnetism in this view is a theory of non-Minkowski spacetime, and one must use covariant derivatives in the field equations of electromagnetism as well as the field equations of gravitation. The same conclusion has been reached independently by Sachs [8] using Clifford algebra in general relativity. O(3) electrodynamics is an example where the covariant derivatives are defined in the complex circular basis [1-5] denoted ((1),(2),(3)).

Consider a region of non-Euclidean, three-dimensional space [6] such that each point is specified by three numbers (u_1, u_2, u_3), the curvilinear coordinates. The transformation equations between the Cartesian and curvilinear coordinates are:

$$\begin{aligned} x &= x(u_1, u_2, u_3), \quad u_1 = u_1(x, y, z), \\ y &= y(u_1, u_2, u_3), \quad u_2 = u_2(x, y, z), \\ z &= z(u_1, u_2, u_3), \quad u_3 = u_3(x, y, z), \end{aligned} \tag{2.18}$$

where the functions are single valued and continuously differentiable. There is therefore a one-to-one correspondence between the point (x, y, z) and (u_1, u_2, u_3). The position vector in general curvilinear coordinates is $\boldsymbol{r}(u_1, u_2, u_3)$, and the arc length is the modulus of the infinitesimal displacement vector:

$$ds = |d\boldsymbol{r}| = \left| \frac{\partial \boldsymbol{r}}{\partial u_1} du_1 + \frac{\partial \boldsymbol{r}}{\partial u_2} du_2 + \frac{\partial \boldsymbol{r}}{\partial u_3} du_3 \right| . \tag{2.19}$$

Here $\partial \boldsymbol{r}/\partial u_i$ is the metric coefficient, whose modulus is the scale factor:

$$h_i = \left| \frac{\partial \boldsymbol{r}}{\partial u_i} \right| . \tag{2.20}$$

The unit vectors of the curvilinear coordinate system are:

$$\boldsymbol{e}_i = \frac{1}{h_i} \frac{\partial \boldsymbol{r}}{\partial u_i}, \tag{2.21}$$

$$e^i = (e^1, e^2, e^3), \tag{2.22}$$

and so we can write

$$d\boldsymbol{r} = h_1 du_1 \boldsymbol{e}_1 + h_2 du_2 \boldsymbol{e}_2 + h_3 du_3 \boldsymbol{e}_3. \tag{2.23}$$

If we write the unit vector in three dimensions as

$$\boldsymbol{e}_i = \partial \boldsymbol{r}/\partial u_i / |\partial \boldsymbol{r}/\partial u_i|, \quad h_i = |\partial \boldsymbol{r}/\partial u_i|, \tag{2.24}$$

then each component is unity, as in the Cartesian coordinates (Eq. (1.2)):

$$\boldsymbol{i} = \frac{\partial \boldsymbol{r}}{\partial x} / \left| \frac{\partial \boldsymbol{r}}{\partial x} \right| , \text{ etc.,} \quad q_x = |\partial \boldsymbol{r}/\partial x|, \text{ etc.} \tag{2.25}$$

The unit vectors $\boldsymbol{e}_i$ are unit tangent vectors to the curve u_i at point P, i.e., the three unit vectors $\mathbf{e}_1, \mathbf{e}_2, \mathbf{e}_3$ of the curvilinear coordinate system are unit tangent vectors to the coordinate curves, are mutually orthogonal, and form the O(3) symmetry cyclic relations:

$$\mathbf{e}_1 \times \mathbf{e}_2 = \mathbf{e}_3{}^*, \quad \mathbf{e}_2 \times \mathbf{e}_3 = \mathbf{e}_1{}^*, \quad \mathbf{e}_3 \times \mathbf{e}_1 = \mathbf{e}_2{}^*. \tag{2.26}$$

This means that each of the unit vectors $\mathbf{e}_1, \mathbf{e}_2, \mathbf{e}_3$ is dual to an antisymmetric rank-two tensor in three-dimensional non-Euclidean space:

$$e^{ij} = -\epsilon^{ijk} e^k, \tag{2.27}$$

$$e_{ij} = \epsilon^{ijk} e_k. \tag{2.28}$$

The rank-two tensors e^{ij} also form an O(3) symmetry cyclic relation. The curvilinear coordinate system is orthogonal if

$$\boldsymbol{e}_1 \cdot \boldsymbol{e}_2 = 0, \quad \boldsymbol{e}_1 \cdot \boldsymbol{e}_3 = 0, \quad \boldsymbol{e}_2 \cdot \boldsymbol{e}_3 = 0. \tag{2.29}$$

The symmetric metric tensor $q^{ij(S)}$ is then defined by the square of the arc length:

$$ds^2 = d\boldsymbol{r} \cdot d\boldsymbol{r} = q^{ij(S)} du_i du_j. \tag{2.30}$$

This is a product of one-forms of differential geometry [7]. The one form du_k is dual to the wedge product $du_i \wedge du_j$, a two-form, and $q^{ij(S)}$ implies $q^{ij(A)}$, where $q^{ij(A)}$ is the antisymmetric metric in curvilinear coordinates. Therefore, the square of the arc length implies the following area zero-form in differential geometry:

$$dA = -\frac{1}{2} g^{ij(A)} du_i \wedge du_j. \tag{2.31}$$

This result means that the existence of a symmetric metric in curvilinear coordinates in three dimensions implies the existence of an antisymmetric metric. Generalization of this result to four dimensions in any non-Minkowski spacetime gives

$$\omega_1 = ds^2 = q^{\mu\nu(S)} du_\mu du_\nu, \tag{2.32}$$

$$\omega_2 = dA = -\frac{1}{2} q^{\mu\nu(A)} du_\mu \wedge du_\nu. \tag{2.33}$$

A graphical summary of the concepts used to derive the important result (2.33) is given below in three space dimensions:

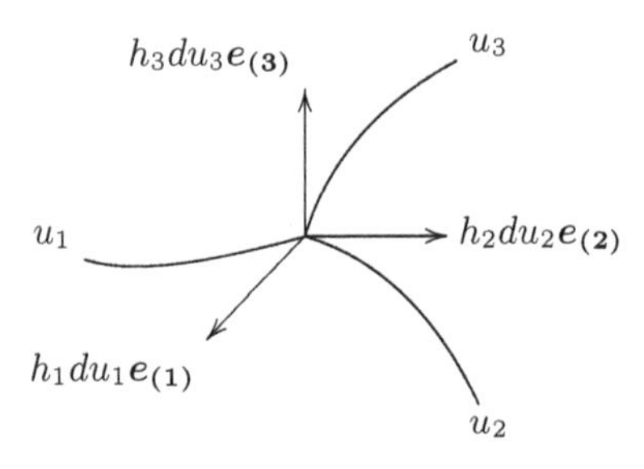

Fig. 2.1. Curvalinear Coordinates

The general vector field in curvilinear coordinates in three space dimensions [6] is

$$\boldsymbol{F} = F_1\boldsymbol{e}_1 + F_2\boldsymbol{e}_2 + F_3\boldsymbol{e}_3, \tag{2.34}$$

and so we may define the metric coefficients in terms of the infinitesimal of the displacement vector in curvilinear coordinates:

$$\begin{aligned} d\boldsymbol{r} &= \frac{\partial \boldsymbol{r}}{\partial u_1}du_1 + \frac{\partial \boldsymbol{r}}{\partial u_2}du_2 + \frac{\partial \boldsymbol{r}}{\partial u_3}du_3 \\ &= \boldsymbol{q}_1 du_1 + \boldsymbol{q}_2 du_2 + \boldsymbol{q}_3 du_3 \,. \end{aligned} \tag{2.35}$$

The scale factors are the moduli of the metric coefficients and the unit vectors are the metric coefficients divided by the scale factors:

$$h_i = |\boldsymbol{q}_i|, \quad \boldsymbol{e}_i = \boldsymbol{q}_i / h_i. \tag{2.36}$$

Therefore the elements of the symmetric metric tensor are

$$q_{ij}{}^{(S)} = \boldsymbol{q}_i \cdot \boldsymbol{q}_j = h_i h_j \boldsymbol{e}_i \cdot \boldsymbol{e}_j, \tag{2.37}$$

and the elements of the antisymmetric tensor are

$$q_{ij}{}^{(A)} = \boldsymbol{q}_i \times \boldsymbol{q}_j = h_i h_j \boldsymbol{e}_i \times \boldsymbol{e}_j. \tag{2.38}$$

These results have been obtained independently by Sachs [8], using Clifford algebra. The square of the arc length in curvilinear coordinates in three dimensions is therefore

$$ds^2 = \boldsymbol{q}_i \cdot \boldsymbol{q}_j du_u du_j \tag{2.39}$$

and can be expressed through the 3×3 symmetric metric tensor:

$$q_{ij}{}^{(S)} = q_i q_j = \begin{pmatrix} h^2{}_1 & h_1 h_2 & h_1 h_3 \\ h_2 h_1 & h^2{}_2 & h_2 h_3 \\ h_3 h_1 & h_3 h_2 & h^2{}_3 \end{pmatrix}. \tag{2.40}$$

The off-diagonal elements of this tensor vanish if and only if the curvilinear coordinates system is orthogonal. In four spacetime dimensions, the metric tensor is the tensor product of two four-dimensional metric vectors whose components are the four scale factors:

$$q^\mu = (h^0, h^1, h^2, h^3), \quad q_\nu = (h_0, h_1, h_2, h_3) \tag{2.41}$$

giving the following 4×4 symmetric metric tensor in spacetime in curvilinear coordinates:

$$q_{\mu\nu}{}^{(S)} = \begin{pmatrix} h^2{}_0 & h_0 h_1 & h_0 h_2 & h_0 h_3 \\ h_1 h_0 & h^2{}_1 & h_1 h_2 & h_1 h_3 \\ h_2 h_0 & h_2 h_1 & h^2{}_2 & h_2 h_3 \\ h_3 h_0 & h_3 h_1 & h_3 h_2 & h^2{}_3 \end{pmatrix}. \tag{2.42}$$

The gravitational field in general relativity is identified through the Einstein equation derived from this symmetric metric tensor in non-Minkowski spacetime. There is a one-to-one relation between any set of curvilinear coordinates (u_1, u_2, u_3) and any other set (u_1', u_2', u_3') so the metric $q_{\mu\nu}^{(S)}$ is generally covariant by definition, i.e., does not change its form under transformation from one set of curvilinear coordinates to another. In general relativity, this means that the equations of gravitation are generally covariant, the same in form under any transformation of coordinates.

The principle of general relativity asserts that all equations of natural philosophy must be generally covariant, including the field equations of electromagnetism. We now follow this well known principle rigorously and derive the generally covariant field equations of electromagnetism from the antisymmetric metric in four-dimensional non-Minkowski spacetime. These equations of electromagnetism derive essentially from the fact that we can form the cross products (2.38) as well as the dot products (2.37), i.e., we can always construct an antisymmetric metric from the same metric vectors are those used to construct the symmetric metric. This result is true for any non-Minkowski spacetime.

Restrict attention to three space dimensions, and note that the antisymmetric metric is built up from the following area elements of curvilinear coordinate analysis [6]:

$$\begin{aligned} dA_1 &= h_2 h_3 du_2 du_3 |\boldsymbol{e}_2 \times \boldsymbol{e}_3|, \\ dA_2 &= h_3 du_3 h_1 du_1 |\boldsymbol{e}_3 \times \boldsymbol{e}_1|, \\ dA_3 &= h_1 du_1 h_2 du_2 |\boldsymbol{e}_1 \times \boldsymbol{e}_2|. \end{aligned} \tag{2.43}$$

The axial surface vector in three dimensions is therefore

$$d\boldsymbol{A} = h_2 h_3 du_2 du_3 \boldsymbol{e}_1 + h_3 du_3 h_1 du_1 \boldsymbol{e}_2 + h_1 du_1 h_2 du_2 \boldsymbol{e}_3, \tag{2.44}$$

and is dual to the antisymmetric surface tensor in three dimensions in curvilinear coordinates:

$$\begin{aligned} dA^{ij} = h_2 h_3 du_2 du_3 &\begin{pmatrix} 0 & 0 & 0 \\ 0 & 0 & -1 \\ 0 & 1 & 0 \end{pmatrix} + h_3 h_1 du_3 du_1 \begin{pmatrix} 0 & 0 & 1 \\ 0 & 0 & 0 \\ -1 & 0 & 0 \end{pmatrix} \\ &+ h_1 h_2 du_1 du_2 \begin{pmatrix} 0 & -1 & 0 \\ 1 & 0 & 0 \\ 0 & 0 & 0 \end{pmatrix}. \end{aligned} \tag{2.45}$$

The antisymmetric metric tensor in three space dimensions is

$$q^{ij(A)} = \begin{pmatrix} 0 & -h_1 h_2 & h_1 h_3 \\ h_2 h_1 & 0 & -h_2 h_3 \\ -h_3 h_1 & h_3 h_2 & 0 \end{pmatrix}. \tag{2.46}$$

In order to generalize this result to four-dimensional spacetime in curvilinear coordinates, note that the existence of the zero-form

$$\omega_1 = ds^2 = q^{\mu\nu(S)} du_\mu du_\nu \tag{2.47}$$

implies the existence of the area zero-form

$$\omega_2 = dA = -\frac{1}{2} q^{\mu\nu(A)} du_\mu \wedge du_\nu, \tag{2.48}$$

where $q^{\mu\nu(A)}$ is the antisymmetric metric in four non-Minkowski spacetime dimensions. The structure of the antisymmetric metric is therefore the same as that of the wedge product $du_\mu \wedge du_\nu$ in four dimensions, i.e., the antisymmetric metric is an antisymmetric 4×4 matrix formed from the antisymmetric product of two metric 4-vectors

$$q_{\mu\nu}{}^{(A)} = \begin{pmatrix} 0 & -h_0h_1 & -h_0h_2 & -h_0h_3 \\ h_1h_0 & 0 & -h_1h_2 & h_1h_3 \\ h_2h_0 & h_2h_1 & 0 & -h_2h_3 \\ h_3h_0 & -h_3h_1 & h_3h_2 & 0 \end{pmatrix} \tag{2.49}$$

The generally covariant electromagnetic field tensor is the metric $q_{\mu\nu}{}^{(A)}$ within a factor that must be negative under charge conjugation. The magnetic field components are defined by

$$B^i = -\frac{1}{2} B^{(0)} \epsilon^{ijk} q^{jk(A)} \tag{2.50}$$

and the electric field components by

$$E^i = E^{(0)} q^{io(A)}. \tag{2.51}$$

Thus the complete electromagnetic field tensor is

$$G^{\mu\nu} = \begin{pmatrix} 0 & -E^1 & -E^2 & -E^3 \\ E^1 & 0 & -B^3 & B^2 \\ E^2 & B^3 & 0 & -B^1 \\ E^3 & -B^2 & B^1 & 0 \end{pmatrix}. \tag{2.52}$$

The generally covariant homogeneous field equation of decoupled electromagnetism is the Bianchi identity [7] defined by the fact that the $\widetilde{G}_{\mu\nu}$ dual two-form is a closed two form whose covariant exterior derivative vanishes identically in any non-Minkowski spacetime:

$$D_\mu \widetilde{G}^{\mu\nu} = 0. \tag{2.53}$$

The homogeneous field equation in form notation is therefore the Bianchi identity

$$D \wedge G = 0, \tag{2.54}$$

which in tensor notation becomes

$$D_\mu \widetilde{G}^{\mu\nu} = 0. \tag{2.55}$$

If we define the covariant derivatives as those of an O(3) symmetry gauge field theory, Eq. (2.55) becomes the homogeneous field equation of O(3) electrodynamics [1-5]:

$$(\partial_\mu - ig\boldsymbol{A}_\mu \times)\widetilde{\boldsymbol{G}}^{\mu\nu} = 0, \tag{2.56}$$

which is also a Jacobi identity of group generators [7]. In differential geometry, the Bianchi identity is [7]

$$D_\rho R^\kappa{}_{\lambda\mu\nu} + D_\mu R^\kappa{}_{\lambda\nu\rho} + D_\nu R^\kappa{}_{\lambda\rho\mu} = 0; \tag{2.57}$$

so the generally covariant homogeneous field equation of electrodynamics is an equation of differential geometry, with Christoffel symbols antisymmetric in their lower two indices. (In gravitation the Christoffel symbols are symmetric in their lower two indices.) The Bianchi identity (2.57) is clearly analogous to the Jacobi identity (2.55) rewritten as

$$D_\rho G_{\mu\nu} + D_\mu G_{\nu\rho} + D_\nu G_{\rho\mu} = 0; \tag{2.58}$$

and the reason for this is that the generally covariant field tensor can be obtained from a round-trip in non-Euclidean spacetime using covariant derivatives [1-5,7]. This procedure is also used to define the Riemann curvature tensor in general relativity [7]. So the Riemann tensor for electromagnetism can be related to the field tensor $G^{\mu\nu}$ through a choice of Christoffel symbol, as illustrated in Eq. (1.64) of Chap. 1.

Therefore electromagnetism exists in general relativity because we can define the four-dimensional surface tensor in non-Euclidean spacetime. Analogously, gravitation exists in general relativity because we can define the square of the line element in four-dimensional non-Euclidean spacetime. The same metric vectors (tetrads) are used to define both fields, so they have a common origin in geometry. This conclusion allows us to unify gravitation and electromagnetism using the concept of tetrad.

The infinitesimal area of surface is the geometrical origin of the electromagnetic field. In Euclidean three-space [1-5], the projection of the area of a parallelogram formed from the infinitesimal line elements $d\boldsymbol{r}$ and $d\boldsymbol{r}'$ on the coordinate planes $x_i x_j$ is

$$dA_{ij} = dx_i dx_j' - dx_j dx_i'. \tag{2.59}$$

This is dual to the axial surface vector $d\tilde{A}_i$:

$$d\tilde{A}_i = \frac{1}{2}\epsilon_{ijk} dA_{jk}, \tag{2.60}$$

which is a pseudo-vector normal to the surface element. In non-Euclidean three-space, the result (2.59) becomes

$$dA_{ij} = h_i h_j' du_i du_j' - h_j h_i' du_j du_i', \tag{2.61}$$

and we may define a generally covariant magnetic pseudo-vector

$$\begin{aligned} B^i &= (B^1, B^2, B^3) \\ &= -\tfrac{1}{2}\epsilon^{ijk}B^{jk}, \quad B^{jk} = \begin{pmatrix} 0 & -B^3 & B^2 \\ B^3 & 0 & -B^1 \\ -B^2 & B^1 & 0 \end{pmatrix} \end{aligned} \tag{2.62}$$

and a generally covariant electric pseudo-vector

$$\widetilde{E}^i = \frac{1}{2}\epsilon^{ijk}\widetilde{E}^{jk}, \quad \widetilde{E}^{jk} = \begin{pmatrix} 0 & E^3 & -E^2 \\ -E^3 & 0 & E^1 \\ E^2 & -E^1 & 0 \end{pmatrix}. \tag{2.63}$$

In four dimensions in Minkowski spacetime, the infinitesimal area of surface is defined by

$$dA^{\mu\nu} = dx^\mu dx^{\nu\prime} - dx^\nu dx^{\mu\prime}, \tag{2.64}$$

and the dual element of surface

$$d\widetilde{A}^{\mu\nu} = \frac{1}{2}\epsilon^{\mu\nu\sigma\rho}dA_{\sigma\rho} \tag{2.65}$$

is normal to $dA^{\mu\nu}$:

$$d\widetilde{A}^{\mu\nu}dA_{\mu\nu} = 0. \tag{2.66}$$

In non-Minkowski spacetime, Eq. (2.64) becomes

$$dA^{\mu\nu} = h^\mu h^{\nu\prime} du^\mu du^{\nu\prime} - h^\nu h^{\mu\prime} du^\nu du^{\mu\prime}, \tag{2.67}$$

and we may define the following generally covariant magnetic and electric field axial vector components:

$$B^\mu = (0, B^i), \tag{2.68}$$

$$\widetilde{E}^\mu = (0, \widetilde{E}^i). \tag{2.69}$$

2.3 Field Equation For Gravitation And Electromagnetism

Both the symmetric and antisymmetric metric tensors are built up of individual metric four vectors q^μ in non-Euclidean spacetime. The metric vector is the inverse tetrad $q^\mu{}_a$ with unwritten index a (see following chapters). It can therefore be inferred that the Einstein field equation can be constructed from the more fundamental field equation

$$R^\mu - \frac{1}{2}Rq^\mu = kT^\mu \tag{2.70}$$

by multiplying Eq. (2.70) on both sides by q^ν. The Ricci tensor of the Einstein field equation is therefore

$$R^{\mu\nu} = R^{\mu} q^{\nu}, \tag{2.71}$$

the symmetric canonical energy-momentum tensor of the Einstein field equation is

$$T^{\mu\nu} = T^{\mu} q^{\nu}, \tag{2.72}$$

and the symmetric metric tensor of the Einstein field equation reads

$$q^{\mu\nu(S)} = q^{\mu} q^{\nu} (= q^{\mu}{}_{a} q^{\nu}{}_{b} \eta^{ab}). \tag{2.73}$$

Equation (2.71) is therefore an equation for the gravitational field, which is described by the field tensor

$$G^{\mu} = R^{\mu} - \frac{1}{2} R q^{\mu}, \tag{2.74}$$

where

$$q^{\mu} = (h^0, h^1, h^2, h^3) \tag{2.75}$$

is the metric vector in non-Euclidean spacetime. The Sachs field equation [8] corresponding to Eq. (2.70) is

$$\frac{1}{4}\left(\kappa_{\mu\nu} q^{\nu} + q^{\nu} \kappa^{+}_{\mu\nu}\right) + \frac{1}{8} R q_{\mu} = k T_{\mu}, \tag{2.76}$$

where $\kappa_{\mu\nu}$ is the Sachs spin curvature tensor.

The generally covariant 4-vector T^{μ} appearing in the novel equation (2.70) is the generalization to non-Minkowski spacetime of the energy-momentum 4-vector p^{μ}, which is covariant under the Lorentz transformation of special relativity. Consider the particular solution

$$R^{\mu} = R q^{\mu}; \tag{2.77}$$

then

$$\frac{1}{2} R q^{\mu} = k p^{\mu} \tag{2.78}$$

is an equation which implies that mass m is defined in general relativity by the scalar curvature R:

$$m = R q^0 / 2c^2 k. \tag{2.79}$$

Newton's second law in general relativity is

$$j^{\mu} = \partial p^{\mu} / \partial \tau, \tag{2.80}$$

where τ is the proper time, and conservation of energy-momentum is expressed through the fact that p^{μ} is an invariant in four-dimensional (spacetime) non-Euclidean geometry:

$$p^{\mu} p_{\mu} = (R/2k)^2 \, q^{\mu} q_{\mu}. \tag{2.81}$$

Similarly, the generally covariant metric vector q^{μ} is an invariant of non-Minkowski spacetime:

$$q^{\mu}q_{\mu} = h^{2}{}_{0} - h^{2}{}_{1} - h^{2}{}_{2} - h^{2}{}_{3}. \tag{2.82}$$

Equation (2.81) in Newton's third law in general relativity. Newton's law of universal gravitation (the inverse square law) is also generalized to non-Minkowski spacetime by Eq. (2.80); the inverse square dependence comes out of the fact that the scalar curvature is inversely proportional to the square of distance.

The metric 4-vector (one-form) q^{ρ} is dual to the wedge product $q^{\mu} \wedge q^{\nu}$ in non-Minkowski spacetime. From this fundamental four-dimensional geometrical property, it is inferred that Eq. (2.70) is dual to the following equation between two-forms:

$$R^{\mu} \wedge q^{\nu} - \frac{1}{2}Rq^{\mu} \wedge q^{\nu} = kT^{\mu} \wedge q^{\nu}, \tag{2.83}$$

which is inferred by multiplying both sides of Eq. (2.70) by the wedge in $\wedge q^{\nu}$; it the antisymmetric metric is defined as

$$q^{\mu\nu(A)} := q^{\mu} \wedge q^{\nu}, \tag{2.84}$$

and antisymmetric Ricci tensor as

$$R^{\mu\nu(A)} := R^{\mu} \wedge q^{\nu}, \tag{2.85}$$

and the antisymmetric energy momentum tensor as

$$T^{\mu\nu(A)} := T^{\mu} \wedge q^{\nu}. \tag{2.86}$$

It can be inferred as follows that (2.83) is an equation of generally covariant electromagnetism, i.e., Eq. (2.83) represents a theory of electromagnetism in general relativity.

Define the antisymmetric field tensor

$$G^{\mu\nu} = -G^{\nu\mu} := G^{(0)} \left(R^{\mu} \wedge q^{\nu} - \frac{1}{2}Rq^{\mu} \wedge q^{\nu} \right). \tag{2.87}$$

The antisymmetry of the tensor implies the following Jacobi identity of general relativity:

$$D_{\rho}G_{\mu\nu} + D_{\mu}G_{\nu\rho} + D_{\nu}G_{\rho\mu} = 0, \tag{2.88}$$

an identity which can be written equivalently as

$$D_{\mu}\widetilde{G}^{\mu\nu} = 0, \tag{2.89}$$

where

$$\widetilde{G}^{\mu\nu} = \frac{1}{2}\epsilon^{\mu\nu\rho\sigma}G_{\rho\sigma} \tag{2.90}$$

is the dual of $G_{\rho\sigma}$. It follows that

$$\widetilde{G}^{\mu\nu} G_{\mu\nu} = 0 \tag{2.91}$$

are invariants of general relativity.

Taking the covariant derivative of both sides of Eq. (2.83) gives

$$D_\mu G^{\mu\nu} = G^{(0)} k D_\mu (T^\mu \wedge q^\nu). \tag{2.92}$$

Define

$$j^\nu := \mu_0 G^{(0)} k D_\mu (T^\mu \wedge q^\nu), \tag{2.93}$$

where μ_0 is the vacuum permeability, a fundamental constant of physics; then Eq. (2.92) can be rewritten as

$$D_\mu G^{\mu\nu} = \mu_0 j^\nu. \tag{2.94}$$

Equations (2.89) and (2.94) are identified as the homogeneous and inhomogeneous field equations of generally covariant electromagnetism. They have been deduced using geometry from the fundamental Eq. (2.70), which is therefore an equation both of gravitation and of electromagnetism, i.e., an equation of unified field theory.

Using the space indices i, j, k, the generally covariant magnetic field tensor is

$$B^{ij} := G^{ij}, \tag{2.95}$$

and the generally covariant electric field tensor is

$$E^{0i} := cG^{0i}. \tag{2.96}$$

Considering the particular solution (2.77), the electromagnetic field tensor becomes

$$G^{\mu\nu} = \frac{1}{2} G^{(0)} R q^\mu \wedge q^\nu \tag{2.97}$$

and thus is directly proportional to the antisymmetric metric as inferred in Chap. 1. Defining

$$E^{(0)} = cB^{(0)} := \frac{1}{2} cG^{(0)} R, \tag{2.98}$$

the generally covariant magnetic field tensor becomes

$$B^{ij} = B^{(0)} q^i \wedge q^j \tag{2.99}$$

and is dual to the generally covariant magnetic field pseudo-vector

$$B^k = -B^{(0)} q^k. \tag{2.100}$$

The generally covariant electric field tensor becomes

$$\widetilde{E}^{ij} = E^{(0)} \tilde{q}^i \wedge \tilde{q}^j \tag{2.101}$$

and is dual to the generally covariant field vector

$$\widetilde{E}^k = E^{(0)}\tilde{q}^k. \tag{2.102}$$

These results are summarized as follows:

$$G^{\mu\nu} = \begin{pmatrix} 0 & -E^1 & -E^2 & -E^3 \\ E^1 & 0 & -cB^3 & cB^2 \\ E^2 & cB^3 & 0 & -cB^1 \\ E^3 & -cB^2 & cB^1 & 0 \end{pmatrix},$$

$$\widetilde{G}^{\mu\nu} = \begin{pmatrix} 0 & -cB^1 & -cB^2 & -cB^3 \\ cB^1 & 0 & E^3 & -E^2 \\ cB^2 & -E^3 & 0 & E^1 \\ cB^3 & E^2 & -E^1 & 0 \end{pmatrix},$$

$$cB^k = -\frac{1}{2}\epsilon^{ijk}G^{ij}, \quad E^k = -G^{0k}, \tag{2.103}$$

$$\widetilde{E}^k = \frac{1}{2}\epsilon^{ijk}\widetilde{G}^{ij}, \quad c\widetilde{B}^k = -\widetilde{G}^{0k},$$

where all quantities are generally covariant. Equation (2.70) therefore leads to a generally covariant theory of electromagnetism as well as generally covariant theory of gravitation. For example, we have inferred that Eq. (2.70) gives the general relativistic equivalents of Newton's second and third laws of dynamics and also the general relativistic equivalents of the homogeneous and inhomogeneous equations of electrodynamics, i.e., the general-relativistic equivalents of the Gauss, Faraday, Coulomb, and Ampère-Maxwell laws of electrodynamics. In electrodynamics the energy momentum 4-vector becomes

$$T^\mu = eA^\mu, \tag{2.104}$$

where e is the charge on the proton and A^μ the potential 4-vector

$$A^\mu = (A^0, A^1, A^2, A^3). \tag{2.105}$$

Using the particular solution (2.77) for the sake of illustration, we obtain the following result from Eq. (2.104):

$$e = Rq^0/2A^{(0)}k, \tag{2.106}$$

which shows that charge e is defined in terms of scalar curvature. This result is the electromagnetic analogue of Eq. (2.79), which shows that mass m is defined in terms of scalar curvature R. Equations (2.79) and (2.106) illustrate the principle of general relativity and Mach's principle [8], i.e., that all equations of physics should be equations of general relativity.

2.4 Parity Violating Fields: Comparison With The Sachs Theory

It is well known that parity-violating fields exist in nature, notably the weak field [1-5,7,8]. Equation (2.70) implies that there may exist parity-violating fields generated by the effect of the parity operator on the metric vector:

$$q_+^\mu := (h_0, h_1^+, h_2^+, h_3^+) \xrightarrow{\hat{P}} q_-^\mu := (h_0, -h_1^-, -h_2^-, -h_3^-). \tag{2.107}$$

It follows that

$$q_+^\mu q_\mu^- = h^2{}_0 - h_1^+ h_1^- - h_2^+ h_2^- - h_3^+ h_3^- \tag{2.108}$$

is generally covariant and is also invariant under parity symmetry. If there is parity violation, Eq. (2.70) becomes two distinct field equations which can be generated from each other by the parity reversal operator $\hat{p}$, defined by [1-5], $\boldsymbol{r} \to -\boldsymbol{r}, \boldsymbol{p} \to -\boldsymbol{p}$, where $\boldsymbol{r}$ is position and $\boldsymbol{p}$ is momentum. The two equations are denoted by $+$ and $-$ subscripts:

$$R_+^\mu - \frac{1}{2} R q_+^\mu = k T_+^\mu, \tag{2.109}$$

$$R_-^\mu - \frac{1}{2} R q_-^\mu = k T_-^\mu. \tag{2.110}$$

The difference of these two equations is a parity-violating field equation

$$(R_+^\mu - R_-^\mu) - \frac{R}{2}(q_+^\mu - q_-^\mu) = k(T_+^\mu - T_-^\mu). \tag{2.111}$$

We may now postulate the existence of a parity-violating field, described by the symmetric metric

$$\begin{aligned} \gamma^{\mu\nu(S)} \neg &:= (R_+^\mu - R_-^\mu) q_\pm^\nu - \tfrac{R}{2}(q_+^\mu - q_-^\mu) q_\pm^\nu \\ &= k(T_+^\mu - T_-^\mu) q_\pm^\nu, \end{aligned} \tag{2.112}$$

and a parity-violating field described by the antisymmetric metric

$$\begin{aligned} \gamma^{\mu\nu(A)} \neg &:= \gamma^{(0)} \left((R_+^\mu - R_-^\mu) \wedge q_\pm^\nu - \tfrac{R}{2}(q_+^\mu - q_-^\mu) \wedge q_\pm^\nu \right) \\ &= \gamma^{(0)} k (T_+^\mu - T_-^\mu) \wedge q_\pm^\nu. \end{aligned} \tag{2.113}$$

The $\gamma^{\mu\nu(A)}$ field can be expressed through the following homogeneous and inhomogeneous field equations:

$$D_\mu \tilde{\gamma}^{\mu\nu(A)} = 0, \tag{2.114}$$

$$D_\mu \gamma^{\mu\nu(A)} = \mu_0 j^\nu; \tag{2.115}$$

and it may be postulated that this is the well-known parity-violating weak field [1-5,7]. The weak field is described through Eq. (2.114) and (2.115) by using an internal SU(2) symmetry gauge group, where there is a two-to-one mapping of the elements of SU(2) onto O(3). The three internal indices of the weak field denote the three parity-violating, massive weak-field bosons. The latter acquire mass by symmetry breaking with the Higgs mechanism [1-5,7], and we will consider this mechanism in more detail later in this book. So, the equations of the weak field are three gauge equations of the type (2.113), one for each weak field boson. The weak field is therefore

$$\gamma^{\mu\nu(A)} = \gamma^{(0)} \left((R_+^\mu - R_-^\mu) \wedge q_\pm^\nu - \frac{R}{2}(q_+^\mu - q_-^\mu) \wedge q_\pm^\nu \right), \tag{2.116}$$

where $\gamma^{(0)}$ is a coefficient akin to $G^{(0)}$ for the O(3) electromagnetic field. It can be inferred that the weak field is closely related to the O(3) electromagnetic field, for which the internal indices (1),(2),(3) denote states of circular polarization. Therefore, the weak field can be thought of as the parity-violating part of the O(3) electromagnetic field [1-5].

The effect of parity on the field equation (2.109) is to generate the parity-reversed matter field described by Eq. (2.110). This parity-reversed matter field is the field of antimatter, which was first predicted by the Dirac equation [7] of special-relativistic quantum field theory. The parity-violating field equation with symmetric metric, Eq. (2.112), may therefore be thought of as the gravitational field equation describing the well-known parity-violating matter that exists in nature [9].

If we define the motion reversal operator $\hat{T}$ by the operations $\boldsymbol{r} \rightarrow \boldsymbol{r}, \boldsymbol{p} \rightarrow -\boldsymbol{p}$, then it is known that there exist $\hat{T}$ violating fields in nature [9]. If so, the operator $\hat{T}$ generates two distinct equations from Eq. (2.70), equations interconvertible by $\hat{T}$. The $\hat{T}$ violating field equation from Eq. (2.70) is therefore a difference of these $\hat{T}$ violating equations, and describes the $\hat{T}$ violating fields of nature. Lastly, there may also exist $\hat{C}$ violating fields, where $\hat{C}$ is the charge conjugation operator defined by charge reversal.

Sachs [8] has developed a theory of antisymmetrized general relativity which is a special case of Eq. (2.70), and which is more complicated in structure. The Sachs theory is also based on a metric vector, but this is a 16-component vector in Clifford algebra, i.e., a 4-component vector each of whose components is described with a 2×2 matrix of Clifford algebra. In the flat spacetime limit, these matrices become the four Pauli matrices. Therefore, the metric vector of the Sachs theory is obtained from Eq. (2.41) by choosing the basis to be a Clifford algebraic basis in non-Minkowski spacetime. Equation (2.41) is true for all well-defined bases in non-Minkowski spacetime and therefore generalizes Sachs's theory.

The Sachs theory is based on the following distinction between the line element ds_+ and parity-reversed line element ds_-:

$$ds_+ = q_+^\mu dx_\mu, \tag{2.117}$$

$$ds_- = q_-^\mu dx_\mu. \tag{2.118}$$

This distinction leads to the symmetric metric defined by the symmetric product of 16-valued metric vectors in the Clifford algebraic basis,

$$ds_+ ds_- = -\frac{1}{2}(q_+^\mu q_-^\nu + q_+^\nu q_-^\mu) dx_\mu dx_\nu = \sigma^0 g^{\mu\nu} dx_\mu dx_\nu, \tag{2.119}$$

and to the two field equations

$$\frac{1}{4}(\kappa_{\rho\lambda} q_+^\lambda + q_+^\lambda \kappa_{\rho\lambda}^T) + \frac{1}{8} R q_\rho^+ = k T_\rho^+, \tag{2.120}$$

$$-\frac{1}{4}(\kappa_{\rho\lambda}^T q_-^\lambda + q_-^\lambda \kappa_{\rho\lambda}) + \frac{1}{8} R q_\rho^- = k T_\rho^-. \tag{2.121}$$

Here $\kappa_{\rho\lambda}$ is the curvature tensor generated from the Riemann curvature tensor through

$$\kappa_{\rho\lambda}^T q_\mu^\nu + q_\mu^- \kappa_{\rho\lambda} = R_{\kappa\mu\rho\lambda} q_-^\kappa, \tag{2.122}$$

and σ^+ denotes the zero'th Pauli matrix

$$\sigma^0 = \begin{pmatrix} 1 & 0 \\ 0 & 1 \end{pmatrix}. \tag{2.123}$$

Sachs therefore does not use the concept of an antisymmetric metric, and the two field equations (2.120) and (2.121) do not lead back to the Einstein field equation. Equation (2.70), on the other hand, is based on the 4-valued metric vector in any non-Euclidean spacetime and in any basis, and Eq. (2.70) leads straightforwardly back to the Einstein field equation, while also straightforwardly generating the equations of electromagnetism. Equation (2.70) contains more information than the Einstein field equation because the latter may be derived from Eq. (2.70) and because Eq. (2.70) describes both gravitation and electromagnetism.

The Sachs theory also generates the equations of gravitation and electromagnetism, but in a considerably more complicated way. The Sachs gravitational field equation reads

$$\begin{aligned} &\tfrac{1}{2}\left(\kappa_{\rho\lambda} q_+^\lambda q_\gamma^\nu - q_\gamma^+ q^{\lambda-} \kappa_{\rho\lambda} + q_+^\lambda \kappa_{\rho\lambda}^T q_\gamma^- - q_\gamma^+ \kappa_{\rho\lambda}^T q_-^\lambda\right) \\ &\quad + \tfrac{1}{4}(q_\rho^+ q_\gamma^- + q_\gamma^+ q_\rho^-) = 2k(T_\rho^+ q_\gamma^- + q_\gamma^+ T_\rho^-), \end{aligned} \tag{2.124}$$

and his electromagnetic field equations are

$$\begin{aligned} &\tfrac{1}{2}\left(\kappa_{\rho\lambda} q_+^\lambda q_\gamma^- + q_\gamma^+ q^{\lambda-} \kappa_{\rho\lambda} + q_+^\lambda \kappa_{\rho\lambda}^T q_\gamma^- + q_\gamma^+ \kappa_{\rho\lambda} q_-^\lambda\right) \\ &\quad + \tfrac{1}{4}(q_\rho^+ q_\gamma^- - q_\gamma^+ q_\rho^-) = 2k(T_\rho^+ q_\gamma^- - q_\gamma^+ T_\rho^-), \end{aligned} \tag{2.125}$$

where the electromagnetic field tensor is defined as

$$\begin{aligned} G_{\rho\lambda\neg} := &\tfrac{G^{(0)}}{2}\left(\kappa_{\rho\lambda} q_+^\lambda q_\gamma^- + q_\gamma^+ q^{\lambda-} \kappa_{\rho\lambda} + q_+^\lambda K_{\rho\lambda}^T q_\gamma^- + q_\gamma^+ \kappa_{\rho\lambda}^T q_-^\lambda\right) \\ &+ \tfrac{G^{(0)}}{4}(q_\rho^+ q_\gamma^- - q_\gamma^+ q_\rho^-) \end{aligned} \tag{2.126}$$

and the 4-current density is

$$j_\gamma = (G^{(0)}k/\mu_0)(T^{;\rho}_{\rho+}q^-_\gamma - q^+_\gamma T^{;\rho}_{\rho-}). \tag{2.127}$$

Therefore, by Ockham's Razor we choose the simpler and the more general of these generally covariant unified field theories, i.e., Eq. (2.70).

There are important concepts shared by both the Sachs theory and Eq. (2.70), notably the concepts of metric vector q^μ and canonical energy-momentum vector T^μ. However, the concept of curvature vector R^μ in Eq. (2.70) does not appear directly in the Sachs theory, in which it is represented by

$$R_\rho = \frac{1}{4}(\kappa_{\rho\lambda}q^\lambda_+ + q^\lambda_+\kappa^T_{\rho\lambda}). \tag{2.128}$$

Both theories are generally covariant, classical, unified field theories of gravitation and electromagnetism and both infer the existence of the $\boldsymbol{B}^{(3)}$ field and O(3) electrodynamics. However, O(3) electrodynamics emerges straightforwardly from Eq. (2.70) by multiplying it on both sides by the wedge $\wedge q^\nu$, while the derivation of the non-Abelian gauge field structure of Sachs's theory is a considerably more complicated problem. Both classical field theories are based on the philosophy of general covariance, i.e., that all theories of physics should be generally covariant by the principle of general relativity and the Mach principle. The special relativistic form of the Sachs theory has met with considerable success [8] in describing a range of phenomena in physics, and this augurs well for the simpler and more powerful Eq. (2.70) in general relativity. One advantage of Eq. (2.70) over Sachs's theory is that Eq. (2.70) produces the Einstein field equation (1.29), which is a well accepted theory of covariant gravitational theory and cosmology. Another advantage is that Eq. (2.70) produces O(3) electrodynamics straightforwardly through duality in differential geometry, i.e., through the well-developed theory of forms [7].

A third advantage of Eq. (2.70) is the introduction into field theory of the curvature 4-vector R^μ defined by

$$R^\mu = (1/h^4)Rq^\mu, \tag{2.129}$$

where

$$h^4 = ({h^2}_0 - {h^2}_1 - {h^2}_2 - {h^2}_3)^2. \tag{2.130}$$

If we multiply Eq. (2.129) by $h^2 q_\mu$, we obtain

$$R = h^2 q^\mu R_\mu = q^{\mu\nu(S)}{R_{\mu\nu}}^{(S)}, \tag{2.131}$$

which is the usual definition of the scalar curvature R in Riemann geometry from the symmetric metric tensor $q^{\mu\nu(S)}$ and symmetric Ricci tensor $R^{\mu\nu(S)}$. Use of Eqs. (2.129) in Eq. (2.70) gives the simple but powerful unified field equation

$$T^\mu = \alpha q^\mu, \tag{2.132}$$

where the proportionality coefficient is defined by

$$\alpha := \frac{R}{k}\left(\frac{1}{h^4} - \frac{1}{2}\right). \tag{2.133}$$

On the unit hypersphere

$$h^2{}_0 - h^2{}_1 - h^2{}_2 - h^2{}_3 = 1, \tag{2.134}$$

Eq. (2.70) assumes the simple form

$$T^\mu = (R/2k)q^\mu. \tag{2.135}$$

This result shows that in the generally covariant unified field theory of gravitation and electromagnetism, the canonical energy momentum vector is proportional to the metric 4-vector. We can surmise at this stage that, in generally covariant grand unified field theory, the same is also true for the weak and strong fields, which are also described by an equation of the type (2.70), or the simplified (2.135) for the special case of the unit hyper-sphere.

We are now in a position to simplify the traditional derivation of Newtonian gravitation from Einstein's general relativity by considering Eq. (2.70). We already know the Eq. (2.70) gives Newtonian gravitation as a limit of the Einstein field equation, which can be obtained from Eq. (2.70) by multiplication on both sides by q^ν. However, by deriving Newton's gravitational field equation directly from Eq. (2.70), we can show that the latter equation is in its own right a generally covariant equation of gravitation that is simpler in structure than the Einstein field equation.

2.5 The Field Equation (2.70) As An Eigenequation Or Wave Equation

The simplest form of Einstein's field equation [10] is the equation relating the scalar curvature R to the scalar quantity T defined by

$$R := -kT. \tag{2.136}$$

This equation is obtained [10] from Eq. (1.29) on multiplying both sides by $q^{\mu\nu(S)}$ and using the result

$$q^{\mu\nu(S)} q_{\mu\nu}{}^{(S)} = 4, \tag{2.137}$$

which is in turn a consequence of the relation [11] between the metric $q_{\mu\nu}{}^{(S)}$ and the inverse metric $q^{\mu\nu(S)}$:

$$q^{\mu\nu(S)} q_{\nu\rho}{}^{(S)} = \delta^\mu_\rho. \tag{2.138}$$

Here δ^μ_ρ is the Kronecker delta in four dimensions:

$$\delta^\mu_\varrho = \begin{cases} 1, & \mu = \rho, \\ 0, & \mu \neq \rho. \end{cases} \tag{2.139}$$

Now, multiply both sides of Eq. (2.136) by the metric vector q_μ to obtain Eq. (2.70) in the form of an *eigenequation* or *wave equation*:

$$\Box q_\mu = R q_\mu = -kT q_\mu, \tag{2.140}$$

where q_μ plays the role of an eigenfunction. From the definitions [10]:

$$R := q^{\mu\nu(S)} R_{\mu\nu}, \tag{2.141}$$

$$T := q^{\mu\nu(S)} T_{\mu\nu}, \tag{2.142}$$

we obtain

$$R = q^\mu q^\nu R_\mu q_\nu = q^\nu q_\nu q^\mu R_\mu, \tag{2.143}$$

$$T = q^\mu q^\nu T_\mu q_\nu = q^\nu q_\nu q^\mu T_\mu. \tag{2.144}$$

Using $\mu = \rho$ in the geometrical definition (2.138), one gets

$$q^{\mu\nu(S)} q_{\mu\nu}{}^{(S)} = \begin{cases} \delta^\mu_\mu = \delta^0_0 + \delta^1_1 + d^2_2 + d^3_3 = 4, \\ q^\mu q^\nu q_\mu q_\nu = (q^\mu q_\mu)^2, \end{cases} \tag{2.145}$$

so

$$q^\mu q_\mu = \pm 2. \tag{2.146}$$

In the flat spacetime limit, it is known that

$$q^\mu q_\mu \to -2, \tag{2.147}$$

because, in this limit,

$$q^\mu \to (1, 1, 1, 1), \quad q_\mu \to (1, -1, -1, -1). \tag{2.148}$$

Therefore, we obtain the following equation, which is valid for all non-Minkowskian spacetimes:

$$q^\mu q_\mu = -2. \tag{2.149}$$

This is a geometrical relation between the metric vector q_μ and the inverse metric vector q^μ.

Therefore, in Eq. (2.133):

$$h^4 = g^{\mu\nu(S)} q_{\mu\nu}{}^{(S)} = 4, \tag{2.150}$$

giving Eq. (2.70) in the form

$$R q^\mu = -4kT^\mu. \tag{2.151}$$

To check this derivation for self-consistency, use Eqs. (2.138), (2.144), and (2.149) to find

$$R_\mu = \frac{1}{4} R q_\mu, \tag{2.152}$$

$$T_\mu = \frac{1}{4} T q_\mu; \tag{2.153}$$

so

$$R_\mu = -kT_\mu. \tag{2.154}$$

This is another form of the unified field Eq. (2.70). Add $-\frac{1}{2}Rq_\mu$ to either side of Eq. (2.154) to obtain Eq. (2.70) self-consistently:

$$R_\mu - \frac{1}{2} R q_\mu = -\frac{1}{4} R q_\mu = \frac{k}{4} T q_\mu = kT_\mu. \tag{2.155}$$

The eigenequation (2.140) can be written as

$$(\Box + kT) q_\mu = 0, \tag{2.156}$$

and it can be seen that it is similar in structure to fundamental eigenequations of physics in the flat spacetime limit, for example the single-particle wave equation [7]:

$$\left(\Box + m^2c^2/\hbar^2\right)\phi = 0. \tag{2.157}$$

Here ϕ is a scalar field, $\Box$ the d'Alembertian operator, $\hbar$ the Dirac constant $h/2\pi$, and m is the particle mass. It is well known [7] that Eq. (2.157) becomes the Klein-Gordon equation after quantization. Equation (2.157) is obtained from the Einstein equation of special relativity,

$$p^\mu p_\mu - m^2c^2 = 0, \tag{2.158}$$

on using the famous [7] operator definition of *wave-mechanics*,

$$p^\mu = i\hbar\partial^\mu. \tag{2.159}$$

The non-relativistic limit of the Klein-Gordon equation is the Schrödinger equation

$$\frac{\hbar^2}{2m}\nabla^2\phi = -i\hbar\frac{\partial\phi}{\partial t}, \tag{2.160}$$

and the non-relativistic limit of the Einstein equation of special relativity is Newton's equation [7]

$$En = \frac{p^2}{2m} = \frac{1}{2}mv^2. \tag{2.161}$$

The Dirac equation of relativistic quantum mechanics is obtained on using Clifford algebra [7] and can be written in the form

$$\left(\Box + m^2c^2/\hbar^2\right)\psi = 0, \tag{2.162}$$

where ψ is a spinor. Finally the Proca equation of quantum electrodynamics is

$$\left(\Box + m_0^2c^2/h^2\right) A_\mu = 0, \tag{2.163}$$

where m_0 is the mass of the photon and where A_μ is the 4-potential of electrodynamics. The d'Alembert equation of electrodynamics is

$$\Box A_\mu = -(1/\epsilon_0)j_\mu, \tag{2.164}$$

where j^μ is the 4-current defined by Eq. (2.94). The d'Alembert equation also becomes an eigenequation if the 4-current is defined in terms of the potential 4-vector

$$j_\mu = \epsilon_0(m^2c^2/\hbar^2)A_\mu. \tag{2.165}$$

The non-relativistic limit of the d'Alembert equation is the Poisson equation of electrodynamics:

$$-\Box A_0 \rightarrow \nabla^2 A_0 = -\frac{1}{\epsilon_0}\rho = -\frac{1}{\epsilon_0}j_0. \tag{2.166}$$

The Poisson equation of non-relativistic classical dynamics is [11]

$$-\Box\Phi \rightarrow \nabla^2\Phi = 4\pi G\rho, \quad \Phi = -GM/r, \tag{2.167}$$

where G is Newton's gravitational constant, related to the Einstein constant k by

$$k = 8\pi G/c^2. \tag{2.168}$$

The relativistic generalization of the Poisson equation in the flat spacetime limit is also an eigenequation which can be written in the form

$$\left(\Box + m^2c^2/\hbar^2\right)\Phi = 0, \quad \Phi = 4\pi\hbar^2G\rho/m^2c^2. \tag{2.169}$$

Therefore many of the fundamental equations of physics are eigenequations in the flat spacetime limit, and it is demonstrated as follows that they can all be obtained from the fundamental eigenequation (2.156).

The eigenequation (2.156) is an equation of non-Minkowskian spacetime, where the metric vector q_μ is the eigenfunction. To obtain the flat spacetime limit of Eq. (2.156), consider the equation of metric compatibility in geometry [11]:

$$D_\rho {q_{\mu\nu}}^{(s)} = D_\rho q^{\mu\nu(S)} = 0. \tag{2.170}$$

Using the definition of the symmetric metric

$${q_{\mu\nu}}^{(S)} = q_\mu q_\nu, \tag{2.171}$$

$$q^{\mu\nu(S)} = q^\mu q^\nu, \tag{2.172}$$

the equation of metric compatibility becomes

$$D_\rho(q_\mu q_\nu) = (D_\rho q_\mu)q_\nu + q_\mu(D_\rho q_\nu) = 0, \tag{2.173}$$

where the Leibnitz rule has been used for the covariant derivative [11]:

$$D(T \otimes S) = (DT) \otimes S + T \otimes (DS). \tag{2.174}$$

Multiply both sides of Eq. (2.173) by q^ν, to obtain

$$-2D_\rho q_\mu + q_\mu q^\nu D_\rho q_\nu = 0. \tag{2.175}$$

Now multiply both sides of this equation by $q^\mu q_\nu$:

$$q^\mu q_\nu D_\rho q_\mu = 2D_\rho q_\nu. \tag{2.176}$$

Similarly, it can be shown that

$$q^\nu q_\mu D_\rho q_\nu = 2D_\rho q_\mu. \tag{2.177}$$

When $\mu = \nu$, these equations become

$$D_\rho q_\mu = -D_\rho q_\mu, \tag{2.178}$$

whose only solution is

$$D_\rho q_\mu = 0. \tag{2.179}$$

Recall that q_μ is the *tetrad* $q^a{}_\mu$ with unwritten index a, and q^μ is the inverse tetrad $q^\mu{}_a$. Equations (2.179) and (2.180) are the *tetrad postulate* of differential geometry [11]. When μ is not equal to ν, Eq. (2.179) is also a solution of the metric compatibility Eq. (2.173).

Equation (2.179) is the equation of metric compatibility for the metric vector q_μ. The equation of metric compatibility for the inverse metric vector follows from Eq. (2.172):

$$D_\rho q^\mu = 0. \tag{2.180}$$

The equation of metric compatibility for the symmetric metric tensor, Eq. (1.173), is used in Riemann geometry [11] to give the famous unique relation between $q^{\mu\nu(S)}$ and the Christoffel symbol:

$$\Gamma^\rho{}_{\mu\alpha} = \frac{1}{2} q^{\rho\lambda(S)} \left(\partial_\mu q_{\lambda\alpha}{}^{(S)} + \partial_\alpha q_{\mu\lambda}{}^{(S)} - \partial_\lambda q_{\alpha\mu}{}^{(S)} \right). \tag{2.181}$$

The equation of metric compatibility for q^μ, written out in terms of the Christoffel symbol, is

$$D_\nu q^\mu = \partial_\nu q^\mu + \Gamma^\mu{}_{\lambda\nu} q^\lambda = 0. \tag{2.182}$$

Multiplication of both sides of this equation by q_λ gives

$$q_\lambda D_\nu q^\mu = q_\lambda \partial_\nu q^\mu + q_\lambda q^\lambda \Gamma^\mu{}_{\lambda\nu} = 0; \tag{2.183}$$

and using Eq. (2.149) gives a new and fundamental definition of the Christoffel symbol in terms of the metric vector:

$$\Gamma^{\mu}{}_{\lambda\nu} = \frac{1}{2} q_{\lambda} \partial_{\nu} q^{\mu}. \tag{2.184}$$

On comparing Eqs. (2.181) and (2.184), it can be seen that the link between the Christoffel symbol and the metric vector q_{ν} is much simpler than the link between the Christoffel symbol and the metric tensor. Furthermore, Eq. (2.184) has been derived without any assumptions about q_{ν}, which can be a polar or axial four-vector. By contrast, the famous Eq. (2.181), on which rests Einstein's theory of general relativity [10], is derived with the little known but restrictive assumption *that the metric tensor be torsion-free* [11]. This means that the Christoffel symbol in the generally covariant theory of gravitation is assumed to be symmetric in its lower two indices:

$$\Gamma^{\rho}{}_{\mu\alpha} = \Gamma^{\rho}{}_{\alpha\mu}. \tag{2.185}$$

In the flat spacetime limit, Eq. (2.182) becomes

$$\partial_{\nu} q^{\mu} = 0, \tag{2.186}$$

a geometrical result which means that objects in flat spacetime move in a straight line. Equation (2.186) is also Newton's first law in classical, non-relativistic, flat spacetime dynamics, a law which states that objects move in a straight line in *vacuum*, i.e., unless acted upon by an external force. In general relativity the concept of force is replaced by the concept of curvature in non-Minkowski spacetime [11], and so the vacuum is defined as flat spacetime. Differentiating Eq. (2.186) gives

$$\Box q^{\mu} = \partial^{\nu}(\partial_{\nu} q^{\mu}) = 0 \tag{2.187}$$

in the flat spacetime limit, where we have used the rule [12]

$$\Box := \partial^{\nu}\partial_{\nu}, \tag{2.188}$$

defining the d'Alembertian operator $\Box$ in flat spacetime. (The d'Alembertian in non-Minkowski spacetime is [11] $D^{\mu}D_{\mu} = \Box + R$.)

The field equation (2.140) in vacuum is obtained when the canonical energy-momentum vector T_{μ} vanishes identically:

$$T q_{\mu} = 0, \tag{2.189}$$

and this is an equation of flat spacetime. Comparing Eqs. (2.189) and (2.187) gives an operator equivalence between R and $\Box$ comparable with the famous operator equivalence that produces wave mechanics (Eq. (2.159)):

$$R q_{\mu} = \Box q_{\mu}. \tag{2.190}$$

Therefore the eigenequation (2.140) becomes the wave equation

$$\Box q_\mu = 0 \tag{2.191}$$

in vacuum. Equation (2.191) is identically true if T_μ is identically zero, because R in vacuum is identically zero and $q_\mu = (1, -1, -1, -1)$. Using the equivalence (2.159), it can be seen that the d'Alembertian also vanishes if energy-momentum ($p_\mu = VT_\mu$, where V has the units of volume) vanishes identically, because

$$p^\mu p_\mu \rightarrow -\hbar^2 \partial^\mu \partial_\mu = -\hbar^2 \Box = 0. \tag{2.192}$$

However, if T_μ vanished identically, the universe would be devoid of matter, so the asymptotic approach to flat spacetime must be considered. This is the *weak-field limit* of general relativity [10], first developed by Einstein and described in hundreds of textbooks. In this limit,

$$T_\mu \rightarrow T_0 = mc^2/V_0, \tag{2.193}$$

i.e., the rest energy is much larger than the other terms, and the curvature R approaches $(mc/\hbar)^2$ asymptotically. Therefore, Eq. (2.140) in this limit becomes

$$\left(\Box + km/V_0\right) q_\mu = 0, \tag{2.194}$$

and we can see the great equations of physics beginning to emerge from Eq. (2.140). The metric vector in the weak-field limit is expanded in the Maclaurin series

$$q_\mu = \epsilon_\mu + \frac{1}{2}\gamma_\mu + \ldots, \tag{2.195}$$

where

$$\epsilon_\mu = (1, -1, -1, -1) \tag{2.196}$$

and where $\gamma_\mu \ll \epsilon_\mu$. The inverse metric vector q^μ is given by Eq. (2.195) as

$$q^\mu = \epsilon^\mu - \frac{1}{2}\gamma^\mu + \ldots. \tag{2.197}$$

Therefore, the wave equations of physics are given by Eq. (2.194) in the weak-field limit:

$$\left(\Box + km/V_0\right) \gamma_\mu = 0. \tag{2.198}$$

It can be seen that the metric perturbation γ_μ "plays the role" of the scalar field in the Klein-Gordon and Schrödinger equations, the spinor in the Dirac equation, and so on.

References

1. M. W. Evans, J.-P. Vigier, S. Roy, and S. Jeffers, *The Enigmatic Photon* (Kluwer Academic, Dordrecht, 1994–2002, hardcover and softcover), in five volumes.
2. M. W. Evans and L. B. Crowell, *Classical and Quantum Electrodynamics and the $B^{(3)}$ Field* (World Scientific, Singapore, 2001).
3. M. W. Evans, ed., *Modern Nonlinear Optics*, 2nd edn., a special topical issue of I. Prigogine and S. A. Rice, series eds., *Advances in Chemical Physics* (Wiley, New York, 2001), Vols. 119(1)–119(3).
4. M. W. Evans and A. A. Hasanein, *The Photomagneton in Quantum Field Theory* (World Scientific, Singapore, 1994).
5. M. W. Evans and S. Kielich, eds., *Modern Nonlinear Optics*, 1st edn., a special topical issue of I Prigogine and S. A. Rice, series eds., *Advances in Chemical Physics* (Wiley Interscience, New York, 1992, 1993, 1997), Vols. 85(1)–85(3).
6. E. G. Milewski, ed., *The Vector Analysis Problem Solver* (Research and Education Assocation, New York, 1987).
7. L. H. Ryder, *Quantum Field Theory*, 2nd edn. (University Press, Cambridge, 1987, 1996).
8. M. Sachs in Ref. 3, Vol. 119(1).
9. J. D. Jackson, *Classical Electrodynamics* (Wiley, New York, 1963).
10. A. Einstein, *The Meaning of Relativity* (Princeton University Press, Princeton, 1921-1953).
11. S. P. Carroll, *Lecture Notes on General Relativity*, a graduate course at Harvard University, University of California at Santa Barbara, and University of Chicago; arXiv:ar-qv/9712019v1-3Dec1977, public domain.

Part II

Developed Theory

3

A Generally Covariant Field Equation For Gravitation And Electromagnetism

Summary. A generally covariant field equation is developed for gravitation and electromagnetism by considering the metric vector q^{μ} in curvilinear, non-Euclidean space-time. The field equation is:

$$R^{\mu} - \frac{1}{2} R q^{\mu} = kT^{\mu} \tag{3.1}$$

where T^{μ} is the canonical energy-momentum four-vector, k the Einstein's constant, R^{μ} is the curvature four-vector, and R the Riemann scalar curvature. It is shown that eqn. (3.1) can be written as:

$$R^{\mu} = \alpha q^{\mu}, \tag{3.2}$$

where α is a coefficient defined in terms of R, k, and the scale factors of the curvilinear coordinate system. Gravitation is described from eqn. (3.2) through the Einstein's field equation, which is recovered by multiplying both sides of eqn. (3.2) by q^{μ}. Generally covariant electromagnetism is described from eqn. (3.2) by multiplying it on both sides by the wedge $\wedge q^{\nu}$. Therefore gravitation is described by the symmetric metric $q^{\mu} q^{\nu}$ and electromagnetism by the anti-symmetric metric defined by the wedge product $q^{\mu} \wedge q^{\nu}$.

Key words: Generally covariant field equation for gravitation and electromagnetism, O(3) electrodynamics, $\boldsymbol{B}^{(3)}$ field.

3.1 Introduction

The Principle of General Relativity states that every theory of physics should be generally covariant, i.e., retain its form under the general coordinate transformation in non-Euclidean space-time defined by any well-defined set of curvilinear coordinates [1]. This is a well known and accepted principle, [2], so a unified field theory should also be generally covariant. At present however, only one out of the four known fields of nature: gravitational, electromagnetic, weak and strong, is described by a generally covariant field equation, the Einstein's field equation of gravitation:

$$R^{\mu\nu(S)} - \frac{1}{2} R q^{\mu\nu(S)} = k T^{\mu\nu(S)} \tag{3.3}$$

Here $q^{\mu\nu(S)}$ is the symmetric metric tensor, $R^{\mu\nu(S)}$ the symmetric Ricci tensor defined in Riemann geometry, R the scalar curvature, k the Einstein's constant and $T^{\mu\nu(S)}$ the symmetric canonical energy-momentum tensor.

In this Letter a generally covariant field equation for gravitation and electromagnetism is inferred through fundamental geometry: in non-Euclidean space-time the existence of a symmetric metric tensor $q^{\mu\nu(S)}$ implies the existence of an anti-symmetric metric tensor $q^{\mu\nu(A)}$. The former is defined by the line element ds^2 formed from the square of the arc length and the latter by the area element dA. In differential geometry the one-form ds^2 is dual to the two-form dA. The symmetric metric tensor is defined by the symmetric tensor product of two metric four- vectors:

$$q^{\mu\nu(S)} = q^{\mu} q^{\nu} \tag{3.4}$$

and the anti-symmetric metric tensor by the wedge product:

$$q^{\mu\nu(A)} = q^{\mu} \wedge q^{\nu} \tag{3.5}$$

where the metric four-vector is:

$$q^{\nu} = (h^0, h^1, h^2, h^3) \tag{3.6}$$

Here h^i are the scale factors of the generally covariant curvilinear coordinate system defining non-Euclidean space-time. Therefore both types of metric tensor are defined by the metric vector q^{μ}. From this result of geometry it is inferred that if gravitation be identified through $q^{\mu\nu(S)}$, through the well known eqn. (3.3), then electromagnetism is identified through $q^{\mu\nu(A)}$. This inference is developed in Section 3.3 into a generally covariant field equation of gravitation and electromagnetism, an equation written in terms of the metric four-vector q^{μ}, which is at the root of both gravitation and electromagnetism. The following section defines the fundamental geometrical concepts needed for the field equation inferred in Section 3.3.

3.2 Fundamental Geometrical Concepts

Restrict attention initially to three non-Euclidean space dimensions. The set of curvilinear coordinates is defined as (u_1, u_2, u_3), where the functions are single valued and continuously differentiable, and where there is a one to one relation between (u_1, u_2, u_3) and the Cartesian coordinates. The position vector is $\boldsymbol{r}(u_1, u_2, u_3)$, and the arc length is the modulus of the infinitesimal displacement vector:

$$ds = |d\boldsymbol{r}| = \left| \frac{\partial \boldsymbol{r}}{\partial u_1} du_1 + \frac{\partial \boldsymbol{r}}{\partial u_2} du_2 + \frac{\partial \boldsymbol{r}}{\partial u_3} du_3 . \right| \tag{3.7}$$

The metric coefficients are $\partial \boldsymbol{r}/\partial u^i$, and the scale factors are:

$$h_i = \left| \frac{\partial \boldsymbol{r}}{\partial u_i} \right| . \tag{3.8}$$

The unit vectors are

$$\boldsymbol{e}_i = \frac{1}{h_i} \frac{\partial \boldsymbol{r}}{\partial u_i} \tag{3.9}$$

and form the O(3) symmetry cyclic relations:

$$\boldsymbol{e}_1 \times \boldsymbol{e}_2 = \boldsymbol{e}_3, \quad \boldsymbol{e}_2 \times \boldsymbol{e}_3 = \boldsymbol{e}_1, \quad \boldsymbol{e}_3 \times \boldsymbol{e}_1 = \boldsymbol{e}_2, \tag{3.10}$$

where O(3) is the rotation group of three dimensional space [3-8]. The curvilinear coordinates are orthogonal if:

$$\boldsymbol{e}_1 \cdot \boldsymbol{e}_2 = 0, \quad \boldsymbol{e}_2 \cdot \boldsymbol{e}_3 = 0, \quad \boldsymbol{e}_3 \cdot \boldsymbol{e}_1 = 0. \tag{3.11}$$

The symmetric metric tensor is then defined through the line element, a one form of differential geometry:

$$\omega_1 = ds^2 = q^{ij(S)} du_i du_j, \tag{3.12}$$

and the anti-symmetric metric tensor through the area element, a two form of differential geometry:

$$\omega_2 = dA = -\frac{1}{2} q^{ij(A)} du_i \wedge du_j. \tag{3.13}$$

These results generalize as follows to the four dimensions of any non-Euclidean space-time:

$$\omega_1 = ds^2 = q^{\mu\nu(S)} du_\mu du_\nu, \tag{3.14}$$

$$\omega_2 =^* \omega_1 = dA = -\frac{1}{2} q^{\mu\nu(A)} du_\mu \wedge du_\nu. \tag{3.15}$$

In differential geometry the element du_σ is dual to the wedge product $du_\mu \wedge du_\nu$.

The symmetric metric tensor is:

$$q^{\mu\nu(S)} = \begin{bmatrix} h^2{}_0 & h_0 h_1 & h_0 h_2 & h_0 h_3 \\ h_1 h_0 & h^2{}_1 & h_1 h_2 & h_1 h_3 \\ h_2 h_0 & h_2 h_1 & h^2{}_2 & h_2 h_3 \\ h_3 h_0 & h_3 h_1 & h_3 h_2 & h^2{}_3 \end{bmatrix}, \tag{3.16}$$

and the anti-symmetric metric tensor is:

$$q^{\mu\nu(A)} = \begin{bmatrix} 0 & -h_0 h_1 & -h_0 h_2 & -h_0 h_3 \\ h_1 h_0 & 0 & -h_1 h_2 & h_1 h_3 \\ h_2 h_0 & h_2 h_1 & 0 & -h_2 h_3 \\ h_3 h_0 & -h_3 h_1 & h_3 h_2 & 0 \end{bmatrix} \tag{3.17}$$

3.3 The Generally Covariant Field Equation

It has been shown that both the symmetric and anti-symmetric metric can be built up of individual metric four vectors q^μ in any non-Euclidean space-time, including the Riemannian space-time used in eqn. (3.3). It can therefore be inferred that the Einstein's field equation (3.3) can be built up from the generally covariant field equation:

$$R^\mu - \frac{1}{2}Rq^\mu = kT^\mu \tag{3.18}$$

Eqn. (3.3) is recovered from eqn. (3.18) by multiplying both sides of the latter by the metric four vector q^ν. We may therefore define the familiar symmetric tensors appearing in the Einstein's field equation of gravitation as follows:

$$R^{\mu\nu(S)} = R^\mu q^\nu, \tag{3.19}$$

$$q^{\mu\nu(S)} = q^\mu q^\nu, \tag{3.20}$$

$$T^{\mu\nu(S)} = T^\mu q^\nu, \tag{3.21}$$

in terms of the more fundamental four vectors R^μ, q^μ, and T^μ.

Eqn. (3.18) gives the generally covariant form of Newtons second law:

$$f^\mu = \frac{\partial T^\mu}{\partial \tau}, \tag{3.22}$$

where f^μ is a force four-vector and τ the proper time. Newtons third Law, and the Noethers Theorem (conservation of energy-momentum) is expressed through the invariant:

$$T^\mu T_\mu = \text{constant}, \tag{3.23}$$

and Newtons Law of universal gravitation in generally covariant form is given from eqn. (3.1) by:

$$f^\mu = \frac{1}{k}\frac{\partial G^\mu}{\partial \tau}, \qquad G^\mu := R^\mu - \frac{1}{2}Rq^\mu. \tag{3.24}$$

The results of generally covariant gravitational theory are also given by equation (3.18) because it is the basis of Einsteins field equation (3.3).

We have argued that the metric four vector q^σ is dual to the wedge product $q^\mu \wedge q^\nu$. From this fundamental result in differential geometry [8] it follows that the Einstein's field equation is dual to the following equation between two forms:

$$R^\mu \wedge q^\nu - \frac{1}{2}Rq^\mu \wedge q^\nu = kT^\mu \wedge q^\nu, \tag{3.25}$$

an equation which is derived by multiplying both sides of eqn. (3.18) by the wedge $\wedge q^\nu$ and in which appear the anti-symmetric Ricci tensor, anti-symmetric metric, and anti-symmetric energy-momentum tensor, respectively defined as follows:

$$R^{\mu\nu(A)} = R^{\mu} \wedge q^{\nu}, \tag{3.26}$$

$$q^{\mu\nu(A)} = q^{\mu} \wedge q^{\nu}, \tag{3.27}$$

$$T^{\mu\nu(A)} = T^{\mu} \wedge q^{\nu}. \tag{3.28}$$

Define the following anti-symmetric field tensor:

$$G^{\mu\nu(A)} = G^{(0)} \left(R^{\mu\nu(A)} - \frac{1}{2} R q^{\mu\nu(A)} \right). \tag{3.29}$$

The anti-symmetry of the tensor implies the following Jacobi identity of non-Euclidean space-time [3-8]:

$$D_{\rho} G_{\mu\nu}^{\ \ (A)} + D_{\mu} G_{\nu\rho}^{\ \ (A)} + D_{\nu} G_{\rho\mu}^{\ \ (A)} := 0, \tag{3.30}$$

where D_{ρ} are generally covariant four derivatives. In Riemannian space-time they can be defined through Christoffel symbols which are anti-symmetric in their lower two indices. The Jacobi identity (3.30) can be rewritten as:

$$D_{\rho} \widetilde{G}^{\mu\nu(A)} := 0, \tag{3.31}$$

where

$$\widetilde{G}^{\mu\nu(A)} = \frac{1}{2} \varepsilon^{\mu\nu\rho\sigma} G_{\rho\sigma}^{\ \ (A)} \tag{3.32}$$

is the dual of $G_{\rho\sigma}^{\ \ (A)}$, and orthogonal to $G_{\rho\sigma}^{\ \ (A)}$:

$$\widetilde{G}^{\mu\nu(A)} G_{\mu\nu}^{\ \ (A)} = 0. \tag{3.33}$$

Taking covariant derivatives either side of eqn. (3.25) gives:

$$D_{\mu} G^{\mu\nu(A)} = G^{(0)} k D_{\mu} T^{\mu\nu(A)}. \tag{3.34}$$

Define the generally covariant four-vector:

$$j^{\nu} = \mu_0 G^{(0)} k D_{\mu} T^{\mu\nu(A)} \tag{3.35}$$

where μ_0 is the vacuum permeability, a fundamental constant [3-8]. Then eqn. (3.34) can be written as:

$$D_{\mu} G^{\mu\nu(A)} = j^{\nu} / \mu_0. \tag{3.36}$$

We define eqns. (3.31) and (3.36) as respectively the homogenous and inhomogeneous field equations of generally covariant electromagnetism.

The generally covariant electric and magnetic fields are defined as:

$$\begin{aligned} cB^{k} &= -\frac{1}{2} \varepsilon^{ijk} G^{ij(A)}, \quad E^{k} = -G^{0k(A)}, \\ \widetilde{E}^{k} &= -\frac{1}{2} \varepsilon^{ijk} \widetilde{G}^{ij(A)}, \quad c\widetilde{B}^{k} = -\widetilde{G}^{0k(A)}, \end{aligned} \tag{3.37}$$

and the generally covariant electromagnetic field tensors as:

$$G^{\mu\nu(A)} = \begin{bmatrix} 0 & -E^1 & -E^2 & -E^3 \\ E^1 & 0 & -cB^3 & cB^2 \\ E^2 & cB^3 & 0 & -cB^1 \\ E^3 & -cB^2 & cB^1 & 0 \end{bmatrix}, \widetilde{G}^{\mu\nu(A)} = \begin{bmatrix} 0 & -cB^1 & -cB^2 & -cB^3 \\ cB^1 & 0 & E^3 & -E^2 \\ cB^2 & -E^3 & 0 & E^1 \\ cB^3 & E^2 & -E^1 & 0 \end{bmatrix}. \tag{3.38}$$

The Jacobi identity (3.31) becomes an identity of a gauge invariant Yang-Mills type field theory [3-8] of electromagnetism if we define the field tensor as a commutator of covariant derivatives:

$$G^{\mu\nu(A)} = \frac{i}{g}[D^\mu, D^\nu], \tag{3.39}$$

where

$$D^\mu = \partial^\mu - igA^\mu, \tag{3.40}$$

and where A^μ is the vector potential. The field tensor $G^{\mu\nu(A)}$ is invariant under the gauge transformation:

$$A^{\mu'} = SA_\mu S^{-1} - \frac{i}{g}(\partial_\mu S)S^{-1} \tag{3.41}$$

for any internal gauge field symmetry. The Jacobi identity (3.31) can be expressed as the identity [3-8]:

$$\sum_{\text{cyclic}} [D_\rho, [D_\mu, D_\nu]] := 0 \tag{3.42}$$

for any gauge group symmetry.

It has therefore been shown that a gauge invariant Yang Mills field theory for electromagnetism can be derived from the generally covariant field equation (3.18), which also produces the gauge invariant and generally covariant theory of gravitation first proposed contemporaneously by Einstein's and Hilbert in late 1915.

3.4 Discussion

The simplest expression of eqn. (3.18) can be obtained from the well known definition of the scalar curvature R in Riemann geometry:

$$R = g^{\mu\nu(S)} R_{\mu\nu}^{\ \ (S)} = ({h^2}_0 - {h^2}_1 - {h^3}_2 - {h^3}_3) q^\mu R_\mu. \tag{3.43}$$

It can be seen that this equation can be obtained from the equation:

$$R^\mu = \frac{1}{h^4} R q^\mu, h^2 := {h^2}_0 - {h^2}_1 - {h^3}_2 - {h^2}_3, \tag{3.44}$$

by multiplying eqn. (3.44) on both sides by $h^2 q_\mu$. Using eqn. (3.44) in eqn. (3.18) gives:

$$T^\mu = \alpha q^\mu, \tag{3.45}$$

where the proportionality coefficient is defined as:

$$\alpha := \frac{R}{k}\left(\frac{1}{h^4} - \frac{1}{2}\right). \tag{3.46}$$

On the unit hyper-sphere:

$$h^2{}_0 - h^2{}_1 - h^2{}_2 - h^2{}_3 = 1, \tag{3.47}$$

the proportionality simplifies to:

$$\alpha = R/2k. \tag{3.48}$$

Using the definition (3.19) for the symmetric Ricci tensor, and multiplying on both sides by q_ν, we also obtain the following useful expression for the curvature four-vector as a contraction of the symmetric Ricci tensor in Riemann geometry:

$$R^\mu = (1/h^2) q_\nu R^{\mu\nu(S)} \tag{3.49}$$

Eqn. (3.45) shows that for both gravitation and electromagnetism the generally covariant energy momentum four-vector T^μ is proportional to the generally covariant metric four-vector q^μ through the metric dependent proportionality coefficient α. It is likely that such a result is also true for the weak and strong fields, because it is known that the electromagnetic field can be unified with the weak field [3-8] and because both the weak and strong fields are gauge fields. It is likely therefore that eqn. (3.18) is a generally covariant field equation of classical grand unified field theory. This result is required by the Principle of General Relativity.

In the special case where te covariant derivatives of the Yang Mills field theory have O(3) internal symmetry, with indices (1), (2) and (3), where ((1), (2) (3)) is the complex circular representation of space, eqns. (3.31) and (3.36) become the field equations of O(3) electrodynamics [3-7]. The latter has been extensively discussed in the literature and tested against experimental data from several sources, and can now be recognised as an example of eqn. (3.18), illustrating the advantages of eqn. (3.18) over the received view of electromagnetism. Therefore O(3) electrodynamics is a theory of general relativity. In the received opinion [3-7] electromagnetism is a theory of special relativity in Euclidean space-time in which the field is an entity superimposed on the frame of reference, a Yang Mills gauge field theory with U(1) internal gauge group symmetry. The several advantages of O(3) electrodynamics over the received opinion have been discussed in the literature [3-7] and it can now be seen that these advantages stem from the fact that eqn. (3.16) gives a theory of generally covariant electromagnetism and also the well known generally covariant theory of gravitation. Using eqn. (3.45) it can be seen that both gravitation and electromagnetism are defined by the metric vector q^μ within the proportionality coefficient α: both fields being essentially the frame of reference itself. We may now conclude that the non-Euclidean nature of space-time gives rise to both gravitation and electromagnetism through eqn. (3.18).

A similar conclusion has been reached by Sachs [9] using Clifford algebra, but The important Sachs unification scheme is based on Clifford algebra and is considerably more complicated than eqn. (3.18), which is therefore preferred by Okhams Razor - choose the simpler of two theories.

Acknowledgments

Extensive discussions are gratefully acknowledged with Fellows and Emeriti of the Alpha Foundation.

References

1. E. G. Milewski (Chief Ed.), *The Vector Analysis Problem Solver*, (Research and Education Association, New York, 1987).
2. A. Einstein, *The Meaning of Relativity* (Princeton Univ. Press, 5th ed., 1955).
3. M. W. Evans, J.-P. Vigier et al., *The Enigmatic Photon* (Kluwer, Dordrecht, 1994 to 2002, hardback and paperback), vols. 1 - 5.
4. M. W. Evans and L. B. Crowell. *Classical and Quantum Electrodynamics and the* $\boldsymbol{B}^{(3)}$ *Field* (World Scientific, Singapore, 2001).
5. M. W. Evans and A. A. Hasanein, *The Photomagneton in Quantum Field Theory* (World Scientific, Singapore, 1994).
6. M. W. Evans and S. Kielich (eds.), *Modern Non-Linear Optics*, in I Prigogine and S. A. Rice (series eds.), *Advances in Chemical Physics* (Wiley Inter-science, New York, 1995 to 1997, hardback and paperback), vol. 85, first edition.
7. M. W. Evans, (ed.) second edition of ref. 6, vol. 119 (Wiley Inter-science, 2001).
8. L. H. Ryder, *Quantum Field Theory* (Cambridge Uniov. Press, papervack, 1987 and 1996, second edition).
9. M. Sachs in ref. 7, vol. 119(1), and references therein.

4

A Generally Covariant Wave Equation For Grand Unified Field Theory

Summary. A generally covariant wave equation is derived geometrically for grand unified field theory. The equation states most generally that the covariant d'Alembertian acting on the vielbein vanishes for the four fields which are thought to exist in nature: gravitation, electromagnetism, weak field and strong field. The various known field equations are derived from the wave equation when the vielbein is the eigenfunction. When the wave equation is applied to gravitation the wave equation is the eigenequation of wave mechanics corresponding to Einstein's field equation in classical mechanics, the vielbein eigenfunction playing the role of the quantized gravitational field. The three Newton laws, Newton's law of universal gravitation, and the Poisson equation are recovered in the classical and nonrelativistic, weak-field limits of the quantized gravitational field. The single particle wave-equation and Klein-Gordon equations are recovered in the relativistic, weak-field limit of the wave equation when scalar components are considered of the vielbein eigenfunction of the quantized gravitational field. The Schrödinger equation is recovered in the non-relativistic, weak-field limit of the Klein-Gordon equation). The Dirac equation is recovered in this weak-field limit of the quantized gravitational field (the non-relativistic limit of the relativistic, quantized gravitational field when the vielbein plays the role of the spinor. The wave and field equations of $O(3)$ electrodynamics are recovered when the vielbein becomes the relativistic dreibein (triad) eigenfunction whose three orthonormal space indices become identified with the three complex circular indices (1), (2), (3), and whose four spacetime indices are the indices of non-Euclidean spacetime (the base manifold). This dreibein is the potential dreibein of the $O(3)$ electromagnetic field (an electromagnetic potential four-vector for each index (1), (2), and (3)). The wave equation of the parity violating weak field is recovered when the orthonormal space indices of the relativistic dreibein eigenfunction are identified with the indices of the three massive weak field bosons. The wave equation of the strong field is recovered when the orthonormal space indices of the relativistic vielbein eigenfunction become the eight indices defined by the group generators of the $SU(3)$ group.

Key words: generally covariant equation, grand unified field theory, gravitation, higher symmetry electromagnetism, $O(3)$ electrodynamics, weak field, strong field.

4.1 Introduction

Recently [1] a generally covariant classical field equation has been proposed for the unification of the classical gravitational and electromagnetic fields by considering the metric four-vector q_μ in non-Euclidean spacetime. In this Letter the corresponding equation in wave (or quantum) mechanics is derived by considering the action of the covariant d'Alembertian operator on the metric four-vector considered as the eigenfunction. By deriving a metric compatibility equation for the metric four-vector a wave equation is obtained from a fundamental geometrical property: the covariant d'Alembertian acting on the metric vector vanishes in non-Euclidean spacetime. This geometrical result is also true when the eigenfunction is a symmetric or anti-symmetric metric tensor [1], and, most generally, when the eigenfunction is a vielbein [2]. The wave equation with symmetric metric tensor as eigenfunction is the direct result of the latter's metric compatibility equation, and the wave equation with vielbein as eigenfunction is the result of the tetrad postulate [2]. The latter is a fundamental result of geometry irrespective of metric compatibility and whether or not the metric tensor is torsion free. The wave equation can therefore be constructed as an eigenequation from geometry with different types of eigenfunction. This is achieved in this Letter by expressing the covariant d'Alembertian operator as a sum of the flat space d'Alembertian operator $\Box$ plus a term dependent on the non-Euclidean nature of spacetime. The latter term is shown to be a scalar curvature R which is identified as the eigenvalue. The eigenoperator is therefore the operator $\Box$, and the wave equation is a fundamental geometrical property of non-Euclidean spacetime [1,2]. Most generally the eigenfunction is the vielbein [2] $e^a{}_\mu$, which relates an orthonormal basis (Latin index) to a coordinate basis (Greek index), and the generally covariant wave equation is the eigenequation

$$(\Box + kT)e^a{}_\mu = 0. \tag{4.1}$$

The Einstein field equation [1-3] of gravitational general relativity can be written in the contracted form [1-4]

$$R = -kT, \tag{4.2}$$

where R and T are obtained [4] from the curvature tensor and the canonical energy momentum tensor by index contraction. If $q_{\mu\nu}{}^{(S)}$ denotes the symmetric metric tensor defined [1] by

$$q_{\mu\nu}{}^{(S)} = q_\mu q_\nu \tag{4.3}$$

then

$$R = q^{\mu\nu(S)} R_{\mu\nu}, \quad T = q^{\mu\nu(S)} T_{\mu\nu}, \tag{4.4}$$

where $R_{\mu\nu}$ and $T_{\mu\nu}$ are also symmetric tensors. Therefore the generally covariant wave equation is

$$(\Box + kT)e^a{}_\mu = 0, \quad kT = -R. \tag{4.5}$$

It can be seen that Eq. (4.5) has the form of the well known second-order wave equations of dynamics and electrodynamics, such as the single particle wave equation, the Klein-Gordon, Dirac, Proca, and d'Alembert [5] and non-relativistic limiting forms, such as the Schrödinger equation, and, in the classical limit, the Poisson and Newton equations. The use of the vielbein as eigenfunction has several well known advantages [2]:

(4.1) The tetrad postulate:

$$D_\nu e^a{}_\mu = 0, \tag{4.6}$$

where D_ν denotes the covariant derivative [2], is true for any connection, whether or not it is metric compatible or torsion free.
(4.2) The use of the vielbein as eigenfunction allows spinors to be analyzed in non-Euclidean spacetime, and this is essential to derive the Dirac equation from Eq. (4.5).
(4.3) The index a of the vielbein can be identified with the internal index of gauge theory [2,5], and this property is essential if Eq. (4.5) is to be an equation of grand unified field theory.
(4.4) Vielbein theory is highly developed and is closely related to Cartan-Maurer theory, a generalization of Riemann geometry [2].

The structure group of the tangent bundle in the four dimensional spacetime base manifold is $GL(4, R)$ [2], the group of real invertible 4×4 matrices. In a Lorentzian metric this reduces to the Lorentz group $SO(3, 1)$. The fibers of the fiber bundle are tied together with ordinary rotations [2] and the structure group of the new bundle is $SO(3)$, the group of rotations in three space dimensions without the assumption of parity symmetry. The electromagnetic potential is defined on this bundle by the dreibein $A^a{}_\mu$, where a is (1), (2) or (3), the indices of the complex circular representation of three dimensional space. The evolution of electrodynamics in this way, as a gauge theory with $O(3)$ gauge group symmetry, where $O(3)$ is the group of rotations in three dimensions with parity symmetry, began with the proposal of the $\boldsymbol{B}^{(3)}$ field [6] as being responsible for the inverse Faraday effect in all materials (phase free magnetization by circularly polarized electromagnetic radiation). Maxwell Heaviside electrodynamics is a gauge field theory with no internal indices, and whose internal gauge group symmetry is $U(1)$ [7-12]. After a decade of development it is known [12] that there are numerous instances in which $O(3)$ electrodynamics surpasses $U(1)$ electrodynamics in its ability to describe experimental data, for example data from interferometry, reflection, physical optics in general, the inverse Faraday effect, and its resonance equivalent, radiatively induced fermion resonance [7-12]. Therefore many data are now known which indicate that the electromagnetic sector of grand unified field theory is described by an $O(3)$ symmetry gauge field theory, not $U(1)$. In the development of $O(3)$ electrodynamics the connection on the internal fiber bundle of gauge theory was identified for the first time as the connection on the tangent bundle of general relativity [2]. This is an essential step towards the evolution of a simple and powerful unified field theory as embodied in Eq. (4.5) of this Letter. The tangent bundle is defined with respect to the base manifold, which is four dimensional non-Euclidean spacetime [1]. Prior to the development of higher symmetry electrodynamics, and generally covariant electrodynamics [1,13,14] the tangent bundle of general relativity [2] was not identified with the fiber bundle of gauge theory, in other words the internal index of gauge theory was thought to be the index of an abstract space unrelated to spacetime [2]. In $O(3)$ electrodynamics the internal index $a = (1), (2), (3)$ represents a physical orthonormal space tangential to the base manifold and in the basis ((1),(2),(3)) it is possible to define unit vectors $\boldsymbol{e}^{(1)}, \boldsymbol{e}^{(2)}, \boldsymbol{e}^{(3)}$ which define a tangent space, a space that is orthonormal to the base manifold (non-Euclidean, four dimensional spacetime). It therefore becomes possible to invoke the dreibein, or triad, as described already, with the three Latin indices (a) representing the orthogonal space and the four Greek indices (μ) the base manifold. The indices $a = (1), (2), (3)$ can be used to define the unit vector system in curvilinear coordinate analysis [1,14,15]. One of the unit vectors, e.g., $\boldsymbol{e}^{(1)}$, is a unit tangent vector to the curve, and the other two, $\boldsymbol{e}^{(2)}$ and $\boldsymbol{e}^{(3)}$, are mutually orthogonal to $\boldsymbol{e}^{(1)}$. This procedure cures two fundamental inconsistencies of field theory as it stands at present:

(1) The gravitational field is non-Euclidean space time in general relativity, while the other three fields (electromagnetic, weak and strong) are entities superimposed on flat or Euclidean spacetime.
(2) The $U(1)$ electromagnetic field has an Abelian and linear character, while the other three fields are non-Abelian and nonlinear[2].

Therefore the internal gauge space of $O(3)$ electrodynamics is identified with a tangent space in the complex circular basis ((1),(2),(3)), a basis chosen to represent circular polarization, a well known empirical property of electromagnetic radiation [5-14]. This allows electromagnetism to be developed as a theory of general relativity [1], using $O(3)$ symmetry covariant derivatives, which become spin affine connections in vielbein theory [2]. The $O(3)$ electromagnetic field tensor becomes a Cartan Maurer torsion tensor [2] which is defined with a spin affine connection on the tangent bundle of general relativity. These properties are contained with Eq. (4.5), together with the ability to describe the gravitational, weak and strong fields. Therefore, Eq. (4.5) is a generally covariant wave equation of grand unified field theory.

In Sec. 2 the wave equation (4.5) is derived for various forms of the eigenfunction using metric compatibility equations for the metric vector q_μ and the symmetric and anti-symmetric metric tensors $q_\mu q_\nu$ and $q_\mu \wedge q_\nu$ [1] and using the tetrad postulate [2] for the vielbein $e^a{}_\mu$. In Sec. 3 the equation of parallel transport and the geodesic equation [2-4] are written in terms of the metric four-vector q_μ, and the equation of metric compatibility of the metric four vector shown to be a solution of the geodesic equation. In Sec. 4

the Poisson equation and the Newtonian equations are derived in the weak field limit of gravitational theory. In Sec. 5 the second order wave equations are derived from Eq. (4.5) in various limits for the four known fields of nature. Finally, Sec. 6 is a discussion of some of the many possible avenues for further work based on Eq. (4.5) and its classical equivalent given in Ref. [1].

4.2 Derivation Of The Generally Covariant Wave Equation

The wave equation is based on the following expression for the covariant d'Alembertian operator:

$$D^\rho D_\rho = \Box + D^\mu \Gamma^\rho{}_{\mu\rho}, \tag{4.7}$$

where

$$D^\mu \Gamma^\rho{}_{\mu\rho} = \partial^\mu \Gamma^\rho{}_{\mu\rho} + \Gamma^{\mu\rho}{}_\lambda \Gamma^\lambda{}_{\mu\rho} \tag{4.8}$$

is the covariant derivative of the index contracted Christoffel symbol $\Gamma^\rho{}_{\mu\rho}$. Equation (4.7) is derived by first considering the commutator [2] $[D_\mu, D_\nu]$ acting on the inverse metric vector q^ρ [1]:

$$\begin{aligned}
&[D_\mu, D_\nu]q^\rho = D_\mu D_\nu q^\rho - D_\nu D_\mu q^\rho \\
&= \partial_\mu(D_\nu q^\rho) - \Gamma^\lambda{}_{\mu\nu} D_\lambda q^\rho + \Gamma^\rho{}_{\mu\sigma} D_\nu q^\sigma - (\mu \leftrightarrow \nu) \\
&= \partial_\mu \partial_\nu q^\rho + (\partial_\mu \Gamma^\rho{}_{\nu\sigma}) q^\sigma + \Gamma^\rho{}_{\nu\sigma} \partial_\mu q^\sigma - \Gamma^\lambda{}_{\mu\nu} \partial_\lambda q^\rho \\
&- \Gamma^\lambda{}_{\mu\nu} \Gamma^\rho{}_{\lambda\sigma} q^\sigma + \Gamma^\rho{}_{\mu\sigma} \partial_\nu q^\sigma + \Gamma^\rho{}_{\mu\sigma} \Gamma^\sigma{}_{\nu\lambda} q^\lambda - (\mu \leftrightarrow \nu) \\
&= (\partial_\mu \Gamma^\rho{}_{\nu\sigma} - \partial_\nu \Gamma^\rho{}_{\mu\sigma} + \Gamma^\rho{}_{\mu\lambda} \Gamma^\lambda{}_{\nu\sigma} - \Gamma^\rho{}_{\nu\lambda} \Gamma^\lambda{}_{\mu\sigma}) q^\sigma \\
&- 2(\Gamma^\lambda{}_{\mu\nu} - \Gamma^\lambda{}_{\nu\mu}) D_\lambda q^\rho \\
&= R^\rho{}_{\sigma\mu\nu} q^\sigma - T^\lambda{}_{\mu\nu} D_\lambda q^\rho,
\end{aligned} \tag{4.9}$$

where $R^\rho{}_{\sigma\mu\nu}$ is the Riemann tensor and $T^\lambda{}_{\mu\nu}$ is the torsion tensor. Consideration of the symmetry of Eq. (4.9), and contracting indices, $\rho = \sigma$, leads to the result

$$D^\mu D_\mu = \partial^\mu \partial_\mu + \partial^\mu \Gamma^\rho{}_{\mu\rho} + \Gamma^{\mu\rho}{}_\lambda \Gamma^\lambda{}_{\mu\rho} + 2\Gamma^\lambda{}_{\mu\mu} D_\lambda. \tag{4.10}$$

In this expression the Christoffel symbols are defined as [2]

$$(\Gamma_\mu)^\rho{}_\rho := \Gamma^\rho{}_{\mu\rho}, \quad (\Gamma_\mu)^\rho{}_\lambda := \Gamma^\rho{}_{\mu\lambda}, \quad \text{etc.}, \tag{4.11}$$

but, by convention [2], the brackets are omitted in the notation. We follow this convention in the rest of this paper. For any vector V^ν,

$$D_\mu V^\nu = \partial_\mu V^\nu + \Gamma^\nu{}_{\mu\lambda} V^\lambda. \tag{4.12}$$

We therefore can write

$$D^{\mu}\Gamma^{\rho}{}_{\mu\rho} = \partial^{\mu}\Gamma^{\rho}{}_{\mu\rho} + \Gamma^{\mu\rho}{}_{\lambda}\Gamma^{\lambda}{}_{\mu\rho}. \tag{4.13}$$

The covariant d'Alembertian operator is therefore in general

$$D^{\mu}D_{\mu} = \Box + D^{\mu}\Gamma^{\rho}{}_{\mu\rho} + 2\Gamma^{\lambda}{}_{\mu\mu}D_{\lambda} \tag{4.14}$$

and can be thought of qualitatively as "half the Riemann tensor plus half the torsion tensor." This is a geometrical result independent of any considerations of field theory.

Equation (4.5), the wave equation for the vielbein as eigenfunction, follows from the tetrad postulate [2]:

$$D_{\rho}e^{a}{}_{\mu} = 0, \tag{4.15}$$

which holds whether or not the connection is metric-compatible or torsion-free. Differentiating Eq. (4.15) covariantly gives Eq. (4.5):

$$\begin{aligned} D^{\rho}(D_{\rho}e^{a}{}_{\mu}) := D^{\rho}D_{\rho}e^{a}{}_{\mu} &= \left(\Box + D^{\mu}\Gamma^{\rho}{}_{\mu\rho} + 2\Gamma^{\lambda}{}_{\mu\mu}D_{\lambda}\right)e^{a}{}_{\mu} \\ &= \left(\Box + D^{\mu}\Gamma^{\rho}{}_{\mu\rho}\right)e^{a}{}_{\mu} = 0. \end{aligned} \tag{4.16}$$

Equation (4.16) is therefore a geometrical result which is independent of any assumptions made concerning the Christoffel symbol and its relation to the metric tensor [2] or metric vector [1]. The covariant d'Alembertian operator appearing in the wave equation (4.16) is the sum of the flat spacetime d'Alembertian $\Box$ and the term $D^{\mu}\Gamma^{\rho}{}_{\mu\rho}$. The latter is identified as *scalar curvature* (R) because it has the units of inverse square meters and is defined by an index contraction [2]. The scalar curvature R is obtained conventionally by contracting indices in the Riemann tensor. Carroll [2], for example, defines R as follows:

$$R := q^{\sigma\nu(S)}R^{\lambda}{}_{\sigma\lambda\nu}, \tag{4.17}$$

where the Ricci tensor is [2]

$$R_{\mu\nu} := R^{\lambda}{}_{\mu\lambda\nu} \tag{4.18}$$

and the Riemann tensor with lowered indices is [2]

$$R_{\rho\sigma\mu\nu} := q_{\rho\lambda}{}^{(S)}R^{\lambda}{}_{\sigma\mu\nu}. \tag{4.19}$$

However, Sachs [16] gives a different definition of the Ricci tensor:

$$R_{\kappa\rho} := q^{\mu\lambda(S)}R_{\mu\kappa\rho\lambda}; \tag{4.20}$$

so, assuming Eq. (4.19) and contracting indices $\alpha = \lambda$:

$$R_{\kappa\rho} = \delta^{\lambda}_{\lambda}R^{\lambda}{}_{\kappa\rho\lambda} = q^{\mu\lambda(S)}q_{\mu\alpha}{}^{(S)}R^{\alpha}{}_{\kappa\rho\lambda} \tag{4.21}$$

Comparing Eqs. (4.18) and (4.21), and it is seen that the definition of the scalar curvature R is a matter of convention, and is not standardized. Different authors give different definitions. Therefore the R that appears in the Einstein equation with contracted indices, Eq. (4.2), is a matter of convention. Furthermore, the minus sign that appears in Eq. (4.2) is also a matter of convention, Einstein himself [4] used the equation in the form $R = kT$, without a minus sign. In the rest of this paper we will use the contemporary [2] convention $R = -kT$. The general rule is that the scalar curvature R is found by a contraction of indices in the Riemann tensor, which has several well-known symmetry properties [2], for example–it is anti-symmetric in its last two indices. Using the following choice of index contraction:

$$R := R^{\mu}{}_{\nu\mu\nu} = \partial_{\mu} \Gamma^{\mu}{}_{\nu\nu} - \partial_{\nu} \Gamma^{\mu}{}_{\mu\nu} + \Gamma^{\mu}{}_{\mu\nu} \Gamma^{\nu}{}_{\nu\nu} - \Gamma^{\mu}{}_{\nu\nu} \Gamma^{\nu}{}_{\mu\nu} , \tag{4.22}$$

it can be seen that in this convention the scalar curvature is

$$\begin{aligned} R &:= \partial_{\mu} \Gamma^{\mu}{}_{\nu\nu} + \Gamma^{\mu}{}_{\mu\nu} \Gamma^{\nu}{}_{\nu\nu} - \left(\partial_{\nu} \Gamma^{\mu}{}_{\mu\nu} + \Gamma^{\mu}{}_{\nu\nu} \Gamma^{\nu}{}_{\mu\nu} \right) \\ &= D_{\mu} \Gamma^{\mu}{}_{\nu\nu} - D_{\nu} \Gamma^{\mu}{}_{\mu\nu} . \end{aligned} \tag{4.23}$$

Comparing Eqs. (4.16) and (4.23), it is deduced that the scalar curvature

$$R := -D^{\mu} \Gamma^{\rho}{}_{\mu\rho} \tag{4.24}$$

that appears in the definition of the covariant d'Alembertian is, qualitatively, "half" of the scalar curvature in Eq. (4.3), obtained directly from the Riemann tensor. This result is consistent with the fact that the covariant d'Alembertian is, qualitatively (or roughly speaking), half the Riemann plus torsion tensors. If the Christoffel symbol is assumed to be symmetric in its lower two indices (as in the convention in standard general relativity [2]) then the scalar curvature R defined in Eq. (4.24) becomes the second term in Eq. (4.23). If the Christoffel symbol is anti-symmetric in its lower two indices, as in the definition of the torsion tensor (Eq. (4.9)), then the second term in Eq. (4.23) is the negative of the definition appearing in Eq. (4.24). Importantly, however, Eq. (4.16) is valid whatever the symmetry of the Christoffel symbol, because Eq. (4.16) is the direct result of the tetrad postulate, Eq. (4.6). Therefore Eq. (4.16) is true for curved spacetime (gravitation) and twisted or torqued spacetime (electromagnetism). Using Eq. (4.2), we deduce the wave equation in the form

$$(\Box + kT) e^{a}{}_{\mu} = 0, \tag{4.25}$$

where

$$D^{\mu} \Gamma^{\rho}{}_{\mu\rho} = kT = -R. \tag{4.26}$$

A less generally valid wave equation can be obtained with the symmetric metric tensor of the Einstein field equation [1-4] as eigenfunction. This wave equation follows from the metric compatibility condition [2]:

$$D_\rho q_{\mu\nu}^{(S)} = 0. \tag{4.27}$$

Differentiating Eqs. (4.17) covariantly leads to the wave equation as the eigen equation:

$$D^\rho D_\rho q_{\mu\nu}^{(S)} = (\Box + kT) q_{\mu\nu}^{(S)} = 0, \tag{4.28}$$

where $q_{\mu\nu}^{(S)}$ is the eigenfunction. A third type of wave equation can be obtained using the definition [1]

$$q_{\mu\nu}^{(S)} = q_\mu q_\nu. \tag{4.29}$$

Covariant differentiation of products is defined by the Leibniz theorem [2]; therefore the metric compatibility of the symmetric metric tensor, Eq. (4.27), implies that

$$D_\rho(q_\mu q_\nu) = q_\mu(D_\rho q_\nu) + (D_\rho q_\mu) q_\nu = 0, \tag{4.30}$$

for the first derivative, and

$$\begin{aligned} D^2(q_\mu q_\nu) &:= (D^\rho D_\rho)(q_\mu q_\nu) \\ &= q_\mu D^2 q_\nu + 2(D_\rho q_\mu)(D_\rho q_\nu) + q_\nu D^2 q_\mu \\ &= 0 \end{aligned} \tag{4.31}$$

for the second derivative. A self consistent solution of Eqs. (4.30) and (4.31) is

$$D_\mu q_\nu = 0, \tag{4.32}$$

which is a metric compatibility condition for the metric vector q_μ. Differentiating Eq. (4.32) covariantly gives the wave equation as an eigenequation with the metric vector q_μ as eigenfunction:

$$D^\rho D_\rho q_\mu = (\Box + kT) q_\mu = 0. \tag{4.33}$$

Finally, it may be shown similarly that there exists a wave equation with the anti-symmetric metric $q_{\mu\nu}^{(A)} = q_\mu \wedge q_\nu$ as eigenfunction, i.e.:

$$(\Box + kT) q_{\mu\nu}^{(A)} = 0. \tag{4.34}$$

4.3 Fundamental Equations In Terms Of The Metric Vector

The equation of metric compatibility (4.32) can be derived independently as a solution of the equation of parallel transport [2] written for the inverse metric four-vector q^μ:

$$\frac{Dq^\mu}{ds} := \frac{dq^\mu}{ds} + \Gamma^\mu{}_{\nu\lambda} \frac{dx^\nu}{ds} q^\lambda = 0, \tag{4.35}$$

where dx^ν/ds is the tangent vector to q^μ. Here

$$(ds)^2 = q^\mu q^\nu dx_\mu dx_\nu \tag{4.36}$$

is the square of the line element in curvilinear coordinates [1]. The geodesic equation for q^μ is

$$\frac{D}{ds}\left(\frac{dq^\mu}{ds}\right) = 0. \tag{4.37}$$

Now use the chain rule [17] if $u = f(x, y)$; then

$$\frac{du}{dt} = \frac{\partial f}{\partial x}\frac{dx}{dt} + \frac{\partial f}{\partial y}\frac{dy}{dt}; \tag{4.38}$$

so, if $q^\mu = q^\mu(x^\nu)$, then

$$\frac{dq^\mu}{ds} = \frac{\partial q^\mu}{\partial x^\nu}\frac{dx^\nu}{ds}. \tag{4.39}$$

Using Eq. (4.39) in Eq. (4.35), one gets

$$\left(\frac{\partial q^\mu}{\partial x^\nu} + \Gamma^\mu{}_{\nu\lambda}\, q^\lambda\right)\frac{dx^\nu}{ds} = 0, \tag{4.40}$$

i.e.,

$$(D_\nu q^\mu)\frac{dx^\nu}{ds} = 0. \tag{4.41}$$

In general $dx^\nu/ds \neq 0$, so Eq. (4.32), the metric compatibility condition for q_μ, is a solution of Eq. (4.41). Therefore the equation of metric compatibility for q_μ can be derived as a solution of the equation of parallel transport for q^μ. The geodesic equation (4.37) follows from the equation of parallel transport (4.35), so the metric compatibility equation for q_μ is a special case of the geodesic equation for q^μ. Using Eq. (4.39), the geodesic equation becomes

$$\left(\frac{D}{ds}\left(\frac{\partial q^\mu}{\partial x^\nu}\right)\right)\frac{\partial x^\nu}{\partial s} + \frac{\partial q^\mu}{\partial x^\nu}\left(\frac{D}{ds}\left(\frac{\partial x^\nu}{\partial s}\right)\right) = 0. \tag{4.42}$$

But the geodesic equation can be written for any vector V^μ, and so

$$\frac{D}{ds}\left(\frac{dx^\nu}{ds}\right) = 0. \tag{4.43}$$

It therefore follows that

$$\frac{D}{ds}\left(\frac{\partial q^\mu}{\partial x^\nu}\right) = 0. \tag{4.44}$$

As shown in Ref. [1], the gravitation and electromagnetism can be described from a novel generally covariant field equation for q_μ:

$$R_\mu - \frac{1}{2}Rq_\mu = kT_\mu. \tag{4.45}$$

4.4 Derivation Of The Poisson And Newton Equations

The Poisson equation of gravitation can be derived straightforwardly in the weak field limit [1-4] from the wave equation for an eigenfunction, for example Eq. (4.33) which can be written as the two equations:

$$(\Box + kT)q_o = 0, \tag{4.46}$$

$$(\Box + kT)q_i = 0, \quad i = 1, 2, 3. \tag{4.47}$$

Using

$$\Box := \frac{1}{c^2}\frac{\partial^2}{\partial t^2} - \nabla^2, \tag{4.48}$$

Eq. (4.46) becomes

$$\nabla^2 q_o = kTq_o \tag{4.49}$$

for a quasi-static q_o. In the weak-field limit [1-4]:

$$q_o = \epsilon + \eta_o = 1 + \eta_o, \quad \eta_o \ll 1, \tag{4.50}$$

where ϵ_μ is the unit four-vector. Therefore Eq. (4.49) becomes

$$\nabla^2 \eta_o = kTq_o \sim kT. \tag{4.51}$$

This is the Poisson equation

$$\nabla^2 \Phi = 4\pi G\rho \tag{4.52}$$

if

$$\Phi = \frac{1}{2}c^2\eta_o \tag{4.53}$$

is the gravitational potential, if

$$\rho = T = m/V \tag{4.54}$$

is the rest energy density, and if G is Newton's gravitational constant, related to Einstein's constant by:

$$k = 8\pi G/c^2. \tag{4.55}$$

Newton's law, his theory of universal gravitation, and the equivalence of inertial and gravitational mass are all contained within the metric compatibility condition

$$\frac{\partial q^\mu}{\partial x^\nu} = -\Gamma^\mu{}_{\nu\lambda}\, q^\lambda. \tag{4.56}$$

Multiplying Eq. (4.56) on both sides by q_λ and using

$$\begin{aligned} \left(\Gamma^\mu{}_{\nu 0}\, q^0\right) q_0 &= \Gamma^\mu{}_{\nu 0}\left(q^0 q_0\right) = \Gamma^\mu{}_{\nu 0}\,, \\ \left(\Gamma^\mu{}_{\nu 1}\, q^1\right) q_1 &= \Gamma^\mu{}_{\nu 1}\left(q^1 q_1\right) = -\Gamma^\mu{}_{\nu 1}\,, \text{ etc.,} \end{aligned} \tag{4.57}$$

a unique equation is obtained for the Christoffel symbol in terms of the metric vector, irrespective or whether or not the metric vector is torsion-free:

$$\Gamma^{\mu}{}_{\nu o} = -q_o \partial_\nu q^\mu, \quad \Gamma^{\mu}{}_{\nu i} = q_i \partial_\nu q^\mu, \quad i = 1, 2, 3. \tag{4.58}$$

(The well-known equation relating the Christoffel symbol to the symmetric metric tensor is more intricate and less useful, because it is derived on the assumption of a torsion-free metric, i.e., that the Christoffel symbol is symmetric in its lower two indices. It is:

$$\Gamma^{\sigma}{}_{\mu\nu} = \frac{1}{2} q^{\sigma\rho(S)} \left(\partial_\mu q_{\nu\rho}{}^{(S)} + \partial_\nu q_{\rho\mu}{}^{(S)} - \partial_\rho q_{\mu\nu}{}^{(S)} \right), \tag{4.59}$$

where $q^{\sigma\rho(S)}$ is the inverse of the symmetric metric tensor. The metric tensors are defined by [2]

$$q^{\mu\nu(S)} q_{\nu\sigma}{}^{(S)} = \delta^\mu_\sigma, \tag{4.60}$$

where

$$\delta^\mu_\sigma = \begin{cases} 1, & \mu = \sigma, \\ 0, & \mu \neq \sigma, \end{cases} \tag{4.61}$$

is the Kronecker delta. In non-Euclidean spacetime, the elements of $q^{\mu\nu(S)}$ and $q_{\mu\nu}{}^{(S)}$ are not the same in general [2].)

In the Newtonian limit, the particle velocities are much smaller than c, so [2-4]:

$$dx^i / d\tau \ll dt / d\tau \sim 1. \tag{4.62}$$

Using the chain rule for the left-hand side of Eq. (4.56), with proper time τ as an affine parameter [2], we obtain

$$\frac{\partial q^\mu}{\partial x^\nu} = \frac{d\tau}{dx^\nu} \frac{dq^\mu}{d\tau} \rightarrow \frac{1}{c} \frac{d\tau}{dt} \frac{dq^\mu}{d\tau} = \frac{1}{c} \frac{dq^\mu}{dt}. \tag{4.63}$$

Consider now the identity obtained from the equation of metric compatibility, Eq. (4.56):

$$\frac{\partial q^\mu}{\partial x^\nu} := \frac{\partial q^\mu}{\partial x^\nu}. \tag{4.64}$$

Using the chain rule in the weak-field limit, the left-hand side of Eq. (4.64) becomes

$$\frac{\partial q^\mu}{\partial x^\nu} \rightarrow \frac{1}{c} \frac{\partial q^\mu}{\partial t}. \tag{4.65}$$

If we consider the four-vector defined by

$$x^\mu = (x^0, x^1, x^2, x^3), \tag{4.66}$$

then the metric vector is defined by ($q^0 := q^\mu(\mu = 0)$, etc.)

$$q^0 = \frac{\partial x^\mu}{\partial x^0}, \quad q^1 = \frac{\partial x^\mu}{\partial x^1}, \quad q^2 = \frac{\partial x^\mu}{\partial x^2}, \quad q^3 = \frac{\partial x^\mu}{\partial x^3}; \tag{4.67}$$

therefore the left-hand side of Eq. (4.64) becomes for $\mu = 0$:

$$\frac{1}{c}\frac{\partial q^0}{\partial t} = \frac{1}{c^2}\frac{\partial^2 x^\nu}{\partial t^2} \tag{4.68}$$

in the weak-field or Newtonian limit. In this limit the metric can be considered as a perturbation of the flat spacetime metric [2]:

$$q^0 = \left(1 - \frac{1}{2}\eta^0\right) \sim 1. \tag{4.69}$$

The gravitational field in the Newtonian limit is quasi-static, and the position vector is dominated by its time-like component, so

$$\frac{\partial q^0}{\partial x^\nu} \to -\frac{1}{2}\frac{\partial \eta^0}{\partial x^\nu}. \tag{4.70}$$

Equating left-hand and right-hand sides of the identity (4.64), gives, in the Newtonian approximation,

$$\frac{d^2 x^i}{dt^2} = -\frac{c^2}{2}\frac{\partial \eta^0}{\partial x^i}, \tag{4.71}$$

which is Newton's second law combined with the Newtonian theory of universal gravitation. It is seen that the equivalence of gravitational and inertial mass implied by Eq. (4.71) is a consequence of geometrical identity (4.64). This is a powerful and original result, obtained from the novel equation of metric compatibility (4.56).

Using the definition (4.53) for the Newtonian potential Φ, Eq. (4.71) can be written in the familiar form

$$\frac{d^2 \boldsymbol{r}}{dt^2} = -\boldsymbol{\nabla}\Phi, \tag{4.72}$$

which is equivalent to the inverse square law of Newton

$$\boldsymbol{F} = m\frac{d^2 \boldsymbol{r}}{dt^2} = -G\frac{mM}{r^2}\boldsymbol{k}. \tag{4.73}$$

From Eq. (4.56) with $\mu = 0$, it can be seen that Eq. (4.72) can be expressed as one Christoffel symbol:

$$\frac{\partial q^0}{\partial x^\nu} = -\Gamma^0{}_{\nu 0}\, q^0 \sim -\Gamma^0{}_{\nu 0}\,, \tag{4.74}$$

an equation which shows that the equivalence of gravitational and inertial mass is a geometrical result, the equation of metric compatibility, (4.56), which also leads to the generally covariant wave equation (4.33), and self-consistently, to the Poisson Eq. (4.52) in the Newtonian approximation. If we write Eq. (4.28) as

$$(\Box + kT)g_{\mu\nu} = 0, \tag{4.75}$$

i.e., using the standard notation $g_{\mu\nu} = g_{\mu\nu}^{(S)}$ for the symmetric metric, the weak-field or Newtonian approximation gives

$$(\Box + km/V)\, g_{oo} = 0, \tag{4.76}$$

where $T = m/V$. If g_{oo} is considered to be quasistatic, Eq. (4.76) reduces to

$$\nabla^2 g_{oo} = kT g_{oo}. \tag{4.77}$$

Using the weak field approximation

$$g_{oo} = 1 - h_{oo} \sim 1 \tag{4.78}$$

for the symmetric metric, we obtain Carroll's Eq. (4.36) [2] (the Einstein field equation in the weak-field limit):

$$\nabla^2 h_{oo} = -kT g_{oo} = -kT_{oo}, \tag{4.79}$$

which is the Poisson equation (4.52) with $h_{oo} = -c^2\Phi/2, \quad k = 8\pi G/c^2, \quad T_{oo} = m/V$. Therefore the wave equation (4.28) is the eigenequation corresponding to the classical Einstein field equation. Einstein [4] arrived at the approximation (4.79) though an intermediate equation (Eq. (4.89b) of Ref. [4]):

$$\Box\gamma_{\mu\nu} = 2kT^*_{\mu\nu}, \quad T^*_{\mu\nu} = T_{\mu\nu} - \frac{1}{2}g_{\mu\nu}T, \tag{4.80}$$

in which the metric tensor was approximated by [4]:

$$g_{\mu\nu} = -\delta_{\mu\nu} + \gamma_{\mu\nu}. \tag{4.81}$$

Using the definition

$$T = g^{\mu\nu}T_{\mu\nu}, \tag{4.82}$$

an expression is obtained for $T_{\mu\nu}$ in terms of T:

$$g_{\mu\nu}T = (g_{\mu\nu}g^{\mu\nu})T_{\mu\nu} = 4T_{\mu\nu}. \tag{4.83}$$

Equation (4.80) can therefore be written as the eigenequation

$$\left(\Box + \frac{1}{2}kT\right) g_{\mu\nu} = 0, \tag{4.84}$$

which is the wave quation (4.28) except for a factor (1/2) coming from the approximation method used by Einstein.

Using the weak-field limit of Eq. (4.33), we obtain

$$(\Box + km/V)\, q_0 = 0, \tag{4.85}$$

where $T = mc^2/V$ is again the rest energy density. Identifying q_0 as a scalar field [5] identifies Eq. (4.85) as the single-particle wave equation, which after quantization can be interpreted as the Klein-Gordon equation [5], whose wavefunction is identified with q_0 in the weak-field approximation. The Klein-Gordon equation is

$$(\Box + m^2c^2/\hbar^2)\, q_0 = 0, \tag{4.86}$$

so

$$E_0 = mc^2 = \frac{m^2c^4V}{\hbar^2 k}. \tag{4.87}$$

Equation (4.87) can be identified as the Planck/de Broglie postulate for any particle:

$$E_0 = \hbar\omega_0 = mc^2, \tag{4.88}$$

where ω_0 is the rest frequency of any particle. The rest frequency is defined by

$$\omega_0 = 8\pi c\ell^2/V, \tag{4.89}$$

where

$$\ell = (G\hbar/c^3)^{1/2} \tag{4.90}$$

is the Planck length. Equation (4.87) means that the product of the rest mass m and the rest volume V of any particle is a universal constant

$$mV = \hbar^2 k/c^2, \tag{4.91}$$

which is an important result of the generally covariant wave equation.

Using the operator equivalence of quantum mechanics [5]

$$p^\mu = i\hbar\partial^\mu,$$

$$p^\mu = (En/c, \boldsymbol{p}), \quad \partial^\mu = \left(\frac{1}{c}\frac{\partial}{\partial t}, -\boldsymbol{\nabla}\right), \tag{4.92}$$

Eq. (4.86) becomes Einstein's equation of special relativity:

$$p^\mu p_\mu = \frac{En^2}{c^2} - p^2 = m^2c^2, \tag{4.93}$$

in which En denotes the total energy (kinetic plus potential) and mc^2 is the rest energy. From the equation [18]

$$\boldsymbol{F} = \gamma m\dot{\boldsymbol{v}} = \dot{\boldsymbol{p}}, \tag{4.94}$$

where $\boldsymbol{p} = \gamma m\boldsymbol{v}$ is the momentum in the limit of special relativity (weak-field limit), an expression is obtained for the kinetic energy in special relativity:

$$T = mc^2(\gamma - 1). \tag{4.95}$$

In the non-relativistic limit $v \ll c$, the Newtonian kinetic energy

$$T = \frac{1}{2}mv^2 \tag{4.96}$$

is obtained from the second Newton law, Eq. (4.73), which is self-consistently the non-relativistic weak-field limit of Eq. (4.33). Using the operator equivalence (4.92) in Eq. (4.96) gives the time-dependent free-particle Schrödinger equation [5,19]:

$$i\hbar\frac{\partial q_0}{\partial t} = -\frac{\hbar^2\nabla^2}{2m}q_0. \tag{4.97}$$

Identifying the Hamiltonian operator as

$$H = -\frac{\hbar^2\nabla^2}{2m} \tag{4.98}$$

transforms Eq. (4.97) into the time-independent free-particle Schrödinger equation

$$Hq_0 = Tq_0, \tag{4.99}$$

which is a weak-field approximation to the wave equation (4.33) when we consider kinetic energy only [5]. The wavefunction of the Schrödinger equation (4.99) is [19]

$$q_0 = 1 + Ae^{iKZ} + Be^{-iKZ}, \tag{4.100}$$

which is the time-like component of the metric eigenfunction of Eq. (4.33) in the weak-field approximation used to recover Eq. (4.99).

These methods illustrate that wave or quantum mechanics can be considered to be an outcome of general relativity, and that the wave-function can be considered to be a deterministic property of general relativity, namely a metric four-vector, a metric tensor, or most generally a vielbein.

The first Newton law is obtained in the weak-field limit of the geodesic equation, or alternatively when the Christoffel symbol $\Gamma^0_{\nu 0}$ in Eq. (4.74) vanishes. These limits correspond to the flat spacetime in which there is no acceleration. Newton's law is contained within the conservation law for q_μ. The latter can be deduced from the Bianchi identity [2]

$$D^\mu G_{\mu\nu} := 0, \tag{4.101}$$

where

$$G_{\mu\nu} = R_{\mu\nu} - \frac{1}{2}Rg_{\mu\nu} \tag{4.102}$$

is the Einstein tensor. Noether's theorem gives the energy conservation law

$$D^\mu T_{\mu\nu} = 0, \tag{4.103}$$

and the metric compatibility assumption of standard general relativity [2] is

$$D^\rho g_{\mu\nu} = 0. \tag{4.104}$$

If we define [1]

$$R_{\mu\nu} := R_{\mu}q_{\nu}, \quad T_{\mu\nu} := T_{\mu}q_{\nu}, \quad g_{\mu\nu} = q_{\mu}q_{\nu}, \tag{4.105}$$

then the Bianchi identity (4.101) becomes

$$\begin{aligned} D^{\mu}G_{\mu\nu} &= (D^{\mu}R_{\mu})q_{\nu} + R_{\mu}(D^{\mu}q_{\nu}) \\ &\quad -\tfrac{1}{2}Rq_{\mu}D^{\mu}q_{\nu} - \tfrac{1}{2}D^{\mu}(Rq_{\mu})q_{\nu} \\ &:= 0. \end{aligned} \tag{4.106}$$

Using the metric compatibility assumption for q_{ν}, Eq. (4.32), gives the result

$$D^{\mu}G_{\mu} = 0, \quad G_{\mu} := R_{\mu} - \frac{1}{2}Rq_{\mu}. \tag{4.107}$$

This is the Bianchi identity for the field tensor:

$$G_{\mu} = kT_{\mu}. \tag{4.108}$$

Using Eq. (4.103) and the Leibniz theorem, the energy conservation law becomes

$$\begin{aligned} D^{\mu}(T_{\mu}q_{\nu}) &= (D^{\mu}T_{\mu})q_{\nu} + T_{\mu}(D^{\mu}q_{\nu}) \\ &= (D^{\mu}T_{\mu})q_{\nu} = 0, \end{aligned} \tag{4.109}$$

and the energy conservation law for T_{μ} is deduced to be

$$D^{\mu}T_{\mu} = 0. \tag{4.110}$$

The unified field equation (4.45) [1] becomes

$$D^{\mu}(G_{\mu} - kT_{\mu}) := 0. \tag{4.111}$$

Using the equations [1]:

$$R_{\mu} = \frac{1}{4}Rq_{\mu}, \tag{4.112}$$

both the energy conservation law (4.110) and the Bianchi identity (4.107) can be expressed as the equation

$$\left(D^{\mu} + \frac{1}{R}D^{\mu}R\right)q_{\mu} = 0. \tag{4.113}$$

4.5 Some Fundamental Equations Of Physics Derived From The Wave Equation

Equation (4.113) is similar in structure to a gauge transformation equation in generic gauge field theory [2,5,7-12]. On using the results [2]

$$D_\mu R = \partial_\mu R, \tag{4.114}$$

$$\begin{aligned} D_\mu q^\mu &= \partial_\mu q^\mu + \Gamma^\mu{}_{\mu\lambda} q^\lambda = \frac{1}{\sqrt{|q|}} \partial_\mu (\sqrt{|q|} q^\mu), \\ &= \partial_\mu q^\mu + \left(\frac{1}{\sqrt{|q|}} \partial_\mu \sqrt{|q|} \right) q^\mu, \end{aligned} \tag{4.115}$$

where $|q|$ is the modulus of the determinant of the symmetric metric $g_{\mu\nu} := q_\mu q_\nu$, Eq. (4.113) becomes

$$\left(\partial_\mu + \frac{1}{\sqrt{|q|}} \partial_\mu \sqrt{|q|} + \frac{1}{R} \partial_\mu R \right) q^\mu = 0, \tag{4.116}$$

a result which has been generated by the Leibniz theorem [2]

$$D_\mu (R q^\mu) = (D_\mu R) q^\mu + R (D_\mu q^\mu) = 0. \tag{4.117}$$

Now consider the definition of gauge transformation in generic gauge field theory [5]:

$$\psi' = S\psi \tag{4.118}$$

where ψ is the generic (n-dimensional) gauge field and S the rotation generator in n dimensions. Application of the Leibniz theorem produces

$$D_\mu (S\psi) = (D_\mu S)\psi + S(D_\mu \psi). \tag{4.119}$$

The covariant derivative in generic gauge field theory is defined through a vielbein [2], the generic gauge potential $A^a{}_\mu$, and a factor g (denoting generic charge):

$$D_\mu := \partial_\mu - igA^\mu, \quad A_\mu := m^a A^a{}_\mu; \tag{4.120}$$

and the gauge transformation (4.118) implies that

$$A'_\mu = A_\mu - \frac{i}{gS} \partial_\mu S, \tag{4.121}$$

i.e.,

$$igA'_\mu = igA_\mu + \frac{1}{S} \partial_\mu S. \tag{4.122}$$

The factor $-i$ in Eq. (4.120) originates in the fact that the gauge group generators m_a in generic gauge field theory are defined as imaginary-valued matrices. This procedure defines the upper index a of the vielbein $A^a{}_\mu$ [2]. However, a basis can always be found for the gauge group generators such that Eq. (4.120) becomes

$$D_\mu = \partial_\mu + gA_\mu. \tag{4.123}$$

Comparing Eqs. (4.123) and (4.116),

$$A_\mu = \frac{1}{g} \cdot \frac{1}{R} \partial_\mu R = \frac{\hbar}{e} \cdot \frac{1}{R} \partial_\mu R = B^{(0)} \partial_\mu R, \tag{4.124}$$

where $\hbar/e$ is the elementary unit of magnetic flux (the fluxon) and where $B^{(0)}$ is a magnetic flux density. Equation (4.124) combines the operator equivalence (4.92) of quantum mechanics with the minimal prescription ($p^\mu = eA^\mu$) in generic gauge field theory, giving the result

$$p^\mu = eA^\mu = i\hbar\partial^\mu. \tag{4.125}$$

This result has been obtained from the wave equation (4.33) and the Bianchi identity (4.107) in general relativity. Comparison of Eqs. (4.117) and (4.119) shows that the scalar curvature $R = -kT$ in general relativity plays the role of the rotation generator S in generic gauge field theory, and that the metric q_μ plays the role of the generic field A_μ or potential. The field can be a scalar field as in the single particle wave equation, Klein-Gordon, and Schrödinger equations (Sec. 4), but can also be a spinor, as in the Dirac equation, and a four-vector as in the Proca, d'Alembert, and Poisson equations of electrodynamics. In gravitation it has been shown in previous sections that the field can be a four-vector, a symmetric and anti-symmetric tensor and, most generally, a vielbein [2]. In $O(3)$ electrodynamics [7-12] the field or potential (Feynman's "universal influence" [5]) is the vielbein $A^a{}_\mu$, where the upper index denotes the Euclidean [2] complex circular basis ((1),(2),(3)) needed for the description of circular polarization in radiation. The lower index denotes non-Euclidean spacetime in general relativity. The upper index a is a basis for the tangent bundle of general relativity; and if we now make the *ansatz*

$$A^a{}_\mu = A^{(0)}q^a{}_\mu = A^{(0)}e^a{}_\mu, \tag{4.126}$$

we identify the internal index of a the field or potential or "universal influence" in gauge field theory with the basis index of the tangent space [2] in general relativity. This identification is the key to field unification in the new wave equation (4.25). In other words field unification is achieved by choosing the eigenfunction of the wave equation to represent the different fields that are presently thought to exist in nature: scalar fields, vector fields, symmetric or anti-symmetric tensor fields, spinor fields, and most generally, vielbeins. The weak field is a vielbein whose $SU(2)$ internal index describes the three massive weak field bosons [5], and the strong field is a vielbein whose $SU(3)$ internal index represents gluons. The internal index of the weak field therefore represents a physical tangent space of general relativity whose structure group is $SU(2)$, homomorphic with the structure group $O(3)$ of $O(3)$ electrodynamics [7-12]. The fiber bundle for both fields is therefore identified with the tangent bundle. In $O(3)$ electrodynamics the fibers are tied together with rotations in three dimensions represented by the structure group $SO(3)$ and the field is defined on this tangent or fiber bundle by the vielbein $A^a{}_\mu$. In weak field theory precisely the same procedure is followed, but the structure group becomes $SU(2)$ and the field becomes the vielbein $W^a{}_\mu$ whose three internal indices represent the three massive weak field bosons. In strong field theory the structure group is $SU(3)$ and the vielbein becomes $S^a{}_\mu$, where there are eight indices a [5]. The *ansatz* (4.126) therefore implies that the massive bosons of the weak field and the gluons of the strong field are different manifestations of the photons with indices (1), (2), and (3) of $O(3)$ electrodynamics. The $O(3)$ photons, the weak field bosons and the gluons are described by Eq. (4.25) in which the eigenfunctions are respectively $A^a{}_\mu, W^a{}_\mu$, and $S^a{}_\mu$, i.e., by the three wave equations of general relativity

$$(\Box + kT)A^a{}_\mu = 0, \tag{4.127}$$

$$(\Box + kT)W^a{}_\mu = 0, \tag{4.128}$$

$$(\Box + kT)S^a{}_\mu = 0. \tag{4.129}$$

In other words, the internal indices of the $O(3)$, weak and strong fields are different representations of the basis used to represent the tangent space in general relativity. The $O(3)$ electromagnetic field is represented by a vielbein in which the tangent space is defined in the $O(3)$ symmetry complex circular basis ((1),(2),(3)) [7-12]. This basis for the vielbein of the weak field becomes the three $SU(2)$ matrices (Pauli matrices), and there are eight $SU(3)$ symmetry matrices (geometrical generalizations [5] of the three complex two by two Pauli matrices to eight complex three by three matrices). These different basis representations are all representations of the same physical tangent space in general relativity.

In the currently accepted convention of the standard model and grand unified field theory the electromagnetic sector is represented by the field or potential A_μ in which there is no internal index, and the abstract fiber bundle of gauge field theory is not identified with the physical tangent bundle of general relativity. Consequently the standard model suffers from the inconsistencies described in the introduction, the most serious of these inconsistencies is that the Principle of General Relativity is not followed in the currently accepted convention known as "the standard model"–the Principle is applied to the gravitational field in the standard model but not to the electromagnetic, weak, and strong fields.

In $O(3)$ electrodynamics the *ansatz* (4.126) implies that

$$A_\mu = A^{(0)} q_\mu = \frac{1}{g} \cdot \frac{1}{R} \partial_\mu R \tag{4.130}$$

(where the scalar magnitude $A^{(0)}$ and the differential operator ∂_μ are the same for all three indices a). If for each index a we assume that

$$q_\mu = \frac{ds}{dx^\mu}, \tag{4.131}$$

then the *ansatz* (4.126) implies that

$$s = \frac{1}{gA^{(0)}} = \frac{1}{\kappa}. \tag{4.132}$$

For each index a the geodesic equation for $O(3)$ electrodynamics [7-12] becomes

$$\frac{d\kappa^\mu}{ds} + \Gamma^\mu{}_{\nu\sigma}\kappa^\nu\kappa^\sigma = 0, \quad \kappa^\mu = \frac{dq^\mu}{ds}, \tag{4.133}$$

an equation which defines the propagation, or path taken in non-Euclidean spacetime, of the three photons (1), (2), and (3) of $O(3)$ electrodynamics.

The wave equation (4.25) becomes the d'Alembert equation of $O(3)$ electrodynamics [7-12]

$$\square A^a{}_\mu = -\frac{1}{\epsilon_0 c^2} j^a{}_\mu \tag{4.134}$$

if we define the four-current density by the vielbein

$$j^a{}_\mu = c^2 \epsilon_0 k T A^a{}_\mu. \tag{4.135}$$

Equation (4.134) represents three wave equations [7-12], one for each photon indexed (1), (2), and (3):

$$\square A^{(1)}{}_\mu = -\frac{1}{\epsilon_0 c^2} j^{(1)}{}_\mu, \tag{4.136}$$

$$\square A^{(2)}{}_\mu = -\frac{1}{\epsilon_0 c^2} j^{(2)}{}_\mu, \tag{4.137}$$

$$\square A^{(3)}{}_\mu = -\frac{1}{\epsilon_0 c^2} j^{(3)}{}_\mu, \tag{4.138}$$

two transverse photons, (1) and (2), and one longitudinal (3). These three equations are evidently equations of general relativity, and are also gravitational wave equations multiplied on each side by the C negative scalar magnitude $A^{(0)}$. It follows from the foregoing discussion that these wave equations are also equations of the weak and strong fields with $A^a{}_\mu$ replaced respectively by $W^a{}_\mu$ and $S^a{}_\mu$. The weak field limit applied to Eq. (4.127) produces three Proca equations [5,7-12], one for each photon (i.e., for each index $a = (1), (2)$ and (3)):

$$\left(\square + m^2c^2/\hbar^2\right) A^{(i)}{}_\mu = 0, \quad i = 1, 2, 3, \tag{4.139}$$

and this procedure also produces the Planck/de Broglie postulate (4.95) applied to the photon, thus identifying the photon as a particle with mass. In the limit of electrostatics we obtain from Eq. (4.127) the Poisson equation

$$\nabla^2 A_0 = -RA_0 = kTA_0, \tag{4.140}$$

which shows that the source of the scalar potential A_0 is the scalar curvature R. This result appears to be an important indication of the fact that electric current can be obtained from the scalar curvature of the non-Euclidean spacetime, i.e., electromagnetic energy can be obtained from non-Euclidean spacetime through devices such as the motionless electromagnetic generator [12].

The identification of the $O(3)$ electromagnetic field as a vielbein implies that the unit vectors $\mathbf{e}^{(1)}, \mathbf{e}^{(2)}, \mathbf{e}^{(3)}$ of the basis described by the upper Latin index a of the vielbein are orthonormal vectors of an Euclidean tangent space to the base manifold (non-Euclidean spacetime) described by the lower Greek index μ of the vielbein. The unit vectors define the $O(3)$ symmetry cyclic equations [7-12]:

$$\begin{aligned} \boldsymbol{e}^{(1)} \times \boldsymbol{e}^{(2)} &= i\boldsymbol{e}^{(3)*}, \\ \boldsymbol{e}^{(2)} \times \boldsymbol{e}^{(3)} &= i\boldsymbol{e}^{(1)*}, \\ \boldsymbol{e}^{(3)} \times \boldsymbol{e}^{(1)} &= i\boldsymbol{e}^{(2)*}, \end{aligned} \tag{4.141}$$

and can be used to define a tangent at any point p of a curve in the non-Euclidean spacetime used to define the base manifold. The basis unit vectors are defined in terms of the Cartesian unit vectors of the tangent space by [7-12]

$$\begin{aligned} \boldsymbol{e}^{(1)} &= (1/\sqrt{2})(\boldsymbol{i} - i\boldsymbol{j}), \\ \boldsymbol{e}^{(2)} &= (1/\sqrt{2})(\boldsymbol{i} + i\boldsymbol{j}), \\ \boldsymbol{e}^{(3)} &= \boldsymbol{k}. \end{aligned} \tag{4.142}$$

It follows that the $O(3)$ electromagnetic field is defined in terms of the metric vectors:

$$\begin{aligned} \boldsymbol{A}^{(1)} &= A^{(0)}/\sqrt{2}(\boldsymbol{i} - i\boldsymbol{j})e^{i\phi} = A^{(0)}\boldsymbol{q}^{(1)}, \\ \boldsymbol{A}^{(2)} &= A^{(0)}/\sqrt{2}(\boldsymbol{i} + i\boldsymbol{j})e^{-i\phi} = A^{(0)}\boldsymbol{q}^{(2)}, \\ \boldsymbol{A}^{(3)} &= A^{(0)}\boldsymbol{k} = A^{(0)}\boldsymbol{q}^{(3)}, \end{aligned} \tag{4.143}$$

where ϕ is the electromagnetic phase. The unit vectors $\boldsymbol{e}^{(1)}$ and $\boldsymbol{e}^{(2)}$ can be thought of as tangent vectors on a circle as illustrated in the following Argand diagram:

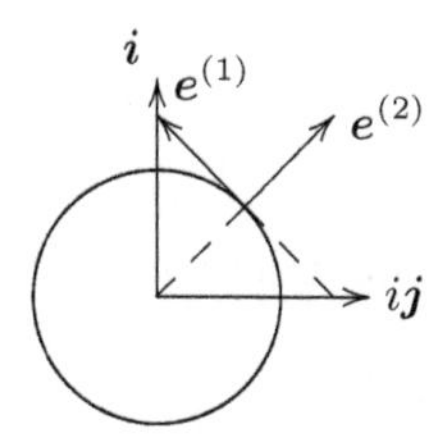

Fig. 4.1. Tangent Vectors

These tangent vectors are vectors of the tangent space to the base manifold. If we write

$$\begin{aligned} \boldsymbol{q}^{(1)'} &= \boldsymbol{q}^{(2)'} = \tfrac{1}{\sqrt{2}}(\boldsymbol{i}\cos\phi + \boldsymbol{j}\sin\phi), \\ \boldsymbol{q}^{(1)''} &= -\boldsymbol{q}^{(2)''} = \tfrac{1}{\sqrt{2}}(\boldsymbol{i}\sin\phi - \boldsymbol{j}\cos\phi), \end{aligned} \tag{4.144}$$

it can be seen in the following diagram that the metric vectors are tangent vectors that rotate around a circle for any given point Z:

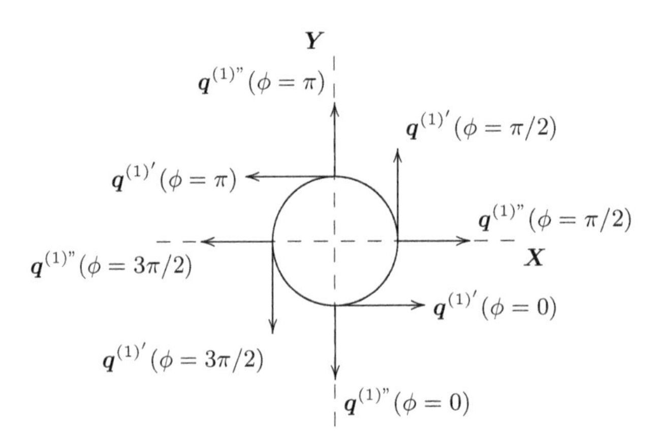

Fig. 4.2. Rotating Tangent Vectors

As we advance along the Z axis, which defines the unit vector $\boldsymbol{e}^{(3)}$ orthonormal to $\boldsymbol{e}^{(1)}$ and $\boldsymbol{e}^{(2)}$, the path drawn out is a helix, and this is the geodesic (propagation path) for $O(3)$ radiation.

Having recognized that the $O(3)$ electromagnetic field is defined by the vielbein in Eq. (4.25), it becomes possible to define scalar-valued components of the electromagnetic field (and scalar fields in general) as scalar-valued vielbein components such as:

$$\begin{aligned} q^{(1)}{}_X &= (1/\sqrt{2})e^{i\phi}, \quad q^{(1)}{}_Y = -(1/\sqrt{2})e^{i\phi}, \\ q^{(2)}{}_X &= (1/\sqrt{2})e^{-i\phi}, \quad q^{(2)}{}_Y = (1/\sqrt{2})e^{-i\phi}, \\ q^{(3)}{}_Z &= 1. \end{aligned} \tag{4.145}$$

These scalar-valued vielbein components are components of the tangent-space vector:

$$\boldsymbol{q}_\mu = q^{(1)}{}_\mu \boldsymbol{e}^{(1)} + q^{(2)}{}_\mu \boldsymbol{e}^{(2)} + q^{(3)}{}_\mu \boldsymbol{e}^{(3)}, \tag{4.146}$$

which is defined by the three four-vectors [7-12] in the base manifold $q^{(1)}{}_\mu$, $q^{(2)}{}_\mu$, $q^{(3)}{}_\mu$, one four-vector for each index $a = (1), (2)$ and (3). The components of the $O(3)$ electromagnetic field are therefore

$$A^{(1)}{}_\mu = A^{(0)} q^{(1)}{}_\mu, \tag{4.147}$$

$$A^{(2)}{}_\mu = A^{(0)} q^{(2)}{}_\mu, \tag{4.148}$$

$$A^{(3)}{}_{\mu} = A^{(0)} q^{(3)}{}_{\mu}, \tag{4.149}$$

two transverse ($a = (1)$ and (2)) and one longitudinal ($a = (3)$).

The vielbein is well defined object in differential geometry [2] and can be used, for example, to generalize Riemann geometry through the Maurer-Cartan structure equations. The close similarity of vielbein theory to gauge theory is also well understood mathematically [2], but in the currently accepted convention of the standard model the vielbein has not been used because the identification of the fiber bundle of gauge field theory is the tangent bundle of general relativity has not been made. In this section we have identified the internal index of $O(3)$ electrodynamics with the tangent space of general relativity by identifying a with the indices $(1), (2)$, and (3). This identification allows results from vielbein theory and differential geometry to be used for unified field theory, i.e., both for general relativity and gauge theory. For example the $O(3)$ gauge field is defined by [2]:

$$\begin{aligned} G^{a}{}_{\mu\nu} &= (dA)^{a}{}_{\mu\nu} + (\omega \wedge A)^{a}{}_{\mu\nu} \\ &= \partial_{\mu} A^{a}{}_{\nu} - \partial_{\nu} A^{a}{}_{\mu} + \omega^{a}{}_{\mu b} A^{b}{}_{\nu} - \omega^{a}{}_{\nu b} A^{b}{}_{\mu}, \end{aligned} \tag{4.150}$$

which is a covariant exterior derivative in differential geometry. In Eq. (4.150) $\omega^{a}{}_{\mu b}$ is a spin affine connection. In gauge field theory the $O(3)$ electromagnetic gauge field is defined by the gauge-invariant commutator of covariant derivatives [5,7-12]:

$$\begin{aligned} G^{a}{}_{\mu\nu} = \tfrac{i}{g}[D_{\mu}, D_{\nu}] &= \partial_{\mu} A^{a}{}_{\nu} - \partial_{\nu} A^{a}{}_{\mu} \\ &+ g(A^{b}{}_{\mu} A^{c}{}_{\nu} - A^{b}{}_{\nu} A^{c}{}_{\mu}). \end{aligned} \tag{4.151}$$

A comparison of Eq. (4.150) and (4.151) defines the spine affine connections in terms of the $O(3)$ fields or vector potentials:

$$\omega_{\mu}{}^{a}{}_{b} A^{b}_{\nu} - \omega_{\mu}{}^{a}{}_{b} A^{b}{}_{\mu} = g(A^{b}{}_{\mu} A^{c}{}_{\nu} - A^{b}{}_{\nu} A^{c}{}_{\mu}) = g \epsilon_{abc} A^{b}{}_{\mu} A^{c}{}_{\nu}. \tag{4.152}$$

Thus, the field or potential or "universal influence" $A^{a}{}_{\mu}$ has been defined in this section in terms of the scalar curvature in general relativity and also in terms of the spin affine connections. The gauge field $G^{a}{}_{\mu\nu}$ is invariant under the gauge transformation (4.128); i.e., if

$$A_{\mu} \rightarrow A_{\mu} - \frac{i}{g} \frac{1}{S} \partial_{\mu} S, \tag{4.153}$$

the gauge field is unchanged. This result is true for all four fields. In gravitation the equivalent of the gauge field is the Riemann tensor, which is covariant under coordinate transformation, while the Christoffel symbol is not covariant under coordinate transformation because it is not a tensor [2].

Some powerful results of vielbein theory may be translated directly into the language of unified field theory developed in this Letter, for example

$O(3)$ electrodynamics. The first of the Maurer-Cartan structure relations [2] of differential geometry is

$$dT^a + \omega^a{}_b \wedge T^b = R^a{}_b \wedge e^b \tag{4.154}$$

and states that the covariant exterior derivative of the torsion form T^a (left hand side of Eq. (4.154) is the wedge product of the Riemann form $R^a{}_b$ and vielbein form e^b (right-hand side of Eq. (4.154)). Equation (4.154) is the inhomogeneous field equation of $O(3)$ electrodynamics:

$$D_\mu G^{\mu\nu,a} = \frac{1}{\mu_0} j^{\nu,a}, \tag{4.155}$$

where the charge-current density vielbein is defined by Eq. (4.135) of this Section. Equations (4.154) and (4.155) are equations of unified field theory– the torsion form T^a represents electromagnetism (or the weak and strong fields), and the Riemann form $R^a{}_b$ represents gravitation. In Ref. (1) the inhomogeneous equation (4.153) was inferred from Eq. (4.45) by multiplying it on both sides by the wedge $\wedge A^\mu{}_\nu$, and by defining the electromagnetic field tensor as

$$G_{\mu\nu} = G^{(0)}(R_\mu \wedge q_\nu - \frac{R}{2} q_\mu \wedge q_\nu) \tag{4.156}$$

and the charge current density as

$$j^\nu = \mu_0 G^{(0)} k D_\mu (T^\mu \wedge q^\nu). \tag{4.157}$$

The gravitational field and Riemann tensor were defined [1] by multiplying the novel field Eq. (4.45) on both sides by q_ν, so Eq. (4.45), the classical analogue of the wave equation (4.25) is an equation of unified field theory.

The second Maurer-Cartan structure relations is the Bianchi identity, and translates into the Bianchi identity of gravitation [2,5], and also into the identity (4.107) used in Sec. 4 to derive the gauge invariance equation (4.113). In $O(3)$ electrodynamics it becomes the homogeneous field equation [1], the Jacobi identity

$$D_\mu \widetilde{G}^{\mu\nu,a} := 0, \tag{4.158}$$

where $\widetilde{G}^{\mu\nu,a}$ is the dual [5,7-12] of $G^a{}_{\mu\nu}$.

The tetrad postulate of vielbein theory, Eq. (4.6), translates into the $O(3)$ symmetry cyclic relations

$$\begin{aligned} \partial_i \boldsymbol{q}^{(1)*}{}_j &= -i\kappa \boldsymbol{q}^{(2)}{}_i \times \boldsymbol{q}^{(3)}{}_j, \\ \partial_i \boldsymbol{q}^{(2)*}{}_j &= -i\kappa \boldsymbol{q}^{(3)}{}_i \times \boldsymbol{q}^{(1)}{}_j, \\ \partial_i \boldsymbol{q}^{(3)*}{}_i &= -i\kappa \boldsymbol{q}^{(1)}{}_i \times \boldsymbol{q}^{(2)}{}_j \end{aligned} \tag{4.159}$$

between space indices of the base manifold ($\mu = i = 1, 2, 3$).

In this section the $O(3)$ electromagnetic gauge field has been identified in three different ways: Eqs. (4.150), (4.151), and (4.156). Self-consistency demands that these three definitions by the same, giving Eq. (4.152), for example. This equation relates the spin affine connection and the vector potential. Comparing equations (4.151) and (4.156) gives the important result

$$G^{(0)}\left(R_\mu \wedge q_\nu - \frac{R}{2} q_\mu \wedge q_\nu\right) = \partial_\mu A^a{}_\nu - \partial_\nu A^a{}_\mu + g\epsilon_{abc} A^b{}_\mu A^c{}_\mu, \qquad (4.160)$$

which indicates that the group structure of generally covariant electrodynamics is non-Abelian and that generally covariant electrodynamics must be a gauge field theory with an internal gauge group such as $O(3)$, of higher symmetry than the conventional $U(1)$ of the standard model. The wedge product $R_\mu \wedge q_\nu$ is accordingly identified as

$$R_\mu \wedge q_\nu = \frac{1}{G^{(0)}}\left(\partial_\mu A^a{}_\nu - \partial_\nu A^a{}_\mu\right) \qquad (4.161)$$

and the wedge product $q_\mu \wedge q_\nu$ as

$$\frac{R}{2} q_\mu \wedge q_\nu = -\frac{g}{G^{(0)}} \epsilon_{abc} A^b{}_\mu A^c{}_\nu. \qquad (4.162)$$

If electrodynamics were a $U(1)$ Abelian theory, then the wedge product $q_\mu \wedge q_\nu$ would be zero:

$$q_\mu \wedge q_\nu = q_\mu q_\nu - q_\nu q_\mu = 0. \qquad (4.163)$$

The electromagnetic field would then disappear because [1]

$$R_\mu = \frac{1}{4} R q_\mu. \qquad (4.164)$$

The tetrad postulate in $U(1)$ symmetry gauge field theory would reduce to

$$D_\mu q_\nu = (\partial_\mu - igA_\mu) q_\nu = (\partial_\mu - igA^{(0)} q_\mu) q_\nu = 0. \qquad (4.165)$$

The $U(1)$ gauge field would then be

$$G_{\mu\nu} = \partial_\mu A_\nu - \partial_\nu A_\mu = igA^{(0)2}(q_\mu q_\nu - q_\nu q_\mu) = 0 \qquad (4.166)$$

and would vanish, a result that is self consistent with Eq. (4.164). It is concluded that *general relativity implies higher symmetry electrodynamics*, a result that is crucial for the development of a unified field theory.

Finally in this section we use another important result of vielbein theory to derive the Dirac equation from the wave equation (4.25): the vielbein allows spinors to be developed in non-Abelian spacetime. Each component of the spinor must obey a Klein-Gordon equation (Ref. [5], p. 45). The Klein-Gordon equation is obtained from the wave equation (4.25) by considering the four scalar components of the vielbein (there are four such components for each index a). The solutions of the Dirac equation for a particle at rest are the positive and negative solutions, respectively,

$$\psi = u(0)\exp(-imt), \quad \psi = v(0)\exp(imt). \tag{4.167}$$

The two positive energy and two negative energy spinors in this limit become

$$u^{(1)}(0) = \begin{pmatrix}1\\0\\0\\0\end{pmatrix}, \quad u^{(2)}(0) = \begin{pmatrix}0\\1\\0\\0\end{pmatrix}, \quad v^{(1)}(0) = \begin{pmatrix}0\\0\\1\\0\end{pmatrix}, \quad v^{(2)}(0) = \begin{pmatrix}0\\0\\0\\1\end{pmatrix}, \tag{4.168}$$

and these are identified as components of the vielbein. The Dirac equation has been obtained from the wave equation (4.25), which uses the vielbein as eigenfunction.

4.6 Discussion

The key to field unification (unification of general relativity and gauge theory) in this Letter is the realization that the internal index (fiber bundle index) of gauge theory is the tangent bundle index of general relativity. Fundamental geometry shows that this internal index is present in basic relations such as the one between Cartesian unit vectors in Euclidean spacetime, $\boldsymbol{i} \times \boldsymbol{j} = \boldsymbol{k}$ in cyclic permutation. This internal index is implicitly assumed to exist in everyday geometry in flat space, but is the key to realizing that the most general eigenfunction for the wave equation (4.25) must be a vielbein. The unit vectors $\boldsymbol{i}, \boldsymbol{j}, \boldsymbol{k}$ (or $\boldsymbol{e}^{(1)}, \boldsymbol{e}^{(2)}, \boldsymbol{e}^{(3)}$ of the complex circular basis) are most generally vielbeins. It follows in generally covariant electrodynamics that the field $A^a{}_\mu$ is also a vielbein and that the gauge group symmetry of electrodynamics must be $O(3)$ or higher. The existence of a $U(1)$ gauge field theory is prohibited by fundamental geometry, because in such a theory the internal index of the vielbein is missing. This is geometrically incorrect. These results are proven as follows.

Consider the displacement vector [1,14,15] in the three dimensions of Euclidean space:

$$\boldsymbol{r} = X\boldsymbol{i} + Y\boldsymbol{j} + Z\boldsymbol{k}. \tag{4.169}$$

The Cartesian unit vectors are

$$\boldsymbol{i} = \frac{\partial \boldsymbol{r}}{\partial X} / \left|\frac{\partial \boldsymbol{r}}{\partial X}\right|, \quad \boldsymbol{j} = \frac{\partial \boldsymbol{r}}{\partial Y} / \left|\frac{\partial \boldsymbol{r}}{\partial Y}\right|, \quad \boldsymbol{k} = \frac{\partial \boldsymbol{r}}{\partial Z} / \left|\frac{\partial \boldsymbol{r}}{\partial Z}\right|, \tag{4.170}$$

and the three metric vectors are [1,18,19]

$$\begin{aligned} \boldsymbol{q}_X &= \boldsymbol{q}^a(a=1) = \left|\tfrac{\partial \boldsymbol{r}}{\partial X}\right| \boldsymbol{i}, \\ \boldsymbol{q}_Y &= \boldsymbol{q}^a(a=2) = \left|\tfrac{\partial \boldsymbol{r}}{\partial Y}\right| \boldsymbol{j}, \\ \boldsymbol{q}_Z &= \boldsymbol{q}^a(a=3) = \left|\tfrac{\partial \boldsymbol{r}}{\partial Z}\right| \boldsymbol{k}. \end{aligned} \tag{4.171}$$

It follows that in Euclidean space that both the unit and metric vector components must be labeled with an upper and lower index:

$$\begin{aligned} &{q^1}_1 = -1, \;\; {q^1}_2 = 0, \;\; {q^1}_3 = 0, \\ &{q^2}_1 = 0, \;\; {q^2}_2 = -1, \;\; {q^2}_3 = 0, \\ &{q^3}_1 = 0, \;\; {q^3}_2 = 0, \;\; {q^3}_3 = -1, \\ &{q^1}_1 = -i_1 = i_X = 1, \text{ etc.} \end{aligned} \tag{4.172}$$

These results extend to Euclidean spacetime on using the index 0

$${q^0}_0 = 1, \quad {q^0}_1 = 0, \quad {q^0}_2 = 0, \quad {q^0}_3 = 0. \tag{4.173}$$

Equations (172) and (4.173) define the vielbein ${q^a}_\mu$ where $a = 0, 1, 2, 3$, and $\mu = 0, 1, 2, 3$. More precisely, the vielbein is a vierbein or tetrad [2] because there are four internal or tangent space indices a and four indices μ of the base manifold. If the tetrad is used in the context of general relativity the a index becomes the tangent space index, and if the tetrad is used in gauge theory a is the index of the internal space that defines the gauge group. *Therefore fundamental geometry shows that the tetrad can be used both in general relativity and gauge theory, and this is the key to field unification.*

In Ref. [1] it has been shown that both the gravitational and electromagnetic field originate in Eq. (4.45): if the gravitational field is described through the symmetric metric tensor $q_\mu q_\nu$ then the electromagnetic field must be described through the anti-symmetric tensor:

$$G_{\mu\nu} = G^{(0)} \left(R_\mu \wedge q_\nu - \frac{1}{2} R q_\mu \wedge q_\nu \right). \tag{4.174}$$

This is again a result of geometry, essentially the result states that there exists a dot product between two vectors (symmetric metric tensor $q_\mu q_\nu$, used to describe the gravitational field) there must exist a cross product between the same two vectors (anti-symmetric metric tensor $q_\mu \wedge q_\nu$, used to describe the electromagnetic field). Taking the definition [1]

$$R = q^{\mu\nu(S)} R_{\mu\nu} = q^\mu q^\nu R_\mu q_\nu = -2 q^\mu R_\mu, \quad q^\nu q_\nu = -2, \tag{4.175}$$

it follows that

$$R_\mu = \frac{1}{4} R q_\mu, \quad G_{\mu\nu} = \frac{1}{4} G^{(0)} R \left(q_\mu \wedge q_\nu - 2 q_\mu \wedge q_\nu \right) \tag{4.176}$$

and that the electromagnetic field can be written in general as the wedge product:

$$G_{\mu\nu} = -\frac{1}{4} G^{(0)} R (q_\mu \wedge q_\nu). \tag{4.177}$$

The minus sign in Eq. (4.177) is a matter of convention and so the electromagnetic field can be succinctly expressed, within a factor $B^{(0)}$, as the wedge product of q_μ and q_ν:

$$G_{\mu\nu} = B^{(0)}(q_\mu \wedge q_\nu), \tag{4.178}$$

where $B^{(0)}$ has the units of magnetic flux density [7-12]:

$$B^{(0)} = \frac{1}{4}G^{(0)}R. \tag{4.179}$$

The wedge product of two one forms in differential geometry is defined [2] by

$$A_\mu \wedge B_\nu = (A \wedge B)_{\mu\nu} = A_\mu B_\nu - A_\nu B_\mu. \tag{4.180}$$

Therefore the wedge product vanishes if q_μ and q_ν are considered as four vectors with no internal index. *It follows that electromagnetism cannot be a gauge theory with no internal index, and therefore cannot be a $U(1)$ gauge field theory.* The wedge product of the two vielbeins $q^a{}_\mu$ and $q^b{}_\nu$ is

$$(q^a \wedge q^b)_{\mu\nu} = q^a{}_\mu q^b{}_\nu - q^a{}_\nu q^b{}_\mu, \tag{4.181}$$

and the electromagnetic field is the differential two-form

$$G^c{}_{\mu\nu} = B^{(0)}(q^a \wedge q^b)_{\mu\nu}. \tag{4.182}$$

In differential geometry, the Greek indices become redundant (i.e., can be assumed implicitly to be always the same on the left and right hand sides of an equation in differential geometry, the theory of differential forms), so the Greek indices can be suppressed [2]. Equation (4.182) can therefore be written as

$$G^c = B^{(0)} q^a \wedge q^b, \tag{4.183}$$

and, within a factor $B^{(0)}$, the electromagnetic field is a torsion two form T^c:

$$G^c = B^{(0)} T^c = B^{(0)} q^a \wedge q^b. \tag{4.184}$$

The first Maurer-Cartan structure relation (Eq. (4.154)) relates the torsion two form to the Riemann form, and so the first Maurer-Cartan structure relation becomes a relation between gravitation (Riemann form) and electromagnetism (torsion form). By adjusting the index a on the torsion form, the Maurer-Cartan structure relation becomes one between the weak field and gravitation, and the strong field and gravitation. This inter-relation between fields is a result of geometry and of the novel grand unified field theory developed in this Letter.

In the language of tetrads and wedge products the geometrical equation $\boldsymbol{i} \times \boldsymbol{j} = \boldsymbol{k}$ becomes

$$(q^1 \wedge q^2)_{12} = q^1{}_1 \wedge q^2{}_2 = q^1{}_1 q^2{}_2 - q^1{}_2 q^2{}_1 = q^1{}_1 q^2{}_2 = (q^3)_{12} = q^3{}_3. \tag{4.185}$$

in Euclidean spacetime in the Cartesian basis the tetrad is non-zero if and only if $a = \mu_1$ so it has been implicitly assumed that $q^a{}_\mu$ can be written as q_μ. This assumption means that the existence of the internal index a in basic geometry has been overlooked. In gauge theory this has led to the incorrect assumption that there can exist a gauge theory (electromagnetism) with no internal index. Careful consideration shows however that the unit vectors $\boldsymbol{i}, \boldsymbol{j}, \boldsymbol{k}$ are the following tetrad components:

$$-\boldsymbol{i} := (0, q^1{}_1, 0, 0), \quad -\boldsymbol{j} := (0, 0, q^2{}_2, 0), \quad -\boldsymbol{k} := (0, 0, 0, q^3{}_3). \tag{4.186}$$

In a non-Euclidean space (base manifold) defined [1,14,15] by the curvilinear coordinate basis (u_1, u_2, u_3) the unit vectors are the orthonormal tangent space vectors

$$\boldsymbol{e}^a = \frac{\partial \boldsymbol{r}}{\partial u^a} / \left| \frac{\partial \boldsymbol{r}}{\partial u^a} \right|, \quad a = 1, 2, 3, \tag{4.187}$$

obeying the $O(3)$ cyclic relations

$$\boldsymbol{e}^1 \times \boldsymbol{e}^2 = \boldsymbol{e}^3, \quad \text{et cyclicum}, \tag{4.188}$$

and the metric vectors are

$$\boldsymbol{q}^a = \frac{\partial \boldsymbol{r}}{\partial u^a}, \quad a = 1, 2, 3, \tag{4.189}$$

i.e., the tetrad components

$$q^a{}_1 = -\frac{\partial X}{\partial u^a}, \quad q^a{}_2 = -\frac{\partial Y}{\partial u^a}, \quad q^a{}_3 = -\frac{\partial Z}{\partial u^a}. \tag{4.190}$$

The tetrad in four-dimensional spacetime is therefore $q^a{}_\mu$. The upper index a of the tetrad denotes a flat, orthonormal tangent spacetime, and the lower index μ the non-Euclidean base manifold (the non-Euclidean spacetime of general relativity). The structure factors [1] are:

$$h^a = (q^a{}_0 q^{a0} - q^a{}_1 q^{a1} - q^a{}_2 q^{a2} - q^a{}_3 q^{a3}). \tag{4.191}$$

In general relativity the metric $q^a{}_\mu$ always has an upper index a, and a lower index μ, and the tetrad $q^a{}_\mu$ is the eigenfunction of the wave equation (4.25) of grand unified field theory. It has been demonstrated in this Letter that this wave equation is the direct result of the tetrad postulate, Eq. (4.6), and so is the direct result of geometry. More generally, it has also been demonstrated in this Letter that there must exist an internal index a in all geometrical relations, such as the relation between Cartesian unit vectors $\boldsymbol{i}, \boldsymbol{j}, \boldsymbol{k}$.

$O(3)$ electrodynamics [7-12] is therefore Eq. (4.178) when the internal index a is (1), (2), and (3), and $O(3)$ *electrodynamics is the direct result of general relativity, and of geometry.* In other words the very existence of gravitation is empirical evidence for the existence of $O(3)$ electrodynamics, because gravitation is described through $q^a{}_\mu q^b{}_\nu$ and $O(3)$ electrodynamics by $q^a{}_\mu \wedge q^b{}_\nu$. Both fields originate in the classical equation (4.45) [1], which is the classical limit of the wave equation (4.25). In $O(3)$ electrodynamics the tetrad postulate (6) becomes the cyclic equations with $O(3)$ symmetry:

$$\begin{aligned} \partial_\mu \boldsymbol{A}^{(3)*}{}_\nu &= -ig\boldsymbol{A}^{(1)}{}_\mu \times \boldsymbol{A}^{(2)}{}_\nu, \\ \partial_\mu \boldsymbol{A}^{(1)*}{}_\nu &= -ig\boldsymbol{A}^{(2)}{}_\mu \times \boldsymbol{A}^{(3)}{}_\nu, \\ \partial_\mu \boldsymbol{A}^{(2)*}{}_\nu &= -ig\boldsymbol{A}^{(3)}{}_\mu \times \boldsymbol{A}^{(1)}{}_\nu, \end{aligned} \tag{4.192}$$

where we have used the relation $A^a{}_\mu = A^{(0)}q^a{}_\mu$. The tetrad postulate (4.6) shows that:

$$\begin{aligned} \partial_\mu \boldsymbol{A}^{(1)*}{}_\nu - \partial_\nu \boldsymbol{A}^{(1)*}{}_\mu &= -ig\boldsymbol{A}^{(2)}{}_\mu \times \boldsymbol{A}^{(3)}{}_\nu, \\ \partial_\mu \boldsymbol{A}^{(2)*}{}_\nu - \partial_\nu \boldsymbol{A}^{(2)*}{}_\mu &= -ig\boldsymbol{A}^{(3)}{}_\mu \times \boldsymbol{A}^{(1)}{}_\nu, \\ \boldsymbol{B}^{(3)*}{}_{\mu\nu} &= -ig\boldsymbol{A}^{(1)}{}_\mu \times \boldsymbol{A}^{(2)}{}_\nu. \end{aligned} \tag{4.193}$$

The gauge field in $O(3)$ electrodynamics is defined by the cyclic relations [7-12]

$$\begin{aligned} \boldsymbol{G}^{(3)*}{}_{\mu\nu} &= \partial_\mu \boldsymbol{A}^{(3)*}{}_\nu - \partial_\nu \boldsymbol{A}^{(3)*}{}_\mu - ig\boldsymbol{A}^{(1)}{}_\mu \times \boldsymbol{A}^{(2)}{}_\nu, \\ \boldsymbol{G}^{(1)*}{}_{\mu\nu} &= \partial_\mu \boldsymbol{A}^{(1)*}{}_\nu - \partial_\nu \boldsymbol{A}^{(1)*}{}_\mu - ig\boldsymbol{A}^{(2)}{}_\mu \times \boldsymbol{A}^{(3)}{}_\nu, \\ \boldsymbol{G}^{(2)*}{}_{\mu\nu} &= \partial_\mu \boldsymbol{A}^{(2)*}{}_\nu - \partial_\nu \boldsymbol{A}^{(2)*}{}_\mu - ig\boldsymbol{A}^{(3)}{}_\mu \times \boldsymbol{A}^{(1)}{}_\nu. \end{aligned} \tag{4.194}$$

But we know from Eq. (4.178) that

$$\begin{aligned} \boldsymbol{G}^{(3)*}{}_{\mu\nu} &= -iB^{(0)}\boldsymbol{q}^{(1)}{}_\mu \times \boldsymbol{q}^{(2)}{}_\nu, \\ \boldsymbol{G}^{(1)*}{}_{\mu\nu} &= -iB^{(0)}\boldsymbol{q}^{(2)}{}_\mu \times \boldsymbol{q}^{(3)}{}_\nu, \\ \boldsymbol{G}^{(2)*}{}_{\mu\nu} &= -iB^{(0)}\boldsymbol{q}^{(3)}{}_\mu \times \boldsymbol{q}^{(1)}{}_\nu, \end{aligned} \tag{4.195}$$

so in $O(3)$ electrodynamics there exist the following three fundamental relations:

$$g\boldsymbol{A}^{(1)}{}_\mu \times \boldsymbol{A}^{(2)}{}_\nu = B^{(0)}\boldsymbol{q}^{(1)}{}_\mu \times \boldsymbol{q}^{(2)}{}_\nu, \quad \text{et cyclicum.} \tag{4.196}$$

Finally the realization that the electromagnetic field must be a tetrad allows the description of the internal space by any appropriate index of the orthonormal tangent space, for example a can be (1),(2),(3) of the complex circular basis, or it can be (X, Y, Z) of the Cartesian basis. So $O(3)$ electrodynamics, or any higher symmetry electrodynamics, can be developed using any well defined index a of the tangent space of general relativity. This means that electrodynamics can be developed as an $SU(2)$ symmetry gauge field theory, or as an $SU(3)$ symmetry gauge field symmetry. This suggests that the weak and strong fields may both be manifestations of the electromagnetic field. Essentially, one field is changed into another by changing the index a. Therefore there emerge many possible inter-relations between fields once it is realized that the index a is always present in the tetrad $q^a{}_\mu$, i.e., in the eigenfunction of the wave equation (4.25).

Acknowledgments

The author gratefully acknowledges many informative discussions with AIAS fellows and others.

References

1. M. W. Evans, *Found. Phys. Lett.* **16**, 367 (2003).
2. S. M. Carroll, *Lecture Notes in General Relativity* (University of California, Santa Barbara Graduate Course, arXiv:gr-gc/9712019 v1 3 Dec., 1997).
3. R. M. Wald, *General Relativity* (University of Chicago Press, 1984).
4. A. Einstein, *The Meaning of Relativity* (Princeton University Press, 1929).
5. L. H. Ryder, *Quantum Field Theory* (Cambridge University Press, 1987, 1996).
6. M. W. Evans, *Physica B* **182**, 227 (1992).
7. M. W. Evans and S. Kielich, eds., *Modern Non-Linear Optics*, in I. Prigogine and S. A. Rice, series eds., *Advances in Chemical Physics*, Vol. 85, 1st edn. (Wiley-Interscience, New York, 1992, 1993, 1997).
8. M. W. Evans and A. A. Hasanein, *The Photomagneton in Quantum Field Theory* (World Scientific, Singapore, 1994).
9. M. W. Evans, J.-P. Vigier, et al., *The Enigmatic Photon*, Vols. 1-5 (Kluwer Academic, Dordrecht, 1994 to 2002).
10. M. W. Evans and L. B. Crowell, *Classical and Quantum Electrodynamics and the $\boldsymbol{B}^{(3)}$ Field* (World Scientific, Singapore, 2001).
11. M. W. Evans, ed., *Modern Non-Linear Optics*, in I. Prigogine and S. A. Rice, series eds., *Advances in Chemical Physics*, Vol. 119, 2nd edn. (Wiley-Interscience, New York, 2001).
12. D. J. Clements and M. W. Evans, *Found. Phys. Lett.* **16** (5) (2003). M. W. Evans *et al.* (AIAS Author Group), *ibid.*, **16**, 195 (2003).
13. M. W. Evans *et al.* (AIAS Author Group), *Found. Phys. Lett.* **16**, 275 (2003).
14. M. W. Evans, *Lecture Notes in O(3) Electrodynamics* (World Scientific, Singapore, in preparation).
15. E. G. Milewski, *The Vector Analysis Problem Solver* (Research and Education Associates, New York, 1987).
16. M. Sachs, in Ref. [11], Vol. 119(1).
17. G. Stephenson, *Mathematical Methods for Science Students* (Longmans & Green, London, 1968).
18. J. B. Marion and S. T. Thornton, 3rd edn., *Classical Dynamics of Particles and Systems* (Harcourt & Brace, New York, 1988).
19. P. W. Atkins, *Molecular Quantum Mechanics*, 2nd edn. (Oxford University Press, 1983).

5

The Equations of Grand Unified Field Theory In Terms Of the Maurer-Cartan Structure Relations of Differential Geometry

Summary. The first and second Maurer-Cartan structure relations are combined with the Evans field equation [1] for differential forms to build a grand unified field theory based on differential geometry. The tetrad or vielbein plays a central role in this theory, and all four fields currently thought to exist in nature can be described by the same equations, the tangent space index of the tetrad in general relativity being identified with the tetrad's internal (gauge group) index in gauge theory.

Key words: grand unified field theory, quantum grand unified field theory, Maurer-Cartan structure relations, differential geometry, $O(3)$ electrodynamics, $\boldsymbol{B}^{(3)}$ field.

5.1 Introduction

It is currently thought that there exist four fundamental force fields in nature: gravitational, electromagnetic, weak and strong. The development of a theory of all known force fields is one which goes back to the mid-nineteenth century, when attempts were made to unify the gravitational and electromagnetic fields. The contemporary theory of all four fields is known as "grand unified field theory" (GUFT). Recently a generally covariant, classical field/matter equation [1] and wave equation of quantum field/matter theory [2] was developed and applied to GUFT. In this theory the tangent bundle of generally relativity [3] is also the fiber bundle of gauge theory [4]. Both the tangent and fiber bundles are described [2] by the tetrad (or vielbein) $q^a{}_\mu$ [3] whose orthonormal tangent space index a becomes the index both of the tangent space in general relativity and of the fiber bundle space in gauge theory. The index μ of the tetrad is that of the non-Euclidean base manifold. In general relativity, the base manifold is non-Euclidean space-time in four dimensions. GUFT is thereby developed [1,2] as a theory of general relativity [3], in which all the equations of physics are generally covariant. In quantum GUFT [2] the tetrad is the eigenfunction of the generally covariant wave equation [2] from which is obtained the equations of the quantized force and matter fields in nature. Examples of such equations in classical field/matter theory are the

Einstein field equation and the equations of generally covariant electrodynamics [5-10], namely the $O(3)$ electrodynamic field/matter equations. These were obtained from the classical equation introduced in Ref. [1]. Equations of quantum field/matter theory such as the single particle wave equation, the Schrödinger equation, the Klein-Gordon and Dirac equations, were obtained from the quantized equation in Ref. [2]. The latter is a generally covariant, second-order wave equation which also gives wave and quantum field equations in higher symmetry electrodynamics [5-10], the $O(3)$ d'Alembert and Proca equations. Electrodynamics is thereby developed into a generally covariant theory, as required by the principle of general relativity [11]. The generally covariant wave equation [2] was used as a theory of the weak and strong fields by recognizing that the internal index of the tetrad is both the fiber bundle index of gauge field theory [3] and the tangent space index of general relativity. The internal gauge group index of the parity-violating weak field [4] therefore becomes a manifestation of geometry and is closely related to the internal $O(3)$ symmetry gauge group index $a = (1), (2), (3)$ of $O(3)$ electrodynamics, where ((1),(2),(3)) denotes the orthonormal, $O(3)$ symmetry, complex circular basis of three-dimensional Euclidean space [5-10]. By developing the basis ((1),(2),(3)) with an $SU(3)$ representation it becomes possible to apply the generally covariant classical [1] and quantum [2] equations into equations in the strong field [4].

In this theory, physics is reduced to geometry. The tetrad is used [2,3] to introduce differential geometry into general relativity, and in this Letter the classical and quantum equations developed in Refs. [1] and [2] are combined with the Maurer-Cartan structure relations to give a GUFT theory based entirely on differential geometry [3]. In so doing, it is seen that all four fields are manifestations of differential geometry and can be inter-related with the structure relations combined with the Evans equations [1,2] expressed as equations in differential forms [3,4]. If the index a represents the tangent space of the base manifold representing gravitation then the Evans equations [1,2] are those of gravitation. If the index a has an $O(3)$ symmetry with the basis ((1),(2),(3)), the Evans equations are those of generally covariant, or higher symmetry, electromagnetism [1,2,5-10]. If the index a has an $SU(2)$ symmetry, the Evans equations are those of the parity-violating weak field; and if the index a has $SU(3)$ symmetry, the Evans equations are those of the strong field. In order to build the theory from first geometrical principles Sec. 5.2 gives a description of the tetrad at work in $O(3)$ electrodynamics, and Sec. 5.3 gives the Evans equations as equations between differential forms and combines them with the Maurer-Cartan structure relations.

5.2 The Tetrad In Generally Covariant (O(3)) Electrodynamics

Define the base manifold in general as the non-Euclidean spacetime indexed by $\mu = (0, 1, 2, 3)$. In this manifold define the infinitesimal displacement vector in the space basis (u_1, u_2, u_3):

$$d\boldsymbol{r} = \frac{\partial \boldsymbol{r}}{\partial u_1} du_1 + \frac{\partial \boldsymbol{r}}{\partial u_2} du_2 + \frac{\partial \boldsymbol{r}}{\partial u_3} du_3 \tag{5.1}$$

and the three metric vectors [2,9,11]:

$$\boldsymbol{q}^1 = \frac{\partial \boldsymbol{r}}{\partial u_1}, \quad \boldsymbol{q}^2 = \frac{\partial \boldsymbol{r}}{\partial u_2}, \quad \boldsymbol{q}^3 = \frac{\partial \boldsymbol{r}}{\partial u_3}. \tag{5.2}$$

Define the orthonormal tangent spacetime by the index a of the $O(3)$ symmetry complex circular basis:

$$\boldsymbol{e}^{(1)} \times \boldsymbol{e}^{(2)} = i\boldsymbol{e}^{(3)*} \text{ et cyclicum,} \tag{5.3}$$

whose unit vectors are related to the orthonormal Cartesian unit vectors $\boldsymbol{i}, \boldsymbol{j}, \boldsymbol{k}$ of the tangent space by

$$\boldsymbol{e}^{(1)} = \frac{1}{\sqrt{2}}(\boldsymbol{i} - i\boldsymbol{j}), \quad \boldsymbol{e}^{(2)} = \frac{1}{\sqrt{2}}(\boldsymbol{i} + i\boldsymbol{j}), \quad \boldsymbol{e}^{(3)} = \boldsymbol{k}. \tag{5.4}$$

In the tangent space, the displacement vector is

$$d\boldsymbol{r} = \frac{\partial \boldsymbol{r}}{\partial e^{(1)}} de^{(1)} + \frac{\partial \boldsymbol{r}}{\partial e^{(2)}} de^{(2)} + \frac{\partial \boldsymbol{r}}{\partial e^{(3)}} de^{(3)}, \tag{5.5}$$

and the three orthonormal metric vectors in the tangent space are

$$\begin{aligned} \boldsymbol{q}^{(1)} &= \partial \boldsymbol{r}/\partial e^{(1)} = \boldsymbol{e}^{(1)}, \\ \boldsymbol{q}^{(2)} &= \partial \boldsymbol{r}/\partial e^{(2)} = \boldsymbol{e}^{(2)}, \\ \boldsymbol{q}^{(3)} &= \partial \boldsymbol{r}/\partial e^{(3)} = \boldsymbol{e}^{(3)}. \end{aligned} \tag{5.6}$$

Let the orthonormal tangent space rotate and translate with respect to the space $i = 1, 2, 3$ by introducing the phase $\phi = \omega t - \kappa Z$ of the wave equation [5-10]. Here ω is an angular frequency at instant t and κ is a wave-vector at point Z. The orthonormal tangent space is thereby defined by the metric vectors

$$\boldsymbol{q}^{(1)} = \boldsymbol{e}^{(1)} e^{i\phi}, \quad \boldsymbol{q}^{(2)} = \boldsymbol{e}^{(2)} e^{-i\phi}, \quad \boldsymbol{q}^{(3)} = \boldsymbol{e}^{(3)} \tag{5.7}$$

whose magnitudes are

$$q^{(1)} = \frac{1}{\sqrt{2}}(1 - i)e^{i\phi}, \quad q^{(2)} = \frac{1}{\sqrt{2}}(1 + i)e^{-i\phi}, \quad q^{(3)} = 1. \tag{5.8}$$

The components of the tangent space and base manifold are inter-related by the tetrad (or vielbein) matrix, whose space-like components form a 3×3 invertible matrix

$$\begin{bmatrix} q^{(1)} \\ q^{(2)} \\ q^{(3)} \end{bmatrix} = \begin{bmatrix} q^{(1)}{}_1 & q^{(1)}{}_2 & q^{(1)}{}_3 \\ q^{(2)}{}_1 & q^{(2)}{}_2 & q^{(2)}{}_3 \\ q^{(3)}{}_1 & q^{(3)}{}_2 & q^{(3)}{}_3 \end{bmatrix} \begin{bmatrix} q^1 \\ q^2 \\ q^3 \end{bmatrix} . \tag{5.9}$$

This expression means that the coordinate system defined by the metric vectors $\boldsymbol{q}^{(1)}, \boldsymbol{q}^{(2)}, \boldsymbol{q}^{(3)}$ is rotating and translating with respect to the coordinate system defined by the metric vectors $\boldsymbol{q}^1, \boldsymbol{q}^2, \boldsymbol{q}^3$.

Let the components of the tetrad matrix be

$$q^a{}_\mu := q^{(1)}{}_1, q^{(1)}{}_2 \dots, \tag{5.10}$$

then these define the $O(3)$ electromagnetic field [1,2,5-10] or potential:

$$A^a{}_\mu := A^{(1)}{}_1, A^{(1)}{}_2, \dots \tag{5.11}$$

through the Evans wave equation [2], which is an eigenequation with the tetrad

$$A^a{}_\mu = A^{(0)} q^a{}_\mu \tag{5.12}$$

as the eigenfunction.

The inverse of the tetrad matrix is defined by [12]

$$A^\mu{}_a := \left(A^a{}_\mu\right)^{-1} = \text{adj } A^a{}_\mu / |A^a{}_\mu|, \tag{5.13}$$

where adj $A^a{}_\mu$ is the adjoint and $|A^a{}_\mu|$ the determinant of the tetrad matrix. (The Evans equations are obtained from the determinant $|A^a{}_\mu|$ using the appropriate Lagrangian and variational methods [3,13]). The adjoint matrix is the matrix of cofactors:

$$\frac{\text{adj} A^a{}_\mu}{A^{(0)}} = \begin{bmatrix} \left(q^{(2)}{}_2 q^{(3)}{}_3 - q^{(2)}{}_3 q^{(3)}{}_3\right) & \left(q^{(2)}{}_1 q^{(3)}{}_3 - q^{(2)}{}_3 q^{(3)}{}_1\right) & \left(q^{(2)}{}_1 q^{(3)}{}_2 - q^{(2)}{}_2 q^{(3)}{}_1\right) \\ \left(q^{(1)}{}_2 q^{(3)}{}_3 - q^{(1)}{}_3 q^{(3)}{}_2\right) & \left(q^{(1)}{}_1 q^{(3)}{}_3 - q^{(1)}{}_3 q^{(3)}{}_1\right) & \left(q^{(1)}{}_1 q^{(3)}{}_2 - q^{(1)}{}_2 q^{(3)}{}_1\right) \\ \left(q^{(1)}{}_2 q^{(2)}{}_3 - q^{(1)}{}_3 q^{(2)}{}_2\right) & \left(q^{(1)}{}_2 q^{(2)}{}_3 - q^{(1)}{}_3 q^{(2)}{}_1\right) & \left(q^{(1)}{}_1 q^{(2)}{}_2 - q^{(1)}{}_2 q^{(2)}{}_1\right) \end{bmatrix} . \tag{5.14}$$

The cofactors can be expressed in terms of the elements of the matrix $q^a{}_\mu$ by using the $O(3)$ cyclic relation that defines the orthonormal, Euclidean, tangent space a:

$$e^{(2)}{}_2 e^{(3)}{}_3 - e^{(2)}{}_3 e^{(3)}{}_2 = -ie^{(1)*}{}_1 = -ie^{(2)}{}_1, \quad \text{etc.} \tag{5.15}$$

Thus,

$$q^{(2)}{}_2 q^{(3)}{}_3 - q^{(2)}{}_3 q^{(3)}{}_2 = -i q^{(2)}{}_1, \quad \text{etc,} \tag{5.16}$$

and the adjoint matrix becomes

$$\text{adj } A^a{}_\mu = -iA^{(0)} \begin{bmatrix} q^{(2)}{}_1 \, q^{(1)}{}_1 \, q^{(3)}{}_1 \\ q^{(2)}{}_2 \, q^{(1)}{}_2 \, q^{(3)}{}_2 \\ q^{(2)}{}_3 \, q^{(1)}{}_3 \, q^{(3)}{}_3 \end{bmatrix} . \tag{5.17}$$

The determinant $|A^a{}_\mu|$ is defined by

$$A \text{ adj} A = |A| I A^{(0)}, \quad |A^a{}_\mu| = -iA^{(0)}. \tag{5.18}$$

The inverse matrix is therefore

$$A^\mu{}_a = A^{(0)} \begin{bmatrix} q^{(2)}{}_1 \, q^{(1)}{}_1 \, q^{(3)}{}_1 \\ q^{(2)}{}_2 \, q^{(1)}{}_2 \, q^{(3)}{}_2 \\ q^{(2)}{}_3 \, q^{(1)}{}_3 \, q^{(3)}{}_3 \end{bmatrix} , \tag{5.19}$$

and we arrive at the equations

$$A^{(0)} \begin{bmatrix} q^{(1)} \\ q^{(2)} \\ q^{(3)} \end{bmatrix} = A^a{}_\mu \begin{bmatrix} q^1 \\ q^2 \\ q^3 \end{bmatrix} , \quad A^{(0)} \begin{bmatrix} q^1 \\ q^2 \\ q^3 \end{bmatrix} = A^\mu{}_a \begin{bmatrix} q^{(1)} \\ q^{(2)} \\ q^{(3)} \end{bmatrix} . \tag{5.20}$$

In the notation of differential geometry [3], these two equations are written as

$$A^{(0)} q^a = A^a{}_\mu q^\mu, \quad A^{(0)} q^\mu = A^\mu{}_a q^a \tag{5.21}$$

and define the tetrad $A^a{}_\mu$ and the inverse tetrad $A^\mu{}_a$. The tetrad and its inverse define the FIELDS in grand unified field theory. In other words, all four fields currently thought to exist are defined by the way in which the tangent space is related to the base manifold. The interrelation of fields is also defined by the tetrad. In the presence of gravitation, for example, the base manifold becomes curved, and the tetrad defining the electromagnetic field in Eq. (5.12) changes, so the electromagnetic field changes due to the presence of gravitation.

These considerations can be extended to the four dimensions of space-time by interrelating the two basis four-vectors

$$\begin{aligned} q^a &:= \left(q^{(0)}, q^{(1)}, q^{(2)}, q^{(3)}\right), \\ q^\mu &:= \left(q^0, q^1, q^2, q^3\right), \end{aligned} \tag{5.22}$$

in which the tetrad is defined by the four by four invertible matrix

$$A^{(0)} \begin{bmatrix} q^{(0)} \\ q^{(1)} \\ q^{(2)} \\ q^{(3)} \end{bmatrix} = A^{a}{}_{\mu} \begin{bmatrix} q^{0} \\ q^{1} \\ q^{2} \\ q^{3} \end{bmatrix}, \quad A^{(0)} \begin{bmatrix} q^{0} \\ q^{1} \\ q^{2} \\ q^{3} \end{bmatrix} = A^{\mu}{}_{a} \begin{bmatrix} q^{(0)} \\ q^{(1)} \\ q^{(2)} \\ q^{(3)} \end{bmatrix} \tag{5.23}$$

whose adjoint (matrix of cofactors) reads

$$\text{adj } A^{a}{}_{\mu} = -i \begin{bmatrix} q^{(0)}{}_{0} \; q^{(2)}{}_{0} \; q^{(1)}{}_{0} \; q^{(3)}{}_{0} \\ q^{(0)}{}_{1} \; q^{(2)}{}_{1} \; q^{(1)}{}_{1} \; q^{(3)}{}_{1} \\ q^{(0)}{}_{2} \; q^{(2)}{}_{2} \; q^{(1)}{}_{2} \; q^{(3)}{}_{2} \\ q^{(0)}{}_{3} \; q^{(2)}{}_{3} \; q^{(1)}{}_{3} \; q^{(3)}{}_{3} \end{bmatrix} \tag{5.24}$$

and whose determinant is

$$|A^{a}{}_{\mu}| = -i. \tag{5.25}$$

The inverse matrix (or inverse tetrad) is therefore defined by

$$A^{a}{}_{\mu} A^{\nu}{}_{a} = \delta^{\nu}{}_{\mu}. \tag{5.26}$$

We have gone through this exercise in detail to illustrate the meaning of the tetrad in $O(3)$ electrodynamics. It is essentially the matrix that inter-relates two frames of reference: that of the base manifold and the orthonormal tangent space. The tetrad is the eigenfunction of the Evans wave equation [2], which for $O(3)$ electrodynamics becomes the eigenequation

$$(\Box + kT) A^{a}{}_{\mu} = 0, \tag{5.27}$$

where k is the gravitational constant and T the contracted energy-momentum tensor [2]. The wave equation (5.27) is a direct consequence of the tetrad postulate [2,3], which for $O(3)$ electrodynamics is

$$D_{\nu} A^{a}{}_{\mu} = 0, \tag{5.28}$$

representing a cyclic relation [2] between tetrad components (components of the $O(3)$ electromagnetic field). The tetrad is centrally important to the theory of general relativity expressed as a theory of differential geometry [3,13] and is used to derive the Maurer-Cartan structure relations [3] between differential forms: the Riemann and torsion forms, valid for all types of connection. By recognizing that the tetrad is the fundamental eigenfunction in the Evans wave equation [2], it follows (Sec. 5.3) that the Maurer-Cartan structure relations become field/matter relations of grand unified field theory, i.e., of all the known force fields and matter waves in nature. The wave equation [2] is the quantized version of the classical field/matter equation introduced in Ref. [1]:

$$R_{\mu} - \frac{1}{2} R q_{\mu} = kT_{\mu}. \tag{5.29}$$

Equation (5.29) is written in terms of four-vectors in a base manifold that is in general a non-Euclidean spacetime. The field tensor in Eq. (5.29) is the four-vector

$$G_\mu = R_\mu - \frac{1}{2} R q_\mu, \tag{5.30}$$

where R_μ is the Ricci four-vector, R the scalar curvature, q_μ the metric four-vector, k is the Einstein constant, and T_μ the canonical energy-momentum four-vector. In the language of differential geometry [2,3,13], Eq. (5.29) becomes a relation between the corresponding differential one-forms. In tetrad notation, Eq. (5.29) is

$$G^a{}_\mu = kT^a{}_\mu . \tag{5.31}$$

In differential geometry, the equation becomes

$$G^a = kT^a, \tag{5.32}$$

where the index μ is implied [3]. The wave equation is differential geometry is the eigenequation with the one form (tetrad) q^a as eigenfunction:

$$(\Box + kT) q^a = 0. \tag{5.33}$$

The $O(3)$ electromagnetic gauge field is then the two-form [2]

$$G^c = G^{(0)} q^a \wedge q^b, \tag{5.34}$$

where the wedge produce is a wedge product between one-forms [3]:

$$\left(q^a \wedge q^b\right)_{\mu\nu} := q^a{}_\mu q^b{}_\nu - q^a{}_\nu q^b{}_\mu . \tag{5.35}$$

A central result of this theory is that electrodynamics is a gauge field (Eq. (5.34)) whose internal index c is the tangent index of the base manifold in general relativity. The tangent space is an orthonormal Euclidean space whose unit vectors form an $O(3)$ symmetry cyclic relation:

$$\boldsymbol{e}^{(1)} \times \boldsymbol{e}^{(2)} = i \boldsymbol{e}^{(3)*}, \quad \text{et cyclicum}. \tag{5.36}$$

Thus, generally covariant electrodynamics in an $O(3)$ symmetry gauge field theory [1,2,5-10]. This is a fundamental result of general relativity and differential geometry.

5.3 GUFT As Differential Geometry: Evans Equations And The Maurer-Cartan Structure Relations

Equation (5.31) is the Evans field equation (5.29) for each index a. The quantities in Eq. (5.29) are defined by [1-3]

$$R_{\mu\nu} = R_\mu q_\nu, \quad q_{\mu\nu} = q_\mu q_\nu, \quad T_{\mu\nu} = T_\mu q_\nu, \tag{5.37}$$

where $R_{\mu\nu}$ is the Ricci tensor, $q_{\mu\nu}$ the symmetric metric tensor, and $T_{\mu\nu}$ the symmetric canonical energy-momentum tensor of Einstein's general relativity [3]. The symmetric metric tensor is defined as

$$q_{\mu\nu} = q_\mu q_\nu \tag{5.38}$$

and [14]

$$q^{\mu\nu} q_{\mu\nu} = 4. \tag{5.39}$$

Considering the flat spacetime limit:

$$q^\mu q_\mu = -2, \quad q^\mu = (1,1,1,1), \quad q_\mu = (1,-1,-1,-1). \tag{5.40}$$

More generally, Eq. (5.40) indicates that, in non-Euclidean spacetime, q^μ is the inverse metric if q_μ is the metric. The scalar curvature is

$$R = q^{\mu\nu} R_{\mu\nu} = q^\mu q^\nu R_\mu q_\nu = -2q^\mu R_\mu; \tag{5.41}$$

and multiplying Eq. (5.41) on both sides by q_μ gives

$$R_\mu = \frac{1}{4} R q_\mu. \tag{5.42}$$

The Evans field equation (5.29) can therefore be developed as

$$G^a{}_\mu = -\frac{1}{4} R q^a{}_\mu = k T^a{}_\mu. \tag{5.43}$$

Similarly, the contracted energy-momentum tensor is defined by Einstein [14] as

$$T = q^{\mu\nu} T_{\mu\nu} = -2q^\mu T_\mu. \tag{5.44}$$

Multiplication of Eq. (5.44) on both sides by q_μ gives

$$T_\mu = \frac{1}{4} T q_\mu. \tag{5.45}$$

Therefore, the Evans field equation becomes the classical equation

$$R q^a{}_\mu = -k T q^a{}_\mu, \tag{5.46}$$

which is the contracted form of Einstein's field equation

$$R = -kT \tag{5.47}$$

multiplied on both sides by the tetrad $q^a{}_\mu$. The Evans wave equation [2]

$$\Box q^a{}_\mu = R q^a{}_\mu = -k T q^a{}_\mu \tag{5.48}$$

is obtained from the tetrad postulate and is the quantized version of Eq. (5.47).

The classical and quantum Evans equations are equations for the fundamental field in grand unified field theory. In both cases the field is recognized as the tetrad. The tetrad represents the components indexed μ of the coordinate basis vectors in terms of the components indexed a of the orthonormal basis defining the vectors o the tangent space in general relativity. In differential geometry the tetrads are also the components of the orthonormal basis one-forms in terms of the coordinate basis one forms [3].

The classical Eq. (5.4) states that

$$R + kT = 0 \tag{5.49}$$

by *ansatz*, the basic postulate of general relativity. The quantum Eq. (5.48) originates in the tetrad postulate [3] of geometry, a postulate which is true for any connection, whether torsion free or not, and Eq. (5.48) gives quantum field/matter theory from general relativity. Gravitation is described by these equations when the field is the tetrad $q^a{}_\mu$. The other three fields are described when the field is the tetrad multiplied by an appropriate scaling factor. For example, the fundamental electromagnetic field is

$$A^a{}_\mu = A^{(0)} q^a{}_\mu, \tag{5.50}$$

where $A^{(0)}$ is the magnitude of the electromagnetic four-potential, as described in Sec. 5.1.

The *gauge invariant* fields, or gauge fields, are defined as follows. The gravitational gauge field is the Riemann form of differential geometry, the dual of the torsion tensor:

$$R_{ab} = \frac{R}{4} \varepsilon_{abcd} [q^c, q^d]. \tag{5.51}$$

The other three fields are defined by the torsion form, the anti-symmetric sum or commutator of tetrads:

$$\tau^c{}_{\mu\nu} := \frac{R}{4} \left[q^a{}_\mu, q^b{}_\nu \right] = \frac{R}{4} \left(q^a{}_\mu q^b{}_\nu - q^a{}_\nu q^b{}_\mu \right). \tag{5.52}$$

The tetrad is a vector valued one form [3]. Define its exterior derivative as the vector valued two-form

$$(dq)^a{}_{\mu\nu} := \partial_\mu q^a{}_\nu - \partial_\nu q^a{}_\mu \tag{5.53}$$

and its covariant exterior derivative as

$$\begin{aligned} (Dq)^a{}_{\mu\nu} &= (dq + \omega \wedge q)^a{}_{\mu\nu} \\ &= \partial_\mu q^a{}_\nu - \partial_\nu q^a{}_\mu + \omega^a{}_{\mu b} q^b{}_\nu - \omega^a{}_{\nu b} q^b{}_\mu, \end{aligned} \tag{5.54}$$

where ω is the spin connection [3].

The first Maurer-Cartan structure relation [3] is then

$$\tau^c = Dq^c, \tag{5.55}$$

and the second Maurer-Cartan structure relation is

$$R^a{}_b = D\omega^a{}_b. \tag{5.56}$$

We therefore arrive at the following equations which combine the two Evans equations with the two Maurer-Cartan structure relations to define the *gauge invariant fields* in grand unified field theory:

$$R^a{}_b = \varepsilon^a{}_{bc}T^c, \tag{5.57}$$

where T^c is the torsion form,

$$\tau^c = \frac{R}{4}\left[q^a, q^b\right] = Dq^c. \tag{5.58}$$

The gravitational field is described by the Riemann form and the other three fields by the torsion form multiplied by the scale factor $G^{(0)}$:

$$G^c = G^{(0)}\tau^c. \tag{5.59}$$

In $O(3)$ electrodynamics the scale factor is the primordial magnetic flux density (in units of tesla, or webers per square meter):

$$B^{(0)} = (1/4)G^{(0)}R. \tag{5.60}$$

Equations (5.57) and (5.58) also define the spin connection in terms of the tetrad for all four fields.

Differential geometry [3] also gives the following identities:

$$DR^a{}_b := 0, \tag{5.61}$$

$$D^*\tau^c := 0, \tag{5.62}$$

$$D\tau^c = R^c{}_b \wedge q^b, \tag{5.63}$$

which can be used to interrelate the four gauge fields. In the following section, several novel results are given by interrelating the gravitational and electromagnetic fields. Equation (5.61) is the Bianchi identity, and Eq. (5.62) and (5.63) are the homogeneous and inhomogeneous gauge field equations. In $O(3)$ electrodynamics Eq. (5.63) shows that the current term is derived from the Riemann form, and therefore from the scalar curvature R multiplied by the anti-commutator of tetrads.

This fundamental result of differential geometry implies that electromagnetic energy can be transmitted from a source to a receiver by scalar curvature R and that electromagnetic energy is available in non-Euclidean spacetime.

Experimental evidence supporting ths result might be found in devices such as the patented motionless electromagnetic generator [15] and Sweet's device [16]. All fields in nature are fundamentally dependent upon, and originate in, scalar curvature R.

5.4 Inter-Relation Of Fields: The Poisson Equation And Other Results

The Evans wave equation [2] is obtained by covariant differentiation of the tetrad postulate (5.28) and reads

$$D^\rho D_\rho q^a{}_\mu = (\Box + kT)q^a{}_\mu = (\Box - R)q^a{}_\mu = 0, \tag{5.64}$$

where $\Box$ is the d'Alembertian in Euclidean spacetime (the spacetime of special relativity). The scalar curvature in this equation is [2]

$$R = -D^\mu \Gamma^\rho{}_{\mu\rho} = -\left(\partial^\mu \Gamma^\rho{}_{\mu\rho} + \Gamma^{\mu\rho}{}_\lambda \Gamma^\lambda{}_{\mu\rho}\right). \tag{5.65}$$

We therefore deduce that

$$\partial^\mu \Gamma^\rho{}_{\mu\rho} + \Gamma^{\mu\rho}{}_\lambda \Gamma^\lambda{}_{\mu\rho} = kT. \tag{5.66}$$

Both Eqs. (5.64) and (5.66) give Poisson's equation in the appropriate approximation. The equivalence principle states that the laws of physics in small enough regions of spacetime reduce to the equations of special relativity [3,4]. In special relativity there exist equations such as

$$(\Box + \kappa^2{}_0)\phi = 0, \tag{5.67}$$

$$(\Box + \kappa^2{}_0)\psi = 0, \tag{5.68}$$

$$(\Box + \kappa^2{}_0)A_\mu = 0, \tag{5.69}$$

where λ_0 is the Compton wavelength of a particle:

$$\lambda_0 = 1/\kappa_0 = \hbar/mc. \tag{5.70}$$

Here m is the particle mass, $\hbar$ is the Dirac constant $h/2\pi$, and c the speed of light in vacuum. Equation (5.67) is the Klein-Gordon equation, in which ϕ is the scalar field; Eq. (5.68) is the Dirac equation, in which ψ is the four spinor; and Eq. (5.69) is the Proca equation, in which A_μ is the electromagnetic wave function, conventionally a four-vector of the Maxwell Heaviside theory [4]. (In generally covariant electrodynamics $A^a{}_\mu$ is a tetrad, as we have already argued.) In order to reduce Eq. (5.64) to Eqs. (5.67), (5.68), and (5.69), we define the latter as limiting forms of the Evans wave equation of general relativity, Eq. (5.64). The limiting forms are obtained when the *rest curvature* R_0 is defined by

$$|R_0| = \kappa^2{}_0 = kT_0, \tag{5.71}$$

where

$$T_0 = m/V = q^0 q^0 T_{00}, \quad q^0 := 1. \tag{5.72}$$

The rest curvature is therefore the inverse square of the Compton wavelength for any particle, including the photon:

$$|R_0| = m^2c^2/\hbar^2. \tag{5.73}$$

The rest curvature is defined by the rest energy:

$$E_0 = mc^2 = \hbar c\sqrt{|R_0|}, \tag{5.74}$$

where

$$\kappa_0 = \sqrt{|R_0|}. \tag{5.75}$$

More generally, the quantum of energy for any particle (not only the photon) is given by

$$E = \hbar c\sqrt{|R|}. \tag{5.76}$$

where

$$|R| = D^{\mu}\Gamma^{\rho}{}_{\mu\rho}. \tag{5.77}$$

Equation (5.76) is therefore a generalization of the Planck postulate for the photon in special relativity to all particles in general relativity.

On writing

$$E = \hbar\omega = \hbar c\sqrt{|R|}, \tag{5.78}$$

it becomes clear the the quantum $\hbar\omega$ is electromagnetic energy, e.g., is available from scalar curvature R, i.e., from any kind of connection $\Gamma^{\rho}{}_{\mu\rho}$, and for any sources of R. Equation (5.63) is a statement of this result in classical electrodynamics.

It is now demonstrated that the well-known Poisson equation for both Newtonian dynamics and for electro-statics can be obtained self consistently both from Eq. (5.64) and Eq. (5.66). Starting from Eq. (5.64), the Poisson equation is obtained from the component $q^0{}_0$:

$$\left(\Box + \frac{km}{V}\right) q^0{}_0 = 0, \tag{5.79}$$

in which $T = m/V$ is the mass density and V is a volume. The rest mass density m/V_0, where V_0 is the rest volume, can be regarded as the zero-point energy of the Evans wave equation of generally covariant quantum field theory. In the weak field or Newtonian limit [1-3], Eq. (5.79) becomes

$$\Box q^0{}_0 = -\frac{km}{V} q^0{}_0 \sim -\frac{km}{V}, \tag{5.80}$$

because in this limit

$$q^0{}_0 = 1 + \eta^0{}_0, \tag{5.81}$$

where $\eta^0{}_0$ is a small perturbation of the tetrad. In the Newtonian limit, $q^0{}_0$ is a quasi-static, so Eq. (5.80) reduces to

$$\nabla^2\Phi = 4\pi G\rho, \tag{5.82}$$

which is the Poisson equation of Newtonian dynamics [3,17] provided that

$$\Phi := \frac{1}{2}c^2 q^0{}_0, \tag{5.83}$$

where G is Newton's gravitational constant and ρ the mass density m/V.

In order to derive Eq. (5.82) from Eq. (5.66), it is assumed that in the weak-field limit the term quadratic in the connection can be neglected and the Eq. (5.72) applies, so that the relevant component of the metric vector q_μ in the non-Euclidean base manifold is q_0. In the Newtonian limit

$$q^0 = 1 + \eta^0, \tag{5.84}$$

where $\eta^0 \ll q^0$. The metric vector is assumed to be metric compatible [2,3], and the equation of metric compatibility for q^μ is

$$D_\nu q^\mu = \partial_\nu q^\mu + \Gamma^\mu{}_{\nu\lambda} q^\lambda = 0. \tag{5.85}$$

The relevant index to consider is $\lambda = 0$, and thus

$$\partial_\nu q^\mu + \Gamma^\mu{}_{\nu 0} q^0 = 0. \tag{5.86}$$

Multiplying by q_0 and using $q_0 q^0 = 1$, one gets

$$\Gamma^\mu{}_{\nu 0} = -q_0 \partial_\nu q^\mu \sim -\partial_\nu q^\mu, \tag{5.87}$$

where Eq. (5.84) has been used. The relevant index in Eq. (5.87) is $\mu = 0$, giving

$$\Gamma^0{}_{\nu 0} = -\partial_\nu q^0, \tag{5.88}$$

so that Eq. (5.66) becomes

$$\partial^\mu \Gamma^0{}_{\mu 0} = -\Box q^0 = (8\pi/c^2) G\rho. \tag{5.89}$$

In the Newtonian limit the field is quasi-static, so Eq. (5.89) reduces to the Poisson Eq. (5.82), provided that

$$\Phi := \frac{1}{2}c^2 q^0. \tag{5.90}$$

These two derivations of the Poisson equation in the Newtonian limit show that Eq. (5.64) is consistent Eq. (5.66) *and that there exists a rest curvature corresponding to rest energy in special relativity.* The rest curvature produces energy which is convertible into any other form of energy, including electromagnetic energy; and, in grand unified field theory, electromagnetic energy originates in general from curvature R. Equations (5.67) and (5.69) are limiting forms of Eq. (5.64) when the curvature is the rest curvature. Therefore the scalar field, the spinor field, and the conventional electromagnetic field of Maxwell Heaviside field theory are all limiting forms of the tetrad.

The Poisson equation in electrostatics is obtained, using the same approximations, from the Evans wave equation (5.64) when its eigenfunction is the electromagnetic tetrad field defined by Eq. (5.50). The Poisson equation for electrostatics is therefore a limiting form of the equation

$$(\Box + kT)A^a{}_\mu = (\Box - R)A^a{}_\mu = 0 \tag{5.91}$$

in general relativity and grand unified field theory; it is

$$\nabla^2(A^{(0)}\Phi) = 4\pi G(A^{(0)}\rho). \tag{5.92}$$

The scalar potential in electrostatics is therefore

$$\phi = (1/c)A^{(0)}\Phi \tag{5.93}$$

and the Poisson equation in electrostatics is the Schrödinger-type eigenequation

$$\nabla^2\phi = (4\pi G\rho/c^2)\phi. \tag{5.94}$$

The units of ϕ are JC^{-1} and the units of $A^{(0)}$ are $JsC^{-1}m^{-1}$, and so

$$\phi = cA^{(0)}. \tag{5.95}$$

The Poisson equation is therefore an eigenequation in the electrostatic potential ϕ, and this result can be thought of as the quantization of charge because the quantity $\varepsilon_0\phi$ where ε_0 is the vacuum permittivity (S.I. Units) has the units of C m^{-1}. Therefore if electrostatics is considered to be an approximation of generally covariant grand unified field theory, we obtain the equation of quantization of charge:

$$\nabla^2\psi_e = (4\pi G\rho/c^2)\psi_e \tag{5.96}$$

where $\psi_e = \varepsilon_0\phi$ and

$$\varepsilon_0 = 8.854188 \times 10^{-12}J^{-1}C^2m^{-1}, \quad G = 6.6726 \times 10^{-11}Nm^2kg^{-2}. \tag{96b}$$

The Poisson equation in S.I. units in electrostatics is usually written as

$$\nabla^2\phi = -\rho_e/\varepsilon_0, \tag{5.97}$$

where ρ_e is the charge density in C m^{-3}. The minus sign in Eq. (5.97) is a matter of convention [18]; it is chosen so that the electric field strength is defined by

$$\boldsymbol{E} := -\boldsymbol{\nabla}\phi, \tag{5.98}$$

i.e., points towards a decrease in potential. The factor 4π is conventionally divided out of the Poisson equation for electrostatics [18] as a matter of convenience in the S.I. system of units. Therefore the Poisson equation for electrostatics can always be written in the same form as that for Newtonian dynamics:

$$\nabla^2\phi = (4\pi/\varepsilon_0)\rho_e, \tag{5.99}$$

where the charge density ρ_e and mass density ρ are, respectively,

$$\rho_e = \int edV = (\varepsilon_0 G/c^2)\phi^{(0)}\rho, \quad \rho = \int mdV. \tag{5.100}$$

Here e is the fundamental charge (defined as the charge on the proton, or the modulus of the charge on the electron) and m the mass of the electron:

$$e = 1.60219 \times 10^{-19}\ \text{C}, \quad m = 9.10953 \times 10^{-31}\ \text{kg}. \tag{5.101}$$

Comparison of Eqs. (5.94) and (5.99) yields

$$\rho_e = (\varepsilon_0 G/c^2)\phi^{(0)}\rho, \tag{5.102}$$

giving the fundamental ratio of charge e to mass m in terms of the fundamental electrostatic potential $\phi^{(0)}$:

$$e = (\varepsilon_0 G/c^2)\phi^{(0)} m. \tag{5.103}$$

Application of (5.103) to the electron gives

$$\phi^{(0)}(\text{one electron}) = \frac{c^2 e}{\varepsilon_0 G m}\text{JC}^{-1}, \tag{5.104}$$

which is a fundamental constant furnishing the number of joules available from the ratio e/m for one electron in Eq. (5.104),

$$\begin{aligned} c &= 2.997925 \times 10^8 \text{m s}^{-1}, \\ e &= 1.60219 \times 10^{-19}\text{C}, \\ \varepsilon_0 &= 8.854188 \times 10^{-12}\text{J}^{-1}\text{C}^2\text{m}^{-1}, \\ G &= 6.6726 \times 10^{-11}\text{Nm}^2\text{kg}^{-2}, \\ m &= 9.10953 \times 10^{-31}\text{kg}, \end{aligned} \tag{5.105}$$

and so

$$\begin{aligned} \phi^{(0)}(\text{one electron}) &= 2.6726 \times 10^{49}\ \text{J C}^{-1}, \\ A^{(0)}(\text{one electron}) &= 8.92473 \times 10^{40}\ \text{J C}^{-1}\text{,s m}^{-1}. \end{aligned} \tag{5.106}$$

It now becomes possible to interrelate the fundamental equations of Newtonian dynamics and electrostatics by using Eqs. (5.102) and (5.103) to interrelate terms in the following well-known equations:

$$\boldsymbol{g} = -\boldsymbol{\nabla}\Phi, \quad \boldsymbol{E} = -\boldsymbol{\nabla}\phi; \tag{5.107}$$

$$\boldsymbol{F} = -(Gm_1m_2/r^2)\boldsymbol{k} = m_1\boldsymbol{g}, \quad \boldsymbol{F} = -(e_1e_2/\varepsilon_0 r^2)\boldsymbol{k} = e_1\boldsymbol{E}, \tag{5.108}$$

$$\boldsymbol{\nabla}\Phi = (Gm_2/r^2)\boldsymbol{k}, \quad \boldsymbol{\nabla}\phi = (e_2/\varepsilon_0 r^2)\boldsymbol{k}, \tag{5.109}$$

in which $\boldsymbol{g}$ is the gravitational acceleration, r the distance between two masses m_1 and m_2 or two charges e_1 and e_2, and $\boldsymbol{F}$ is the force between the two masses of the two charges.

All these well-known equations originate in the wave equation (5.64) of generally covariant grand unified field theory. Both the gravitational and the electrostatic fields originate in the tetrad when approximated in the weak-field limit in which the tetrad is a perturbation of Euclidean spacetime and in which the field is quasistatic.

All forms of energy are inter-convertible, and it follows from the equation

$$R = -kT \tag{5.110}$$

that all forms of curvature R are interconvertible. Newtonian dynamics is the weak-field limit of the wave equation (5.64), and electrostatics is the weak-field limit of the same equation multiplied by $A^{(0)}$. It follows that, given a charge e, there is an equivalent amount of mass m. If the force in Newton's between two masses a distance r apart is numerically the same as that between two charges the same distance r apart, then

$$e^2 = \varepsilon_0 G m^2. \tag{5.111}$$

From this equation,

$$e = \pm 2.430647 \times 10^{-11}\text{m}, \quad m = \pm 4.11413 \times 10^{10}|e|, \tag{5.112}$$

which shows that if there is only one sign of mass (as observed experimentally). Charge can be thought to originate in "symmetry breaking" of mass into two different signs, and mass originates in curvature of non-Euclidean spacetime, something which can be loosely described as a "symmetry breaking of the vacuum." Therefore there can be neither charge nor mass in Euclidean spacetime.

The force between two charges of one coulomb each, one meter apart, is

$$1/\varepsilon_0 = 1.12941 \times 10^{11}\ \text{N}, \tag{5.113}$$

and the force between two masses of one kilogram each, one meter apart, is

$$G = 6.6726 \times 10^{-11}\ \text{N}. \tag{5.114}$$

The electrostatic force is therefore

$$1/\varepsilon_0 G = 1.69261 \times 10^{21} \tag{5.115}$$

times greater than the gravitational force, which is why the gravitational force is conventionally referred to as "weak" in comparison with the electrostatic force in the laboratory. However, both forces originate in rest curvature R_0. From Eq. (5.102), the charge density is given by

$$\rho_e = -(\varepsilon_0 \phi^{(0)}/8\pi)R_0, \tag{5.116}$$

and the mass density is

$$\rho = -R/k. \tag{5.117}$$

The Einstein constant is

$$k = 8\pi G/c^2 = 1.86595 \times 10^{-26} \text{ N s}^2\text{kg}^{-2}, \tag{5.118}$$

and thus

$$\rho_e = -3.5229 \times 10^{-13}\phi^{(0)}R_0, \quad \rho = -5.35920 \times 10^{25}R_0. \tag{5.119}$$

If the fundamental potential (5.106) is used in Eq. (5.119), the following results are obtained:

$$\rho_e = -9.41530 \times 10^{36}R_0, \quad \rho = -5.35920 \times 10^{25}R_0. \tag{5.120}$$

The charge obtainable from a given curvature R in a given volume is about twelve orders of magnitude greater than the mass obtainable from the same curvature R for the same volume.

From Eq. (5.79) it is possible to define the rest volume of a particle:

$$V_0 = \frac{8\pi G\hbar^2}{mc^4} = \frac{8\pi {\lambda^2}_p}{\sqrt{|R_0|}}, \tag{5.121}$$

where

$$\lambda_c = 1/\sqrt{|R_0|} = \hbar/mc \tag{5.122}$$

is the Compton wavelength, and where λ_p denotes the Planck length

$$\lambda_p = \sqrt{G\hbar/c^3}. \tag{5.123}$$

For the electron,

$$\lambda_c = 2.42631 \times 10^{-12}\text{m}, \quad \lambda_p = 4.05087 \times 10^{-35}\text{m}, \tag{5.124}$$

and the rest volume is therefore

$$V_0 = 8\pi\lambda_c {\lambda^2}_p = 1.00065 \times 10^{-79}\text{m}^3. \tag{5.125}$$

Equation (5.122) interrelates wave-particle dualism and general relativity.

The Planck and Compton wavelengths can be expressed in terms of the rest curvature R_0 as follows:

$$\lambda_c = (c^2/Gm){\lambda^2}_p = 1/\sqrt{|R_0|}. \tag{5.126}$$

The classical radius of the electron (λ_e) can also be expressed in terms of rest curvature R_0:

$$\lambda_e = \alpha\lambda_0, \quad \lambda_0 = 1/\sqrt{|R_0|}, \tag{5.127}$$

where

$$\alpha = e^2/4\pi\varepsilon_0\hbar c \tag{5.128}$$

is the fine structure constant of electrodynamics:

$$\alpha = 7.297351 \times 10^{-3}. \tag{5.129}$$

Equation (5.127) derives quantum electrodynamics from generally covariant grand unified field theory.

The *rest wave-number*

$$\kappa_0 = 1/\lambda_0 = \sqrt{|R_0|} \tag{5.130}$$

is fundamental to the interrelation of general relativity with quantum mechanics and classical and quantum electrodynamics. The Klein-Gordon, Dirac, and Proca equations can each be written in terms of the rest wavenumber, which is the inverse of the Compton wavelength:

$$(\Box - R_0)\phi = 0, \tag{5.131}$$

$$(\Box - R_0)\psi = 0, \tag{5.132}$$

$$(\Box - R_0)A_\mu = 0. \tag{5.133}$$

The Evans wave equation shows that

$$D^\mu \Gamma^\rho{}_{\mu\rho} \rightarrow -R_0 \tag{5.134}$$

in the limit of special relativity.

The equation of quantization of mass is

$$m = \hbar\kappa_0/c := \hbar\omega_0/c^2. \tag{5.135}$$

The ratio of charge to mass can always be expressed as

$$e^2/m^2 = \zeta\varepsilon_0 G, \tag{5.136}$$

where ζ is a dimensionless coefficient. Equation (5.136) is the result of the Newton and Coulomb inverse square laws, which we have derived from the Evans equation in the weak-field limit. Therefore the Newton and Coulomb inverse square laws become the same. Using Eq. (5.135) in Eq. (5.136),

$$e = \pm\hbar\kappa_0\sqrt{\zeta\varepsilon_0 G/c^2}. \tag{5.137}$$

Comparing Eqs. (5.136) and (5.102), one gets

$$\zeta = (e/mc^2)\phi, \tag{5.138}$$

and thus Eq. (5.137) is the charge quantization equation.

All these results emerge from the Evans wave equation in the weak-field approximation.

Equation (5.135) can be generalized to

$$m = \hbar\kappa_0/c = \hbar\sqrt{|R|}/c, \tag{5.139}$$

i.e., to

$$E = \hbar c\sqrt{|R|} = \hbar c\kappa = \hbar\omega, \tag{5.140}$$

which is Eq. (5.76) for the quantization of energy in terms of the wavenumber $\sqrt{|R|}$. Similarly, Eq. (5.139) can be generalized to

$$e = \pm(\hbar/c)\sqrt{\zeta\varepsilon_0 G|R|}. \tag{5.141}$$

Equations (5.140) and (5.141) are generally covariant equations for the quantization of charge and mass in terms of the wave number $\sqrt{|R|}$, where $R + kT = 0$. *They show that mass and charge are quantized in terms of the fundamental wavenumbers:*

$$\kappa = \sqrt{|R|}. \tag{5.142}$$

5.5 Discussion

It has been shown that all force and matter fields in nature are determined and interrelated by the two Evans equations [1,2] combined with the Maurer-Cartan structure relations of differential geometry [3]. The homogeneous and inhomogeneous gauge field equations then follow from geometrical considerations, and interrelate the torsion and Riemann forms. The gauge field equations show how the force and matter fields are interrelated. Some consequences of this deduction are worked out in Sec. 5.4, and there are many other consequences yet to be inferred and tested experimentally. One of the major predictions of the Evans equations, the existence of higher symmetry electrodynamics, has been extensively developed and tested experimentally [5-10]; for example, the fundamental $\boldsymbol{B}^{(3)}$ field of generally covariant electrodynamics has been observed experimentally in the inverse Faraday effect, in the phase factor of physical optics and interferometry, and in several other ways. Another major prediction of the Evans equations is the availability of electromagnetic energy from non-Euclidean spacetime. This prediction appears to be verified experimentally in devices such as the patented electromagnetic generator [15] and in Sweet's device [16]. Development of such devices could lead to the general availability of electromagnetic energy, energy which originates in the wavenumber $\sqrt{|R|}$. The Evans equations show that the acquisition of such energy does not violate Noether's Theorem. Energy is obtained from kT in a source situated anywhere in the universe and is transmitted to the receiver by the scalar curvature $R = -kT$ of non-Euclidean spacetime, sometimes referred to loosely as "the vacuum." The popular phrase "energy from the vacuum" [15] does not imply "energy from nothing at all."

The most important aspect of a grand unified field theory is its ability to interrelate fields, and in Sec. 5.4 we have illustrated this by interrelating, and thus identifying, charge and mass through the fundamental potential. Since all forms of scalar curvature are inter-convertible, electromagnetic energy is obtainable from *any* type of scalar curvature anywhere in the universe, given the existence of the primordial fluxon $\hbar/e$ anywhere in the universe. Therefore electromagnetic energy from non-Euclidean spacetime is available to the Earth-bound engineer in usable form and originates in the curvatures inherent in the rest of the universe. The Evans field and wave equations describe how the energy propagates from source to Earth-bound engineer in usable form, and originates in the curvatures inherent in the rest of the universe. The Evans field and wave equations describe how the energy propagates from source to Earth-bound observer. It has been shown in Sec. 5.4 that the charge generated by a given curvature R is about twelve orders of magnitude greater than the mass generated by the same curvature R in a given volume. *This means that the charge and concomitant electromagnetic energy available from curvature induced by mass is amplified by about twelve orders of magnitude.* Loosely writing, a small amount of mass results in a very large amount of charge, and this augurs well for the design of devices which can trap and use this electromagnetic energy.

In general the Evans field and wave equations are nonlinear in the connection and require numerical methods for solution, but in Sec. 5.4 it has been shown that approximate analytical methods of solution can lead to powerful results on the interrelation of fields. The field is the tetrad, and the influence of one type of field (e.g., gravitational), on another (e.g., electromagnetic) is measured through changes in tetrad components as discussed in Sec. 5.2. It is important to develop the interrelation of the $O(3)$ electromagnetic field with the weak field and the gravitational field with the strong field by developing and interrelating the bases used for the tangent space and fiber bundle space. This will be the subject of further work. It is already clear however that the four fields thought to exist in nature have a common origin, the scalar curvature R.

Acknowledgments

Many formative discussions are gratefully acknowledged with the Fellows and Emeriti of the AIAS, and funding from the Ted Annis Foundation and Craddock, Inc. is gratefully acknowledged.

References

1. M. W. Evans, *Found. Phys. Lett.* **16**, 369 (2003).
2. M. W. Evans, *Found. Phys. Lett.*, in press (2003); www.aias.us
3. S. M. Carroll, *Lecture Notes in General Relativity* (University of California, Santa Barbara, graduate course on arXiv:gr-qc/9712019 v1 3 Dec., 1997).
4. L. H. Ryder, *Quantum Field Theory* (Cambridge University Press, Cambridge, 1987).
5. M. W. Evans, J.-P. Vigier, *et al. The Enigmatic Photon* (Kluwer Academic, Dordrecht, 1994 to 2002, hardback and paperback; Japanese translation in preparation, 2003; www.aias.us).
6. M. W. Evans and L. B. Crowell, *Classical and Quantum Electrodynamics and the* $\boldsymbol{B}^{(3)}$ *Field* (World Scientific, Singapore, 2001); www.aias.us
7. M. W. Evans, ed., *Modern Non-linear Optics*, in I. Prigogine and S. A. Rice, series eds., *Advances in Chemical Physics*, Vol. 119 (Wiley-Interscience, New York, 2001); www.aias.us; see also P. K. Anastasowski and D. Hamilton, Vol. 119 (3).
8. D. J. Clements and M. W. Evans, *Found. Phys. Lett.*, in press, July, 2003; www.aias.us
9. M. W. Evans, *Lecture Notes in* $O(3)$ *Electrodynamics* (World Scientific, Singapore, in preparation); first two lectures available on www.aias.us
10. D. J. Clements and M. W. Evans, *Higher Symmetry Electrodynamics in Special and General Relativity* (Kluwer Academic, Dordrecht, in preparation); www.aias.us
11. R. M. Wald, *General Relativity* (University of Chicago Press, 1984).
12. G. Stephenson, *Mathematical Methods for Science Students* (Longmans & Green, London, 1968).
13. B. F. Schutz, *A First Course in General Relativity* (Wiley, New York, 1972).
14. A. Einstein, *The Meaning of Relativity* (Princeton University Press, 1921).
15. T. E. Bearden, in Ref. 7, Vol. 119(2), and www.cheniere.org
16. DVD available from A. Craddock; see www.cheniere.org of the Sweet device.
17. J. B. Marion and S. T. Thornton, *Classical Dynamics of Particles and Systems* (HBJ, New York, 1988).
18. D. Corson and P. Lorrain, *Introduction to Electromagnetic Fields and Waves* (Freeman, San Francisco, 1962).

6

Derivation Of Dirac's Equation From The Evans Wave Equation

Summary. The Evans wave equation [1] of general relativity is expressed in spinor form, thus producing the Dirac equation in general relativity. The Dirac equation in special relativity is recovered in the limit of Euclidean or flat spacetime. By deriving the Dirac equation from the Evans equation it is demonstrated that the former originates in a novel metric compatibility condition, a geometrical constraint on the metric vector q_μ used to define the Einstein metric tensor. Contrary to some claims by Ryder, it is shown that the Dirac equation cannot be deduced unequivocally from a Lorentz boost in special relativity. It is shown that the usually accepted method in Clifford algebra and special relativity of equating the outer product of two Pauli spinors to a three-vector in the Pauli basis leads to the paradoxical result $X = Y = Z = 0$. The method devised in this paper for deriving the Dirac equation from the Evans equation does not use this paradoxical result.

Key words: Evans wave equation, general relativity, Dirac equation in general and special relativity, metric four-vector.

6.1 Introduction

A novel and fundamental wave equation of general relativity can be deduced [1] from the metric compatibility condition

$$D_\mu q^\nu = \partial_\mu q^\nu + \Gamma^\nu{}_{\mu\lambda} q^\lambda = 0 \tag{6.1}$$

on the metric four-vector q_μ used in the standard definition [2,3] of the metric tensor of the Einstein field equation

$$q_{\mu\nu} = q_\mu q_\nu \tag{6.2}$$

as the outer product of two metric vectors. Equation (6.1) states that the covariant derivative of the metric vector q^ν vanishes. It has been shown that the metric compatibility condition (6.1) leads to the usual metric compatibility condition of the metric tensor

$$D_\sigma q_{\mu\nu} = 0 \tag{6.3}$$

used in Riemann geometry [3] to relate the Christoffel symbol $\Gamma^{\nu}{}_{\mu\lambda}$ and metric tensor $q_{\mu\nu}$. Covariant differentiation [4] of the compatibility condition (6.1) leads to the Evans wave equation [1,4]

$$(\square + kT)q^\mu = 0. \tag{6.4}$$

In Eq. (6.4), T is the standard contracted form of the canonical energy momentum tensor

$$T = q^{\mu\nu}T_{\mu\nu}, \tag{6.5}$$

k is the Einstein constant, and $\square$ the flat spacetime d'Alembertian operator.

In Sec. 6.2 the Dirac equation in general relativity is obtained from Eq. (6.4) by expressing the metric three vector in non-Euclidean spacetime as a two-component metric spinor, then applying the parity operation to the metric spinor to obtain a metric spinor with four components. This procedure produces the Dirac equation in general relativity. In Sec. 6.3 the well known Dirac equation of special relativity is obtained as a limiting form of Eq. (6.4) defined by

$$kT \to m^2c^2/\hbar^2, \tag{6.6}$$

where m denotes mass, c the speed of light in vacuo, and $\hbar$ the Dirac constant. In Sec. 6.4 it is shown that the Dirac equation in general relativity can also be written as

$$(\square + kT)\gamma^\mu = 0, \tag{6.7}$$

where γ^μ is the Dirac matrix generalized to non-Euclidean spacetime, a 4×4 matrix related to the tetrad. The most general form of Eq. (6.4) is [1,2]

$$(\square + kT)q^a{}_\mu = 0, \tag{6.8}$$

where $q^a{}_\mu$ is the tetrad [3].

In Sec. 6.5 it is shown that the accepted method [5] in Clifford algebra in special (and also general) relativity of relating the components of a three-vector to those of a two-spinor leads to a fundamental paradox of Clifford algebra

$$X = Y = Z = 0. \tag{6.9}$$

Therefore we do not use this paradoxical but unfortunately well accepted method in our derivations in this paper. We devise a simple and correct method Eq. (6.28) for relating the three vector components to the two Pauli spinor components.

Finally it is shown that the Dirac equation cannot be derived unequivocally from a Lorentz boost in special relativity. The correct way to derive the Dirac equation is from non-Euclidean geometry in general relativity, using the metric four vector q_μ. We refer to Eq. (6.4) as the "Evans equation" for ease of reference.

6.2 Derivation of the Dirac Equation of General Relativity

We start the derivation by considering the ordinary position vector $\boldsymbol{R}$ in three dimensional space and correctly deriving the Pauli spinor from $\boldsymbol{R}$. The method thus developed is used then to construct the metric spinor from the metric three-vector in non-Euclidean spacetime of general relativity. This allows the metric three-vector $\boldsymbol{q}$ in non-Euclidean spacetime to be expressed as a metric two-spinor

$$q = \begin{pmatrix} q_1 \\ q_2 \end{pmatrix}. \tag{6.10}$$

Finally the parity operation is applied to the two-spinor to produce a metric four-spinor ψ:

$$\psi = \begin{pmatrix} {q_1}^{(R)} \\ {q_2}^{(R)} \\ {q_1}^{(L)} \\ {q_2}^{(L)} \end{pmatrix} = \begin{pmatrix} q^{(R)} \\ q^{(L)} \end{pmatrix}, \tag{6.11}$$

where the superscripts (R) and (L) denote right-handed and left-handed two-spinors, respectively. The Dirac equation in general relativity is then deduced to be

$$(\Box + kT)\psi = 0. \tag{6.12}$$

The position vector $\boldsymbol{R}$ is defined in the Cartesian basis in standard notation [2]

$$\boldsymbol{R} = X\hat{\boldsymbol{i}} + Y\hat{\boldsymbol{j}} + Z\hat{\boldsymbol{k}}. \tag{6.13}$$

In the Pauli basis [5], the position vector is

$$r = \boldsymbol{\sigma} \cdot \boldsymbol{r} = X\sigma_1 + Y\sigma_2 + Z\sigma_3 = \begin{pmatrix} Z & X - iY \\ X + iY & -Z \end{pmatrix} \tag{6.14}$$

and the square of the position vector is the invariant

$$\begin{aligned} r^2 = (\boldsymbol{\sigma} \cdot \boldsymbol{r})(\boldsymbol{\sigma} \cdot \boldsymbol{r}) &= \begin{pmatrix} Z & X - iY \\ X + iY & -Z \end{pmatrix} \begin{pmatrix} Z & X - iY \\ X + iY & -Z \end{pmatrix} \\ &= (X^2 + Y^2 + Z^2) \begin{pmatrix} 1 & 0 \\ 0 & 1 \end{pmatrix}. \end{aligned} \tag{6.15}$$

The four Pauli matrices

$$\begin{aligned} \sigma_0 &= \begin{pmatrix} 1 & 0 \\ 0 & 1 \end{pmatrix}, \quad \sigma_1 = \begin{pmatrix} 0 & 1 \\ 1 & 0 \end{pmatrix}, \\ \sigma_2 &= \begin{pmatrix} 0 & -i \\ i & 0 \end{pmatrix}, \quad \sigma_3 = \begin{pmatrix} 1 & 0 \\ 0 & -1 \end{pmatrix}, \end{aligned} \tag{6.16}$$

are all Hermitian and unitary. However, while the determinant of the unit matrix σ_0 is $+1$, the determinants of the other three Pauli matrices are -1 in each case. The latter therefore do not belong to the group $SU(2)$. But this does not affect their $SU(2)$ commutation symmetry. The $SU(2)$ group is commonly defined as consisting of all unitary, unimodular 2×2 matrices:

$$UU^{\dagger} = 1, \quad \det U = 1. \tag{6.17}$$

They all have the general form

$$U = \begin{pmatrix} a & b \\ -b^* & a^* \end{pmatrix}, \tag{6.18}$$

subject to the constraint

$$aa^* + bb^* = 1. \tag{6.19}$$

Under an $SU(2)$ transformation, the following occurs:

$$\xi' = U\xi, \quad \text{i.e.,} \quad \begin{pmatrix} \xi_1' \\ \xi_2' \end{pmatrix} = \begin{pmatrix} a & b \\ -b^* & a^* \end{pmatrix} \begin{pmatrix} \xi_1 \\ \xi_2 \end{pmatrix}, \tag{6.20}$$

$$\xi^{+'} = \xi^{+}U^{+}, \quad \text{i.e.,} \quad (\xi_1^{*'}, \xi_2^{*'}) = (\xi_1^*, \xi_2^*) \begin{pmatrix} a^* & -b \\ b^* & a \end{pmatrix}, \tag{6.21}$$

where

$$\xi = \begin{pmatrix} \xi_1 \\ \xi_2 \end{pmatrix}, \quad \xi^{+} = (\xi_1^*, \xi_2^*), \tag{6.22}$$

is the two-component spinor with complex-valued elements. The two-spinor is often referred to as the Pauli spinor but was devised in the nineteenth century by Clifford. Therefore, under a $SU(2)$ transformation, we obtain four equations

$$\xi_1' = a\xi_1 + b\xi_2, \tag{6.23}$$

$$\xi_2' = -b^*\xi_1 + a^*\xi_2, \tag{6.24}$$

$$\xi_1^{*'} = \xi_1^* a^* + \xi_2^* b^*, \tag{6.25}$$

$$\xi_2^{*'} = -b\xi_1^* + a\xi_2^*. \tag{6.26}$$

Consider the inner products

$$\begin{aligned} R^2 &= (X, Y, Z) \begin{pmatrix} X \\ Y \\ Z \end{pmatrix} = X^2 + Y^2 + Z^2, \\ R^2 &= (\xi_1^*, \xi_2^*) \begin{pmatrix} \xi_1 \\ \xi_2 \end{pmatrix} = \xi_1\xi_1^* + \xi_2\xi_2^* \end{aligned} \tag{6.27}$$

to obtain the simple invariant

$$R^2 = X^2 + Y^2 + Z^2 = \xi_1 \xi_1^* + \xi_2 \xi_2^*. \tag{6.28}$$

This is a relation between the three real-valued components X, Y and Z of the three-vector, and the two complex-valued components ξ_1 and ξ_2 of the two-spinor. Define $O(3)$ as the group of rotations in three-dimensional space [5]. Under an $O(3)$ transformation,

$$R^2 = X^2 + Y^2 + Z^2 = X'^2 + Y'^2 + Z'^2. \tag{6.29}$$

Similarly, under an $SU(2)$ transformation,

$$R^2 = \xi_1 \xi_1^* + \xi_2 \xi_2^* = \xi_1' \xi_1^{*\prime} + \xi_2' \xi_2^{*\prime} \tag{6.30}$$

If we choose $b = 0$ in the constraint condition (6.19), then

$$aa^* = 1; \tag{6.31}$$

and, from Eqs. (6.31), (6.30), (6.24), and (6.23):

$$R^2 = aa^* \xi_1 \xi_1^* + a^* a \xi_2 \xi_2^* = \xi_1 \xi_1^* + \xi_2 \xi_2^*, \tag{6.32}$$

as required. Therefore

$$aa^* = 1, \quad bb^* = 0 \tag{6.33}$$

and

$$U = \begin{pmatrix} a & 0 \\ 0 & a^* \end{pmatrix} \tag{6.34}$$

is one possible representation of U, the $SU(2)$ transformation matrix.

Therefore, Eqs. (6.20) and (6.21) become

$$\begin{pmatrix} \xi_1' \\ \xi_2' \end{pmatrix} = \begin{pmatrix} a & 0 \\ 0 & a^* \end{pmatrix} \begin{pmatrix} \xi_1 \\ \xi_2 \end{pmatrix}, \quad (\xi_1^{*\prime}, \xi_2^{*\prime}) = (\xi_1^*, \xi_2^*) \begin{pmatrix} a^* & 0 \\ 0 & a \end{pmatrix}. \tag{6.35}$$

If $b = b^* = 0$, Eq. (6.35) gives Eqs. (6.23) and (6.24), as required.

In order to build up the Dirac equation in general relativity from the Evans equation the above derivations based on the position vector $\boldsymbol{R}$ in flat (or Euclidean) spacetime must be repeated for the metric vector $\boldsymbol{q}$ of non-Euclidean spacetime in its spinor representation in non-Euclidean spacetime.

First, consider the metric three-vector in Euclidean spacetime:

$$\boldsymbol{q} = q_X \boldsymbol{i} + q_Y \boldsymbol{j} + q_Z \boldsymbol{k}. \tag{6.36}$$

Then

$$q^2 = q^2{}_X + q^2{}_Y + q^2{}_Z \tag{6.37}$$

is invariant under an $O(3)$ transformation. Define the metric spinor

$$q = \begin{pmatrix} q_1 \\ q_2 \end{pmatrix}, \quad q^+ = (q_1^*, q_2^*), \tag{6.38}$$

such that

$$q^2 = {q^2}_X + {q^2}_Y + {q^2}_Z = q_1 q_1^* + q_2 q_2^* \tag{6.39}$$

is an invariant under an $SU(2)$ transformation. In Euclidean space:

$$q_Z = q_Y = q_Z = 1 \tag{6.40}$$

and

$$q_1 q_1^* + q_2 q_2^* = 3, \tag{6.41}$$

a result which is approached asymptotically in the weak-field limit of general relativity [1-5].

Under a $SU(2)$ transformation:

$$\begin{pmatrix} q_1' \\ q_2' \end{pmatrix} = \begin{pmatrix} a & 0 \\ 0 & a^* \end{pmatrix} \begin{pmatrix} q_1 \\ q_2 \end{pmatrix}, \tag{6.42}$$

and this is a transformation of the metric spinor in the weak field limit. From Eqs. (6.39) and (6.42) it can be seen that the metric three-vector $\boldsymbol{q}$ can always be represented as the metric two-spinor q in three-dimensional Euclidean space. Therefore, the corresponding metric four-vector [1,4]

$$q^\mu = (q^0, \boldsymbol{q}) \tag{6.43}$$

can always be represented as the metric three-spinor

$$q^\mu = (q^0, q) \tag{6.44}$$

in four-dimensional Euclidean spacetime, in which is defined the invariant

$$\begin{aligned} q^\mu q_\mu &= q^{02} - {q^2}_X - {q^2}_Y - {q^2}_Z \\ &= q^{02} - q_1 q_1^* - q_2 q_2^*. \end{aligned} \tag{6.45}$$

The Evans wave equation (6.4) in component form is

$$(\Box + kT) q^0 = 0, \tag{6.46}$$

$$(\Box + kT)\boldsymbol{q} = \boldsymbol{0}, \tag{6.47}$$

and thus there exists the equation

$$(\Box + kT)\begin{pmatrix} q_1 \\ q_2 \end{pmatrix} = 0, \tag{6.48}$$

which the Evans equation for the two-spinor q.

In Euclidean spacetime, the metric three-vector components are defined as [2,6]

$$q_X = \left|\frac{\partial \boldsymbol{R}}{\partial X}\right|, \quad q_Y = \left|\frac{\partial \boldsymbol{R}}{\partial Y}\right|, \quad q_Z = \left|\frac{\partial \boldsymbol{R}}{\partial Z}\right|. \tag{6.49}$$

Consider now a region of non-Euclidean three-dimensional space such that each point is specified by three numbers (u_1, u_2, u_3), the curvilinear coordinates [2]. The transformation equations between Cartesian and curvilinear coordinates are

$$\begin{aligned} X &= X(u_1, u_2, u_3), u_1 = u_1(X, Y, Z), \\ Y &= Y(u_1, u_2, u_3), u_2 = u_2(X, Y, Z), \\ Z &= Z(u_1, u_2, u_3), u_3 = u_3(X, Y, Z), \end{aligned} \tag{6.50}$$

where the functions are single-valued and continuously differentiable. There is therefore a one-to-one correspondence between (X, Y, Z) and (u_1, u_2, u_3). The position vector in curvilinear coordinates in $\boldsymbol{R}(u_1, u_2, u_3)$. Define the arc length as the modulus of the infinitesimal displacement vector:

$$ds = |d\boldsymbol{R}| = \left|\frac{\partial \boldsymbol{R}}{\partial u_1} du_1 + \frac{\partial \boldsymbol{R}}{\partial u_2} du_2 + \frac{\partial \boldsymbol{R}}{\partial u_3} du_3\right|, \tag{6.51}$$

where $\partial \boldsymbol{R}/\partial u_i$ is the metric coefficient whose modulus is the scale factor

$$h_i = \left|\frac{\partial \boldsymbol{R}}{\partial u_i}\right|. \tag{6.52}$$

The unit vectors of the curvilinear coordinate system may now be defined as

$$\boldsymbol{e}_{(i)} = \frac{1}{h_i}\frac{\partial \boldsymbol{R}}{\partial u_i}, \quad i = 1, 2, 3; \tag{6.53}$$

and we can write

$$d\boldsymbol{R} = h_1 du_1 \boldsymbol{e}_{(1)} + h_2 du_2 \boldsymbol{e}_{(2)} + h_3 du_3 \boldsymbol{e}_{(3)}. \tag{6.54}$$

The unit vectors $\boldsymbol{e}_{(i)}$ are unit tangent vectors to the curve u_i at point P, i.e., the three unit vectors of the curvilinear coordinate system are unit tangent vectors to the coordinate curves, are mutually orthogonal, and cyclically symmetric with $O(3)$ symmetry:

$$\boldsymbol{e}_{(1)} \cdot \boldsymbol{e}_{(2)} = 0, \quad \boldsymbol{e}_{(1)} \cdot \boldsymbol{e}_{(3)} = 0, \quad \boldsymbol{e}_{(2)} \cdot \boldsymbol{e}_{(3)} = 0, \tag{6.55}$$

$$\boldsymbol{e}_{(1)} \times \boldsymbol{e}_{(2)} = \boldsymbol{e}_{(3)}, \quad \boldsymbol{e}_{(2)} \times \boldsymbol{e}_{(3)} = \boldsymbol{e}_{(1)}, \quad \boldsymbol{e}_{(3)} \times \boldsymbol{e}_{(1)} = \boldsymbol{e}_{(2)}. \tag{6.56}$$

In the Euclidean limit the unit vectors $\boldsymbol{e}_{(i)}$ become the Cartesian unit vectors and the scale factors h_i become the metric coefficients q_i. Therefore we arrive at the definition of the general three-vector field $\boldsymbol{F}$ in non-Euclidean three-space [2]:

$$\boldsymbol{F} = F_{(1)}\boldsymbol{e}_{(1)} + F_{(2)}\boldsymbol{e}_{(2)} + F_{(3)}\boldsymbol{e}_{(3)}. \tag{6.57}$$

This definition is sketched below:

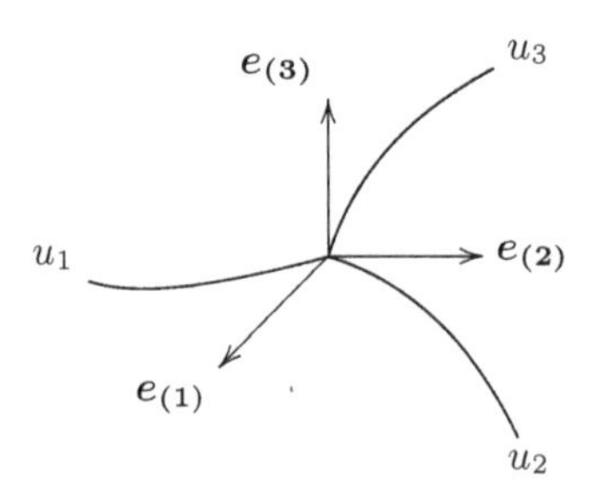

Fig. 6.1. Curvalinear Coordinates

(6.58)

The metric three-vector in non-Euclidean three-space is therefore

$$\boldsymbol{q} = q_{(1)}\boldsymbol{e}_{(1)} + q_{(2)}\boldsymbol{e}_{(2)} + q_{(3)}\boldsymbol{e}_{(3)}, \tag{6.59}$$

and similarly the metric four-vector in non-Euclidean four-dimensional space-time is

$$q^\mu = (q^0, \boldsymbol{q}). \tag{6.60}$$

The metric spinor in non-Euclidean three-space can always be defined through the relation

$$q^2 = q^2_{(1)} + q^2_{(2)} + q^2_{(3)} = q_1 q_1^* + q_2 q_2^*, \tag{6.61}$$

which is Eq. (6.39) in non-Euclidean three-space. We denote this metric spinor by

$$q = \begin{pmatrix} q_1 \\ q_1 \end{pmatrix}, \tag{6.62}$$

and it obeys the Evans equation

$$(\square + kT)q = 0 \tag{6.63}$$

in non-Euclidean three-space. Similarly the metric three spinor in four-dimensional non-Euclidean space is $(q^0, \boldsymbol{q})$.

Define the parity operation in Euclidean three-space by

$$\hat{P}(\boldsymbol{R}) = -\boldsymbol{R} = -X\boldsymbol{i} - Y\boldsymbol{j} - Z\boldsymbol{k}, \tag{6.64}$$

i.e., the parity operation reverses the signs of $\boldsymbol{i}, \boldsymbol{j}, \boldsymbol{k}$, and therefore reverses the signs of the position vector $\boldsymbol{R}$ and the metric three-vector $\boldsymbol{q}$:

$$\hat{P}(\boldsymbol{q}_X) = -\left|\frac{\partial \boldsymbol{R}}{\partial X}\right| \boldsymbol{i} = -\boldsymbol{q}_X, \quad \text{etc.} \tag{6.65}$$

Therefore, under parity,

$$\hat{P}(q^{(R)}) = q^{(L)}. \tag{6.66}$$

Similarly, in non-Euclidean three-space, parity is defined as reversing the signs of the unit vectors $\boldsymbol{e}_{(1)}, \boldsymbol{e}_{(2)}, \boldsymbol{e}_{(3)}$ reversing the signs of $\boldsymbol{R}$ and $\boldsymbol{q}$ in non-Euclidean three-space, in which the parity operation produces, again:

$$\hat{P}(q^{(R)}) = q^{(L)}. \tag{6.67}$$

Therefore Eq. (6.63) is extended under parity to an equation in the four-spinor:

$$(\Box + kT)\psi = 0, \tag{6.68}$$

and this is the Evans equations for the four-spinor in non-Euclidean spacetime.

6.3 Derivation Of The Dirac Equation of Special Relativity

In the weak-field limit, Eq. (6.68) asymptotically approaches [1,4]

$$kT \to \frac{km}{V_0} = \frac{m^2c^2}{\hbar^2}, \quad \left(\Box + \frac{m^2c^2}{\hbar^2}\right)\psi = 0, \tag{6.69}$$

and the Dirac equation of special relativity is deduced. In Eq. (6.69), m is the rest mass of a particle and V_0 its rest volume. Equation (6.69) generalizes the Planck/de Broglie postulate to any particle (not only the photon), and also implies that the photon has a rest mass and rest volume. The rest frequency is defined by

$$E_0 = \hbar\omega_0 = mc^2, \quad \omega_0 = 8\pi c\ell^2/V, \tag{6.70}$$

where

$$\ell = \sqrt{G\hbar/c^2} \tag{6.71}$$

is the Planck length. The product of the rest mass m and the rest volume V_0 is a universal constant

$$mV_0 = \hbar^2 k/c^2, \tag{6.72}$$

so every particle occupies a rest volume in nature, and there can be no point particles. This deduction is consistent with the fact that there are no singularities in nature, and no singularities in general relativity, a classical field theory. The Dirac Eq. (6.69) is well known to be an equation of relativistic quantum mechanics, which we have therefore deduced from classical general relativity using geometry. No probabilistic assumptions have been made in this derivation. However, it is well known [5] that the Dirac equation, unlike the Klein-Gordon equation, is consistent with the probabilistic interpretation of the wave function proposed by Bohr, Born, Heisenberg, and the Copenhagen School in natural philosophy. Neither the probabilistic interpretation of the wave function, nor the positivist speculations [6] of this School are required to deduce the Dirac equation from the novel Evans equation of classical general relativity. We see in Eq. (6.70) that the Evans equation also produces the wave particle dualism of de Broglie from geometry. Having deduced the Dirac equation, it is a straightforward to deduce from the Dirac equation the Schrödinger equation in the former's non-relativistic limit. Having deduced the Schrödinger equation it then becomes possible in turn to deduce Newtonian dynamics in the classical limit of the Schrödinger equation [5,7].

Therefore all the equations of linear quantum and classical dynamics can be deduced from non-linear Evans equation in well defined limits [1,4]. Noether's theorem can also be deduced [1,3,4] from the non-Euclidean geometry used to construct the Evans equation, and therefore conservation laws of which the archetypical example is Newton's third law, can be deduced from the metric compatibility Eq. (6.1) and the Evans wave Eq. (6.4).

Another fundamental concept introduced by the Evans equation is that of rest curvature, which is the minimum curvature associated with any particle, such as the electron. The rest curvature is the inverse of the Compton wavelength squared:

$$|R_0| = 1/\lambda^2 = m^2c^2/\hbar^2. \tag{6.73}$$

The rest curvature is therefore related to the rest energy in special relativity by

$$E_0 = mc^2 = \hbar c\sqrt{|R_0|}. \tag{6.74}$$

More generally, the quantum of energy for any particle (not only the photon) is given by

$$E = \hbar c\sqrt{|R_0|}, \tag{6.75}$$

where

$$|R| = \partial_\mu \Gamma^{\mu\nu}{}_{\nu} - \Gamma^{\mu\nu}{}_{\nu}\Gamma^{\nu}{}_{\mu\nu} \tag{6.76}$$

in general relativity [1,4,8]. Here $\Gamma^{\nu}{}_{\mu\nu}$ is the Christoffel symbol. Therefore we arrive at the novel and fundamental law for all particles in general relativity there exists the relation

$$E = \hbar\omega = \hbar c\sqrt{|R|}, \tag{6.77}$$

and this is a generalization of both the Planck-Law and the de Broglie wave-particle dualism to all particles in general relativity.

This new law is another consequence of the Evans wave Eq. (6.4), and is therefore a consequence of the geometrical constraint (6.1).

6.4 The Dirac Matrices as Eigenfunctions

Dirac constructed his equation in about 1927 because the Schrödinger equation did not satisfy the requirements of relativity [7], in which the space and time coordinates must be developed as parts of a single spacetime. In relativity the relativistic momentum γmv and the energy are related by the Einstein equation

$$E^2 - p^2c^2 = m^2c^4, \tag{6.78}$$

and this equation reduces to the Newtonian kinetic energy

$$E = p^2/2m, \tag{6.79}$$

when p is small compared with mc [7]. The operator replacement [1,4,5]

$$p^\mu = i\hbar\partial^\mu, \quad p^\mu = (E/c, \boldsymbol{p}), \quad \partial^\mu = \left(\frac{1}{c}\frac{\partial}{\partial t}, -\boldsymbol{\nabla}\right), \tag{6.80}$$

in Eq. (6.78) produces the Klein-Gordon equation

$$\left(\Box + m^2c^2/\hbar^2\right) q^\mu = 0, \tag{6.81}$$

to which the Evans Eq. (6.4) reduces in the limit (6.69). The Klein-Gordon equation produces a fundamental and well-known paradox [5,7] if its eigenfunction is assumed to have a probabilistic ontology. The probability density from the Klein-Gordon equation can be negative (and therefore unphysical), whereas the probability density of the Schrödinger equation is always positive definite. Dirac therefore constructed an equation of the form

$$\frac{\partial\psi}{\partial t} = \alpha_X\frac{\partial\psi}{\partial X} + \alpha_Y\frac{\partial\psi}{\partial Y} + \alpha_Z\frac{\partial\psi}{\partial Z} + \beta mc^2, \tag{6.82}$$

where $\alpha_X, \alpha_Y, \alpha_Z, \beta$ are constants [7]. The equation leads to a positive definite probability density. The four components of the Dirac equation must also be four Klein-Gordon equations, and the constants $\alpha_X, \alpha_Y, \alpha_Z, \beta$ must be 4×4 matrices, known as the Dirac matrices γ^μ. The solutions of the Dirac equation separate at low kinetic energies into two doubly degenerate sets, one set is associated with positive kinetic and the other with negative kinetic energy, giving rise to anti-particles and the Dirac sea [5,7]. The degeneracy is removed by a magnetic field, giving rise from half integral spin to fermion spin resonance.

In order to derive the original form of the Dirac equation from the equivalent form (6.69) (the Klein-Gordon form of the Dirac equation), the d'Alembertian is expressed in terms of the anti-commutator of Dirac matrices [5]:

$$\Box = \frac{1}{2}\left\{\gamma^\mu, \gamma^\nu\right\}\partial_\mu\partial_\nu. \tag{6.83}$$

This equation implies

$$\{\gamma^\mu, \gamma^\nu\} = 2q^{\mu\nu}, \tag{6.84}$$

i.e., in Euclidean spacetime the symmetric metric tensor $q^{\mu\nu}$ is half the commutator of the Dirac matrices. However we also know that the symmetric metric tensor is defined by [2]

$$q^{\mu\nu} = \frac{1}{2} \{q^\mu, q^\nu\}, \tag{6.85}$$

and so

$$\{\gamma^\mu, \gamma^\nu\} = \{q^\mu, q^\nu\}. \tag{6.86}$$

This equation shows that the Evans wave equation may be expressed as an eigenequation in the Dirac matrix generalized to non-Euclidean spacetime:

$$(\Box + kT)\gamma^\mu = 0. \tag{6.87}$$

In the Euclidean limit

$$\left(\Box + m^2c^2/\hbar^2\right) \gamma^\mu = 0. \tag{6.88}$$

In non-Euclidean spacetime, the elements of the 4×4 Dirac matrices are no longer unity but become space- and time-dependent quantities and the Dirac matrices in non-Euclidean spacetime are related to the tetrads [3] $q^a{}_\mu$. This result is consistent with the fact that the most general form of the Evans equation is an eigenequation in tetrads [1,4]:

$$(\Box + kT)q^a{}_\mu = 0. \tag{6.89}$$

The Dirac matrix and spinor forms of the Dirac equation in general relativity are related by Eqs. (6.84) and (6.86).

The original form of the Dirac equation is obtained by substituting Eq. (6.83) into Eq. (6.69), which has been deduced in this paper from the Evans Eq. (6.4). This substitution produces:

$$\left(\gamma^\mu\gamma^\nu\partial_\mu\partial_\nu + m^2c^2/\hbar^2\right) \psi = 0. \tag{6.90}$$

Equation (6.90) may be obtained from the Dirac equation in its original form

$$(i\gamma^\mu\partial_\mu - mc/\hbar)\psi = 0. \tag{6.91}$$

The proof of this result is as follows:

Using the operator equivalence (6.80), Eq. (6.91) becomes

$$\left(\gamma^\mu p_\mu - mc/\hbar\right) \psi = 0. \tag{6.92}$$

Application of the operator $i\gamma^\mu\partial_\mu$ to Eq. (6.91) produces

$$\left(-\gamma^\mu\partial_\mu\gamma^\nu\partial_\nu - i(\gamma^\mu\partial_\mu)(mc/\hbar)\right) \psi = 0. \tag{6.93}$$

However, we know that

$$i\gamma^{\mu}\partial_{\mu}\psi = (mc/\hbar)\psi \tag{6.94}$$

and

$$\partial_{\mu}\left[(mc/\hbar)\psi\right] = (mc/\hbar)\partial_{\mu}\psi, \tag{6.95}$$

and so we obtain Eq. (6.90) from Eq. (6.91).

Equation (6.92), the original form of the Dirac equation, may now be expressed in the form

$$(E + c\boldsymbol{\sigma}\cdot\mathbf{p})q^{(L)}(\mathbf{p}) = mc^2q^{(R)}(\mathbf{p}), \tag{6.96}$$

$$(E - c\boldsymbol{\sigma}\cdot\mathbf{p})q^{(R)}(\mathbf{p}) = mc^2q^{(L)}(\mathbf{p}), \tag{6.97}$$

which consists of two simultaneous equations [5] with

$$q^{(L)}(\mathbf{0}) = q^{(R)}(\mathbf{0}) \tag{6.98}$$

in the space and time dependent metric two-spinors. In the discussion to this paper it is shown that these spinors are obtained from the rest frame spinors by a Lorentz boost in special relativity.

Therefore all the major properties of the Dirac equation have been obtained from the Evans Eq. (6.4), which is therefore a generalized and nonlinear Dirac equation. The Evans equation also contains physics not present in the Dirac equation and therefore supplants the latter in natural philosophy.

6.5 A Fundamental Paradox In Clifford Algebra

This paradox of Clifford algebra occurs in the standard method [5] of identifying the outer product of two Pauli spinors with the position vector $\boldsymbol{R}$ in the Pauli basis:

$$H = \boldsymbol{\sigma}\cdot\boldsymbol{r} = \xi\xi^{+}, \tag{6.99}$$

where

$$\boldsymbol{\sigma}\cdot\boldsymbol{r} = \begin{pmatrix} Z & X - iY \\ X + iY & -Z \end{pmatrix} \tag{6.100}$$

and

$$\xi\xi^{+} = \begin{pmatrix} \xi_1\xi_1^{*} & \xi_1\xi_2^{*} \\ \xi_2\xi_1^{*} & \xi_2\xi_2^{*} \end{pmatrix}. \tag{6.101}$$

The standard method in Clifford algebra [5] expresses the outer product $\xi\xi^{+}$ as a traceless unitary matrix by noting that Eqs. (6.23) to (6.26) can be written as

$$\xi_1' = a\xi_1 + b\xi_2, \tag{6.102}$$

$$\xi_2' = -b^{*}\xi_1 + a^{*}\xi_2, \tag{6.103}$$

$$-\xi_2^{*\prime} = a(-\xi_2^{*}) + b\xi_1^{*}, \tag{6.104}$$

$$\xi_1^{*\prime} = -b^{*}(-\xi_2^{*}) + a^{*}\xi_1^{*}. \tag{6.105}$$

Therefore the Pauli spinor $\begin{pmatrix} \xi_1 \\ \xi_2 \end{pmatrix}$ transform in the same way as $\begin{pmatrix} -\xi_2^* \\ \xi_1^* \end{pmatrix}$. In shorthand notation,

$$\xi \sim \zeta\xi^*, \quad \zeta = \begin{pmatrix} 0 & -1 \\ 1 & 0 \end{pmatrix}. \tag{6.106}$$

In other words, we can think of

$$\xi_1 \to -\xi_2^*, \quad \xi_1' \to -\xi_2^{*\prime}, \quad \xi_2 \to \xi_1^*, \quad \xi_2' \to \xi_1^{*\prime}. \tag{6.107}$$

Similarly, we can write Eqs. (6.9) to (6.12) as

$$\xi_1^{*\prime} = \xi_1^* a^* + \xi_2^* b^*, \tag{6.108}$$
$$\xi_2^{*\prime} = -\xi_1^* b + \xi_2^* a, \tag{6.109}$$
$$\xi_2' = \xi_2 a^* - \xi_1 b^*, \tag{6.110}$$
$$-\xi_1' = -\xi_2 b - \xi_1 a, \tag{6.111}$$

so we can think of

$$\xi_1^* \to \xi_2, \quad \xi_1^{*\prime} \to \xi_2', \quad \xi_2^* \to -\xi_1, \quad \xi_2^{*\prime} \to -\xi_1'. \tag{6.112}$$

From Eqs. (6.107) and (6.112),

$$\xi\xi^+ \sim \begin{pmatrix} -\xi_2^* \\ \xi_1^* \end{pmatrix} (-\xi_1, \xi_2) = \begin{pmatrix} \xi_2^*\xi_1 & -\xi_2^*\xi_2 \\ -\xi_1\xi_1^* & \xi_1^*\xi_2 \end{pmatrix}. \tag{6.113}$$

The standard method [5], however, asserts that

$$\xi^+ \sim (\zeta\xi)^T = (-\xi_2, \xi_1). \tag{6.114}$$

This method does not agree with the fact that ξ^+ transforms as $(-\xi_1, \xi_2)$, from Eqs. (6.112) and (6.107). However, if, for the sake of continuity, one accepts the incorrect Eq. (6.114), one obtains

$$H = h = \xi\xi^+ = \begin{pmatrix} -\xi_1\xi_2 & {\xi^2}_1 \\ -{\xi^2}_2 & \xi_1\xi_2 \end{pmatrix} = \begin{pmatrix} Z & X - iY \\ X + iY & -Z \end{pmatrix}, \tag{6.115}$$

that is,

$$X - iY = -{\xi^2}_1, \tag{6.116}$$
$$X + iY = {\xi^2}_2, \tag{6.117}$$
$$Z = \xi_1\xi_2. \tag{6.118}$$

Multiplication of Eqs. (6.116) and (6.117) produces

$$X^2 + Y^2 = (X - iY)(X + iY) = -\xi^2{}_1\xi^2{}_2; \tag{6.119}$$

and, adding Eqs. (6.118) and (6.119),

$$X^2 + Y^2 + Z^2 = 0, \tag{6.120}$$

an equation that produces the paradox

$$X = Y = Z \stackrel{?}{=} 0 \tag{6.121}$$

because X, Y and Z are all real-valued.

This fundamental paradox is also present for any type of outer product of Pauli spinors. For example, if we attempt to identify

$$\begin{pmatrix} Z & X - iY \\ X + iY & -Z \end{pmatrix} = \begin{pmatrix} \xi_1\xi_2^* & -\xi_2\xi_2^* \\ -\xi_1\xi_1^* & \xi_2\xi_1^* \end{pmatrix}, \tag{6.122}$$

we find

$$X - iY = -\xi_2\xi_2^*, \tag{6.123}$$

$$X + iY = -\xi_1\xi_1^*, \tag{6.124}$$

$$Z = \xi_1\xi_2^* = -\xi_2\xi_1^*. \tag{6.125}$$

Multiplication of Eqs. (6.123) and (6.124) gives

$$X^2 + Y^2 = \xi_1\xi_2^*\xi_2\xi_1^*; \tag{6.126}$$

and, adding Eqs. (6.125) and (6.126), we obtain Eqs. (6.120) and (6.121) again.

Therefore this paradox is a basic paradox of Clifford algebra because it occurs in the basic relation between three-vector elements and two-spinor elements. We avoid using the paradox in this paper by devising Eq. (6.28) by identifying the inner product of Pauli spinors with the inner product of three vectors.

6.6 Discussion

In this paper we have derived the fundamental properties of the Dirac equation from the metric compatibility condition (6.1) of non-Euclidean geometry. There remains only the problem of relating the Dirac equation to the Lorentz boost of special relativity [5]. The original method, due to Ryder [5], relies solely on the Lorentz boosts:

$$q^{(R)}(\boldsymbol{p}) = \exp(-i\frac{\boldsymbol{\sigma}}{2} \cdot \boldsymbol{\phi})q^{(R)}(\boldsymbol{0}), \tag{6.127}$$

$$q^{(L)}(\boldsymbol{p}) = \exp(i\frac{\boldsymbol{\sigma}}{2} \cdot \boldsymbol{\phi})q^{(L)}(\boldsymbol{0}). \tag{6.128}$$

These are generated from each other by the parity operator. Solving Eqs. (6.127) and (6.128) simultaneously leads to the result

$$\begin{aligned}(E - c\boldsymbol{\sigma}\cdot\boldsymbol{p})q^{(R)}(\boldsymbol{p}) - mc^2 q^{(L)}(\boldsymbol{p})\\ = (E + c\boldsymbol{\sigma}\cdot\boldsymbol{p})q^{(L)}(\boldsymbol{p}) - mc^2 q^{(R)}(\boldsymbol{p}).\end{aligned} \tag{6.129}$$

Comparison with Eqs. (6.96) and (6.97) shows that the Dirac equation is a special case of Eq. (6.129). The Ryder method does not, however, unequivocally obtain the Dirac equation from Eq. (6.129). In order to try to achieve this result we have to consider the extra condition

$$\begin{aligned}q^{(R)}(\boldsymbol{p})q^{(L)T}(\boldsymbol{p}) &= q^{(R)}(0)q^{(L)T}(0) = q^{(R)}(0)q^{(R)T}(0)\\ &= q^{(L)}(0)q^{(R)T}(0) = q^{(L)}(0)q^{(L)T}(0),\end{aligned} \tag{6.130}$$

which is obtained from Eq. (6.98). Equation (6.130) implies

$$\begin{aligned}q^{(R)}(\boldsymbol{p})q^{(L)T}(\boldsymbol{p}) = q^{(R)}(\boldsymbol{p})q^{(R)T}(\boldsymbol{p})\\ = q^{(L)}(\boldsymbol{p})q^{(R)T}(\boldsymbol{p}) = q^{(L)}(\boldsymbol{p})q^{(L)T}(\boldsymbol{p}).\end{aligned} \tag{6.131}$$

Multiplying Eq. (6.129) on both sides by $q^{(L)T}$, we obtain

$$\begin{aligned}(E - c\boldsymbol{\sigma}\cdot\boldsymbol{p})q^{(R)}q^{(L)T} - mc^2 q^{(L)}q^{(L)T}\\ = (E + c\boldsymbol{\sigma}\cdot\boldsymbol{p})q^{(L)}q^{(L)T} - mc^2 q^{(R)}q^{(L)T};\end{aligned} \tag{6.132}$$

and, using Eq. (6.131),

$$E - c\boldsymbol{\sigma}\cdot\boldsymbol{p} - mc^2 = E + c\boldsymbol{\sigma}\cdot\boldsymbol{p} - mc^2, \tag{6.133}$$

which implies

$$\boldsymbol{p} = \boldsymbol{0}, \tag{6.134}$$

$$E - mc^2 = E - mc^2. \tag{6.135}$$

This is the maximum amount of information obtainable from the Lorentz boost, so the Dirac Eq. (6.92) cannot be obtained unequivocally from a Lorentz boost. The correct way of deducing the Dirac equation unequivocally is from general relativity, as in this paper.

Acknowledgments

The Ted Annis Foundation and Craddock Inc. are acknowledged for funding and the Fellows and Emeriti of AIAS for many interesting discussions.

References

1. M. W. Evans, *Found Phys. Lett.* **16**, 367 (2003); in press (December 2003); www.aias.us.
2. E. G. Milewski, ed., *Vector Analysis Problem Solver* (Research and Education Association, New York, 1987).
3. S. M. Carroll, *Lecture Notes in General Relativity* (University of California, Santa Barbara, graduate course on arXiv:gr-qe/9712019vl, 3 December 1997).
4. M. W. Evans, *Found Phys. Lett.*, in press (2004).
5. L. H. Ryder, *Quantum Field Theory* (University Press, Cambridge, 1987, 1996).
6. M. Sachs and A. R. Roy, eds., *Mach's Principle and the Origin of Inertia* (Apeiron, Montreal, 2003).
7. P. W. Atkins, *Molecular Quantum Mechanics* (University Press, Oxford, 1983).

7

Unification Of Gravitational And Strong Nuclear Fields

Summary. Using the recently derived Evans wave equation of unified field theory, the strong nuclear field is described with an SU(3) representation of the gravitational field and the Gell-Mann color triplet is derived from general relativity as a three-spinor eigenfunction of the Evans wave equation.

Key words: Evans wave equation, strong field, gravitational field, Gell-Mann color triplet, field unification.

7.1 Introduction

Recently a wave equation for unified field theory has been derived [1-3] from a lemma of Cartan-Riemann differential geometry [4, 5] and used to give the first self consistent unified description of the gauge invariant gravitational and electromagnetic fields. The former is the Riemann form and the latter is the torsion form. One form is the Hodge dual of the other. The theory has been shown [1-3] to reduce to all the main equations of physics in the appropriate limits, these equations include: the four Newton laws; the Schrödinger equation; the Dirac equation; the d'Alembert equation; the Poisson equations of dynamics and electrostatics; and the correct generally covariant form of the Maxwell-Heaviside field equations. The latter are referred to as "O(3) electrodynamics" because the symmetry group of the underlying gauge field theory is O(3). The theory of O(3) electrodynamics has been extensively tested against experimental data, and found experimentally to have numerous advantages over the Maxwell-Heaviside field theory [6-8]. It has been inferred that in order to unify gravitation and electromagnetism within the theory of general relativity, electrodynamics must have a gauge symmetry higher than U(1) [6-8]. This inference produces the fundamental Evans-Vigier magnetic field, which reaches mega-gauss in the inverse Faraday effect [9] of under dense plasma, and is now a routine observable. The theory reduces to two Maxwell-Heaviside equations for transverse plane waves, but in addition gives the observable Evans-Vigier field, governed by a third field equation [6-8] not present in Maxwell-Heaviside theory. The unified field theory therefore has all the hallmarks of a major paradigm shift in physics. The inter-relation between the gravitational and electromagnetic field is given, for example, by the first and second Bianchi identities of differential geometry. These identities inter-relate the Riemann and torsion forms, which are defined in terms of the spin connection and the tetrad by the second and first Maurer-Cartan structure relations, respectively. It has been inferred that the Riemann form is the Hodge dual of the torsion form, and that spin connection is the Hodge dual of the tetrad. In the condensed notation of differential geometry the first and second Maurer-Cartan structure relations are, respectively:

$$T^a = D \wedge q^a, \tag{7.1}$$

$$R^a{}_b = D \wedge \omega^a{}_b, \tag{7.2}$$

where $R^a{}_b$ is the Riemann form and T^a the torsion form. The symbol $D\wedge$ denotes exterior covariant derivative, and $\omega^a{}_b$ and q^a are, respectively, the spin connection and the tetrad. The first and second Bianchi identities are the homogeneous field equations of unified field/matter theory and are, respectively:

$$D \wedge T^a := 0, \tag{7.3}$$

$$D \wedge R^a{}_b : -0. \tag{7.4}$$

The first and second Evans duality equations state that there exist the Hodge duality relations

$$R^a{}_b = \varepsilon^a{}_{bc} T^c, \tag{7.5}$$

$$\omega^a{}_b = \varepsilon^a{}_{bc} q^c, \tag{7.6}$$

where $\varepsilon^a{}_{bc}$ is the appropriate Levi-Civita symbol in the well defined [4,5] orthonormal space of the tetrad.

In this notation, the Maxwell-Heaviside field theory is described by

$$F = d \wedge A, \tag{7.7}$$

$$d \wedge F := 0, \tag{7.8}$$

where F is the gauge invariant electromagnetic field (a scalar-valued two-form), and where A is the potential field (a scalar-valued one-form). In generally covariant unified field theory [1-3] F becomes directly proportional to the torsion form T^a (a vector valued two form) and A becomes directly proportional to the tetrad q^a (a vector-valued one-form). The exterior derivative $d\wedge$ becomes the covariant exterior derivative $D\wedge$. The inhomogeneous Maxwell-Heaviside field equation is

$$d \wedge^* F = J, \tag{7.9}$$

where $^*F$ is the Hodge dual of F and where J is a scalar-valued three-form, the current form. In unified field theory, Eq. (7.9) is developed into the first and second Evans inhomogeneous field equations

$$D \wedge^* T^a = J^a, \tag{7.10}$$

$$D \wedge^* R^a{}_b = J^a_b, \tag{7.11}$$

where J^a is a vector valued three form, and $J^a{}_b$ is a tensor-valued three form, where $^*T^a$ is the Hodge dual of T^a, defined by

$$^*T^{a\mu\nu} = \frac{1}{2}\varepsilon^{\mu\nu\rho\sigma}T^a{}_{\rho\sigma}, \tag{7.12}$$

and where $^*R^a{}_b$ the Hodge dual of $R^a{}_b$, defined by

$$^*R^{a\ \mu\nu}{}_b = \frac{1}{2}\varepsilon^{\mu\nu\rho\sigma}R^a{}_{b\rho\sigma}. \tag{7.13}$$

Equations (7.10) and (7.11) govern the interaction of the gauge invariant fields $^*T^a$ and $^*R^a{}_b$, respectively, with matter fields in unified field theory.

It can be seen that there is no longer a distinction between the gravitational and electromagnetic fields and they are unified with differential geometry. The potential field (Feynman's "universal influence" [10]) is developed into the tetrad, which is in turn governed by the Evans lemma of differential geometry:

$$DD = dD := \Box - R, \tag{7.14}$$

where $\Box$ is the d'Alembertian operator in Euclidean spacetime, and where R is scalar curvature in differential geometry. The lemma (7.14) is the subsidiary proposition which leads to the Evans wave equation of differential geometry, the eigenequation

$$\Box q^a = R q^a, \tag{7.15}$$

$$(\Box + kT)q^a = 0. \tag{7.16}$$

Combining Eq. (7.15) with the Evans field equation [1-3]

$$R q^a = -kTq^a \tag{7.17}$$

gives the Evans wave equation (7.16) of generally covariant unified field theory. Equation (7.16) completes Einstein's theory of general relativity of the gravitational field and develops it into a generally covariant unified field theory of all known radiation and matter fields.

The torsion form is the commutator of tetrads [1-3]:

$$T^c = Rq^a \wedge q^b; \tag{7.18}$$

and, using Eq. (7.5), the Riemann form is also defined in terms of a commutator of tetrads:

$$R^a{}_b = \epsilon^a{}_{bcd} R q^c \wedge q^d. \tag{7.19}$$

In this paper, we use Eq. (7.16) to derive the Gell-Mann quark color triplet from differential geometry in non-Euclidean spacetime, thus unifying the theory of the strong field and the theory of the gravitational field.

7.2 Derivation Of The Quark Color Triplet

It is well known that the quark Lagrangian is invariant under the SU(3) transformation [10], where q is the quark color triplet

$$q \to Mq, \tag{7.20}$$

$$q = \begin{pmatrix} q_R \\ q_W \\ q_B \end{pmatrix} \tag{7.21}$$

and a three spinor. In unifying the gravitational and strong fields it is shown in this section that q is a manifestation of differential geometry, an eigenfunction of Eq. (7.16). Therefore the equation governing the strong force is

$$(\Box + kT)q = 0, \tag{7.22}$$

and q is the gluon wavefunction, the eigenfunction of the quantized strong field. The corresponding wave equation for the gravitational field is

$$(\Box + kT)q^a{}_\mu = 0, \tag{7.23}$$

where $q^a{}_\mu$ is the tetrad. Therefore, if we can express Eq. (7.22) in terms of Eq. (7.23), we can unify the strong and gravitational fields.

In order to achieve this unification, note that M in Eq. (7.20) is a unitary, orthogonal matrix of the SU(3) group representation of space, a 3×3 generalization of the 2×2 complex Pauli matrices [10] of the SU(2) group representation of space. Only unitary, orthogonal matrices form groups; Hermitian matrices do not form groups [10]. Therefore M must be either an O(3) or SU(3) representation of space [10]. The two representations are representations of the same base manifold, so there must be relations between elements of SU(3) (components of the three spinor) and those of O(3) (components of the tetrad). In the unified field theory summarized in Sec. 1, this geometrical relation unifies the strong and gravitational fields, and shows that the strong field is simply the gravitational field in an SU(3) representation of space. Quarks have not yet been observed individually, so it remains to be seen whether the SU(3) representation has a physical meaning akin to the half integral spin of a fermion in the SU(2) representation. In other words the gravitational field can always be represented mathematically in SU(3) form, but it remains to be seen whether this has any meaning in physics, because individual quarks have never been observed. Indeed the standard model goes so far as to suggest that quarks cannot be observable, and "self-destructs" because quarks cannot be physical if they are unobservable. The major inference in Eq. (7.22) is that the quark color triplet is derived from a tetrad, and it should be noted that Eq. (7. 22) is valid in non-Euclidean spacetime, and is a generally covariant description of the strong field.

In the standard model the SU(3) group is chosen to represent the strong field for experimental reasons [10], but these reasons do not include the observation of individual quarks. The SU(3) symmetry matrix M is the unitary complex 3×3 matrix with unit determinant defined by

$$MM^{\dagger} = 1, \quad \det M = 1. \tag{7.24}$$

It follows that M is 3×3 matrix with eight independent parameters, which are the eight group generators denoted [10] by $\lambda_n/2$. The corresponding generators in the SU(2) representation are the three Pauli matrices. Elementary particles in the standard model are thought to be composite states of quarks, and there are thought to be six quarks: u, d, s, c, t, and b. Each quark has spin half. Baryons and mesons are composite states of quarks, and in the standard model only these composite states are thought to be observable as elementary particles. Pions, for example, are thought to be three members of a supermultiplet of eight, and the strong interaction is thought to be an interaction between quarks mediated by a massless gluon field. All this is in the context of special relativity. There is no concept in the standard model of the strong field being derivable from the gravitational field, and so the strong field is not generally covariant, a major weakness of the standard model. In the unified field theory represented by Eqs. (7.22) and (7.23) the strong field is a manifestation of the gravitational field using SU(3) representation for the base manifold. The quark color triplet is simply a three spinor equivalent of the tetrad used to describe the gravitational field. The index of the orthonormal tangent space of the tetrad becomes the internal index of the SU(3) gauge group representation of the strong field. In the standard model the tangent space is a physical space but the internal space of gauge theory is an abstract space. In the unified field theory outlined in Sec. 1 the hitherto abstract internal space of gauge theory is given a physical meaning for the first time. Despite the fact that quarks are unobservable in the standard model, they are thought to possess a quantum number with three degrees of freedom, and these are the components of the three spinor (7.21): red, white and blue.

(Similarly the weak nuclear field in the standard model is thought to derive from a SU(2) symmetry gauge field theory. However the theory of the weak nuclear field is again a theory of special relativity, and spontaneous symmetry breaking is needed to provide the bosons of the weak field with mass. Unification of the weak and electromagnetic fields is standard-modeled in such a way that the electromagnetic field photons remain massless. This is essentially an empirical model with adjustable parameters. There is no experimental evidence for the Higgs boson for example, and so there is no evidence for spontaneous symmetry breaking, a cornerstone of the standard model. In the unified field theory of Sec. 1, the weak field is the electromagnetic field in SU(2) representation, and the mass of the three weak field bosons is represented by the limit of T in special relativity, a mass density. Each boson of the weak field has its individual mass, determined experimentally.)

Therefore, in order to describe quarks in general relativity, non-Euclidean space-time must be expressed in SU(3) representation using the tetrad as a starting point. Similarly the tetrad is used [4,5] as the starting point for the recent derivation [11] of the Dirac equation from the more general Evans wave Eq. (7.16). The quarks (albeit unobservable by hypothesis in the standard model) of the strong nuclear field and the bosons of the weak nuclear field both become quanta of the unified field from the Evans wave equation. Ultimately the quarks are SU(3) representations of quantized gravitation from Eq. (7.16), and the weak nuclear field bosons are SU(2) representations of the electromagnetic field. Both the gravitational and the electromagnetic field spring from the tetrad, so we have built field unification with differential geometry.

The eight SU(3) matrices [10] are generalizations of the familiar Pauli matrices

$$\sigma^1 = \begin{pmatrix} 0 & 1 \\ 1 & 0 \end{pmatrix}, \quad \sigma^2 = \begin{pmatrix} 0 & -i \\ i & 0 \end{pmatrix}, \quad \sigma^3 = \begin{pmatrix} 1 & 0 \\ 0 & -1 \end{pmatrix} \tag{7.25}$$

as follows:

$$\sigma^1 \rightarrow \lambda_1 = \begin{pmatrix} 0 & 1 & 0 \\ 1 & 0 & 0 \\ 0 & 0 & 0 \end{pmatrix}, \lambda_4 = \begin{pmatrix} 0 & 0 & 1 \\ 0 & 0 & 0 \\ 1 & 0 & 0 \end{pmatrix}, \lambda_6 = \begin{pmatrix} 0 & 0 & 0 \\ 0 & 0 & 1 \\ 0 & 1 & 0 \end{pmatrix}, \tag{7.26}$$

$$\sigma^2 \rightarrow \lambda_2 = \begin{pmatrix} 0 & -i & 0 \\ i & 0 & 0 \\ 0 & 0 & 0 \end{pmatrix}, \lambda_5 = \begin{pmatrix} 0 & 0 & -i \\ 0 & 0 & 0 \\ i & 0 & 0 \end{pmatrix}, \lambda_7 = \begin{pmatrix} 0 & 0 & 0 \\ 0 & 0 & -i \\ 0 & i & 0 \end{pmatrix}, \tag{7.27}$$

$$\sigma^3 \rightarrow \lambda_3 = \begin{pmatrix} 1 & 0 & 0 \\ 0 & -1 & 0 \\ 0 & 0 & 0 \end{pmatrix}, \lambda_8 = \frac{1}{\sqrt{3}} \left(\begin{pmatrix} 1 & 0 & 0 \\ 0 & 0 & 0 \\ 0 & 0 & -1 \end{pmatrix} + \begin{pmatrix} 0 & 0 & 0 \\ 0 & 1 & 0 \\ 0 & 0 & -1 \end{pmatrix} \right). \tag{7.28}$$

The matrices are characterized by:

$$[(1/2)\lambda_a, (1/2)\lambda_b] = (i/2) j_{abc} \lambda_c, \tag{7.29}$$

where

$$C_{abc} := i j_{abc} \tag{7.30}$$

are the group structure constants of SU(3) [10]. The structure constants of SU(2) are

$$C_{abc} = i \epsilon_{abc}. \tag{7.31}$$

It can be seen that λ_2, λ_5, and λ_7 are O(3) rotation generator matrices, cyclically related by

$$[(1/2)\lambda_2, (1/2)\lambda_5] = (i/2) j_{257} \lambda_7; \tag{7.32}$$

and this is a representation of three-dimensional Euclidean space with O(3) group rotation generators. Similarly, $-i\lambda_1, -i\lambda_4, -i\lambda_6$ are three O(3) symmetry rotation generator matrices related cyclically by

$$[(1/2)\lambda_1, (1/2)\lambda_4] = (i/2)j_{146}\lambda_6. \tag{7.33}$$

Finally, the matrices λ_3 and λ_8 are also rotation generator matrices related cyclically by

$$[(1/2)\lambda_3, (1/2)\lambda_8] = (i/2)j_{38c}\lambda_c, \tag{7.34}$$

where $c = 1, 2, 4, 5, 6,$ or 7.

Therefore a rotation in three-dimensional space in SU(3) representation is given by

$$q' = \exp\left(i\frac{\lambda_a}{2}\epsilon_a\right)q, \tag{7.35}$$

an SU(3) transformation with

$$q = \begin{pmatrix} q_R \\ q_W \\ q_B \end{pmatrix}. \tag{7.36}$$

Similarly a rotation in Euclidean, three-dimensional space (the base manifold) can be represented by the SU(2) transformation

$$q' = \exp\left(i\frac{\sigma^i}{2}\epsilon_i\right)q, \tag{7.37}$$

where

$$q = \begin{pmatrix} q_1 \\ q_2 \end{pmatrix} \tag{7.38}$$

is the Pauli two-spinor representation of the base manifold using a two-dimensional representation space. In Eq. (7.36), q is a three-spinor representation of Euclidean three dimensional space.

In order now to represent the strong nuclear field in general relativity, the group indices a, b, c in Eq. (7.29) are recognized as indices of the orthonormal space of the tetrad, so that Eq. (7.29) becomes an equation of differential geometry, valid in any spacetime, and independent of the details of the base manifold geometry [4,5]. Similarly, in order to represent the weak nuclear field in general relativity, the indices a, b, c of Eq. (7.31) become indices of differential geometry. Thus both equations (7.29) and (7.31) can be written as

$$(1/2)\lambda_a \wedge (1/2)\lambda_b = (i/2)j_{abc}\lambda_c, \tag{7.39}$$

that is,

$$(1/2)q_a \wedge (1/2)q_b = (i/2)j_{abc}q_c, \tag{39a}$$

i.e., become valid in non-Euclidean spacetime as equations of differential geometry. In wedge product notation of differential geometry, Eqs. (7.30) and (7.31) define cyclically symmetric relations between wedge products in respectively an SU(3) and SU(2) representation of the orthonormal space of the tetrad. In the SU(3) representation for the strong nuclear field in general relativity the tetrad is defined by the equation

$$q^a = q^a{}_\mu q^\mu : SU(3) \tag{7.40}$$

between a metric three-spinor q^a in the orthonormal space labeled a and a metric three-spinor q^μ in the non-Euclidean base manifold, labeled μ. Similarly the tetrad in the SU(2) representation for the weak nuclear field in general relativity is defined by

$$q^a = q^a{}_\mu q^\mu : SU(2), \tag{7.41}$$

where q^a is a metric two-spinor in the orthonormal space and where q^μ is a metric two-spinor in the base manifold. In both cases the Evans wave equation is an equation in the tetrad

$$(\Box + kT)q^a{}_\mu = 0; \tag{7.42}$$

and in this notation takes the same form for gravitation and for electromagnetism. For both gravitation and electromagnetism, the tetrad is defined by

$$q^a = q^a{}_\mu q^\mu : O(3), \tag{7.43}$$

where q^a is a metric four vector in the orthonormal space and where q^μ is a metric four-vector in the base manifold.

It is seen that the strong and weak nuclear fields and the gravitational and electromagnetic field share a common origin in the concept of tetrad, and that the Evans wave equation in all cases is an equation in the tetrad. The nuclear weak field is essentially an SU(2) representation for the electromagnetic field, and the nuclear strong field an SU(3) representation of the gravitational field. The Dirac equation is derived [11] from the Evans equation with an SU(2) representation of the gravitational field.

Finally the quark color triplet is a field, so must be a manifestation of the tetrad field. In the standard model only an empirical representation of the color triplet is given, i.e., it is not derived from general relativity and is not identified as a manifestation of a tetrad, the matrix that relates the metrics in the orthonormal space and the base manifold. For each label a of the tetrad, it is a generally covariant object, i.e., a generally covariant four-vector, three-spinor or two-spinor, so the equation governing the gluon field (and all other fields) in general relativity is the Evans equation for each upper index a.

Acknowledgments

Craddock, Inc., ADAS and the Ted Annis Foundation are thanked for funding, and the Fellows and Emeriti of the AIAS are thanked for many interesting discussions.

References

1. M. W. Evans, *Found. Phys. Lett.* **16**, 367 (2003).
2. M. W. Evans, "A generally covariant wave equation for grand unified field theory," *Found. Phys. Lett.* **16**, 507 (2003); www.aias.us
3. M. W. Evans, "Equations of unified field theory in terms of the Maurer-Cartan structure relations of differential geometry," *Found. Phys. Lett.* **17**, 25 (2004); www.aias.us
4. S. M. Carroll, *Lecture Notes on General Relativity* (University of California, Santa Barbara, graduate course on arXiv:gr-qe / 9712019 v1 3 Dec, 1997).
5. R. M. Wald, *General Relativity* (Chicago University Press, 1984).
6. M. W. Evans, "O(3) Electrodynamics," in M. W. Evans, ed., *Modern Nonlinear Optics*, a special topical issue in three parts of I. Prigogine and S. A. Rice, series eds, *Advances in Chemical Physics* (Wiley-Interscience, New York, 2001, second and e-book editions), Vol. 119(2), pp. 79-269.
7. M. W. Evans and S. Jeffers, "The present status of the quantum theory of light," in Ref. 6, Vol. 119(3), pp. 1-197.
8. M. W. Evans, J.-P. Vigier, *et al., The Enigmatic Photon* (Kluwer Academic, Dordrecht, 1994 to 2002, hardback and softback editions), Vols 1 to 5.
9. M. Tatarikis, K. Krushelnik, Z. Najmudin, E. L. Clark, M. Salvati, A. E. Dangor, V. Malka, D. Neely, R. Allott, and C. Danson, "Measurements of the inverse Faraday effect in high intensity laser produced plasmas," CLF Annual Report, Science, High Power Laser Programme, Short Pulse Plasma Physics, Rutherford Appleton Laboratories of the UK EPSRC, 1998/1999, m.tatarikis@ic.ac.uk (Imperial College, London).
10. L. H. Ryder, *Quantum Field Theory* (Cambridge University Press, 1996).
11. M. W. Evans, "Derivation of the Dirac equation from the Evans field equation," *Found. Phys. Lett.* **17**(2) (2004); www.aias.us

8

Derivation Of The Evans Wave Equation *From* The Lagrangian And *Action*: Origin Of The Planck Constant In General Relativity

Summary. The Evans wave equation is derived from the appropriate Lagrangian and action, identifying the origin of the Planck constant $\hbar$ in general relativity. The classical Fermat principle of least time, and the classical Hamilton principle of least action, are expressed in terms of a tetrad multiplied by a phase factor $\exp(iS/\hbar)$, where S is the action in general relativity. Wave (or quantum) mechanics emerges from these classical principles of general relativity for all matter and radiation fields, giving a unified theory of quantum mechanics based on differential geometry and general relativity. The phase factor $\exp(iS/\hbar)$ is an eigenfunction of the Evans wave equation and is the origin in general relativity and geometry of topological phase effects in physics, including the Aharonov-Bohm class of effects, the Berry phase, the Sagnac effect, related interferometric effects, and all physical optical effects through the Evans spin field $\boldsymbol{B}^{(3)}$ and the Stokes theorem in differential geometry. The Planck constant $\hbar$ is thus identified as the least amount possible of action or angular momentum or spin in the universe. This is also the origin of the fundamental Evans spin field $\boldsymbol{B}^{(3)}$, which is always observed in any physical optical effect. It originates in torsion, spin and the second (or spin) Casimir invariant of the Einstein group. Mass originates in the first Casimir invariant of the Einstein group. These two invariants define any particle.

Key words: Evans wave equation, Lagrangian, action, Fermat principle, Hamilton principle, topological phase effects, Evans spin field, unified theory, origin of the Planck constant and $\boldsymbol{B}^{(3)}$ in general relativity.

8.1 Introduction

The Evans wave equation of unified theory [1-3] is derived from the appropriate Lagrangian density and action in general relativity. The action S so defined is shown to be the origin both of the Planck constant $\hbar$ and of the Evans spin field $\boldsymbol{B}^{(3)}$ [4-8]. The latter is always observed in any physical optical or topological phase effect involving electromagnetism. The latter, in turn, is part of the Evans unified field theory [1-3] in general relativity and differential geometry [9,10], and electromagnetism is the gauge invariant component of unified field theory originating in the torsion form of differential geometry. Gravitation is the gauge invariant component of the unified field theory originating in the Riemann form. The potential fields for both gravitation and electromagnetism originate in the tetrad $q^a{}_\mu$ [1-3], which is most generally defined as the invertible matrix linking two frames of reference. Section 8.2 identifies the Lagrangian density and action which lead to the Evans wave equation through the Euler Lagrange equation. In Sec. 3 the action defined in Sec. 2 is used to construct the phase factor

$$\Phi = \exp(iS/\hbar), \tag{8.1}$$

which is the origin of all topological phase effects in physics. By defining the Evans spin field $\boldsymbol{B}^{(3)}$ in terms of the action S it is shown that the phase factor Φ is also the origin of all physical optical effects [4-8]. Inter alia, the latter are defined by $\boldsymbol{B}^{(3)}$ and serve to show the presence of $\boldsymbol{B}^{(3)}$ in any physical optical effect, the reason being that $\boldsymbol{B}^{(3)}$ always defines the electromagnetic phase through S and the Stokes theorem of differential geometry applied to general relativity. An elegant and powerful understanding of natural philosophy is therefore possible through the use of differential geometry applied to unified field theory [1-3], and through the development of electromagnetism as a non-Abelian gauge field theory [4-8]. Section 8.4 unifies the Fermat principle of least time in optics and the Hamilton principle of least action in dynamics by showing that these principles of classical physics can both be defined as the tetrad multiplied by the phase factor Φ. The propagation of a wave in optics or a particle in dynamics is governed essentially by Φ, and the wave and particle become conceptually unified in general relativity, thus leading to the de Broglie principle of duality in general relativity and unified theory, and thus to quantum mechanics in general relativity.

8.2 The Lagrangian Density And Action Of The Evans Wave Equation

The action S is the integral [9-14] with respect to the four-volume d^4x over the Lagrangian density:

$$S = (1/c) \int \mathcal{L} d^4x, \tag{8.2}$$

where c is the speed of light in vacuo, a universal constant of general relativity. The four volume d^4x is used because the integration takes place in a non-Euclidean spacetime of four-dimensions. The Lagrangian density is defined to be a function of the tetrad $q^a{}_\mu$, the derivative $\partial_\nu q^a{}_\mu$, and of x_μ:

$$\mathcal{L} = \mathcal{L}\left(q^a{}_\mu, \partial_\nu q^a{}_\mu, x_\mu\right). \tag{8.3}$$

It follows [9-14] from Hamilton's principle of least action and the variational principle that the Lagrangian density is governed by the Euler Lagrange equation of motion:

$$\frac{\partial \mathcal{L}}{\partial q^{\nu}{}_{a}} = \partial^{\mu}\left(\frac{\partial \mathcal{L}}{\partial(\partial^{\mu} q^{\nu}{}_{a})}\right). \tag{8.4}$$

The Evans wave equation of motion [1-3] is

$$(\Box + kT) q^{a}{}_{\mu} = 0, \tag{8.5}$$

where

$$T = -R/k \tag{8.6}$$

is the index contracted energy-momentum tensor defined through Eq. (6) by the negative of the scalar curvature R divided by k, the Einstein constant. In Eq. (8.5) $\Box$ is the d'Alembertian operator with respect to Euclidean spacetime [1-3]:

$$\Box = \frac{1}{c^2}\frac{\partial^2}{\partial t^2} - \frac{\partial^2}{\partial X^2} - \frac{\partial^2}{\partial Y^2} - \frac{\partial^2}{\partial Z^2}. \tag{8.7}$$

Equation (8.5) follows from the Eq. (8.4) given the Lagrangian density

$$\mathcal{L} = -\frac{c^2}{k}\left[\frac{1}{2}(\partial_{\mu} q^{a}{}_{\nu})(\partial^{\mu} q^{\nu}{}_{a}) + \frac{R}{2} q^{a}{}_{\nu} q^{\nu}{}_{a}\right] = -\frac{R}{k} c^2. \tag{8.8}$$

Using Eq. (8.8), the left-hand side of Eq. (8.4) becomes

$$\partial \mathcal{L}/\partial q^{\nu}{}_{a} = -\frac{Rc^2}{k} q^{a}{}_{\nu}, \tag{8.9}$$

and the right-hand side is

$$\partial \mathcal{L}/\left(\partial\left(\partial^{\mu} q^{\nu}{}_{a}\right)\right) = -\frac{c^2}{k}\partial_{\mu} q^{a}{}_{\nu}. \tag{8.10}$$

The Euler-Lagrange equations (8.4) therefore gives the Evans lemma [15] or subsidiary proposition of differential geometry:

$$\Box q^{a}{}_{\nu} = R q^{a}{}_{\nu}, \tag{8.11}$$

given the Lagrangian density (8.8). The Evans wave equation (8.5) follows from the lemma (8.11) using Eq. (8.6). The Lagrangian density (8.8) also defines the scalar curvature in terms of the derivative of the tetrad:

$$R = \partial_{\mu} q^{a}{}_{\nu} \partial^{\mu} q^{\nu}{}_{a}. \tag{8.12}$$

The Evans wave equation and lemma are derived [1-3] from the tetrad postulate of differential geometry [9,10]:

$$D_{\mu} q^{a}{}_{\lambda} = \partial_{\mu} q^{a}{}_{\lambda} + \omega^{a}{}_{\mu b} q^{b}{}_{\lambda} - \Gamma^{\nu}{}_{\mu\lambda} q^{a}{}_{\nu} = 0. \tag{8.13}$$

Using Eq. (8.13) in Eq. (8.12) defines the scalar curvature R (and thus T) in terms of the Christoffel connection $\Gamma^{\nu}{}_{\mu\lambda}$ and the spin connection $\omega^{a}{}_{\mu b}$ for any spacetime:

$$R = \left(\Gamma^{\nu}{}_{\mu\lambda} q^{a}{}_{\nu} - \omega^{a}{}_{\mu b} q^{b}{}_{\lambda}\right) \left(\Gamma^{\mu\lambda}{}_{\nu} q^{a}{}_{\nu} - \omega^{\mu b}{}_{a} q^{\lambda}{}_{b}\right). \tag{8.14}$$

This definition of the scalar curvature follows from the assumption that the Lagrangian density is a function of three variables of the type (8.3). It shows that for any non-zero T the connections must also be non-zero, meaning that the spacetime is never Euclidean, and that the spacetime always contains both curvature and torsion. This is a major advance in understanding over the original theory of general relativity [15], which applies only to gravitation, and in which the torsion tensor is set to zero. For this reason the original theory of general relativity [15] cannot define electromagnetism.

In Eq. (8.5) T is the total energy in general. When there is no potential energy, the total energy and the Lagrangian are identical and given by the kinetic energy:

$$T = L = E_{\text{kin}}. \tag{8.15}$$

More generally, the Hamiltonian and Lagrangian are defined by

$$H = E_{\text{kin}} + E_{\text{pot}}, \tag{8.16}$$

$$L = E_{\text{kin}} - E_{\text{pot}}. \tag{8.17}$$

When considering the Evans lemma and wave equation, the effect of potential energy or interaction energy is always to change R. This means that Feynman diagrams, for example, could be replaced by a theory depending on changes of R, a theory which contains no singularities. This would be a major advance in areas such as quantum electrodynamics. In general the Evans equation would be solved numerically for any given problem in physics and general relativity using powerful contemporary code and code libraries for the solution of partial, second-order, differential equations (wave equation).

8.3 Phase Factor, Topological Effects, And Origin Of The Planck Constant In General Relativity

For a free particle (or field) there is only kinetic energy present and so total energy T and Lagrangian density originate in kinetic energy. This means that the Lagrangian density is defined in terms of $R(= -kT)$ using Eq. (8.8). The action for a free particle is therefore

$$S = -\frac{c}{k} \int R d^4 x. \tag{8.18}$$

The action is an integral over the scalar curvature. This result remains true in the presence of potential or interaction energy provided the scalar curvature R is changed, and provided the Lagrangian is defined correctly by Eq. (8.17). This result may be summarized in the following theorem:

If the scalar curvature is defined by

$$T = -R/k, \tag{8.19}$$

then

$$\int R d^4 x = -\frac{k}{c} S. \tag{8.20}$$

The phase factor of the Evans unified field/matter theory can therefore always be defined by a scalar curvature for any situation in physics:

$$\Phi = \exp(iS/\hbar) = \exp\left(\frac{-ic}{\hbar k} \int R d^4 x\right). \tag{8.21}$$

As shown in Sec. 4, the classical Fermat and Hamilton principles are defined in turn by Eq. (8.21), thus providing a powerful basis for the development of quantum mechanics from general relativity and differential geometry. The origin of these principles of classical physics is found from the fact that the eigenfunction of the Evans wave equation can always be written as the tetrad

$$q^a{}_\mu(x) = \Phi q^a{}_\mu(0). \tag{8.22}$$

Application of the Leibnitz theorem [16] to Eq. (8.22) shows that

$$(\Box + kT)\Phi = 0, \tag{8.23}$$

and so the phase factor Φ *itself* gives curvature and energy eigenvalues from the wave equation (8.26). This is the origin of the well observed topological phase effects [17,18] in physics. Examples are the Aharonov Bohm class of effects, the Berry phase in quantum mechanics, the Sagnac effect in physical optics, the closely related Tomita Ciao effect [4-8] and in general all physical optical effects, because the latter all depend on light propagation, and so they all depend on the phase defined in Eq. (8.21) through the Fermat principle of least time [19]. A gauge transformation can also be understood as a phase effect dependent on Φ, and so Φ is related to the rotation operator in the gauge transformation of the general n-dimensional field [22]. Therefore all these effects become understandable as effects of general relativity, as required by the principle of relativity of Einstein, and not of special relativity. The principle of relativity requires that all theories of physics be theories of general relativity, and ultimately, all topological phase effects can be traced, through the Evans wave equation, to the non-trivial topology of spacetime itself. This topology is summarized in $R(= -kT)$, and the phase (8.21) is defined by an integral over R.

The global or gauge-invariant phase factor is defined through the Dirac or Wu-Yang phase factor [17,18]:

$$\Phi^a = \exp\left(ie \oint A^a{}_\mu dx^\mu\right). \tag{8.24}$$

The physical quantity in Wu-Yang-Dirac phase is e multiplied by the gauge invariant contour integral over the electromagnetic potential $A^a{}_\mu$. The latter is related through the Stokes theorem to the surface integral over the gauge invariant electromagnetic field. The gauge invariant and non-oscillatory (i.e., phase free) electromagnetic field component in the Wu-Yang-Dirac phase is the Evans spin field $\boldsymbol{B}^{(3)}$ [4-8]. This means that the action can be defined through the Stokes theorem and in general relativity and unified field theory the action is defined in this way for any radiated or matter field.

The conventional [14] definition of a matter or radiated wave is

$$\psi = \psi_0 \exp(i\kappa_\mu x^\mu), \tag{8.25}$$

but this definition is incorrect, because it is invariant under parity inversion

$$\hat{P}(x_\mu \kappa^\mu) = x_\mu \kappa^\mu, \tag{8.26}$$

and for this reason cannot provide even a qualitative description of ordinary effects in physical optics such as reflection (i.e., parity inversion) and interferometry [8]. If the phase is invariant under reflection there cannot be an observed reflection. This simple fact appears to have been overlooked in conventional physical optics. In order to construct a Michelson interferogram, for example [4-8], the phase must change sign under $\hat{P}$ at the point of beam reflection from the mirrors of the Michelson interferometer, and this property is given by the contour integral definition of phase used in the Wu-Yang-Dirac phase factor of electrodynamics:

$$\psi = \psi_0 \exp\left(i \oint \kappa_\mu dx^\mu\right). \tag{8.27}$$

This is an important development in physical optics because the Stokes theorem always relates the contour integral to a surface integral, showing that spin is always present in the phase of a matter wave (such as an electron beam) or a radiated wave (such as an electromagnetic beam). In the conventional description of the phase only the energy-momentum is present, represented by $\hbar\kappa^\mu$. The conventional description is incomplete essentially because there are two invariants of the Einstein group, mass and spin, and both are needed to define a particle such as an electron or photon. The fundamental spin field in the electromagnetic phase in general relativity is the $\boldsymbol{B}^{(3)}$ field, which appears in the surface or area integral of the Stokes theorem. In the notation of differential geometry [9,10] the latter means that

$$\oint \kappa_\mu dx^\mu = \int (d \wedge \kappa)_{\mu\nu} d\sigma^{\mu\nu} \tag{8.28}$$

or, in condensed notation,

$$\oint_{\delta S} \kappa = \int_S d \wedge \kappa, \tag{8.29}$$

where $d\wedge$ is the exterior derivative. Eq. (8.29) is true for all geometries of the base manifold, and so is true in general relativity, which is the geometrical theory of all physics.

If q^a is the tetrad form (a vector valued one form), then the Stokes theorem implies that

$$\oint_{\delta S} q^a = \int_S \tau^a, \tag{8.30}$$

where

$$\tau^a = d \wedge q^a \tag{8.31}$$

is the torsion form [1-3,9,10] as defined by the first Maurer-Cartan structure relation of differential geometry [9,10]. So the Stokes theorem relates the contour integral over the tetrad form to the surface integral over the torsion form for any geometry of the base manifold:

$$\oint q^a{}_\mu dx^\mu = \int \tau^a{}_{\mu\nu} dv^{\mu\nu}. \tag{8.32}$$

The torsion form in unified field theory [1-3] is the electromagnetic field within a factor $A^{(0)}$ with the units of weber m^{-1} = volts $s^{-1}m^{-1}$ and the tetrad form is the electromagnetic potential field [1-3]:

$$A^a{}_\mu = A^{(0)} q^a{}_\mu, \tag{8.33}$$

where $A^{(0)}$ is a potential magnitude. Therefore in unified field theory the Stokes theorem (8.30) inter-relates the potential and gauge-invariant fields:

$$\oint_{\delta S} A^a = \int_S B^a = \int_S d \wedge A^a. \tag{8.34}$$

The potential field itself is not gauge invariant, but the contour integral over the potential field is gauge invariant. In the Aharonov-Bohm effect [14], gauge transformation of the potential field produces a physical effect, so both the potential and the gauge invariant fields are physically meaningful in classical general relativity as well as quantum mechanics. The unified field theory [1-3] has shown that quantum mechanics emerges from differential geometry, (see Sec. 4), so the gauge transformation of the second kind produces effects such as entanglement, which are conventionally explained in terms of quantum mechanical wavefunctions without classical meaning. Unified field theory [1-3] therefore traces the origin of entanglement [8] to the gauge transformation of the second kind. Action at a distance is therefore not needed to explain entanglement.

The Wu-Yang-Dirac phase developed for general relativity then follows from Eq. (8.34):

$$\exp\left(i\frac{e}{\hbar}\oint A^a\right) = \exp\left(i\frac{e}{\hbar}\int B^a\right) \tag{8.35}$$

and completely defines the electromagnetic field because this phase is gauge invariant for all geometries of the base manifold. In contrast, the conventional phase factor of Maxwell-Heaviside theory is not gauge-invariant:

$$\Phi_{\mathrm{MH}} = \exp\left[i(\omega t - \boldsymbol{\kappa} \cdot \boldsymbol{r} + \alpha)\right], \tag{8.36}$$

so an arbitrary factor α can be added [17] to it without affecting physical optics. This description of nature is therefore over determined. It also fails to give a qualitative account of reflection and interferometry as argued already.

This development suggests that the phase for radiated and matter waves is defined in general relativity by the tetrad form of differential geometry:

$$\exp\left(i\kappa \oint_{\delta S} q^a\right) = \exp\left(i\kappa \int_S d \wedge q^a\right) \tag{8.37}$$

for all possible geometries of the base manifold. The index a defines the tangent space basis, which is orthogonal and normalized, i.e., an Euclidean space. If we make the gauge transformation of the second kind [8]:

$$q^a \to q^a + \frac{1}{\kappa} d\chi^a, \tag{8.38}$$

where d denotes the ordinary (not exterior) derivative, then

$$d \wedge q^a \to d \wedge q^a + \frac{1}{\kappa} d \wedge d\chi^a = d \wedge q^a, \tag{8.39}$$

on using the Poincaré lemma

$$d \wedge d := 0. \tag{8.40}$$

This result shows that the right-hand side of Eq. (8.37) is gauge invariant. Therefore the left-hand side must also be gauge invariant, implying that

$$\oint d\chi^a := 0 \tag{8.41}$$

for all geometries of the base manifold. Eq. (8.41) is true for all χ^a, irrespective of whether χ^a is single valued or periodic (integral or multi-valued). Therefore the Aharonov-Bohm effect, for example, cannot be described by the Stokes theorem (8.34), and the Aharonov-Bohm effect cannot be understood with the conventional phase (8.36), showing in another way that that phase is unphysical. Recall that the phase (8.36) cannot be used to understand reflection or interferometry in physical optics, and cannot be used to understand interferometry in matter waves, such as the Sagnac effect in electron beams [8]. The conventional description of the Aharonov-Bohm effect [14] contradicts the fundamental result (8.41) from gauge invariance, and the conventional description is therefore incorrect.

This fundamental paradox of conventional electromagnetic theory is resolved, however, by developing the Stokes theorem with the covariant exterior derivative [9,10,18] denoted by $D\wedge$:

$$\oint_{\Delta S} q^a = \int_S D \wedge q^a, \tag{8.42}$$

where

$$D \wedge q^a = d \wedge q^a + \kappa q^b \wedge q^c. \tag{8.43}$$

Equation (8.43) is the non-Abelian Stokes theorem [8,17] in the notation of differential geometry. The covariant exterior derivative acts on a differential n form to produce a differential $n+1$ form which transforms properly under general coordinate transformation [9,10], i.e., as a proper tensor. The ordinary exterior derivatives produces a vector valued $n+1$ form from a vector valued n form. The $n+1$ form transforms correctly under the transformation law for (0,2) tensors for general coordinate transformations, but does not transform correctly as a vector under linear Lorentz transformations [9,10]. This flaw in the $d\wedge$ operator is remedied by replacing it by the $D\wedge$ operator, the covariant exterior derivative operator of differential geometry. The latter acts on the vector valued one form $X^a{}_\mu$, for example the tetrad, to produce [9,10] the vector-valued two-form

$$\begin{aligned}(D \wedge X)^a{}_{\mu\nu} &= (d \wedge X)^a{}_{\mu\nu} + (\omega \wedge X)^a{}_{\mu\nu} \\ &= \partial_\mu X^a{}_\nu - \partial_\nu X^a{}_\mu + \omega^a{}_{\mu b} X^b{}_\nu - \omega^a{}_{\nu b} X^b{}_\mu,\end{aligned} \tag{8.44}$$

where $\omega^a{}_{\mu b}$ is the spin connection. Recent work [3] has shown that the spin connection is the Hodge dual of the tetrad, and that the gauge invariant Riemann form of gravitation is the Hodge dual of the gauge invariant torsion form of electromagnetism [1-3]. The object $D \wedge X^a$ always transforms as a proper tensor [9,10] for all spacetime geometries of the base manifold, and therefore for all situations in general relativity and unified field theory. The unified field theory [1-3] traces the origin of both gravitation and electromagnetism to differential geometry, thus implying the use of $D\wedge$ for basic mathematical correctness. The replacement of $d\wedge$ by $D\wedge$ produces the $\boldsymbol{B}^{(3)}$ field [1-8,17], produces the correct explanation of reflection and interferometry through the phase constructed from (8.42), and clearly explains the Aharonov-Bohm effect as follows.

The electromagnetic potential field in the unified field theory is defined as the tetrad [1-3]

$$A^a{}_\mu = A^{(0)} q^a{}_\mu, \tag{8.45}$$

where $A^{(0)}$ is the scalar magnitude of the fundamental, C negative, potential field, whose origin can be traced to primordial magnetic flux in units of volts s^{-1} [1-3]. So, in generally covariant electrodynamics [4-8],

$$D \wedge A^a = d \wedge A^a + g A^b \wedge A^c, \tag{8.46}$$

with

$$g = e/\hbar = \kappa / A^{(0)} \tag{8.47}$$

as the fundamental charge e on the proton divided by the Planck constant $\hbar$. Thus g is a universal constant [4-8] that defines the photon momentum by

$$p = \hbar\kappa = eA^{(0)}, \tag{8.48}$$

and the origin of g can be traced to differential geometry, the need for a $D\wedge$ operator instead of a $d\wedge$ operator as argued already. Thus, general relativity and unified field theory show the need for g in electrodynamics. In conventional Maxwell-Heaviside theory (a theory of special relativity) g is missing, essentially because spacetime in the Maxwell Heaviside theory is flat or Euclidean and the spin connection and R are both zero. In general relativity this implies that the universe is devoid of all matter fields, all radiation fields and all energy-momentum (T is zero for all k if R is zero). General relativity and differential geometry imply the $\boldsymbol{B}^{(3)}$ field, defined as the differential form

$$B^{(3)} = -igA^{(1)} \wedge A^{(2)}. \tag{8.49}$$

(If g were zero (flat spacetime) $\boldsymbol{B}^{(3)}$ would be zero, but as just argued, $\boldsymbol{B}^{(3)}$ is responsible for all physical optics.) The correctly covariant definition of the Wu-Yang phase in general relativity is therefore

$$\Phi = \exp\left(ig\oint_{DS} A^a\right) = \exp\left(ig\int_S B^a\right) = \exp\left(ig\int_S D\wedge A^a\right). \tag{8.50}$$

In order to understand the Aharonov-Bohm class of effects, consider the property of the covariant Stokes theorem

$$\oint_{DS} A = \int_S D\wedge A, \tag{8.51}$$

under the local gauge transformation

$$A' = \sigma A, \tag{8.52}$$

where σ is rotation operator [9]. In Eqs. (8.51) and (8.52) we have used the generic notation of differential geometry. Covariant differentiation of the gauge transform relation (8.52) produces

$$D'A' = \sigma DA, \tag{8.53}$$

where

$$D = d + gA, \tag{8.54}$$

$$D' = d + gA', \tag{8.55}$$

while ordinary differentiation leads to

$$dA' = d(\sigma A) = \sigma dA + Ad\sigma. \tag{8.56}$$

From Eqs. (8.54) and (8.55) inserted in (8.53),

$$dA' + gA'A' = \sigma dA + g\sigma AA. \tag{8.57}$$

Using Eq. (8.56) in (8.57),

$$\begin{aligned} \sigma dA + Ad\sigma + gA'A' &= \sigma dA + g\sigma AA, \\ Ad\sigma + gA'A' &= g\sigma AA, \end{aligned} \tag{8.58}$$

and, using Eq. (8.52) in (8.58),

$$Ad\sigma + gA'\sigma A = g\sigma AA, \tag{8.59}$$

$$A' = \sigma A\sigma^{-1} - \frac{1}{g}d\sigma\sigma^{-1}. \tag{8.60}$$

This equation defines the change in the tetrad form $A^a{}_\mu$ (the electromagnetic potential form) under the local gauge transformation (8.52), also known as the gauge transform of the second kind [8,9]. This is the relativistically correct gauge transform and is the basis for all gauge field theory in physics because it leaves the action invariant, and leads to fundamental conservation theorems such as the Noether theorem.

In conventional electromagnetic theory [9] the electromagnetic field is defined by the ordinary exterior derivative acting on the scalar-valued one-form A:

$$F = d \wedge A, \tag{8.61}$$

and the field is invariant under the gauge transform (8.6):

$$F' = d \wedge A' = d \wedge A - (1/g\sigma)d \wedge d\sigma = d \wedge A. \tag{8.62}$$

It follows from the Stokes theorem of conventional electromagnetic theory,

$$\oint_{\delta S} A = \int_S d \wedge A, \tag{8.63}$$

that there is no Aharonov-Bohm effect in conventional electromagnetic theory. The reason is that the right-hand side of Eq. (8.63) is unchanged (i.e., invariant) under the gauge transform (8.60), so

$$\oint_{\delta S} d\sigma = \int_S d \wedge d\sigma := 0 \tag{8.64}$$

for all σ, irrespective of whether σ is a single-valued or periodic (multi-valued) function. The conventional explanation of the Aharonov-Bohm effect [9] incorrectly [8] asserts however that

$$\oint_{\delta S} d\sigma \neq 0(?). \tag{8.65}$$

The correctly and generally covariant electromagnetic field [1-3] transforms as

$$G' = D' \wedge A' = d \wedge A' + gA' \wedge A' \tag{8.66}$$

under the frame rotation induced by σ as described in Eq. (8.52). Using the result

$$d \wedge A = d \wedge A', \tag{8.67}$$

it is seen that the rotation produces the effect

$$A \wedge A \rightarrow A' \wedge A', \tag{8.68}$$

which means that the magnetic field defined in Eq. (8.49) transforms under local gauge transformation, i.e., the rotation (8.52), as

$$B = -igA \wedge A \rightarrow -igA' \wedge A'. \tag{8.69}$$

This transformation is the origin of the Aharonov-Bohm effect, which is therefore due to general relativity, not special relativity as in the (incorrect) conventional theory.

The definition (8.69) of the magnetic field in terms of the cross product of two potentials was first inferred [4-8] for the radiated Evans-Vigier field $\boldsymbol{B}^{(3)}$ (now known to be the fundamental spin Casimir invariant of the Einstein group in electrodynamics), but the definition has been extended [4-8] to a static (non-radiated) magnetic field because the correctly and generally covariant definition of the electric and magnetic fields must always be

$$G = D \wedge A, \tag{8.70}$$

in which $d\wedge$ has been replaced by $D\wedge$. As argued already, this replacement is a fundamental requisite of differential geometry. The exterior derivative of the potential $(d \wedge A)$ as in conventional electromagnetic theory (special relativity) is replaced by the covariant exterior derivative of the tetrad $(D \wedge A)$ in general relativity. In the Aharonov-Bohm effect, regions of the experimental set up are considered where the original magnetic flux density and original potential are both zero. In these regions an electromagnetic effect is nevertheless observed, for example as a shift in an electron diffraction pattern. The Aharonov-Bohm effect originates in the magnetic flux

$$B' = -ig\left(\frac{1}{g^2\sigma^2} d\sigma \wedge d\sigma\right), \tag{8.71}$$

which is purely geometrical or topological in nature and which does not depend on the original potential A. The magnetic flux that is observed in the effect is

$$\phi = \int B' dAr, \tag{8.72}$$

and so the original magnetic flux density has been shifted by the local gauge transform (8.52) into other regions of spacetime. This effect is due to the structure of spacetime itself, and therefore due to general relativity. It is obviously not due to action at a distance, a concept which violates relativity and therefore violates the Noether theorem, which as argued already is derived from relativity through local gauge transformation. Anything that violates the Noether theorem violates conservation of energy, momentum, charge and current. The gauge transform (8.52) is defined by the complex-valued rotation operator

$$\sigma = \exp\left(i\Lambda^a(x^\mu)M^a\right), \tag{8.73}$$

and so

$$|A' \wedge A'| = |A \wedge A|, \tag{8.74}$$

by the property of complex conjugation.

It is worth emphasizing that the basic error in the conventional theory of the Aharonov-Bohm effect [9] can be understood clearly in the notation of differential geometry as follows:

$$\begin{aligned} A &\to A + \frac{1}{g}d\chi, \\ \oint_{\delta S} d\chi &= \int_S d \wedge d\chi := 0. \end{aligned} \tag{8.75}$$

However, it is incorrectly asserted in the original theory [9] that

$$\oint_{\delta S} d\chi \neq 0(?). \tag{8.76}$$

This mathematical error was first pointed out in Ref. [8] and is now easily understood using differential geometry as argued already.

A generally covariant theory of electrodynamics is therefore needed to correctly explain the Aharonov-Bohm effect in terms of $D \wedge A$. In the correctly covariant Stokes theorem (8.51), gauge transformation (8.52) produces the result

$$\oint_{DS} d\sigma = \frac{i}{\sigma}\int_S d\sigma \wedge d\sigma \neq 0 \tag{8.77}$$

in regions where the original magnetic flux density and potential are both zero. There is therefore an electromagnetic effect in these regions as observed. The unified field theory [1-3] means that there must be a gravitational analogy of the Aharonov-Bohm effect, and such an effect is indeed observed – the well-known Coriolis and centripetal acceleration due to a rotating frame of reference. The rotation of the frame is a local, or relativistically correct, gauge transformation of type two observed in the weak field limit. These accelerations are observed where there is no Newtonian (or central or linearly directed) acceleration, and therefore no Newtonian gravity and no Newtonian force. It is nevertheless clear that the centripetal and Coriolis accelerations are physical accelerations that exist in regions where there is no Newtonian acceleration, and which are generated by a gauge or frame transformation in the weak field limit of general relativity. They therefore meet all the requirements of an Aharonov-Bohm effect.

The fundamental geometrical reason for the Aharonov-Bohm effect is that the Stokes theorem must be generally covariant, as in Eq. (8.51). Such a Stokes theorem is mathematically non-Abelian [8], and has properties in general relativity and unified field theory that are not present in special relativity. One of these is the Aharonov-Bohm effect as just argued, others include the topological phase effects, the Sagnac effect, and indeed all physical optics, because the phase in electrodynamics can only be explained correctly with the non-Abelian or generally covariant Stokes theorem [8] (Eq. (8.50)). The latter can always be written as

$$\oint_{DS} A^a = \int_S d \wedge A^a + g \int_S A^b \wedge A^c. \tag{8.78}$$

If $a = (1), (2), (3)$, the labels of the complex circular basis of circularly polarized electromagnetic radiation [4-8], then

$$\oint_{DS} A^{(i)} = \int_S d \wedge A^{(i)}, \quad i = 1, 2, \tag{8.79}$$

for the transverse components (1) and (2), and

$$\oint_{DS} A^{(3)} = -ig \int_S A^{(1)} \wedge A^{(2)} = \int_S B^{(3)} \tag{8.80}$$

for the longitudinal component (3). In the conventional theory of electromagnetism the transverse components are radiated plane waves and the longitudinal component is missing. The fundamental reason for this is that the conventional theory is an incorrectly Abelian gauge field theory in which the phase is incorrectly defined. In the generally covariant theory of electrodynamics [1-3] however, the (3) component of gauge transformation from a region 1 to a region 2 can always be written (From Eq. (8.80) [4-8]) as

$$A^{(3)}{}_1 \rightarrow A^{(3)'}{}_2. \tag{8.81}$$

Under this gauge transformation the magnetic flux density is invariant, but is shifted from region 1 to region 2, so the effect can be traced to a property of spacetime itself in general relativity. In special relativity, the spacetime is flat or Euclidean, and the Aharonov-Bohm effect cannot be explained with flat spacetime. Similarly the action

$$S = e \oint_{DS} A^{(3)}{}_1 \rightarrow e \oint_{DS} A^{(3)'}{}_2 \tag{8.82}$$

is invariant under the longitudinal, local gauge transform (8.52) but the action is shifted from region 1 to region 2. This shift in the action is accompanied by a shift in the generally covariant electromagnetic phase factor

$$\Phi = \exp(iS/\hbar). \tag{8.83}$$

The local gauge transformation responsible for this shift in the phase factor is a frame rotation defined by the rotation operator in Eq. (8.73). In the Sagnac effect, a corresponding shift in phase factor is brought about by a physical rotation of the platform of the Sagnac interferometer [8] and the Sagnac effect can be understood [8] as a phase shift brought about by an increase or decrease in the wave number ($\kappa = \omega/c$) which is related to the longitudinal potential component in the phase factor (8.50) by

$$\kappa^a = gA^a. \tag{8.84}$$

Therefore the shift (8.82) can be understood in the Sagnac effect as a frequency shift

$$\omega \to \omega \pm \Omega. \tag{8.85}$$

In Eq. (8.85) Ω is the angular frequency of the rotating platform and the plus or minus signs originate from clockwise and anticlockwise platform rotation respectively. In the Aharonov-Bohm effect the area Ar in the surface integral on the right-hand side of Eq. (8.50) is defined by the area enclosed by diffracting beams in a Young or two-slit interferometer. In the Sagnac effect the corresponding area is defined by the area enclosed by the Sagnac interferometer, i.e., by the paths of the electromagnetic beams or matter beams such as an electron beam or molecular beam on the platform, either at rest or rotating at angular frequency, Ω. The Sagnac effect is explained as follows from the generally covariant phase factor (8.50).

The magnitude of B^a is defined by [4-8]

$$B^{(0)} = \kappa A^{(0)}. \tag{8.86}$$

The phase factor (8.50) therefore becomes

$$\Phi = \exp\left(i\frac{e}{\hbar}\oint \boldsymbol{A}^{(0)} \cdot d\boldsymbol{r}\right) = \exp\left(i\frac{e}{\hbar}\kappa A^{(0)} \int dAr\right) \tag{8.87}$$

and can be written as

$$\Phi = \exp\left(i \oint \boldsymbol{\kappa}^{(3)} \cdot d\boldsymbol{r}\right) = \exp(i\kappa^2 Ar). \tag{8.88}$$

The Sagnac effect with platform at rest is given by the area integral on the right-hand side, which is equal to the contour integral around the boundary of this area on the left-hand side. If the area is a circle the boundary is the circumference. The interferogram or diffraction pattern of the Sagnac effect with platform at rest is therefore

$$\mathrm{Re}(\Phi) = \cos(\kappa^2 Ar) = \cos(\omega^2/c^2 Ar) \tag{8.89}$$

as observed experimentally to one part in 10^{23} precision [8]. The Sagnac effect from Eq. (8.89) depends on the magnitude of the area enclosed by the boundary, but not to the shape of the boundary, and this is again as observed experimentally. Thirdly, the Sagnac effect with platform at rest originates in the generally invariant phase factor

$$\mathrm{Re}(\Phi) = \cos\left(\omega^2/c^2 Ar\right) \tag{8.90}$$

(a scalar frequency squared divided by c^2 and multiplied by a scalar magnitude of area), and so the Sagnac effect with platform at rest is the same for an observer on and off the platform, again as observed experimentally [8]. Similarly the contour integral on the left-hand side of Eq. (8.88) is the same to an observer on and off the platform. The Sagnac effect with platform in motion is given by the interferogram

$$\mathrm{Re}(\Phi_m) = \cos\left(\left((\omega+\Omega)^2 - (\omega-\Omega)^2\right)\frac{Ar}{c^2}\right) = \cos\left(4\frac{\omega\Omega}{c^2}Ar\right), \tag{8.91}$$

as observed experimentally to one part in 10^{23}.

In conventional electrodynamics (Maxwell-Heaviside theory) there is no Sagnac effect [8], because that theory is invariant under motion reversal T, and is also metric invariant, i.e., a theory of flat spacetime. In generally covariant electrodynamics [1-4] the Sagnac effect is described by the generally covariant phase factor (8.50), which is also the origin of the Wu-Yang phase factor. The Sagnac effect can only be described if it is recognised that the torsion form in gravitational theory is the origin of the Coriolis and centripetal accelerations, and may be the origin of dark matter in the universe.

The electromagnetic phase in general relativity and unified field theory is therefore

$$\Phi = \exp\left(i\frac{e}{\hbar}\oint A^{(3)}dZ\right) = \exp\left(i\frac{e}{\hbar}\int B^{(3)}dAr\right), \tag{8.92}$$

and the following equations define the magnitudes $B^{(0)}$ and $A^{(0)}$, respectively, in terms of R and κ:

$$eB^{(0)} = \hbar R, \tag{8.93}$$

$$eA^{(0)} = \hbar\kappa, \tag{8.94}$$

$$B^{(3)} = \frac{\hbar}{e}R. \tag{8.95}$$

Similarly the mass is defined by

$$m = -\int\frac{R}{\kappa}dV. \tag{8.96}$$

These equations show that mass m is the first (mass) Casimir invariant of the Einstein group, and the $\boldsymbol{B}^{(3)}$ field is the second (spin) Casimir invariant of the Einstein group within the scalar magnitude $B^{(0)}$. The corresponding conclusions hold in special relativity if the Einstein group is replaced by the Poincaré group, as first demonstrated by Wigner [8].

The $\boldsymbol{B}^{(3)}$ or Evans-Vigier field originates therefore in the fundamental spin invariant of general relativity, mass being the other fundamental invariant.

The magnitude of the photon momentum is defined by

$$\rho = \hbar\kappa = eA^{(0)}, \tag{8.97}$$

and so e is a primordial charge (with the magnitude of the charge on the proton) present within the radiated electromagnetic field. So the photon is both a particle and a field. The g of non-Abelian gauge field theory [4-8] applied to the electromagnetic field is $e/\hbar$. With these definitions, the electromagnetic phase factor in general relativity is therefore

$$\begin{aligned} \Phi &= \exp(iS/\hbar) = \exp\left(i\int \mathcal{L}d^4x\right) \\ &= \exp\left(-\frac{i}{k}\int Rd^4x\right) \\ &= \exp\left(i\kappa\oint dZ\right) = \exp\left(i\kappa^2\int dAr\right), \\ R &= \kappa^2. \end{aligned} \tag{8.98}$$

The Planck constant is the least amount of action or angular momentum present in the universe, and so is defined in terms of scalar curvature by

$$\hbar = -\frac{c}{r}\int R_0 d^4x. \tag{8.99}$$

The local gauge transformation (8.52) is the generation of one tetrad, $A'^{a}{}_{\mu}$, from another, $A^{a}{}_{\mu}$,

$$A'^{a}{}_{\mu} = \sigma A^{a}{}_{\mu} \tag{8.100}$$

and is therefore a form of the Fermat principle of least time or Hamilton principle of least action combined into one equation (8.100) of unified field theory expressed by differential geometry. In Eq. (8.100) the action S is automatically invariant under the gauge transform because the latter is defined in terms of rotation generators of a given gauge group, so we obtain an equation of general relativity defining the invariant action in terms of the rotation generator

$$S = \hbar M^a \Lambda^a \tag{8.101}$$

for any given gauge group and any given representation space of the base manifold and for any geometry of the base manifold.

It is well known [19] that Fermat's principle of least time governs all physical optics, so it must also govern the Aharonov-Bohm and Sagnac effects, as just shown. Using Eq. (8.98) the Sagnac effect for example can now be understood in general relativity and unified field theory [1-3] as a change in scalar curvature of spacetime produced by rotating the platform in the Sagnac experiment. The Sagnac effect observed [8] in electron beams can be understood in the same way; it is fundamentally a change in phase produced by the scalar curvature of spacetime, and can be observed both in matter waves (electrons or molecules for example) and radiated waves (visible frequency light for example). The Sagnac effect observed in radiated and matter fields is therefore proof of the fundamental relations (8.98), and shows that all physical optical effects of the electromagnetic field are produced by the fundamental $\boldsymbol{B}^{(3)}$ field [8]. In matter waves the $\boldsymbol{B}^{(3)}$ field becomes the fundamental spin of the particle making up the matter field (for example the electron). The Sagnac effect therefore constitutes proof of the Evans unified field theory, in which particle spin from the Dirac equation is deduced from the Evans wave equation [1-3]. The latter applies to both radiated and matter fields. The Sagnac effect in matter fields is also an experimental demonstration of quantum mechanics, and unifies the Fermat and Hamilton principles through Eq. (8.22), producing the de Broglie wave particle duality. In the closely related Tomita-Chao effect [8] an increment of the phase (8.98) or (8.92) is detected after an electromagnetic beam propagates through several loops of a helix using an optical fiber. In the Berry effects [14] a similar type of geometrically generated phase (8.98) is detected after an electromagnetic beam propagates through several loops of a helix using an optical fiber. In both experiments the phase increment is described by a contour integral, so is additive if the light propagates through several loops. The Bohm-Aharonov effect is detected with a two-aperture interferometer and is described by the same equation (8.98) as the Sagnac effect. The only difference between the two effects is geometry and the way in which an observable phase shift is induced: the former by a frame rotation, the latter by a physical rotation. In a physical optical effect such as reflection or in an interferometric effect such as Michelson interferometry [8], the effect being observed is described in the same way, mathematically, as in the Sagnac and Aharonov-Bohm effects. All these effects [8] are manifestations of Eq. (8.98), in which the phase factor and action are defined with the Stokes theorem, and manifestation of the Fermat principle of least time, Eq. (22). The existence of Michelson interferometry (and all types of interferometry) depends [8] on the following property of the contour integral within the exponent of the phase factor:

$$\oint_{0A} \boldsymbol{A} \cdot d\boldsymbol{r} = - \oint_{A0} \boldsymbol{A} \cdot d\boldsymbol{r}. \tag{8.102}$$

In an electromagnetic wave propagating along the Z axis, this contour integral is valued along the following closed boundaries defining an area. If the distance $0A$ is n times the wavelength λ, then the area enclosed [8] is $n\lambda^2/\pi$.

The change in phase on normal reflection is then [8]

$$\exp\left(i\oint\kappa dZ\right)=\exp\left(i\int_0^Z\kappa dZ-i\int_Z^0\kappa dZ\right)=\exp(2i\kappa Z),\tag{8.103}$$

and this is observed in the interferogram of a device such as a Michelson interferometer, Young (two-slit) interferometer, Sagnac, or Mach/Zehnder interferometer.

It is not possible to describe interferometry in the Maxwell-Heaviside theory because the phase factor in that theory is

$$\exp\left(i(\boldsymbol{\kappa}\cdot\boldsymbol{r}-\omega t)\right)\xrightarrow{\hat{P}}\exp\left(i(\boldsymbol{\kappa}\cdot\boldsymbol{r}-\omega t)\right).\tag{8.104}$$

Normal reflection is parity inversion, and under the parity inversion operator $\hat{P}$:

$$\boldsymbol{\kappa}\to-\boldsymbol{\kappa},\quad \boldsymbol{r}\to-\boldsymbol{r},\tag{8.105}$$

so the phase (8.104) does not change under normal reflection in Maxwell-Heaviside theory, implying that there is no observable interferogram, contrary to experiment. The effect of parity inversion on the phase (8.103) however, is as follows:

$$\begin{aligned}\hat{P}\left(\oint\kappa dZ\right)&=\hat{P}\left(\int_0^Z\kappa dZ-\int_Z^0\kappa dZ\right)\\&=\int_0^{-Z}\kappa dZ-\int_{-Z}^0\kappa dZ=-\oint\kappa dZ,\end{aligned}\tag{8.106}$$

and the interferogram is predicted and observed experimentally. This is experimental evidence in favor of electrodynamics as a unified field theory in general relativity, in which the $\boldsymbol{B}^{(3)}$, Evans spin, field is well defined as the following wedge product of tetrads:

$$B^{(3)}{}_{\mu\nu}=-ig\left(A^{(1)}\wedge A^{(2)}\right)_{\mu\nu}.\tag{8.107}$$

The phase factor in physical optics follows from the Wu-Yang-Dirac phase as the surface integral over the commutator of covariant derivatives [8]:

$$\Phi=\exp\left(\int\left[D_\mu,D_\nu\right]d\sigma^{\mu\nu}\right).\tag{8.108}$$

Only the $\boldsymbol{B}^{(3)}$ field

$$\Phi=\exp\left(\mp\frac{g^2}{2}\int\left(A^{(1)}{}_Z A^{(2)}{}_Y-A^{(2)}{}_X A^{(1)}{}_Y\right)dS^{XY}\right)\tag{8.109}$$

contributes to the phase factor in optics and interferometry because the other terms are either zero or oscillatory, averaging to zero over many cycles [8]. The expression of this result in differential geometry is

$$\oint_{DS}q^a=\int_S q^b\wedge q^c=\kappa\int_S D\wedge q^a.\tag{8.110}$$

8.4 Unification Of The Fermat And Hamilton Principles In The Principle of Least Curvature, Derivation Of Wave Particle Duality And Quantum Mechanics From General Relativity

It has been argued that in general relativity the least action in the universe is the Planck constant, a universal constant and the archetypical signature of quantum mechanics. This inference suggests that the well known Fermat principle of least time, which governs optics, and the Hamilton principle of least action, which governs dynamics, can be derived from one principle, to which we refer as the Principle of Least Curvature. The Principle asserts that scalar curvature R is minimized in the equations of motion which govern natural philosophy.

A mathematical expression of the principle is the equation

$$q^a{}_\mu(x^\mu) = e^{iS(x^\mu)/\hbar} q^a{}_\mu(0), \tag{8.111}$$

which is the equation of motion of the tetrad in general relativity in terms of the spacetime dependent action $S(x^\mu)$. The Evans wave equation [1-3] is obtained from Eq. (8.111) by applying the operator $\Box$ to both sides to give

$$\Box q^a{}_\mu(x^\mu) = \Box\left(e^{iS(x^\mu)/\hbar} q^a{}_\mu(0)\right) = R\, e^{iS(x^\mu)/\hbar} q^a{}_\mu(0). \tag{8.112}$$

Using Eq. (8.111), we find the Evans lemma and wave equation [1-3],

$$\Box q^a{}_\mu(x^\mu) = R q^a{}_\mu(x^\mu) = -kT\left(q^a{}_\mu(x^\mu)\right) \tag{8.113}$$

and identify the scalar curvature as

$$R = \Box\left(e^{iS(x^\mu)/\hbar}\right). \tag{8.114}$$

The least possible curvature associated with any particle is

$$|R_0| = 1/\lambda^2{}_0, \tag{8.115}$$

where λ_0 is its Compton wavelength

$$\lambda_0 = \hbar/mc. \tag{8.116}$$

The principle of least curvature means that a particle never travels in a precise straight line, because the scalar curvature of a straight line is zero. The least curvature of the particle is defined by the least action $\hbar$ in Eq. (8.99). This inference means that a particle always has a wave-like nature (observed in diffraction and interferometry of matter waves, for example), and so we have derived the de Broglie wave-particle duality from general relativity. The principle of least curvature also means that the phase in optics and dynamics (radiated and matter waves respectively) is an always defined by a Stokes theorem as in Eq. (8.103) for example, because the phase itself must also have a rotational as well as a translational nature. As we have argued, this inference leads to the first correct explanation of physical optics.

The Evans wave equation is therefore the fundamental equation of quantum mechanics, and is derived from the principle of least curvature in general relativity.

Under local gauge transformation the action is invariant, and so R is invariant. This means that the Aharonov-Bohm effect is a shift from one region of spacetime with a given curvature R to another region of spacetime with the same curvature R. This shift is not action at a distance, but could be interpreted as "non-locality" in the sense that one (local) region of spacetime has the same curvature as another region. This is also a possible explanation of entanglement in quantum mechanics. Therefore both entanglement and non-locality are reconciled with the local nature of general relativity. No such reconciliation is possible in special relativity, where there is no concept of R, and where there is no principle of least curvature. This means that general relativity is a general theory of natural philosophy. All theories of physics are theories of general relativity. The Dirac equation, for example, is

$$\begin{aligned} q^a{}_{\mu R}(x^\mu) &= \exp(iS/\hbar)q^a{}_{\mu L}(0), \\ &\downarrow \hat{P} \\ q^a{}_{\mu L}(x^\mu) &= \exp(iS^*/\hbar)q^a{}_{\mu R}(0), \end{aligned} \tag{8.117}$$

and the time-dependent Schrödinger equation is obtained from

$$q^a{}_\mu(t,\boldsymbol{r}) = e^{iS/\hbar}q^a{}_\mu(0,\mathbf{0}) \tag{8.118}$$

by the following differentiation:

$$\partial q^a{}_\mu/\partial t = (i/\hbar)Hq^a{}_\mu. \tag{8.119}$$

The Hamiltonian is defined [19] by

$$H = -\partial S/\partial \hbar. \tag{8.120}$$

Finally, the Heisenberg commutator equation is a cyclic relation between tetrad forms in differential geometry:

$$q^c{}_{\mu\nu} = (q^a \wedge q^b)_{\mu\nu}. \tag{8.121}$$

Defining the angular momentum as the tetrad form

$$J^a{}_\mu = J^{(0)}q^a{}_\mu, \tag{8.122}$$

the Heisenberg commutator relation for angular momentum is obtained:

$$J^a \wedge J^b = \hbar J^c, \tag{8.123}$$

the starting point for molecular quantum mechanics [19].

Acknowledgments

Craddock Inc., the Ted Annis Foundation, and the Association of Distinguished American Scientists are thanked for funding, and Fellows and Emeriti of AIAS are thanked for several interesting discussions.

References

1. M. W. Evans, *Found. Phys. Lett.* **16**, 367 (2003).
2. M. W. Evans, *Found. Phys. Lett.* **16**, 507 (2003); preprint on www.aias.us.
3. M. W. Evans, *Found. Phys. Lett.* **17**, 25 (2004); **17**, 149 (2004); these and other preprints on www.aias.us.
4. L. Felker, ed., *The Evans Equations* (World Scientific, 2004, in preparation).
5. M. W. Evans, J.-P. Vigier, *et al.*, *The Enigmatic Photon* (Kluwer Academic, Dordrecht, 1994 to 2002, hardback and softback), in five volumes.
6. M. W. Evans and A. A. Hasanein, *The Photomagneton in Quantum Field Theory* (World Scientific, Singapore, 1994).
7. M. W. Evans and L. B. Crowell, *Classical and Quantum Electrodynamics and the $\boldsymbol{B}^{(3)}$ Field* (World Scientific, Singapore, 2001).
8. M. W. Evans, ed., *Modern Nonlinear Optics*, a special topical issue in three parts of I. Prigogine and S. A. Rice, series eds., *Advances in Chemical Physics* (Wiley Interscience, New York, 2001, second edn. and e-book edn.), Vol. 119(1) to 119(3).
9. S. M. Carroll, *Lecture Notes in General Relativity*, arXiv:gr-qe/971200019 vl 3 Dec 1997.
10. R. M. Wald, *General Relativity* (University of Chicago Press, 1994).
11. J. B. Marion and S. T. Thornton, *Classical Dynamics of Particles and Systems* (HBJ, New York, 1988).
12. L. D. Landau and E. M. Lifshitz, *The Classical Theory of Fields* (Pergamon, New York, 1975).
13. A. O. Barut *Electrodynamics and Classical Theory of Fields and Particles* (Macmillan, New York, 1964).
14. L. H. Ryder, *Quantum Field Theory*, 2nd edn. (Cambridge University Press, 1996).
15. A. Einstein, *The Meaning of Relativity* (Princeton University Press, 1921).
16. G. Stephenson, *Mathematical Methods for Science Students* (Longmans & Green, London, 1968).
17. T. W. Barrett and D. M. Grimes, eds., *Advanced Electromagnetism* (World Scientific, Singapore, 1996), pp. 297 ff.
18. T. W. Barrett in A. Lakhtakia, ed., *Essays on the Formal Aspects of Electromagnetic Theory* (World Scientific, Singapore, 1993), pp. 6 ff.

19. P. W. Atkins, *Molecular Quantum Mechanics*, 2nd edn. (Oxford University Press, 1983).
20. L. O'Raigheartaigh, *Rep. Prog. Phys.* **42**, 159 (1979).

9

The Evans Lemma Of Differential Geometry

Summary. A rigorous proof is given of the Evans lemma of general relativity and differential geometry. The lemma is the subsidiary proposition leading to the Evans wave equation and proves that the eigenvalues of the d'Alembertian operator, acting on any differential form, are scalar curvatures. The Evans wave equation shows that the eigenvalues of the d'Alembertian operator, acting on any differential form, are eigenvalues of the index-contracted canonical energy momentum tensor T multiplied by the Einstein constant k. The lemma is a rigorous and general result in differential geometry, and the wave equation is a rigorous and general result for all radiated and matter fields in physics. The wave equation reduces to the main equations of physics in the appropriate limits, and unifies the four types of radiated fields thought to exist in nature: gravitational, electromagnetic, weak and strong.

Key words: Evans lemma, Evans wave equation, differential geometry, general relativity, unified field theory, generally covariant electrodynamics, Evans-Vigier field $\boldsymbol{B}^{(3)}$.

9.1 Introduction

The original theory of general relativity [1] used Riemann geometry to develop a generally covariant theory of gravitation, i.e., a theory which is objective in all frames of reference, so that the equations of gravitation take the same form in any frame of reference, and no frame is a preferred frame. It did not prove to be possible to unify this theory with that of the electromagnetic field for several reasons, prominent among which is the fact that the Maxwell-Heaviside theory is one of special relativity. The nineteenth century Maxwell Heaviside theory was the precursor to special relativity, and was extended from electrodynamics to other areas of physics by Einstein in 1905, after work by Michelson, Morley, Fitzgerald, Lorentz, Poincaré and others. It is not sufficient to replace the ordinary derivative by the covariant derivative in Maxwell-Heaviside theory, because in Riemann geometry the torsion tensor is not used, and there is no concept of orbital or intrinsic spin in Einstein's original general relativity.

In contemporary differential geometry, however [2], non-Euclidean spacetime is defined by two differential forms for any connection: the torsion and Riemann forms. The former is a vector-valued two-form and the latter a tensor-valued two-form, and the tetrad postulate of differential geometry [2-6] proves that the covariant derivative of the tetrad vanishes for all possible types of spacetime. Recently the tetrad postulate has been developed into the most general known equation of physics, the Evans wave equation [3-7], which can be used to describe and inter-relate all known radiated and matter fields, thus opening up many new areas of physics hitherto unknown.

In this paper the subsidiary proposition (a lemma of differential geometry) leading to the Evans wave equation is proven with differential geometry. The lemma states that scalar curvature R always occurs in nature as eigenvalues of the d'Alembertian operator $\Box$ acting on any differential form. The Evans wave equation states that the product $kT = -R$ always occurs in nature as eigenvalues generated by $\Box$ acting on any differential form. Here k is the Einstein constant and T is the index contracted energy-momentum tensor. Both the lemma and the wave equation apply in general relativity without any restriction, and therefore unify the theory of radiated and matter fields. This is a major advance in physics because the electromagnetic and gravitational fields become parts of the same theory of differential geometry, and so there exist hitherto unknown inter-relations between fields which could be of fundamental importance in future applications.

If the differential form is the tetrad $q^a{}_\mu$, the lemma is

$$\Box q^a{}_\mu = R q^a{}_\mu, \tag{9.1}$$

and the wave equations reads

$$(\Box + kT) q^a{}_\mu = 0. \tag{9.2}$$

In Sec. 2 a rigorous proof of the lemma is given from the tetrad postulate, and it is shown that the lemma always leads to the wave equation given that

$$-R = kT, \tag{9.3}$$

which must be interpreted as an equation valid for *all* radiated and matter fields, as first inferred, but not proved, by Einstein [1]. Equation (9.3) is the result of the Evans field equation [3-7]

$$(R + kT) q^a{}_\mu = 0. \tag{9.4}$$

Usually Eq. (9.3) is given a restricted interpretation as the contracted form of the Einstein field equation for gravitation. The more general Evans field equation (9.4) reduces to the Einstein field equation [1,2] and also gives field equations for the gauge-invariant electromagnetic field.

In Sec. 3 it is emphasised that both the lemma and the wave equation are valid for any differential form, thus giving an entire class of Evans wave equations valid for all radiated and matter fields.

9.2 Proof Of the Evans Lemma

The tetrad can be defined by the equation

$$V^a = q^a{}_\mu V^\mu, \tag{9.5}$$

where V^a is any contravariant vector in the orthonormal basis, indexed a, and V^μ is any contravariant vector in the base manifold indexed μ. For example, V can represent position vectors x, so the tetrad becomes

$$x^a = q^a{}_\mu x^\mu \tag{9.6}$$

and has sixteen independent components, the sixteen irreducible representations of the Einstein group [8-12]. The vector V can also represent metric vectors q [3-7], so the tetrad can be defined by

$$q^a = q^a{}_\mu q^\mu. \tag{9.7}$$

This means that the symmetric metric [2-7] can be defined in general by the dot product of tetrads:

$$q_{\mu\nu}{}^{(S)} = q^a{}_\mu q^b{}_\nu \eta_{ab}, \tag{9.8}$$

where η_{ab} is the metric diag$(-1,1,1,1)$ of the orthogonal space [2]. The anti-symmetric metric is defined in general by the wedge product of tetrads:

$$q^c{}_{\mu\nu}{}^{(A)} = q^a{}_\mu \wedge q^b{}_\nu, \tag{9.9}$$

and the general metric tensor with sixteen independent components is defined by the outer product of tetrads:

$$q^{ab}{}_{\mu\nu} = q^a{}_\mu q^b{}_\nu = q^a{}_\mu \otimes q^b{}_\nu. \tag{9.10}$$

The dot product is the gauge invariant gravitational field, the cross product the gauge invariant electromagnetic field, and the outer product combines the two fields and defines the way in which one influences the other.

The tetrad can also be defined [2] by basis vectors; for example,

$$\hat{e}_{(\mu)} = q^a{}_\mu \hat{e}_{(a)}, \tag{9.11}$$

for covariant basis vectors, or

$$\hat{\theta}^{(a)} = q^a{}_\mu \hat{\theta}^{(\mu)} \tag{9.12}$$

for contravariant, or orthogonal, basis vectors. It can be seen that the tetrad always links quantities defined in the orthonormal space indexed a to quantities defined in the base manifold, indexed μ. As a simplified, less abstract, aid to understanding, the orthonormal space [2] can be thought of as the tangent space, and the basis vectors of the orthonormal space can be thought of as the Cartesian unit vectors in a system of axes defining a tangent to a curve in ordinary three dimensional space (the base manifold). The curve is defined in the base manifold, the tangent to the curve is defined in the tangent space. In general the orthonormal space can be [2] any abstract space whose coordinate system is always orthogonal (i.e., Euclidean) and whose unit vectors are normalized, as in gauge field theory. In general relativity, the orthonormal space is the tangent space to the base manifold. The development and experimental verification of O(3) electrodynamics [8-12] showed that the orthonormal space of gauge field theory can be thought of as a physical space, and this is the key to field unification with differential geometry. All indices a in the new unified field theory [3-7] are indices of a physical space, tangent to a physical base manifold. The base manifold is four-dimensional non-Euclidean spacetime.

The tetrad can also be defined in terms of:
Pauli matrices,

$$\hat{\sigma}_{(\mu)} = q^a{}_\mu \hat{\sigma}_{(a)}, \tag{9.13}$$

Pauli two-spinors,

$$\phi^a = q^a{}_\mu \phi^\mu, \tag{9.14}$$

Dirac four-spinors,

$$\psi^a = q^a{}_\mu \psi^\mu, \tag{9.15}$$

or Dirac matrices,

$$\gamma^a = q^a{}_\mu \gamma^\mu. \tag{9.16}$$

The tetrad is also the generalization of the Lorentz transformation to general relativity:

$$x^{a\prime} = q^a{}_\mu{}' x^\mu. \tag{9.17}$$

This equation represents a generally covariant transformation [2]. The tetrad can also be a generally covariant transformation matrix between gauge fields:

$$\psi^{a\prime} = q^a{}_\mu{}' \psi^\mu. \tag{9.18}$$

So

$$q^a{}_\mu = \frac{\partial V^a}{\partial V^\mu}, \frac{\partial q^a}{\partial q^\mu}, \frac{\partial x^a}{\partial x^\mu}, \dots. \tag{9.19}$$

The tetrad is also the keystone to the representation and generalization of the Dirac equation in general relativity [15]. In terms of two-spinors, the generally covariant Dirac equation is

$$\phi^a_R = (q^a{}_\mu)_R \phi^\mu_L(0), \tag{9.20a}$$

$$\phi_L^a = (q^a{}_\mu)_L \phi_R^\mu(0). \tag{9.20b}$$

In terms of four-spinors, the same equation is

$$\psi^a(p) = q^a{}_\mu \psi^\mu(0). \tag{9.21}$$

This illustrates some of the ways in which the tetrad can be used. In general it is an invertible matrix (9.2) with a well-defined inverse:

$$q^a{}_\mu q^\nu{}_a = \delta^\nu_\mu, \tag{9.22}$$

where δ^ν_μ is the Kronecker delta function. The dimensionality of the tetrad matrix depends on the way in which it is defined; for example, using Eqs. (9.6),(9.7),(9.11), or (9.12), the tetrad is a 4×4 matrix; using Eq. (9.13), it is a 2×2 complex matrix. The name "tetrad" is used generically [2].

The tetrad is a vector-valued one-form, i.e., is a one form q_μ with labels a. If a takes the values 1, 2, or 3 of a Cartesian representation of the tangent space, for example, the vector

$$\boldsymbol{q}_\mu = q^1{}_\mu \boldsymbol{i} + q^2{}_\mu \boldsymbol{j} + q^3{}_\mu \boldsymbol{k} \tag{9.23}$$

can be defined in this space. Each of the components q^1_μ, q^2_μ, or $q^3{}_\mu$ are scalar-valued one-forms of differential geometry [2], and each of $q^1{}_\mu, q^2{}_\mu$, and $q^3{}_\mu$ is therefore a covariant four-vector in the base manifold. The three scalar-valued one-forms are therefore the three components of the vector-valued one-form $q^a{}_\mu$, the tetrad form [2].

The power of differential geometry is that it inter-relates differential forms (and thus gives us new equations of geometry, topology and physics) without needing to know the details of the geometry of the base manifold.

The well-known tetrad postulate [2-7] is

$$D_\mu q^a{}_\nu = \partial_\mu q^a{}_\nu + \omega^a{}_{\mu b} q^b{}_\nu - \Gamma^\lambda{}_{\mu\nu} q^a{}_\lambda = 0 \tag{9.24}$$

and is the basis of the Evans lemma and Evans wave equation [3-7] of differential geometry, the most powerful and general wave equation known in general relativity, and thus in physics. This statement is based on the fact that differential geometry is the most powerful and general form of non-Euclidean geometry. In order to help understanding of the lemma and wave equation, we first give the detailed proof of the tetrad postulate.

The proof inter-relates the definitions of the covariant derivative D of a vector X in the base manifold and tangent space and shows that the tetrad postulate is the direct consequence of these definitions. In the base manifold, the vector X is defined by its components X^ν in the basis set ∂_ν and the covariant derivative D by its components D_μ in the basis set dx^μ. Thus

$$DX = (D_\mu X^\nu) dx^\mu \otimes \partial_\nu = (\partial_\mu X^\nu + \Gamma^\nu{}_{\mu\lambda} X^\lambda) dx^\mu \otimes \partial_\nu, \tag{9.25}$$

where $\Gamma^{\nu}{}_{\mu\lambda}$ is the Chistoffel connection [2]. Similarly, in the tangent space,

$$DX = (D_\mu X^a)dx^\mu \otimes \hat{e}_{(a)} = (\partial_\mu X^a + \omega^a{}_{\mu b} X^b)dx^\mu \otimes \hat{e}_{(a)}, \tag{9.26}$$

where $\omega^a{}_{\mu b}$ is the spin connection [2]. Using the following definitions:

$$X^a = q^a{}_\nu X^\nu, \tag{9.27}$$

$$\partial_a = q^\sigma{}_a \partial_\sigma, \tag{9.28}$$

Eq. (9.26) can be rewritten as

$$DX = q^\sigma{}_a \left(q^a{}_\nu \partial_\mu X^\nu + X^\nu \partial_\mu q^a{}_\nu + \omega^a{}_{\mu b} q^b{}_\lambda X^\lambda \right) dx^\mu \otimes \partial_\sigma, \tag{9.29}$$

where the commutator rule for matrices has been used. Replacing the dummy (i.e., repeated) indices σ by ν and using

$$q^\nu{}_a q^a{}_\nu = 1, \tag{9.30}$$

we obtain

$$DX = \left(\partial_\mu X^\nu + q^\nu_a \partial_\mu q^a{}_\lambda X^\lambda + q^\nu{}_a q^b{}_\lambda \omega^a{}_{\mu b} X^\lambda \right) dx^\mu \otimes \partial_\nu. \tag{9.31}$$

Comparing Eqs. (9.25) and (9.31), one gets

$$\Gamma^\nu{}_{\mu\lambda} = q^\nu{}_a \partial_\mu q^a{}_\lambda + q^\nu{}_a q^b{}_\lambda \omega^a{}_{\mu b}. \tag{9.32}$$

Multiply both sides of this equation by q^a_μ to obtain the tetrad postulate

$$q^a{}_\nu \Gamma^\nu{}_{\mu\lambda} = \partial_\mu q^a{}_\lambda + q^b{}_\lambda \omega^a{}_{\mu b}. \tag{9.33}$$

It is seen that the postulate is valid [2] for any Christoffel or spin connection and for any base manifold. The importance of this is that the postulate is true for manifolds with non-zero torsion form as well as non-zero Riemann form. In the Evans unified field theory [3-7] the torsion form represents the gauge invariant electromagnetic field, and the Riemann form represents the gauge invariant gravitational field. The tetrad form represents the potential fields in both cases, and the torsion form is the Hodge dual of the Riemann form. These concepts are summarised in Table 1 of the Appendix.

In contrast, in Einstein's generally covariant theory of gravitation, nineteenth century Riemann geometry is still used and in this theory the torsion tensor is defined as zero by the metric compatibility condition [2] of Riemann geometry used to define the Christoffel connection in terms of the symmetric, ten component, metric tensor. Therefore electromagnetism cannot be represented by Riemannian geometry, and this is the reason why Einstein was not able to unify gravitation and electromagnetism. The absence of the torsion tensor is also the reason why electromagnetism cannot be made into a theory of Einsteinian general relativity simply by replacing the derivatives of the Maxwell Heaviside field equations by the covariant derivatives with Christoffel connection. This is a self-inconsistent procedure because it attempts to represent electromagnetism with spacetime curvature without defining spacetime torsion, or spin. When one self-consistently includes the torsion tensor, electromagnetism becomes a theory of general relativity defined essentially by the wedge product of tetrads

$$T^c = Rq^a \wedge q^b \tag{9.34}$$

and by the Evans lemma and wave equation. The same theory [3-7] also describes gravitation, and the inter-relation between gravitation and electromagnetism. It can be seen from Eq. (9.34) that the torsion is a vector valued two form of differential geometry [2]. The fact that it is vector valued means that the internal index c must have more than one component, and this means that electromagnetism cannot be a U(1) gauge field theory. The latter defines the electromagnetic potential with a scalar-valued one-form A, and the gauge-invariant electromagnetic field by the scalar valued two form F. Unfortunately the U(1) representation of electromagnetism is accepted as part of the standard model, but this is inconsistent with differential geometry (as well as with general relativity) for the above reasons. The first theory of electromagnetism that was found to be consistent with differential geometry was O(3) electrodynamics, which was based in turn on the Evans Vigier field $\boldsymbol{B}^{(3)}$ observed in the inverse Faraday effect [13]. The $\boldsymbol{B}^{(3)}$ field can now be recognized as the wedge product

$$B^{(3)} = -igA^{(1)} \wedge A^{(2)} \tag{9.35}$$

in the complex circular basis [8-12]. The theory of O(3) electrodynamics is therefore the precursor of the Evans unified field theory based on the Evans lemma and wave equation, and O(3) electrodynamics has been extensively verified experimentally [8-12]. It can be made to take the form of two Maxwell Heaviside equations under certain well-defined conditions but it is a gauge field theory of fundamentally different symmetry, a theory in which the electromagnetic field is a torsion form as required, with well defined index c. The theory of O(3) electrodynamics is consistent with differential geometry because the torsion form is well defined, the index c takes the values (1), (2), and (3) of the complex circular representation [8-12] of the tangent space. This is a physical representation of circular polarization, and is an illustration of the more general finding [3-7] that gauge theory can be unified with general relativity provided with the internal gauge space is a physical space.

In contrast, in the standard model, the internal space of a gauge field is abstract, and is the fiber bundle space, unrelated to the tangent space of general relativity. This leads to serious (and well known) internal inconsistencies in the standard model, in that only the gravitational field is generally covariant. The other three fields are manifestations of special relativity and an abstract gauge theory whose internal, or fiber bundle, index has no physical meaning of definition [2]. In the standard model the internal index of the gravitational field represents a physical tangent space, but the internal index of the other three fields represents an unphysical space. The standard model is also internally inconsistent, it does not define the torsion tensor as described already, i.e., the internal index for electromagnetism is missing.

The Evans lemma is a direct consequence of the tetrad postulate. The proof of the lemma starts from covariant differentiation of the postulate:

$$D^{\mu}\left(\partial_{\mu}q^{a}{}_{\lambda}+\omega^{a}{}_{\mu b}q^{b}{}_{\lambda}-\Gamma^{\nu}{}_{\mu\lambda}q^{a}{}_{\nu}\right)=0. \tag{9.36}$$

Using the Leibniz rule, we have

$$\begin{aligned}(D^{\mu}\partial_{\mu})q^{a}{}_{\lambda} &+\partial_{\mu}(D^{\mu}q^{a}{}_{\lambda})+(D^{\mu}\omega^{a}{}_{\mu b})q^{b}{}_{\lambda}+\omega^{a}{}_{\mu b}(D^{\mu}q^{b}{}_{\lambda})\\ &-(D^{\mu}\Gamma^{\nu}{}_{\mu\lambda})q^{a}{}_{\nu}-\Gamma^{\nu}{}_{\mu\lambda}(D^{\mu}q^{a}{}_{\nu})=0,\end{aligned} \tag{9.37}$$

and so

$$(D^{\mu}\partial_{\mu})q^{a}{}_{\lambda}+(D^{\mu}\omega^{a}{}_{\mu b})q^{b}{}_{\lambda}-(D^{\mu}\Gamma^{\nu}{}_{\mu\lambda})q^{a}{}_{\nu}=0, \tag{9.38}$$

because

$$D^{\mu}q^{a}{}_{\lambda}=D^{\mu}q^{b}{}_{\lambda}=D^{\mu}q^{a}{}_{\nu}=0. \tag{9.39}$$

The covariant divergence of a vector V^{μ} is defined [2] as

$$D_{\mu}V^{\mu}=\partial_{\mu}V^{\mu}+\Gamma^{\mu}_{\mu\lambda}V^{\lambda}; \tag{9.40}$$

thus, using Eq. (9.40), it is found that

$$D_{\mu}\partial^{\mu}=\partial_{\mu}\partial^{\mu}+\Gamma^{\mu}{}_{\mu\lambda}\partial^{\lambda}. \tag{9.41}$$

Rewriting the dummy indices inside the connection,

$$D_{\mu}\partial^{\mu}=\partial_{\mu}\partial^{\mu}+\Gamma^{\nu}{}_{\nu\mu}\partial^{\mu}=\Box+\Gamma^{\nu}{}_{\nu\mu}\partial^{\mu}, \tag{9.42}$$

and the wave equation (9.38) therefore becomes

$$(\Box+\Gamma^{\nu}{}_{\nu\mu}\partial^{\mu})q^{a}{}_{\lambda}-R_{1}q^{a}{}_{\nu}=0, \tag{9.43}$$

where the scalar curvature R_1 is defined by

$$-R_{1}q^{a}{}_{\lambda}:=(D^{\mu}\omega^{a}{}_{\mu b})q^{b}{}_{\lambda}-(D^{\mu}\Gamma^{\nu}{}_{\mu\lambda})q^{a}{}_{\nu}. \tag{9.44}$$

Using the tetrad postulate,

$$\partial_{\mu}q^{a}{}_{\lambda}=-\omega^{a}{}_{\mu b}q^{b}{}_{\lambda}+\Gamma^{\nu}{}_{\mu\lambda}q^{a}{}_{\nu}, \tag{9.45}$$

and thus

$$\partial^{\mu}q^{a}{}_{\lambda}=-\omega^{\mu a}{}_{b}q^{b}{}_{\lambda}+\Gamma^{\mu\nu}{}_{\lambda}q^{a}{}_{\nu}. \tag{9.46}$$

Therefore, in Eq. (9.43),

$$\Box q^a{}_\lambda - \Gamma^\nu{}_{\nu\mu}\omega^{\mu a}{}_b q^b{}_\lambda + \Gamma^\nu{}_{\nu\mu}\Gamma^{\mu\nu}{}_\lambda q^a{}_\nu - R_1 q^a{}_\lambda = 0. \tag{9.47}$$

Define the scalar curvature R_2 by

$$-R_2 q^a{}_\lambda := -\Gamma^\nu{}_{\nu\mu}\omega^{\mu a}{}_b q^b{}_\lambda + \Gamma^\nu{}_{\nu\mu}\Gamma^{\mu\nu}{}_\lambda q^a{}_\nu, \tag{9.48}$$

to obtain the Evans lemma

$$\Box q^a{}_\lambda = R q^a{}_\lambda, \tag{9.49}$$

where

$$R = R_1 + R_2. \tag{9.50}$$

Given the tetrad postulate, the lemma shows that scalar curvature R is always an eigenvalue of the wave equation (9.49) for all spacetimes, in other words, R is quantized. The eigenoperator is the d'Alembertian operator $\Box$, and the eigenfunction in this case is the tetrad. In Sec. 3 we will show that the eigenfunction can be any differential form, thus introducing a powerful class of wave equations to differential geometry and physics. The lemma is the subsidiary proposition leading to the Evans wave Eq. (9.2) through Eq. (9.3). The lemma is an identity of differential geometry, and so is comparable in generality and power to the well known Poincaré lemma [14]. In other words, new theorems of topology can be developed from the Evans lemma in analogy with topological theorems [2,14] from the Poincaré lemma. This can be the subject of future work in mathematics, work which may lead in turn to new findings in physics based on topology. The immediate importance of the lemma to physics is that it is the subsidiary proposition leading to the Evans wave equation, which is valid for all radiated and matter fields. Equation (9.49) can be solved for R given tetrad components, or vice versa, solved for tetrad components for a given R. The equation is non-linear in the spin and Christoffel connections, but for a given R it is a linear second order partial differential wave equation, or eigenequation. In this sense it is an equation of wave mechanics and therefore of quantum mechanics, and so unifies quantum mechanics, unified field theory and general relativity. Its power is therefore apparent and the wave equation (9.49) reduces to known equations of physics [3-7] in the appropriate limits. These include the four Newton equations, the Poisson equations of gravitation and electrostatics, the Schrödinger, Klein-Gordon and Dirac equations, and the equations of O(3) electrodynamics. Equation (9.49) produces the quark color triplet through a choice of eigenfunction (a three-spinor of the SU(3) representation), and so unifies the gravitational and strong fields. In the limit of special relativity [3-7],

$$R \to m^2c^2/\hbar^2, \tag{9.51}$$

where m is the particle mass, $\hbar$ Planck's constant, and c the speed of light in vacuo. Equation (9.51) means that there exists a minimum mass density defined by a rest volume V:

$$\rho_{\min} = m/V = m^2c^2/\hbar^2 k. \tag{9.52}$$

The product mV is a universal constant:

$$mV = \hbar^2 k/c^2. \tag{9.53}$$

Equation (9.52) has the important consequence that there must be a minimum amount of mass/energy in the universe, in other words the universe cannot be "empty" ($R = -kT = 0$). This minimum amount of energy is the zero point energy of the Evans wave equation of quantum physics, and sometimes this zero point energy is referred to as "vacuum energy." It is observed in the Casimir effect and in the Lamb shift for example. Electromagnetic energy can therefore be obtained from the vacuum [15], interpreted as spacetime with torsion, and such energy may have been observed in devices such as the motionless electromagnetic generator [16]. In the limit of classical special relativity this minimum amount of energy is the rest energy mc^2, and so the Evans wave equation gives a new meaning to the Einstein rest energy mc^2 – it is the special-relativistic limit of the scalar curvature R. In nonrelativistic equations such as the Schrödinger equation the zero point energy (for example of a harmonic oscillator) does not have a classical meaning, but in relativistic equations such as the Evans equation it has a clear classical meaning, zero point energy originates in R. This is another important aspect of the Evans equation, it derives well-known relativistic equations from differential geometry. The existence of a minimum density of mass/energy, defined by Eq. (9.52) also means that the volume V is there at a maximum, the universe cannot have infinite extent, and can never, therefore be devoid entirely of matter or energy. The scalar curvature R can therefore never be zero. In coming to these conclusions T is recognized as the index contracted form of *total* energy/momentum, the sum of potential and kinetic terms. Total energy/momentum is always conserved through the Noether theorem [2,14]

$$D_\mu T^{\mu\nu} = 0. \tag{9.54}$$

9.3 Gauge Invariant Fields, Duality Equations, Inhomogeneous Field Equation, And The Class Of Evans Equations For All Differential Forms

The wave equation (9.2) defines the potential fields [14] (Feynman's "universal influence") of gravitation and electromagnetism. The former is the tetrad $q^a{}_\mu$ and the latter is

$$A^a{}_\mu = A^{(0)} q^a{}_\mu, \tag{9.55}$$

where $A^{(0)}$ is a C negative coefficient [3-7]. The wave equation also shows how the metrics and gauge invariant fields of gravitation and electromagnetism emerge from the Evans field equation [3]:

$$-Rq^a{}_\mu = kTq^a{}_\mu. \tag{9.56}$$

Equation (9.56) can be expressed as $kT = -R$ and is the equation which transforms the lemma (9.49) into the wave equation (9.2). Using Eqs. (9.8) to (9.10), it can be seen that the Evans field equation (9.56) can be developed in at least three ways, to give

$$-Rq_{\mu\nu}{}^{(S)} = kTq^a{}_\mu q^b{}_\nu \eta_{ab}, \tag{9.57}$$

$$-Rq^c{}_{\mu\nu}{}^{(A)} = kTq^a{}_\mu \wedge q^b{}_\nu, \tag{9.58}$$

$$-Rq^{ab}{}_{\mu\nu} = kTq^a{}_\mu \otimes q^b{}_\nu, \tag{9.59}$$

which are equations in the symmetric, antisymmetric and general metric tensors, respectively. Equation (9.57) has been shown [3-7] to the Einstein field equation, and Eq. (9.58) defines the gauge invariant electromagnetic field through the torsion tensor (9.34). Equation (9.59) defines the gauge invariant electromagnetic field through the torsion tensor (9.34). Equation (9.59) is an equation for the general metric, i.e., the general outer product of tetrads, and defines the way in which gravitation effects electromagnetism, and vice versa. A complete description of all radiated and matter fields requires the gauge invariant as well as the potential fields.

The gauge invariant gravitational field is defined [2] by the Riemann form of differential geometry through the second Maurer-Cartan structure relation

$$R^a{}_b = D \wedge \omega^a{}_b. \tag{9.60}$$

The symbol $D\wedge$ denotes the covariant exterior derivative of differential geometry (see Table 1), (The spin connection $\omega^a{}_b$ does not transform, however, as a tensor [2] so $D\wedge$ is defined in this equation is an operator which is identical to, and has the same effect as, the covariant exterior derivative.) The gauge invariant electromagnetic field is defined by the torsion form of differential geometry through the first Maurer-Cartan structure relation:

$$T^a = D \wedge q^a, \tag{9.61}$$

in which q^a transforms as a tensor. The gauge invariant electromagnetic field is then

$$G^a = D \wedge A^a = gA^b \wedge A^c, \tag{9.62}$$

where G^a is a C negative magnetic flux density (tesla = weber m^{-2}) and $1/g$ is a fundamental or primordial magnetic flux (weber). The quantum (or minimum possible value) of magnetic flux is the universal constant $\hbar/e$ and is the magnetic fluxon [8-14]. This means that a minimum amount of magnetic flux is always present in the universe. In analogy a minimum amount of $T, m, R,$ and V is always present in the universe, as discussed already. The equivalent of Eqs. (9.61) and (9.62) in U(1) or Maxwell-Heaviside electrodynamics is

$$F = d \wedge A, \tag{9.63}$$

where $d\wedge$ is the exterior derivative (see Table 1). Comparing the two available definitions (9.34) and (9.61) of the electromagnetic field gives the cyclically symmetric equation of differential geometry

$$D \wedge q^a = \kappa q^b \wedge q^c, \quad R = \kappa^2, \tag{9.64}$$

which generalizes the B cyclic theorem of O(3) electrodynamics [8-12]. The cyclic equation (9.64) is generally covariant and gauge invariant. All the equations of the Evans unified field theory are generally covariant, in other words they are all equations of differential geometry. This means that in the limit of special relativity, they are all Lorentz covariant. Indeed, as we have seen in Eq. (9.17), the tetrad, the eigenfunction of the Evans lemma and wave equation, can be defined as the equivalent of Lorentz transform matrix in general relativity.

The homogeneous unified field equations of the Evans theory are the Bianchi identities (see Table 1)

$$D \wedge R^a{}_b := 0, \tag{9.65}$$

$$D \wedge T^a := 0. \tag{9.66}$$

In the Maxwell-Heaviside theory, the equivalent equation is

$$d \wedge F := 0. \tag{9.67}$$

The Bianchi identities follow from the Poincaré lemma

$$D \wedge D := 0, \tag{9.68}$$

and Eq. (9.67) follows from the Poincaré lemma

$$d \wedge d := 0. \tag{9.69}$$

The inhomogeneous field equations of the Evans theory are

$$D \wedge \; ^*R^a{}_b = J^a{}_b, \tag{9.70}$$

$$D \wedge \; ^*T^c = J^c, \tag{9.71}$$

where $^*R^a{}_b$ is the dual of $R^a{}_b$, defined as in Table 1, and where $^*T^c$ is the dual of T^c. In Eq. (9.70), $J^a{}_b$ is a tensor-valued three-form, the current form of gravitation, and J^c is a vector-valued three-form, the current form of electromagnetism. The equivalent of Eqs. (9.70) and (9.71) in the Maxwell-Heaviside field theory is

$$d \wedge \; ^*F = J, \tag{9.72}$$

where $^*F$ is the dual of F and J is a scalar-valued three-form, the current form of this theory [2,14].

In the development of the Evans field theory, novel and important duality relations of differential geometry have been discovered (9.1) between the torsion and Riemann forms and (9.2) between the spin connection and tetrad. These are fundamental to differential geometry, to topology, and to generally covariant physics. They are proven from the fact that the tangent space is an orthonormal Euclidean space, so in this space there exists the relation [2-7] between any axial vector V_a and its dual antisymmetric tensor V_κ:

$$V_a = \epsilon_{abc} V_{bc}, \tag{9.73}$$

where ϵ_{abc} is the Levi Civita symbol or three-dimensional totally antisymmetric unit tensor (see Table 1). Raising one index gives

$$\epsilon^a{}_{bc} = \eta^{cd} \epsilon_{dbc}, \tag{9.74}$$

where $\eta^{ad} = \text{diag}(-1, 1, 1, 1)$ is the metric of the orthonormal space. The first Evans duality equation of differential geometry is

$$R^a{}_b = \kappa \epsilon^a{}_{bc} T^c, \tag{9.75}$$

showing that the spin connection is dual to the Riemann form, and the second Evans duality equation of differential geometry is

$$\omega^a{}_b = \kappa \epsilon^a{}_{bc} q^c, \tag{9.76}$$

showing that the spin connection is dual to the tetrad. The duality equations define the symmetry of the Riemann form because they imply

$$R_{ab} = R \epsilon_{abcd} q^c \wedge q^d, \tag{9.77}$$

where ϵ_{abcd} is the four-dimensional totally antisymmetric unit tensor. Therefore the Riemann form is antisymmetric in its indices a and b and also antisymmetric in the indices μ and ν of the base manifold [2]. The symmetries of the Riemann form and Riemann tensor are explained in more detail in Table 1. The duality relations greatly simplify the unified field theory and show that gravitation and electromagnetism are closely related by duality symmetry in differential geometry.

It has been shown that the tetrad can be defined and understood in a number of different ways (9.5) to (9.23), for example, and Table 1. From these and other relations it follows straightforwardly that the Evans lemma and wave equation apply to *any* differential form, so there exists a class of new wave equations in differential geometry, topology and general relativity. For example, the tetrad can be defined in terms of the metric vectors as in Eq. (9.7); and, using the Leibnitz theorem, it follows that

$$(\square + kT) q^\mu = 0. \tag{9.78}$$

Similarly,

$$(\square + kT)T^a = 0, \tag{9.79}$$

$$(\square + kT)R^a{}_b = 0, \tag{9.80}$$

and so on.

Acknowledgments

Craddock Inc., the Ted Annis Foundation, and the Association of Distinguished American Scientists are thanked for funding this work, and AIAS colleagues for interesting discussions.

APPENDIX

Table 1. Glossary of new results and fundamental definitions

1	$D_\mu q^a{}_\nu = 0$	The tetrad postulate
2	$Dq^a = 0$	The tetrad postulate in the notation of differential geometry
3	$d(\ ) = \partial_\mu(\ )$ $D(\ ) = D_\mu(\ )$ $d \wedge (\ )$ $D \wedge (\ )$	Partial derivative Covariant derivative Exterior derivative Exterior covariant derivative
4	$D_\mu V^\nu = \partial_\mu V^\nu + \Gamma^\nu{}_{\mu\lambda} V^\lambda$	Covariant derivative of a vector V^ν
5	$(dX)^a{}_{\mu\nu} = (\partial \wedge X^a)_{\mu\nu}$ $= (d \wedge X^a)_{\mu\nu}$ $= \partial_\mu X^a{}_\nu - \partial_\nu X^a{}_\mu = [\partial_\mu, X^a{}_\mu]$	The exterior derivative in component notation
6	$d \wedge X^a$	The exterior derivative in the notation of differential geometry
7	$D \wedge X^a$ $= d \wedge X^a + \omega^a_b \wedge X^b,$ $(D \wedge X^a)_{\mu\nu}$ $= \partial_\mu X^a{}_\nu - \partial_\nu X^a{}_\mu$ $+\omega^a{}_{\mu b} X^b{}_\nu - \omega^a{}_{\nu\ b} X^b{}_\mu$	The exterior covariant derivative, where $\omega^a{}_b$ is i.e., the spin connection
8	$D_\mu q^a{}_\lambda = \partial_\mu q^a{}_\lambda$ $+\omega_{\mu b}{}^a{}^b q^b{}_\lambda - \Gamma^\nu{}_{\mu\lambda} q^a{}_\nu = 0$	Covariant derivative of the tetrad, denoted by Dq^a
8	$\partial_\mu \phi = D_\mu \phi$	Definition when ϕ is a scalar
9	If $\phi = 0, D_\mu 0 = \partial_\mu 0 := 0$	Identity of geometry
10	$d(Dq^a) := 0,$ i.e., $\partial_\mu (D^\mu q^a{}_\nu) := 0$	Wave equation of differential geometry
11	$DD = dD = \Box - R$	The Evans lemma of differential geometry
12	$\Box q^a = Rq^a$	The Evans lemma as an eigenequation
13	$R = -kT$	Index contracted form of the Einstein field equation

14	$(\square + kT)q^a := 0$	The Evans wave equation in the notation of differential geometry
15	$(\square + kT)q^a{}_\mu := 0$	The Evans wave equation in component notation
16	$Rq^a = -kTq^a$	The Evans field equation in differential geometry
17	$A^a = A^{(0)}q^a$	The electromagnetic potential field (a tetrad one-form)
18	$(\square + kT)A^a := 0$	Evans wave equation for electrodynamics
19	$(A \wedge B)_{\mu\nu} = A_\mu B_\nu - A_\nu B_\mu$	The wedge product in component notation
20	$G^c = G^{(0)}q^a \wedge q^b$	The gauge invariant e/m field
21	$T^c = Rq^a \wedge q^b$	The torsion form, where R is the scalar curvature
22	$R^a{}_b = \epsilon^a{}_{bc}T^c$	The gauge invariant gravitational field and the second Evans duality relation
23	$\epsilon^a{}_{bc} = \eta^{ad}\epsilon_{dbc}$	Definition of the Levi-Civita symbol $\varepsilon^a{}_{bc}$ in the orthonormal space of the tetrad
24	$\eta^{ad} = \mathrm{diag}(1, -1, -1, -1)$	Metric in the orthonormal space of the tetrad
25	$\epsilon_{abc} = \begin{cases} 1, & \text{even} \\ -1, & \text{odd} \\ 0, & \text{otherwise} \end{cases}$	Levi-Civita symbol, the rank three totally antisymmetric unit tensor
26	$\omega^a{}_b = \kappa\epsilon^a{}_{bc}q^c$	The first Evans duality equation, the spin connection is the dual of the tetrad
27	$T^a = D \wedge q^a$	The first Maurer-Cartan structure relation

28	$R^a{}_b = D \wedge \omega^a{}_b$	The second Maurer-Cartan structure relation
29	$D \wedge T^a := R^a{}_b \wedge q^b$	The first Bianchi identity
30	$D \wedge R^a{}_b := 0$	The second Bianchi identity
31	$D \wedge D = d \wedge d := 0$	The Poincaré lemma
32	$T^a = D \wedge q^a; D \wedge T^a = R^a{}_b \wedge q^b$	The generally covariant theory of electromagnetism; gauge invariant
33	$R^a{}_b = D \wedge \omega^a{}_b; D \wedge R^a{}_b := 0$	The generally covariant theory of gravitation; gauge invariant
34	$F = d \wedge A; d \wedge F := 0$	Maxwell-Heaviside theory
35	$D \wedge ({}^*T^a) = 0$	First homogenous Evans field equation
36	$D \wedge ({}^*R^a{}_b) = 0$	Second homogenous Evans field equation
37	$d \wedge ({}^*F) = J$	The inhomogenous Maxwell-Heaviside field equation
38	$({}^*T^a)^{\mu\nu} = \frac{1}{2}\epsilon^{\mu\nu\rho\sigma}(T^a)_{\rho\sigma}$	Definition of ${}^*T^a$
39	$({}^*R^a{}_b)^{\mu\nu} = \frac{1}{2}\epsilon^{\mu\nu\rho\sigma}(R^a{}_b)_{\rho\sigma}$	Definition of ${}^*R^a{}_b$
40	$\boldsymbol{E} = \left(\phi^{(0)}/c^2\right)\boldsymbol{g}$	Evans electrogravitic equation in the weak-field limit
41	$e^2 + m^2 = 1$ $e^2 - m^2 = 0$ $e^2 = m^2 = 1/2$	The Evans symmetry condition of unified field theory (in reduced units)
42	$e^2 + m^2 \neq 1$ $e^2 - m^2 \neq 0$ $e^2 \neq m^2$	The Evans symmery-breaking rule for hybrid effects of gravitation on electro-magnetion (in reduced units)
43	$D \wedge q^a = \kappa q^b \wedge q^c$	The Evans cyclic equation of differential geometry (reduced units)
44	$\boldsymbol{B}^{(1)} \times \boldsymbol{B}^{(2)} = iB^{(0)}\boldsymbol{B}^{(3)*}$	The Evans B cyclic theorem of $O(3)$ electrodynamics

45	$q_{\mu\nu}{}^{(S)} = q^a{}_\mu q^b{}_\nu \eta_{ab}$	The symmetric diagonal metric (gravitation)
46	$q^c{}_{\mu\nu}{}^{(A)} = q^a{}_\mu \wedge q^b{}_\nu$	The antisymmetric metric (electromagnetism)
47	$q^{ab}{}_{\mu\nu} = q^a{}_\mu q^b{}_\nu = q^a{}_\mu \otimes q^b{}_\nu$	The outer product, combined gravitational and e/m field
48	$q^a = q^a_\mu q^\mu$	Definition of metric vectors q^a and q^μ
49	$x^a = q^a{}_\mu x^\mu$	Definition of position vectors
50	$q^a{}_\mu = \partial x^a / \partial x^\mu$	The sixteen independent components $q^a{}_\mu$ of the tetrad, i.e., the 16 irreps of the Einstein group
51	$q^0{}_\mu = \partial x^0 / \partial x^\mu$ $\vdots$ $q^3{}_\mu = \partial x^3 / \partial x^\mu$	The generally covariant metric four-vectors
52	$q^a{}_\mu = \frac{\partial V^a}{\partial V^\mu} = \frac{\partial q^a}{\partial q^\mu} = \frac{\partial x^a}{\partial x^\mu}$ $= \ldots\ldots$	Generally covariant definitions of the tetrad in terms of vectors
53	$\hat{e}_{(\mu)} = q^a{}_\mu \hat{e}_{(a)}$	Basis vector definition of the tetrad for any basis set
54	$q^\mu{}_a q^a{}_\nu = \delta^\mu{}_\nu$	Invertible matrix rule for the tetrad
55	$\hat{\theta}^{(a)} = q^a{}_\mu \hat{\theta}^{(\mu)}$	Orthogonal basis vector definition of the tetrad
56	$\hat{\sigma}_{(\mu)} = q^a{}_\mu \hat{\sigma}_{(a)}$	Pauli matrix definition of the tetrad in Euclidean limit
57	$\hat{q}_{(\mu)} = q^a{}_\mu \hat{q}_{(a)}$	Sachs's basis-vector definition of the tetrad
58	$\psi^a = q^a{}_\mu \psi^\mu$	Dirac spinor definition of the tetrad
59	$\gamma^a = q^a{}_\mu \psi^\mu$	Dirac matrix definition of the tetrad
60	$\Lambda^a{}_\mu \rightarrow q^a{}_\mu$	Tetrad as generalization of the Lorentz transform matrix $\Lambda^a{}_\mu$
61	$x^{a'} = q^{a'}{}_\mu x^\mu$	Tetrad as the generally covariant coordinate transformation

62	$\psi^{a'} = q_\mu{}^{a'}\psi^\mu$	Tetrad as the generally covariant gauge transformation between field vectors ψ^μ and $\psi^{a'}$
63	$\phi^a = q^a{}_\mu\phi^\mu$	Tetrad defined by two spinors
64	$\phi^a_R = (q^a{}_\mu)_R\phi^\mu_L(0)$ $\phi^a_L = (q^a{}_\mu)_L\phi^\mu_R(0)$	The generally covariant Dirac equation in terms of two-spinors
65	$\psi^a(p) = q^a{}_\mu\psi^\mu(0)$	The generally covariant Dirac equation as a tetrad definition
66	$(\Box + kT)\psi^a = 0$	The Evans equation with spinor eigenfunction, a generally covariant Dirac equation
67	$V = V^\mu\hat{e}_{(\mu)}$	Vector
68	V^μ	Vector components, contravariant
69	$\hat{e}_{(\mu)}$	Basis vectors
70	$V^* = V_\mu\hat{\theta}^{(\mu)}$	Vector in the dual vector space
71	V_μ	Vector components, covariant
72	$\hat{\theta}^{(\mu)}$	Basis vectors of the dual space
73	$\delta^\nu_\mu = 1, \nu = \mu$ $= 0, \nu \neq \mu$	Kronecker delta function
74	$\hat{\theta}^{(\nu)}\hat{e}_{(\mu)} = \delta^\nu_\mu$ $V^\mu V_\nu = \delta^\mu_\nu$ $q^\nu q_\mu = \delta^\nu_\mu$ $q^\nu{}_a q_\mu{}^a = \delta^\nu_\mu$ $\vdots$	Defining relations
75	$g^{\mu\nu}g_{\nu\rho} = g_{\rho\nu}g^{\nu\mu} = \delta^\mu_\rho$	Definition of the inverse metric tensor given in terms of the metric tensor $g_{\nu\rho}$
77	$g^{\mu\nu}g_{\mu\nu} = 4$	Trace of the product in (9.75)
78	V_μ	Components of the scalar-valued one-form V.
79	$V_\mu = [V_0 \ldots V_n]$	Row vector representation of the one form

80	$V^\mu = \begin{pmatrix} V^0 \\ \vdots \\ V^n \end{pmatrix}$	Column vector representative of the contravariant vector
81	$V_\mu V^\mu = (V_0 \dots V_n) \begin{pmatrix} V^0 \\ \vdots \\ V^n \end{pmatrix}$ $= V_0V^0 + \dots + V_nV^n$ $= \mathbf{V} \cdot \mathbf{V} = g_{\mu\nu}V^\mu V^\nu$ $= g^a_\mu V^\mu V_a = (g^a_\mu V_a)V^\mu$	Inner or dot product of covariant and contravariant vectors, where $g_{\mu\nu}$ is the metric tensor and q^a_μ is the tetrad
82	$V^a_{\ \mu}$	Components of the vector-valued one-form V^a
83	$V^a_{\ \mu\nu} = -V^a_{\ \nu\mu}$	Components of the vector-valued two-form
84	$V^a_{\ b\mu\nu} = -V^a_{\ b\nu\mu}$	Components of the tensor-valued two-form
85	ϕ	Scalar component of the zero-form
86	$\epsilon_{\mu\nu\rho\sigma} = \begin{cases} 1, & \text{even} \\ -1, & \text{odd} \\ 0, & \text{otherwise} \end{cases}$	Totally antisymmetric tensor in 4 D, the same in all spacetimes
87	$d\phi = (\partial\phi/\partial x^\mu)\,\hat{\theta}^\mu$	Exterior derivative of a scalar
88	$(d\phi)_\mu = \partial_\mu\phi = D_\mu\phi$	Relation between exterior, partial and covariant derivative of a scalar ϕ
89	$^*(U \wedge V)_i = \epsilon^{jk}_{\ \ i}U_jV_k$	Hodge dual definition of the cross product in 3-space
90	$d \wedge F := 0$ $\partial_{[\mu}F_\nu\lambda_{]} := 0$ $\partial_\mu F_{\nu\lambda} + \partial_\lambda F_{\mu\nu} + \partial_\nu F_{\lambda\mu} := 0$ $\partial^*_\mu F^{\mu\nu} := 0$ $^*F^{\mu\nu} = \frac{1}{2}\epsilon^{\mu\nu\rho\sigma}F_{\rho\sigma}$	Equivalent representation of the Jacobi identity, or first Bianchi identity, the homogeneous field equation
91	$(dV)_{\mu_1 \cdots \mu_{p+1}}$ $= (p+1)\partial_{[\mu_1}V_{\mu_2 \cdots \mu_{p1}]}$	Exterior derivative of the scalar-valued $(p+1)$-form.

92	$T_{[\mu\nu\rho]\sigma} = \frac{1}{6}(T_{\mu\nu\rho\sigma} - T_{\mu\rho\nu\sigma} + T_{\rho\mu\nu\sigma} - T_{\nu\mu\rho\sigma} + T_{\nu\rho\mu\sigma} - T_{\rho\nu\mu\sigma})$	Example of the general rule for the antisymmetric tensor; odd number of exchanges give a minus sign
93	$U^\mu = dx^\mu/d\tau$	Velocity four-vector in special relativity
94	$p^\mu = mU^\mu$	Momentum four-vector
95	$T^{\mu\nu} = (\rho + p)u^\mu u^\nu + p\eta^{\mu\nu}$	Energy-momentum tensor
96	$D_\mu V_\nu = \partial_\mu V_\nu - \Gamma^\lambda{}_{\mu\nu} V_\lambda$	Covariant derivative of a one-form
97	$T^\lambda{}_{\mu\nu} = \Gamma^\lambda{}_{\mu\nu} - \Gamma^\lambda{}_{\nu\mu}$	Torsion tensor
98	$R^\rho{}_{\sigma\mu\nu} = -R^\rho{}_{\sigma\nu\mu}$	Riemann tensor
99	$[D_\mu, D_\nu]V^\rho = R^\rho{}_{\sigma\mu\nu}V^\sigma - T^\lambda{}_{\mu\nu} D_\lambda V^\rho$	Commutator of covariant derivatives
100	$R_{\rho\sigma\mu\nu} = g_{\rho\lambda}R^\lambda{}_{\sigma\mu\nu}$ $R_{\rho\sigma\mu\nu} = -R_{\sigma\rho\mu\nu} = -R_{\rho\sigma\nu\mu} = R_{\mu\nu\rho\sigma}$	Symmetries of the Riemann tensor
101	$R_{\rho[\sigma\mu\nu]} := 0 = R_{\rho\sigma\mu\nu} + R_{\rho\mu\nu\sigma} + R_{\rho\nu\sigma\mu}$	First Bianchi identity
102	$R_{ab} = g_{ac}R^c{}_b$ $g_{ac} = \eta_{ac} = \text{diag}(1, -1, -1, -1)$	Relation between Riemann forms of differential geometry
103	$\xi^\mu = \begin{pmatrix} \xi^1 \\ \xi^2 \end{pmatrix}$	The spinor, a vector in 2-D rep. space (Barut, (1.42))
104	$\xi_\mu = C_{\mu\nu}\xi^\nu = [\xi_1, \xi_2]$	The covariant spinor (Barut, Eq. (1.48)). This is a scalar-valued one-form
105	$\xi^\alpha\eta_\alpha = \xi_\alpha\eta^\alpha$	Scalar invariant dot product of two spinors (Barut, (1.45′))
106	$\psi_\mu = [\xi_1^{(R)}, \xi_2^{(R)}, \xi_1^{(L)}, \xi_2^{(L)}]$	Dirac four-spinor, a one form of differential geometry

107	$X = \sigma_\mu x^\mu$ $X_{\alpha\dot{\alpha}} = (\sigma_\mu x^\mu)_{\alpha\dot{\alpha}}$	Relation between spinors and vectors. Both spinors and vectors describe the same base manifold, so we must be able to interrelate them (Barut, p. 30)
108	$\xi_\mu \xi^{\mu *} = \lvert\xi_1^2\rvert + \lvert\xi_2^2\rvert$ $= \xi_1\xi_1^* + \xi_2\xi_2^*$ $= \xi_1\xi^{1*} + \xi_2\xi^{2*}$	Dot product of two 2-spinors, a scalar invariant
109	$x_i x^i = {x^2}_1 + {x^2}_2 + {x^2}_3$	Dot product of two 3-vectors, a scalar invariant
110	${x^2}_1 + {x^2}_2 + {x^2}_3 = \xi_1\xi_1^* + \xi_2\xi_2^*$ (three-dimensional space)	Equating two scalar invariants in the same base manifold
111	$\sigma_0 = \begin{pmatrix} 1 & 0 \\ 0 & 1 \end{pmatrix}, \sigma_1 = \begin{pmatrix} 0 & 1 \\ 1 & 0 \end{pmatrix}$ $\sigma_2 = \begin{pmatrix} 0 & -i \\ i & 0 \end{pmatrix}, \sigma_3 = \begin{pmatrix} 1 & 0 \\ 0 & -1 \end{pmatrix}$	The set of Pauli matrices of the SU(2) group
112	$\sigma_0 \boldsymbol{i} \cdot \boldsymbol{i} = \sigma_1\sigma_1 = 1_2$ $\sigma_0 \boldsymbol{j} \cdot \boldsymbol{j} = \sigma_2\sigma_2 = 1_2$ $\sigma_0 \boldsymbol{k} \cdot \boldsymbol{k} = \sigma_3\sigma_3 = 1_2$ $1_2 := \begin{pmatrix} 1 & 0 \\ 0 & 1 \end{pmatrix}$	Another example of inter-relating scalars in the O(3) and SU(2) reps. of 3-D space, in this case, the scalar components, 1, of σ_0

References

1. A. Einstein, *The Meaning of Relativity* (Princeton University Press, 1921).
2. S. M. Carroll, *Lecture Notes on General Relativity* (University of California, Santa Barbara, Graduate Course on arXiv:gr-qe/9712019 vl 3 Dec. 1997).
3. M. W. Evans, *Found. Phys. Lett.* **16**, 367 (2003).
4. M. W. Evans, *Found. Phys. Lett.* **16**, 507 (2003).
5. M. W. Evans, *Found. Phys. Lett.* **17**, 25 (2004).
6. M. W. Evans, *Found. Phys. Lett.* **17**, 149 (2004).
7. M. W. Evans, *Found. Phys. Lett.* **17**, 301 (2004).
8. M. W. Evans, *Physica B* **182**, 227, 237 (1992).
9. M. W. Evans, J.-P. Vigier, *et al.*, *The Enigmatic Photon* (Kluwer Academic, Dordrecht, 1994 to 2002, hardback and softback, in five volumes).
10. M. W. Evans and L. B. Crowell, *Classical and Quantum Electrodynamics and the* $\boldsymbol{B}^{(3)}$ *Field* (World Scientific, Singapore, 2001).
11. M. W. Evans, ed., *Modern Nonlinear Optics*, a special topical issue of I. Prigogine and S. A. Rice, series eds., *Advances in Chemical Physics*, Vols. 119(1) to 119(3) (Wiley Interscience, New York, 2001, second and e-book editions).
12. L. Felker, ed., *The Evans Equations* (World Scientific, Singapore, 2004, in preparation).
13. M. Tatarikis *et al.*, "Measurements of the inverse Faraday effect in high intensity laser produced plasmas," CFL Annual Report 1998/1999, EPSRC Rutherford Appleton Laboratories, m.tatarikis@ic.ac.uk.
14. L. H. Ryder, *Quantum Field Theory*, 2nd edn. (Cambridge University Press, 1996).
15. P. W. Atkins, *Molecular Quantum Mechanics*, 2nd edn. (Oxford University Press, 1983).
16. T. E. Bearden, in Ref. (12), Vol. 119(2), and www. cheniere.org.

10

Physical Optics, The Sagnac Effect, And The Aharonov-Bohm Effect In The Evans Unified Field Theory

Summary. A generally covariant and gauge invariant description of physical optics, the Sagnac effect, and the Aharonov-Bohm (AB) effect is developed using the appropriate phase factor for electrodynamics. The latter is a generally covariant development of the Dirac-Wu-Yang phase factor based on the generally covariant Stokes theorem. The Maxwell-Heaviside (MH) field theory fails to describe physical optics, interferometry, and topological phase effects in general because the phase factor in that theory is under-determined. A random number from a U(1) gauge transformations can be added to the MH phase factor. The generally covariant phase factor of the Evans unified field theory is gauge invariant and has the correct property under parity inversion to produce observables such as reflection, interferograms, the Sagnac and AB effects, the Tomita-Chao effect and topological phase effects in general. These effects are described simply and self consistently with the generally covariant Stokes theorem in which ordinary exterior derivative is replaced by the covariant exterior derivative of differential geometry.

Key words: Evans unified theory, $\boldsymbol{B}^{(3)}$ field, physical optics, reflection, interferometry, Sagnac effect, AB effect, Tomita-Chao effect, topological phase effects.

10.1 Introduction

In generally covariant electrodynamics [1-8] the potential field is the tetrad within a C negative coefficient, and the gauge invariant electromagnetic field is a wedge product of tetrads. The covariant exterior derivative is used to describe the electromagnetic field using the first Maurer-Cartan structure relation of differential geometry [9]. The generally covariant field theory leads to $O(3)$ electrodynamics and the Evans-Vigier field $\boldsymbol{B}^{(3)}$. The latter is observed in several experiments, including the inverse Faraday effect [10]. The experimental evidence for the $\boldsymbol{B}^{(3)}$ field, and thus for generally covariant electrodynamics, is reviewed in the literature [11].

In this paper a generally covariant Dirac-Wu-Yang phase is defined within Evans unified field theory by using a generally covariant exterior derivative to develop the Stokes theorem of differential geometry [12]. The electromagnetic phase factor is described by this generally covariant Stokes theorem. This procedure produces the correct parity inversion symmetry needed to describe fundamental physical optical phenomena such as reflection and interferometry. The MH theory of electrodynamics is not generally covariant, i.e., is a theory of special relativity, and for this reason is unable to describe physical optics and interferometry, and unable to describe the Sagnac and AB effects. In Sec. 2, these shortcomings of the MH theory are summarized. In Sec. 3 the Evans unified field theory is applied to physical optics, which is correctly described with a phase factor constructed from the appropriate contour and area integrals of the generally covariant Stokes theorem. This procedure gives the correct parity inversion symmetry of the electromagnetic phase factor, and so is able to correctly describe physical optical phenomena such as reflection, interferometry, the Sagnac effect and the AB effect. In each case the generally covariant theory is simpler than the MH theory and at the same time is the first correct theory of physical optics as a theory of electrodynamics.

10.2 Shortcomings Of The Maxwell-Heaviside Field Theory In Physical Optics

In the MH field theory the phase factor is defined by

$$\Phi_{\mathrm{MH}} = \exp[i(\omega t - \kappa Z)], \tag{10.1}$$

where ω is the angular frequency of radiation at instant t and κ is the wave number of radiation at point Z in the propagation axis. Under a U(1) gauge transformation [13] the phase factor (10.1) becomes

$$\Phi_{\mathrm{MH}} \rightarrow \exp(i\alpha)\Phi_{\mathrm{MH}} = \exp[i(\omega t - \kappa Z + \alpha)], \tag{10.2}$$

where α is arbitrary. In consequence the phase factor is not gauge invariant, a major failure of MH field theory. In other words the phase factor (10.2) is undetermined theoretically, because any number α can be added to it without affecting the description of experimental data. In Michelson interferometry [14], for example, an interferogram is formed by displacing a mirror in one arm of the interferometer, thus changing Z in Eq. (10.1) for constant ω, t and κ. The path of a beam of light from the beam-splitter to the mirror and back to the beam-splitter is increased by $2Z$. The conventional description of Michelson interferometry incorrectly asserts that this increase $2Z$ in the path of the light beam results in a change of phase factor

$$\Delta\Phi_{\mathrm{MH}} = \exp(2i\kappa Z) \tag{10.3}$$

producing the observed interferogram, $\cos(2\kappa Z)$, for a monochromatic light beam in Michelson interferometry.

This basic error in the conventional theory has been reviewed in detail in Ref. [11]. The main purpose of this paper is to show that the error is corrected when the theory of electrodynamics is developed into a generally covariant unified field theory.

The origin of the error is found when parity inversion symmetry is examined. Consider the parity inversion symmetry of the product $\boldsymbol{\kappa} \cdot \boldsymbol{r}$ of wave vector $\boldsymbol{\kappa}$ and position vector $\boldsymbol{r}$:

$$\hat{P}(\boldsymbol{\kappa} \cdot \boldsymbol{r}) = \boldsymbol{\kappa} \cdot \boldsymbol{r} \tag{10.4}$$

because

$$\hat{P}(\boldsymbol{\kappa}) = -\boldsymbol{\kappa}, \tag{10.5}$$

$$\hat{P}(\boldsymbol{r}) = -\boldsymbol{r}. \tag{10.6}$$

Parity inversion is equivalent to reflection, and so κZ does not change under reflection in the MH phase factor (10.1). This means that the reflected phase factor is the same as the phase factor of the beam before reflection. In terms of the Fermat principle of least time in optics [15], this means

$$\Phi_2 = e^0 \Phi_1 = \Phi_1, \tag{10.7}$$

where Φ_1 is the phase factor of the wave before reflection and Φ_2 the factor after reflection.

The experimentally observed interferogram in the Michelson interferometer [14] and in reflection in general must be described, however, by

$$\Phi_2 = e^{2i\kappa Z} \Phi_1, \tag{10.8}$$

and so the MH field theory of electrodynamics fails at a fundamental level to describe physical optics. Various hand-waving arguments have been used over the years to address this problem, but there has never been a solution. The reason is that there cannot be a solution in special relativistic gauge field theory because, as we have seen, the phase factor is not gauge invariant; a random number α can be added to it [13]. This number is random in the same sense that χ is random in the usual gauge transform of the MH potential, i.e.,

$$\boldsymbol{A} \rightarrow \boldsymbol{A} + \boldsymbol{\nabla}\chi. \tag{10.9}$$

The U(1) gauge transformation is a rotation of the phase factor (10.2), a rotation of a gauge field brought about by multiplication by the rotation generator of U(1) gauge field theory, the exponential $e^{i\alpha}$. It is always argued in U(1) gauge transformation that χ is unphysical; so, for self-consistency, α must be unphysical. However, an unphysical factor α cannot appear in the phase factor [13] of a physical theory, i.e., a theory of physics aimed at describing data. So the MH field theory fails at a fundamental level to describe physical optics. In other words, MH gauge field theory fails to describe reflection and interferometry.

The Sagnac interferometer [13] with platform at rest produces an observable interferogram both in electromagnetic waves and matter waves (interfering electron beams [11]):

$$\gamma = \cos(2\omega^2/c^2 Ar), \tag{10.10}$$

where Ar is the area enclosed by the interfering beams (electromagnetic or electron beams). When the platform is rotated at an angular frequency Ω, a shift is observed in the interferogram:

$$\Delta\gamma = \cos(4\omega\Omega/c^2 Ar). \tag{10.11}$$

The interferograms (10.10) and (10.11) are independent of the shape of the area Ar, and are the same to an observer on and off the rotating platform. There have been many attempts to explain the Sagnac effect since it was first observed, all have their shortcomings, as reviewed by Barrett [13]. In particular the MH field theory (a U(1) gauge field theory) is wholly incapable of explaining the effect [13], either with platform at rest or in motion. In Sec. 3 it is shown that the Sagnac effect is described straightforwardly in the Evans unified field theory as a change in the Cartan tetrad [1-8]. It is well known that the tetrad plays the role of metric in differential geometry. This gets to the core of the problem with MH theory – it is a theory of flat spacetime and so is metric-invariant [13]. The phase factor (10.1) of the MH field theory is a number, and so is invariant under both parity and motion reversal symmetry. The phase factor (10.1) has no sense of parity, of handedness or of chirality (of being clockwise of anticlockwise) and does not change under frame rotation or platform rotation. The MH theory cannot describe any feature of the Sagnac effect, and again fails at a fundamental level.

There has been a forty-year controversy over the AB effect [12]. The only way to describe it with MH field theory is to shift the problem to the nature of spacetime itself. It is asserted conventionally [12] that the latter is multiply connected and that this feature (rather magically) produces an AB effect by U(1) gauge transformation into the vacuum. It is straightforward to show as follows that this assertion is incorrect because it violates the Poincaré lemma. In Sec. 3 we show that the Evans unified field theory explains the AB effect with the correct generally covariant phase factor. This is the same in mathematical structure as for the Sagnac effect and physical optics, only it interpretation is different. The Evans unified theory is therefore much simpler, as well as much more powerful, than the earlier MH field theory. To prove that the MH theory of the AB effect is incorrect, consider the Stokes theorem in the well-known [9] notation of differential geometry:

$$\exp\left(ig\oint_{\delta S} A\right) = \exp\left(ig\oint_{S} d\wedge A\right), \tag{10.12}$$

where A is the potential one-form and $d \wedge A$ is the exterior derivative of A, a two-form. The factor g in the AB effect is $e/\hbar$, where $-e$ is the charge on the electron and $\hbar$ is the Dirac constant ($h/2\pi$). Under the U(1) gauge transformation

$$A \rightarrow A + d\chi \tag{10.13}$$

the two-form becomes

$$d \wedge A \rightarrow d \wedge A + d \wedge d\chi. \tag{10.14}$$

However, the Poincaré lemma asserts that, for any topology,

$$d \wedge d := 0. \tag{10.15}$$

Therefore the two-form $d \wedge A$ is unchanged under the U(1) gauge transformation (13). The Stokes theorem (10.12) then shows that

$$\oint_{\delta S} d\chi := 0. \tag{10.16}$$

However, the conventional description of the AB effect [12] relies on the incorrect assertion

$$\oint_{\delta S} d\chi \neq 0. \tag{10.17}$$

This is incorrect because it violates the Poincaré lemma (10.15). The lemma is true for multiply-connected as well as simply-connected regions. In ordinary vector notation the lemma states that, for any function χ,

$$\nabla \times \nabla\chi := 0, \tag{10.18}$$

and this is true for a periodic function because it is true for any function. A standard freshman textbook [16] will show that the Stokes theorem and the Green theorem are both true for multiply-connected regions as well as for simply-connected regions.

There is therefore no correct explanation of the AB effect in MH theory and special relativity. Similar arguments show that MH theory cannot be used to describe a topological phase effect such as the Tomita-Chao effect, a shift in phase brought about by rotating a beam of light around a helical optical fibre. In Sec. 3 we show that the Tomita-Chao effect is a Sagnac effect with several loops, and is a shift in the Cartan tetrad of the Evans unified field theory. Similarly, the Berry phase of matter wave theory is a shift in the tetrad of the Evans unified field theory. Many more examples could be given of the advantages of the Evans unified field theory over the MH theory.

10.3 Generally Covariant Phase Factor From The Evans Unified Field Theory

The first correct description of the phase factor in electrodynamics was given in Ref. [11]. Since then, a unified field theory has been developed [1-8] which in turn gives a further insight to the phase factor of electrodynamics and physical optics. In order to construct a phase factor that has the correct symmetry under parity inversion, which is generally covariant and valid for all topologies, and which furthermore is correctly gauge invariant, we use the generally covariant Stokes theorem in differential geometry within the exponent of the electromagnetic phase factor. The phase factor is therefore an application of Eq. (10.12) for the propagating electromagnetic field, which is considered to be part of the generally covariant unified field theory [1-8]. Therefore the phase factor of electrodynamics and physical optics is

$$\Phi = \exp\left(ig \oint_{DS} A\right) = \exp\left(ig \int_{S} D \wedge A\right) \tag{10.19}$$

under all conditions (free or radiated field and field matter interaction).

We shall show in this section that the phase factor (10.19) produces fundamental phenomena of physical optics such as reflection and interferometry through a difference in contour integrals of Eq. (10.19)

$$\Delta\Phi = \exp\left(i\left(\oint_{0}^{Z} \boldsymbol{\kappa} \cdot d\boldsymbol{r} - \oint_{Z}^{0} \boldsymbol{\kappa} \cdot d\boldsymbol{r}\right)\right). \tag{10.20}$$

It is also shown that the phase factor (10.19) can be written as

$$\Phi = \exp\left(i \oint \boldsymbol{\kappa} \cdot d\boldsymbol{r}\right) = \exp\left(i \int \kappa^2 dAr\right) \tag{10.21}$$

and automatically gives the Sagnac effect from the right-hand side of Eq. (10.21). Equation (10.21) follows from the generally covariant Stokes formula applied to the one-form κ of differential geometry, the one-form that represents the wave number. In Eq. (10.19), $D \wedge A$ is the covariant exterior derivative [1-9] necessary to make the theory correctly covariant in general relativity. Equation (10.19) is the generally covariant Stokes theorem, i.e., the Stokes theorem defined for the non-Euclidean geometry of the base manifold and therefore in general relativity [11], and it follows from differential geometry that $D \wedge A$ is a two-form. The electromagnetic phase can be written in general

$$\Phi = \exp\left(i \oint_{DS} \kappa\right) = \exp\left(i \int_{S} D \wedge \kappa\right); \tag{10.22}$$

and Eq. (10.19) and (10.22) are equivalent definitions provided that we recognise the following duality between the electromagnetic field as a potential one-form, denoted by A in differential geometry [9], and the wave number one-form κ:

$$\kappa = gA. \tag{10.23}$$

Here g is a constant of proportionality between the A and κ one-forms. From gauge theory [11] g is given by the wave-particle dualism represented by

$$p = \hbar\kappa = eA, \tag{10.24}$$

where p is a momentum one-form. The duality (10.24) asserts that the photon is a particle with momentum $\hbar\kappa$ and also a field with momentum eA, where $-e$ is the charge on the electron. The meaning of the duality (10.24) has been explained extensively in the literature on $O(3)$ electrodynamics [11]. The charge e is defined as the ratio of magnitudes:

$$e = |p|/|A|, \tag{10.25}$$

and so Eq. (10.23) means that the wave number in physical optics and electrodynamics is multiplied by the potential and divided by the magnitude of the potential to inter-relate the field as particle with momentum $\hbar\kappa$ and the field as wave, with momentum eA. This is the de Broglie wave-particle duality applied to the electromagnetic field.

The electromagnetic phase in unified field theory and general relativity is therefore

$$\Phi = \exp\left(i \oint_{DS} \kappa\right) = \exp\left(\frac{i}{\hbar} \oint_{DS} p\right) = \exp\left(ig \oint_{DS} A\right), \tag{10.26}$$

which is the generally covariant Dirac phase or Wilson loop for electrodynamics as part of a unified field theory [1-8]. The covariant derivative appearing in Eq. (10.19) is defined in the notation of differential geometry by

$$D \wedge A = d \wedge A + gA \wedge A. \tag{10.27}$$

The symbol A is convenient shorthand notation for the tetrad $A^a{}_\mu$ [1-9], so Eq. (10.27), for example, denotes

$$(D \wedge A^c)_{\mu\nu} = (d \wedge A^c)_{\mu\nu} + gA^a{}_\mu \wedge A^b{}_\nu. \tag{10.28}$$

For $O(3)$ electrodynamics with orthonormal space index

$$a = (1), (2), (3) \tag{10.29}$$

the complex circular basis, then:

$$D \wedge A^{(1)*} = d \wedge A^{(1)*} - igA^{(2)} \wedge A^{(3)}, \quad \text{et cyclicum.} \tag{10.30}$$

In Eq. (10.30) the indices μ and ν of the base manifold have been suppressed, as is the custom in differential geometry [9], because Eqs. (10.27) to (10.30) are equations of differential geometry valid for all geometries of the base manifold. So, the indices μ and ν are always the same on both sides and can be suppressed for convenience of notation [9]. The shorthand notation of Eq. (10.27) is therefore a summary of the essential features of Eqs. (10.28) to (10.30). The $B^{(3)}$ form of differential geometry

$$B^{(3)} = D \wedge A^{(3)} = -igA^{(1)} \wedge A^{(2)}, \tag{10.31}$$

is the expression of the fundamental Evans-Vigier field [11] in differential geometry. The $B^{(3)}$ form is a component of a torsion tensor (a vector valued two-form or antisymmetric tensor), and is the spin Casimir invariant of the Einstein group. It will be demonstrated in this section that it is responsible for and therefore observed in all physical optical effects through the generally covariant and gauge invariant phase factor of electrodynamics

$$\Phi = \exp\left(ig \oint_{DS} A^{(3)}\right) = \exp\left(ig \int_S B^{(3)}\right). \tag{10.32}$$

Equation (10.32) is the result of a generally covariant theory of electrodynamics and it is necessary to be precise about the meaning of the contour integral on the left-hand side and the area integral on the right-hand side of Eq. (10.32). In particular, $A^{(3)}$ is an irrotational function, whose curl vanishes. It follows that Eq. (10.32) cannot be the result of the ordinary Stokes theorem

$$\oint_{\delta S} A = \int_S B = \int_S d \wedge A, \tag{10.33}$$

in which $d\wedge$ is the ordinary exterior derivative [11] of differential geometry. In vector notation, Eq. (10.33) reads

$$\oint \boldsymbol{A} \cdot d\boldsymbol{r} = \int \boldsymbol{B} \cdot \boldsymbol{k} dAr = \int \boldsymbol{\nabla} \times \boldsymbol{A} \cdot \boldsymbol{k} dAr, \tag{10.34}$$

and for an irrotational function $\boldsymbol{A}$ both sides of Eq. (10.34) vanish. Therefore, in Eq. (10.32), $\oint A^{(3)}$ is not the result of an integration around a closed loop as defined in the ordinary Stokes theorem [16].

Another way of seeing this is that, if we attempt to integrate the ordinary plane wave

$$\boldsymbol{A} = \frac{A^{(0)}}{\sqrt{2}}(\boldsymbol{i} - i\boldsymbol{j})e^{i\phi} \tag{10.35}$$

around a closed loop (such as a circle) using the ordinary Stokes theorem, we obtain the result

$$\oint \boldsymbol{A} \cdot d\boldsymbol{r} = \int \boldsymbol{\nabla} \times \boldsymbol{A} \cdot \boldsymbol{k} dAr = 0. \tag{10.36}$$

The right-hand side is true because $\boldsymbol{k}$ is perpendicular to $\boldsymbol{i}$ and $\boldsymbol{j}$, and the left-hand side is proven by parameterizing the circle as

$$dx = -x_0 \sin\theta d\theta, \quad dy = y_0 \cos\theta d\theta \tag{10.37}$$

and using

$$\int_0^{2\pi} \sin\theta d\theta = \int_0^{2\pi} \cos\theta d\theta = 0. \tag{10.38}$$

This means that the ordinary Stokes theorem of special relativity cannot be used to represent the phase factor of electrodynamics. The essential reason for this is that electrodynamics must be a theory of general relativity in which the exterior derivative is replaced by the covariant exterior derivative [1-9] under all conditions (free field and field-matter interaction). Only in this way can a unification of gravitation and electromagnetism be achieved [1-8] within general relativity.

This type of field unification is the result of the principle of general relativity, i.e., that all theories of physics must be theories of general relativity. (The MH theory is the archetypical theory of special relativity, and is not generally covariant. This leads to the fundamental problems summarized in Sec. 2.)

The method of interpreting Eq. (10.32) is found by considering integration around the transverse part of a helix. The transverse part of a helix is a position vector which coils around the Z axis as follows:

$$\boldsymbol{r} = (X\boldsymbol{i} - iY\boldsymbol{j})e^{i\phi} \tag{10.39}$$

and defines a non-Euclidean (curling) base-line in the non-Euclidean base manifold of general relativity. The product

$$r^2 = \boldsymbol{r} \cdot \boldsymbol{r}^* = X^2 + Y^2 \tag{10.40}$$

for the helix is the same as for a circle of radius r. Therefore integration around the transverse part of a helix is the same as integration around a circle of circumference $2\pi r$ and area πr^2 provided that the arc length of the helix is also $2\pi r$. The arc length is the distance along the helix from A to B in Fig. 1. This diagram summarizes the process of taking a circle of diameter $2r$ and drawing it out into a helix coiled along the Z axis. The distance from A to B around the circumference of the circle is the same as the distance from A to B along the transverse part of the helix (its arc length [16]). The difference between the helix and the circle is that there is a longitudinal component of the helix, the distance along the Z axis from A to B. The arc length of the helix is along its transverse component from A to B, therefore the distance along Z from A to B is defined by

$$Z \leq 2\pi r. \tag{10.41}$$

When this distance is equal $2\pi r$, the helix becomes a straight line of length $2\pi r$ along the Z axis.

The difference between the helix and the circle illustrates the difference between the generally covariant Stokes theorem and the ordinary Stokes theorem, and is also the essential difference between generally covariant electrodynamics and MH electrodynamics in which radiation is the plane wave (10.35), with no longitudinal component. In generally covariant electrodynamics there is a longitudinal component, the Z component of the helix. The latter can be parameterized [16] by

$$x = x_0 \cos\theta, \quad y = y_0 \sin\theta, \quad z = z_0\theta, \tag{10.42}$$

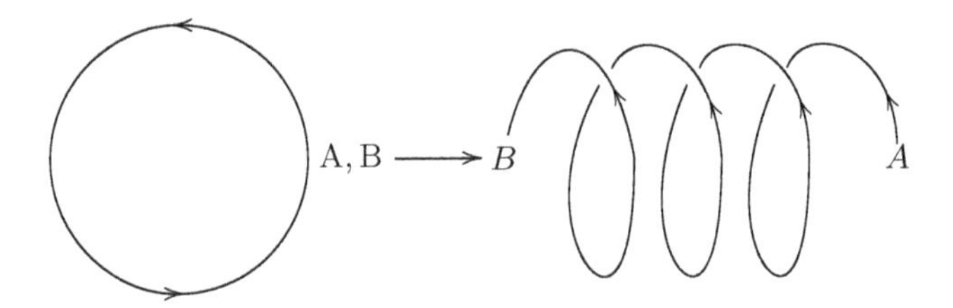

Fig. 10.1. Representation of the circle whose circumference is the same as the arc length of the helix AB.

and so contains both longitudinal and transverse components. The circle can be parameterized [16] by

$$x = x_0 \cos\theta, y = y_0 \sin\theta \tag{10.43}$$

and contains only components which are perpendicular, or transverse, to the Z axis. The latter is evidently undefined in the circle, but well defined in the helix.

It is well accepted that electromagnetic radiation propagates and so contains both rotational and translational components of emotion. These must be described by the helix, and not by the plane wave of the MH theory in special relativity. There must be longitudinal components of the free electromagnetic field, represented by $\boldsymbol{A}^{(3)}$ and $\boldsymbol{B}^{(3)}$.

The ordinary Stokes theorem cannot be applied to the helix, because the latter does not define a closed curve, i.e., the path from A to B of the helix is not closed. In the circle the path around the circumference from A to B brings us back to the starting point, and defines the area of the circle. In order to close the path in the helix the path from A to B along its transverse part must be followed by the path back from B to A along its Z axis, as in Fig. 2. The closed path on the right-hand side of this diagram defines the contour integral of the generally covariant Stokes theorem appearing on the left-hand side of Eq. (10.32). Furthermore, as we have seen, integration around the transverse part of the helix is equivalent to integration around a circle whose circumference is equal to the arc length of the helix. Therefore the plane wave contributes nothing to the generally covariant Stokes theorem because the integration of the plane wave around a circle is zero, as shown in Eq. (10.34) to (10.38).

It is concluded that the electromagnetic phase in generally covariant electrodynamics and physical optics is completely described by Eq. (10.32). The contour integral on the left-hand side of this equation is along the Z axis of the helix whose arc length is the same as the circumference $2\pi r$ of a circle of radius r. The area integral on the right-hand side of Eq. (10.32) is an integral around the area πr^2 of this circle. The contour integral is an integral over the irrotational and longitudinal potential field $\boldsymbol{A}^{(3)}$, and the area integral is an

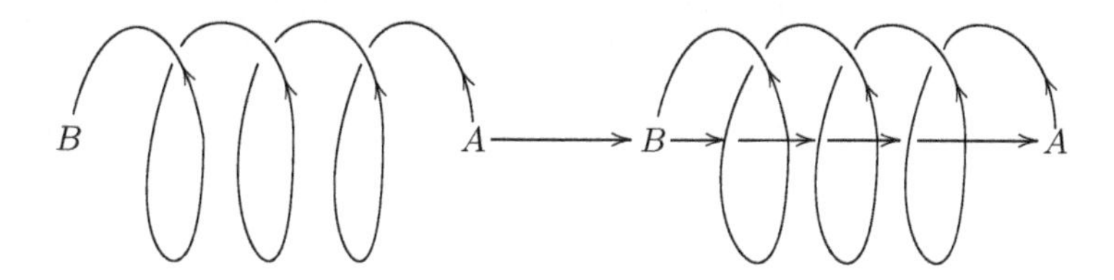

Fig. 10.2. Representation of the line over which the integral is evaluated in the non-Abelian Stokes theorem and Evans phase law.

integral over the Evans-Vigier field $\boldsymbol{B}^{(3)}$, the spin invariant of the Einstein group. Therefore all of the physical optics and electrodynamics are described by a phase factor which is completely defined by $\boldsymbol{A}^{(3)}$ and $\boldsymbol{B}^{(3)}$.

In the following sections, applications of this law are given for reflection, interferometry, the Sagnac effect and the AB effect. The catastrophic failings of the MH field theory of special relativity (summarized in Sec. 2) are corrected in each case. The easiest way to see that the MH theory is not generally covariant is to note that the plane wave defines a finite scalar curvature

$$R = (\partial^2 \boldsymbol{A}^{(1)}/\partial Z^2)/|\boldsymbol{A}^{(1)}| = \kappa^2. \tag{10.44}$$

However, in special relativity, the scalar curvature $R = 0$ by definition, because the spacetime of a theory of special relativity is Euclidean (or "flat"). In the generally covariant description of electrodynamics, the scalar curvature κ^2 of the helix is consistent with the fact that we are reconsidering a baseline that coils around the Z axis in the base manifold.

10.4 Reflection And Michelson Interferometry

The generally covariant phase law (10.32) has the correct parity inversion symmetry for the description of reflection in physical optics, and thus of Michelson interferometry. Considering the Z axis as the propagation axis, the parity inversion symmetry is

$$\oint \kappa^{(3)} dZ \xrightarrow{\hat{P}} -\oint \kappa^{(3)} dZ, \tag{10.45}$$

and so parity inversion is equivalent to traversing the path along the Z axis in the opposite direction

$$\int_0^Z \kappa^{(3)} dZ \xrightarrow{\hat{P}} \int_Z^0 \kappa^{(3)} dZ. \tag{10.46}$$

Traversing the path in the opposite direction therefore produces the following change in phase factor:

$$\exp\left(i\int_0^Z \kappa^{(3)}dZ\right) \to \exp\left(i\int_Z^0 \kappa^{(3)}dZ\right), \tag{10.47}$$

producing the experimentally observed phase change upon normal reflection of electromagnetic radiation from a perfectly reflecting mirror:

$$\exp\left(i\int_0^Z \kappa^{(3)}dZ\right)\exp\left(i\int_Z^0 \kappa^{(3)}dZ\right) = \exp(2i\kappa^{(3)}Z). \tag{10.48}$$

In Michelson interferometry [14] the phase change (10.48) is observed as the cosinal interferogram for a monochromatic beam of radiation:

$$\Delta\gamma = \cos(2\kappa^{(2)}Z). \tag{10.49}$$

In MH electrodynamics (special relativity) normal reflection produces no phase change, because normal reflection, by definition, is equivalent to the parity inversion (10.4). The latter produces no change in phase in MH electrodynamics, and no Michelson interferogram, contrary to experimental observation. In generally covariant electrodynamics, the phase factor is chiral (or handed) in nature, because it is defined by the following cyclic equation:

$$\oint \boldsymbol{A}^{(3)}\cdot d\boldsymbol{r} = -ig\int \boldsymbol{A}^{(1)}\times\boldsymbol{A}^{(2)}\cdot \boldsymbol{k}dAr, \tag{10.50}$$

an expression of the generally covariant Stokes theorem.

10.5 The Sagnac Effect

The generally covariant phase law (10.32) can be exemplified by integration around a circle of radius r. The area on the right-hand side of Eq. (10.32) in this case is πr^2, so we obtain

$$\int \boldsymbol{B}^{(3)}\cdot \boldsymbol{k}dAr = \pi r^2 B_Z^{(3)} = \pi r^2 B^{(0)}. \tag{10.51}$$

The phase factor observed in the Sagnac effect therefore originates in the phase law (10.32) as follows:

$$\Phi = \exp(igB^{(0)}\pi r^2) = \exp(i\kappa^2 Ar) = \exp(i\omega^2/c^2 Ar), \tag{10.52}$$

where we have used

$$B^{(0)} = \kappa A^{(0)}, \quad g = \kappa/A^{(0)}, \quad Ar = \pi r^2. \tag{10.53}$$

The Sagnac effect with platform at rest is an interferogram formed from interference of clockwise and counter-clockwise waves, one with phase $e^{i\kappa^2 Ar}$ and one with phase factor $e^{-i\kappa^2 Ar}$ giving the difference in phase factor $e^{2i\kappa^2 Ar}$ and interferogram:

$$\Delta\gamma = \cos(2\kappa^2 Ar) = \cos(2\omega^2/c^2 Ar), \tag{10.54}$$

as observed experimentally (10.13) with great precision in the ring laser gyro and similar devices. In the MH theory there is no explanation for the Sagnac effect, as we have argued, because the MH phase is parity invariant with no sense of chirality. In MH theory there is no counterpart of the beam area, as on the right-hand side of the generally covariant phase law (10.32).

When the platform of the Sagnac effect is rotated at angular frequency Ω, there is a frequency shift

$$\omega \to \omega \pm \Omega, \tag{10.55}$$

giving rise to an extra interferogram from the phase law (10.32):

$$\Delta\Delta\gamma = \cos\left(4\omega\Omega/c^2\, Ar\right) \tag{10.56}$$

from the difference

$$(\omega+\Omega)^2 - (\omega-\Omega)^2 = 4\omega\Omega. \tag{10.57}$$

This is precisely as observed experimentally to great precision in the ring laser gyro [13].

The left-hand side of the phase law (10.32) for the Sagnac effect in a circle of radius r is

$$\int_0^Z A^{(3)}{}_Z dZ = A^{(0)} Z = \pi r^2 B^{(0)}, \tag{10.58}$$

where we have used

$$\int_0^Z dZ = Z, \quad A^{(3)}{}_Z = A^{(0)}. \tag{10.59}$$

From Eq. (10.53) and (10.58)

$$2\pi r \geq Z = \kappa\pi r^2, \quad \kappa r \geq 2. \tag{10.60}$$

This equation defines the distance from B to A in Fig. 2 in terms of the area πr^2 of a circle whose circumference $2\pi r$ is the same as the arc length of the helix. The latter must always be greater than or equal to the distance from A to B along Z, so:

$$\kappa r \geq 2. \tag{10.61}$$

Using

$$\kappa = \omega/c, \tag{10.62}$$

$$\omega = 2\pi f, \tag{10.63}$$

$$f\lambda = c, \tag{10.64}$$

we find that

$$\kappa = \frac{2\pi}{\lambda} \tag{10.65}$$

and in Eq. (10.60)

$$Z = \frac{2\pi}{\lambda} Ar. \tag{10.66}$$

This equation is a result of the phase law (10.32) and defines the propagation length of electromagnetic radiation in terms of its wavelength and area.

Equation (10.66) is the result of generally covariant electrodynamics and shows that a beam of light has a finite area. This result is obviously consistent with experimental data, but in special relativity (MH theory) the plane wave has infinite lateral extent, and the area of a beam of light is ill defined by the plane wave. In general relativity as we have seen the area is well defined by the phase law (10.32).

We can cross check this result for the Sagnac effect by integrating the wave number vector $\boldsymbol{\kappa}$ around the helix parameterized [16] by:

$$X = X_0 \cos\theta, \quad Y = Y_0 \sin\theta, \quad Z = Z_0\theta, \tag{10.67}$$

$$\frac{dX}{d\theta} = -X_0 \sin\theta, \quad \frac{dY}{d\theta} = Y_0 \cos\theta, \quad \frac{dZ}{d\theta} = Z_0. \tag{10.68}$$

The wave number vector is

$$\boldsymbol{\kappa} = \kappa_X \boldsymbol{i} + \kappa_Y \boldsymbol{j} + \kappa_Z \boldsymbol{k}, \tag{10.69}$$

and the contour integral on the left-hand side of the phase law (10.32) is

$$\oint \boldsymbol{\kappa} \cdot d\boldsymbol{r} = -\kappa_X X_0 \oint \sin\theta d\theta + \kappa_Y Y_0 \oint \cos\theta d\theta + \kappa_Z Z_0 \oint \theta d\theta. \tag{10.70}$$

The transverse part of the helix gives no contribution to this contour integral because:

$$\int_0^{2\pi} \sin\theta d\theta = \int_0^{2\pi} \cos\theta d\theta = 0. \tag{10.71}$$

The longitudinal component along the Z axis from B to A in Fig. 2 gives the only non-zero contribution to the contour integral, a contribution originating in

$$\int_0^{2\pi} \theta d\theta = 2\pi^2. \tag{10.72}$$

So,

$$\oint \boldsymbol{\kappa} \cdot d\boldsymbol{r} = 2\pi^2 \kappa_Z Z_0. \tag{10.73}$$

This result is Eq. (10.58) after identifying

$$Z = 2\pi^2 Z_0, \quad \kappa Z = g A^{(0)} Z. \tag{10.74}$$

The whole of Eq. (10.73) comes from the contribution along the Z axis in Fig. 2 through the axis of the helix. This contribution is not present in special relativity and the conventional Stokes theorem. Our generally covariant unified field theory [1-8] gives the observed phase factor of the Sagnac effect with platform at rest from

$$\Phi = \exp(i\kappa Z) = \exp(i\omega^2/c^2 Ar) \tag{10.75}$$

and also with platform in motion, as argued already.

The topological fundament of this result is that a circle can be shrunk continuously to a point and is simply connected, but a helix shrinks continuously to a line, and in this sense is not simply connected. This is the essential difference between special and general relativity in optics and electrodynamics.

10.6 The Aharonov-Bohm Effect

The AB effect in generally covariant electrodynamics is also described straightforwardly by the phase law (10.32):

$$\begin{aligned}\exp\left(ig\int \boldsymbol{B}^{(3)}\cdot \boldsymbol{k}dAr\right) &= \exp(ig\oint \boldsymbol{A}^{(3)}\cdot d\boldsymbol{r})\\ &= \exp\left(g^2\int \boldsymbol{A}^{(1)}\times \boldsymbol{A}^{(2)}\cdot \boldsymbol{k}dAr\right)\end{aligned} \tag{10.76}$$

The interpretation of Eq. (10.76) for the AB effect is as follows:

(a) The magnetic field $\boldsymbol{B}^{(3)}$ is the static magnetic field (iron whisker or solenoid) placed between the two openings of a Young interferometer which measures the interferogram formed by two electron beams.

(b) Any magnetic field (including a static magnetic field) in generally covariant electrodynamics is the two-form [1-9]

$$B = d\wedge A + gA\wedge A, \tag{10.77}$$

and is also the torsion form or wedge product of tetrads:

$$B^c = -igA^a \wedge A^b. \tag{10.78}$$

Equation (10.78) follows from the fact that a magnetic field must be a component of a torsion two-form (or antisymmetric second rank tensor) that is the signature of spin in general relativity. The concept of spin is missing completely from Einstein's generally covariant theory of gravitation [9]. In the Evans unified field theory, spin (or torsion) gives rise to the generally covariant electromagnetic field. Curvature of spacetime gives rise to the gravitational field, and the latter is the symmetric product of tetrads [1-8]. In the complex circular basis [11],

$$c = (3), \quad a = (1), \quad b = (2). \tag{10.79}$$

(c) The factor g in Eq. (10.76) for the AB effect is

$$g = e/\hbar \tag{10.80}$$

and is the ratio of the modulus of the charge on the electron (situated in the electron beams) to the reduced Planck constant.

(d) The potentials $\boldsymbol{A}^{(1)} = \boldsymbol{A}^{(2)*}$ are defined by the cyclic relation (10.76) and extend outside the area of the solenoid because they are components of a tetrad multiplied by the scalar $A^{(0)}$. The tetrad is the metric tensor in differential geometry, so $\boldsymbol{A}^{(1)} = \boldsymbol{A}^{(2)*}$ are properties of non-Euclidean (i.e., spinning) spacetime. This spacetime, evidently, is not restricted to the solenoid.

(e) The area Ar is the area enclosed [12] by the electron beams of the Young interferometer.

The AB effect is therefore

$$\Delta\Phi = \exp\left(ie/\hbar\int \boldsymbol{B}^{(3)}\cdot\boldsymbol{k}dAr\right) = \exp(ie/\hbar\phi), \tag{10.81}$$

where ϕ is the magnetic flux produced by the solenoid of magnetic flux density $\boldsymbol{B}^{(3)}$. The magnetic flux (weber = volts $^{-1}$) is $\boldsymbol{B}^{(3)}$ (tesla) multiplied by the area Ar enclosed by the electron beams.

The effect (10.81) is observed experimentally [12] as a shift in the interferogram of the Young interferometer, a shift caused by the iron whisker. It is seen that the AB effect is closely similar to the Sagnac effect and to Michelson interferometry and reflection in physical optics. All effects are described straightforwardly by the same phase law (10.76), which is therefore verified with great precision in these experiments. Recall that the MH, or U(1), phase factor (10.1) fails qualitatively in all four experiments (Sec. 2).

When we use the correctly covariant phase law of general relativity, Eq. (10.76), the AB effect becomes a direct interaction of the following potentials with the electron beam:

$$\boldsymbol{A}^{(1)} = \boldsymbol{A}^{(2)*} = \left(A^{(0)}/\sqrt{2}\right)(\boldsymbol{i} - i\boldsymbol{j})e^{i\omega t}. \tag{10.82}$$

These potentials define the static magnetic field of the iron whisker or solenoid as follows:

$$\boldsymbol{B}^{(3)} = -ig\boldsymbol{A}^{(1)} \times \boldsymbol{A}^{(2)}, \tag{10.83}$$

where ω is an angular frequency defining the rate at which the potentials spin around the Z axis of the solenoid. The potential

$$\boldsymbol{A}^{(1)} = \boldsymbol{A}^{(2)*} \tag{10.84}$$

is defined completely by the scalar magnitude $\boldsymbol{A}^{(0)}$, the frequency ω, and the complex unit vectors

$$\boldsymbol{e}^{(1)} = \boldsymbol{e}^{(2)*}. \tag{10.85}$$

The magnitude of the static magnetic field is

$$B^{(0)} = gA^{(0)2} \tag{10.86}$$

and

$$\boldsymbol{B}^{(3)} = B^{(0)}\boldsymbol{e}^{(3)}, \tag{10.87}$$

where

$$\boldsymbol{e}^{(1)} \times \boldsymbol{e}^{(2)} = i\boldsymbol{e}^{(3)*} = i\boldsymbol{k}, \quad \text{et cyclicum.} \tag{10.88}$$

Therefore the essence of the AB effect in general relativity is that it is an effect of spinning spacetime itself, the spinning potential $\boldsymbol{A}^{(1)} = \boldsymbol{A}^{(2)*}$ extends outside the solenoid (to whose Z axis $\boldsymbol{B}^{(3)}$ is confined) and $\boldsymbol{A}^{(1)} = \boldsymbol{A}^{(2)*}$ interacts directly with the electron beam, giving the observed phase shift (10.81). The spin of the spacetime is produced by the iron whisker, or static magnetic field. Conversely, a static magnetic field is spacetime spin as measured by the torsion form of the Evans theory [1-8] of generally covariant electrodynamics. This theory completes the earlier theory of gravitation [9].

This explanation has the advantage of simplicity (Ockham Razor), and there is no need for the obscure, and mathematically incorrect, assumptions used in the received view of the AB effect (Sec. 2). The explanation shows that the AB effect originates in general relativity with spin (the Evans unified field theory [1-8]). In analogy the AB effect is a whirlpool effect, the whirlpool is created by a stirring rod (the iron whisker) and the effect at the edges of the whirlpool is evidently measurable even though the stirring rod is not present there. The water of the whirlpool is the analogy to spinning spacetime. In the same type of analogy the Sagnac effect is a rotational effect caused by a physical rotation of the platform with respect to a fixed reference frame, whereas in the AB effect the magnetic field is a rotation or spin of spacetime (the reference frame itself). The Evans unified field theory gives a quantitative explanation of these whirlpool effects with a precision of up to 1 part on 10^{23} (contemporary ring laser gyro [11,13]).

10.7 Gauge Invariance Of The Phase Law (10.32)

Finally in this paper a brief discussion is given of the gauge invariance of the generally covariant phase law (10.32). The Evans theory [1-8] is a theory of general relativity, in which the base manifold is non-Euclidean in nature, therefore covariant derivatives are always used from the outset instead of ordinary derivatives [9]. In both gravitation and electrodynamics the use of a covariant derivative implies that the Evans theory is intrinsically gauge invariant. The reason is that local gauge invariance in relativity theory is defined as the replacement of the ordinary derivative by a covariant derivative [12,13]. In the MH (or U(1)) electrodynamics, the covariant derivative is the *result* of a local gauge transformation of the generic gauge field ψ, a transformation under which the Lagrangian is invariant [12]. This local (or type two) gauge transformation is defined only in special relativity (flat spacetime) in MH theory and introduces the covariant derivative by changing the ordinary derivative to the U(1) covariant derivative:

$$\partial_\mu \rightarrow \partial_\mu - igA_\mu. \tag{10.89}$$

The introduction of the U(1) potential A_μ in this way is equivalent to the well-known minimal prescription in the received view [12], and so is equivalent to the interaction of field with matter (i.e., of A_μ with e). The shortcomings of this point of view, i.e., of the MH theory, are reviewed briefly in Sec. 2, and elsewhere [1-8,11]. One of several problems with the received view is that A_μ itself is not gauge invariant (i.e., assumed not to be a physical quantity), yet is used in the minimal prescription to represent the physical electromagnetic field. The same type of problem appears in the U(1) phase through the introduction of the random factor α (Sec. 2). In consequence there has been a long, confusing, and misleading debate within U(1) electrodynamics as to whether or not A_μ is physical [13]. This type of debate has also bedeviled the understanding of the AB effect.

In the Evans unified field theory the debate is resolved using general relativity by recognising that A^a_μ is a tetrad and therefore an element of a metric tensor in a spinning spacetime. The gauge invariance of the Evans theory is evident through the fact that it always uses covariant derivatives. In the language of differential geometry, this is the covariant exterior derivative $D\wedge$ [9], without which the geometry of spacetime is incorrectly defined. A magnetic field for example is always (as we have seen) the covariant derivative of a potential form, which is a vector valued one form. The correctly and generally covariant magnetic field is therefore a vector valued two-form. The latter is the torsion form of spacetime within a C negative scalar. This concept is missing both from the Einstein theory of gravitation and the MH theory of electrodynamics. It is the concept needed for a unified field theory of gravitation and electromagnetism in terms of general relativity and differential geometry.

In this paper we have developed the unified field theory into a novel phase law (10.32) that is gauge invariant and generally covariant. The phase law gives the first correct description of physical optics, interferometry, and related effects such as the Sagnac and AB effects. If we use many loops of the Sagnac effect (10.11), we obtain the Tomita-Chao effect; and applying the phase law (10.32) to matter waves, we obtain the Berry phase effects. This will be the subject of future communications.

Acknowledgements

Craddock Inc. and the Ted Annis Foundation are acknowledged for funding, and the staff of AIAS and others for many interesting discussions.

A

The Sagnac Effect As A Change In Tetrad

As argued in the text, the Sagnac effect is a change in frequency caused by rotating the platform of the interferometer:

$$\omega \rightarrow \omega \pm \Omega. \tag{A1}$$

It is a shift in wave number:

$$\kappa \rightarrow \kappa \pm \Omega / c \tag{A2}$$

and thus in the following component of the potential:

$$A^{(3)}{}_{Z} \rightarrow A^{(3)}{}_{Z} \pm (\Omega/\omega) A^{(3)}{}_{Z}. \tag{A3}$$

In the unified field theory,

$$A^{a}{}_{\mu} = A^{(0)} q^{a}{}_{\mu}, \tag{A4}$$

and in the Sagnac effect is a shift in a tetrad component:

$$q^{(3)}{}_{Z} \rightarrow (1 \pm \Omega/\omega) q^{(3)}{}_{Z} \tag{A5}$$

brought about by rotating the platform. The Sagnac effect is therefore one of general relativity applied to physical optics, and so cannot be explained by MH field theory, which is a theory of special relativity and so metric invariant [13]. Recall that the tetrad is the equivalent of the metric matrix in differential geometry. The components of the tetrad in the Evans unified field theory are clearly not those of special relativity.

References

1. M. W. Evans, "A unified field equation for gravitation and electromagnetism," *Found. Phys. Lett.* **16**, 367 (2003).
2. M. W. Evans, "A generally covariant wave equation for grand unified field theory," *Found. Phys. Lett.* **16**, 507 (2003).
3. M. W. Evans, "The equations of grand unified field theory in terms of the Maurer-Cartan structure relations of differential geometry," *Found. Phys. Lett.* **17**, 25 (2004).
4. M. W. Evans, "Derivation of Dirac's equation from the Evans wave equation," *Found. Phys. Lett.* **17**, 149 (2004).
5. M. W. Evans, "Unification of the gravitational and strong nuclear fields," *Found. Phys. Lett.* **17**, 267 (2004).
6. M. W. Evans "The Evans lemma of differential geometry," *Found. Phys. Lett.*, submitted for publication; preprint on www.aias.us.
7. M. W. Evans, "Derivation of the Evans wave equation from the Lagrangian and action. Origin of the Planck constant in general relativity," *Found. Phys. Lett.*, in press; preprint on www.aias.us.
8. M. W. Evans, "Development of the Evans equation in the weak-field limit. The electrogravitic equation," *Found. Phys. Lett.*, in press.
9. S. M. Carroll, *Lecture Notes in General Relativity*, a graduate course in Univ. California, Santa Barbara, arXiv:gr-gq/9712019 vl 3 Dec 1997.
10. M. Tatarikis *et al.*, "Measurements of the inverse Faraday effect in high intensity laser produced plasmas" (CFL Annual Report, 1998/1999, EPSRC, Rutherford Appleton Laboratories, m.tatarikis@ic.ac.uk).
11. M. W. Evans, "O(3)electrodynamics," in M. W. Evans, ed., *Modern Nonlinear Optics*, series eds. I. Prigogine and S. A. Rice, *Advances in Chemical Physics*, Vol. 119 (2), 2nd edn. (Wiley Interscience, New York, 2001; e-book cross link on www.aias.us. L. Felker, ed., *The Evans Equations* (World Scientific, in press, 2004). M. W. Evans, J.-P. Vigier, et al., *The Enigmatic Photon* (Kluwer Academic, Dordrecht, 1994 to 2002, hardback and paperback), 10 volumes. M. W. Evans and L. B. Crowell, *Classical and Quantum Electrodynamics and the* $B^{(3)}$ *Field* (World Scientific, Singapore, 2001). M. W. Evans and A. A. Hasanein, *The Photomagneton in Quantum Field Theory* (World Scientific, Singapore, 1994).
12. L. H. Ryder, *Quantum Field Theory*, 2nd edn. (University Press, Cambridge, 1996).

13. T. W. Barrett, in T. W. Barrett, and D. M. Grimes, *Advanced Electrodynamics* (World Scientific, Singapore, 1996). T. W. Barrett, in A. Lakhtakia, ed., *Essays on the Formal Aspects of Electromagnetic Theory* (World Scientific, Singapore, 1992).
14. M. W. Evans, G. J. Evans, W. T. Coffey, and P. Grigolini, *Molecular Dynamics* (Wiley Interscience, New York, 1982).
15. P. W. Atkins, *Molecular Quantum Mechanics*, 2nd. edn. (University Press, Oxford, 1983).
16. E. M. Milewski, ed., *The Vector Analysis Problem Solver* (Research and Education Association, New York, 1987).

11

Derivation Of The Geometrical Phase From The Evans Phase Law Of Generally Covariant Unified Field Thoery

Summary. The phase law of generally covariant electrodynamics is used to explain straightforwardly the origin of the geometrical and Berry phase effects, exemplified by the Tomita-Chiao effect. Both effects are described by a phase factor that is constructed from the generally covariant Stokes formula of differential geometry, a phase factor in which the contour integral over the potential field $\boldsymbol{A}^{(3)}$ is equated to the area integral over the gauge invariant field $\boldsymbol{B}^{(3)}$, the Evans-Vigier field. The latter is the fundamental spin Casimir invariant of the Einstein group of general relativity applied to electrodynamics. General relativity as extended in the Evans unified field theory is needed for a correct understanding of all phase effects in physics, an understanding that is forged through the Evans phase law, the origin both of the Berry phase and the geometrical phase of electrodynamics observed in the Sagnac and Tomita-Chiao effects.

Key words: geometrical phase, Tomita-Chiao effect, Berry phase, Evans-Vigier field, generally covariant electrodynamics, $\boldsymbol{B}^{(3)}$ field.

11.1 Introduction

Phase factors such as the Berry phase or that observed in the Tomita-Chiao effect are due in general [1,2] to parallel transport in the presence of a gauge field. This inference suggests that such phase factors should properly be described by covariant derivatives in general relativity. Recently a unified field theory has been developed in which electrodynamics is generally covariant [3-9], i.e., becomes understandable with general relativity as required by the Einsteinian principle that all theories of natural philosophy be theories of general relativity. One of the outcomes of this theory is the generally covariant Evans phase law of the unified field, a phase factor which is constructed [10,11] from the generally covariant Stokes formula of differential geometry [12,13]. The field theory is made generally covariant by replacing the exterior derivative of differential geometry, denoted $d\wedge$, by the covariant exterior derivative, denoted by $D\wedge$. Therefore the magnetic field, for example, is defined by the first Maurer Cartan structure relation [12]

$$B = D \wedge A = d \wedge A + gA \wedge A, \tag{11.1}$$

where A is the potential field and g is a proportionality factor with the units of inverse magnetic flux $(e/\hbar)$. In the older Maxwell Heaviside (MH) field theory the magnetic field is defined by

$$B = d \wedge A, \tag{11.2}$$

and it is therefore not a generally covariant field theory. For electrodynamics the Evans phase law is

$$\Phi = \exp\left(ig \oint \boldsymbol{A}^{(3)} \cdot d\boldsymbol{r}\right) = \exp\left(ig \oint \boldsymbol{B}^{(3)} \cdot \boldsymbol{k} dAr\right) := \exp(i\Phi_E), \tag{11.3}$$

where $\boldsymbol{A}^{(3)}$ and $\boldsymbol{B}^{(3)}$ are directed in the propagation axis of the electromagnetic beam and where Ar is the area enclosed by the beam. The Z axis in Eq. (11.3) is the propagation axis of the beam. For matter fields the phase law (11.3) becomes

$$\Phi = \exp\left(i \oint \boldsymbol{\kappa} \cdot d\boldsymbol{r}\right) = \exp\left(i \int \boldsymbol{\kappa}^2 dAr\right) := \exp(i\Phi_E), \tag{11.4}$$

where κ is the wave number [10]. The phase law (11.3) and (11.4) is the first correct phase law of field theory, and gives the first correct explanation [10] of well known phenomena of physical optics such as reflection, interferometry, the Sagnac and Aharonov Bohm (AB) effects. The MH field theory is unable to describe these effects because it is an incomplete theory [2-10] of special relativity.

In Sec. 2 of the Tomita-Chiao effect [1,2] is derived from the Evans phase law (11.4), and in Sec. 3 the Berry phase of matter fields [1,2] is derived from the equivalent phase law (11.4). It is concluded that the origin of the Berry phase is general relativity as developed in the Evans unified field theory [3-10].

11.2 Derivation Of The Tomita-Chiao Effect From The Evans Phase Law

The phase law (11.4) results in a rotation of plane polarized radiation (50% right- and 50% left-circularly polarized) upon propagation in a helical path:

$$\boldsymbol{I}_e = (\boldsymbol{i} - i\boldsymbol{j})(e^{i\phi} + e^{-i\phi}) = 2\boldsymbol{i} \cos\phi. \tag{11.5}$$

To see this, consider initially plane-polarized light, defined as a sum of 50% right- and 50% left-circularly polarized radiation:

$$\begin{aligned} \boldsymbol{I}_L &= \mathrm{Re}(\boldsymbol{i} - i\boldsymbol{j})e^{i\phi} = \cos\phi\boldsymbol{i} + \sin\phi\boldsymbol{j}, \\ \boldsymbol{I}_R &= \mathrm{Re}(\boldsymbol{i} - i\boldsymbol{j})e^{-i\phi} = \cos\phi\boldsymbol{i} - \sin\phi\boldsymbol{j}. \end{aligned} \tag{11.6}$$

After light has propagated along the arc length s of a helix, the phase factor in Eq. (11.5) becomes

$$\boldsymbol{I}'_e = (\boldsymbol{i} - i\boldsymbol{j})e^{i(\phi+\Phi_E)} + (\boldsymbol{i} - i\boldsymbol{j})e^{-i(\phi-\Phi_E)} = \exp(i\Phi_E)\boldsymbol{I}_e \tag{11.7}$$

as a result of the Evans phase law. The angle Φ_E is the Evans phase. Equation (11.7) is generally covariant, and an example of the principle of least curvature [9]. The Tomita-Chiao effect [1,2] is therefore the observation of the Berry phase [14] by rotation of the plane of linearly polarized light due to the scalar curvature R of the helical optical fibre through which the light propagates. After propagation over a distance s along the helix, the polarization changes to

$$\boldsymbol{I}'_e = 2\cos\phi\left(\cos\Phi_e\boldsymbol{i} - \sin\Phi_e\boldsymbol{j}\right), \tag{11.8}$$

where we have used the angle formulas

$$\begin{aligned} \cos(A \pm B) &= \cos A\cos B \mp \sin A\sin B, \\ \sin(A \pm B) &= \sin A\cos B \pm \cos A\sin B. \end{aligned} \tag{11.9}$$

Initially the plane polarized light was polarized as

$$\boldsymbol{I}_e = 2\cos\phi\boldsymbol{i}; \tag{11.10}$$

so, after a propagation distance Z, the plane of polarization has changed from (11.5) to (11.8) due to the Evans phase law (11.4). In MH electrodynamics there is no such effect because the spacetime of the theory is flat, so $R = 0$. In MH theory the phase is purely dynamical, and given by

$$I_{MH} = \exp\left(i(\omega t - \boldsymbol{\kappa}\cdot\boldsymbol{r})\right), \tag{11.11}$$

so there is no mechanism available in special relativity with which to change the plane of polarization of light propagating through a helical optical fibre. Therefore, the Tomita-Chiao effect proves experimentally that the spacetime of electrodynamics is non-Euclidean, with non-zero scalar curvature R. The angle through which the plane of the light is rotated is given from Eq. (11.8) as

$$\tan\theta = \frac{\sin\phi_S}{\cos\phi_S}, \tag{11.12}$$

from which it is inferred that

$$\theta = \Phi_E. \tag{11.13}$$

Therefore the angle through which the plane of light is rotated in the Tomita-Chiao effect, or in any Berry phase, originates in the Evans phase (11.4) of unified field theory and is given in general by

$$\theta = \kappa \oint ds = R \int dAr. \tag{11.14}$$

In this equation, R is the scalar curvature of a given spacetime, or base manifold, and κ is a wavenumber (inverse wavelength) associated with the wave nature of the spacetime or base manifold. Thus matter waves and electromagnetic wave are manifestations of spacetime itself as required in general relativity. In Eq. (11.14), ds is an infinitesimal line element of the non-Euclidean spacetime. Significantly, Einstein's original theory of general relativity originated from the fact that the square ds^2 of the line element is an invariant of general coordinate transformation [12]. The helix defines a rotating and translating baseline in a non-Euclidean manifold. Similarly, the rotating and translating transverse electromagnetic potential vector, defined by

$$\boldsymbol{A}^{(1)} = \boldsymbol{A}^{(2)*} = \frac{A^{(0)}}{\sqrt{2}}(\boldsymbol{i} - i\boldsymbol{j})e^{i\phi}, \tag{11.15}$$

describes the helical baseline of non-Euclidean geometry in the Evans unified field theory [3-10]. The vector $\boldsymbol{A}^{(1)}$ is the spacetime geometry itself multiplied by $A^{(0)}$. The helix in the Tomita-Chiao effect is a physical object that guides $A^{(0)}$ along a helical path, creating a base manifold, or spacetime, with a helical baseline. So the Evans phase law (11.14) applies equally well to both situations. Similarly, the Evans phase law describes straightforwardly [10] the Sagnac and AB effects. The latter is a type of Berry phase [1,2,15].

In the original Tomita-Chiao effect [1,2[it was observed experimentally that the plane of polarization of light is changed after propagation through a helical optical fiber. If the helix is parameterized [10,16] by

$$x = x_0 \cos\theta, \quad y = y_0 \sin\theta, \quad z = z_0\theta, \tag{11.16}$$

then

$$\frac{dx}{d\theta} = -x_0 \sin\theta, \quad \frac{dy}{d\theta} = y_0 \cos\theta, \quad \frac{dz}{d\theta} = z_0. \tag{11.17}$$

The wavenumber vector in general is

$$\boldsymbol{\kappa} = \kappa_x \boldsymbol{i} + \kappa_y \boldsymbol{j} + \kappa_z \boldsymbol{k}, \tag{11.18}$$

and so the contour integral appearing on the right-hand side of Eq. (11.14) is

$$\oint \boldsymbol{k} \cdot d\boldsymbol{r} = -\kappa_x x_0 \oint \sin\theta d\theta + \kappa_y y_0 \oint \cos\theta d\theta + \kappa_z z_0 \oint \theta d\theta. \tag{11.19}$$

Integration from 0 to 2π of Eq. (11.19) produces the result

$$\kappa \oint ds = \oint \phi\boldsymbol{\kappa} \cdot d\boldsymbol{r} = z_0 \int_0^{2\pi} \kappa_z \theta d\theta = 2\pi^2 \kappa_z z_0 \tag{11.20}$$

because

$$\int_0^{2\pi} \sin\theta d\theta = \int_0^{2\pi} \cos\theta d\theta = 0. \tag{11.21}$$

Therefore the angle of rotation of plane polarized light in the Tomita-Chiao effect is given by

$$\theta = 2\pi^2 \kappa_z z_0 = \kappa_z^2 Ar, \tag{11.22}$$

where Ar is the area of a circle whose circumference $2\pi r$ is equal to the arc length of the helical fibre. (A helix can always be constructed from a circle by cutting the circle, pulling it out into a line, and winding the line on a cylinder. The arc length of the helix so constructed must be the same as the circumference of the original circle. If the circle is pulled out into a straight line, the length of the line is the circumference of the original circle.) By integrating from 0 to 2π in Eq. (11.19), we have considered a special case for simplicity of argument.

Equation (11.22) is therefore the result of the general formula (11.14) applied to a helix of arc length $s = 2\pi r$. On writing $\kappa_z = z_1^{-1}$, the Tomita-Chiao phase can finally be expressed as

$$\theta = 2\pi^2 z_0 / z_1. \tag{11.23}$$

The Tomita-Chiao phase is conventionally described [1,2] by

$$\begin{aligned} \exp(-i\theta) &= \exp\left(-i2\pi(1 - \cos\lambda)\right) \\ &= \exp\left(-2\pi\right) i \exp(2\pi i \cos\lambda) \\ &= \exp(2\pi i \cos\lambda), \end{aligned} \tag{11.24}$$

and this is the same as Eq. (11.23) upon identifying

$$\cos\lambda = \pi z_0 / z_1. \tag{11.25}$$

Therefore we have derived the experimentally observed Tomita-Chiao phase from the Evans phase law of generally covariant unified field theory [10]. In so doing the Tomita-Chiao phase is recognized as a phenomenon of general relativity in which spacetime itself has a given non-Euclidean geometry, in this case helical in nature.

11.3 Derivation Of The Berry Phase From The Evans Phase

The Tomita-Chiao effect [1,2] is considered to be the first experimental observation of the Berry phase [14], and is sometimes known as the optical Berry phase or Hannay angle [17]. Such a phase shift occurs whenever a physical phenomenon is defined by a closed path in state space or parameter space [15].

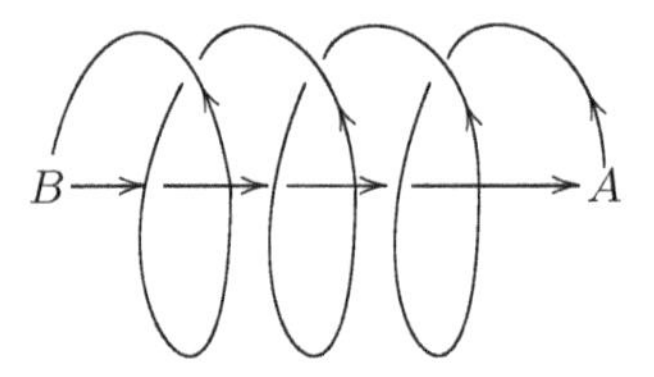

Fig. 11.1.

The closed path in the Evans phase law is defined by the generally covariant Stokes law [3-10] of differential geometry, in which the exterior derivative $d\wedge$ is replaced wherever it occurs with the covariant exterior derivative $D\wedge$. Thus, a magnetic field of any type is always defined in generally covariant electrodynamics in a non-Euclidean spacetime by

$$B = D \wedge A \tag{11.26}$$

and not by

$$B = d \wedge A, \tag{11.27}$$

as in MH field theory in flat (i.e., Euclidean) spacetime. This realization gives rise [18] to the Evans-Vigier field $\boldsymbol{B}^{(3)}$ and to $O(3)$ electrodynamics [19], an example of generally covariant electrodynamics [3-10] in which the orthonormal space of differential geometry (indexed a [12]) is described by the complex circular basis (1),(2), (3)) with $O(3)$ symmetry. The complex circular basis is a natural description of circular polarization [18].

The closed path for the helix is defined as follows, from A to B along the helix and back from B to A along the axis of the helix; see Fig. (11.1). In terms of topology [13] the Evans phase (and therefrom the Berry phase) originates in the fact that a helix cannot be shrunk to a point, and so is not simply connected. A circle can be shrunk to a point, a helix shrinks to a line, the axis of the helix along Z. This procedure is analogous to drawing a helix out into a straight line along Z. Therefore a covariant Stokes theorem must be used to describe the closed path in Fig. (11.1), because the path back from B to A along the axis Z of the helix cuts through the center of the path from A to B along the helix. The generally covariant and non-Abelian Stokes theorem needed to describe this path is [10]

$$\oint \boldsymbol{\kappa}^{(3)} \cdot d\boldsymbol{r} = -i \int \boldsymbol{\kappa}^{(1)} \times \boldsymbol{\kappa}^{(2)} \cdot \boldsymbol{k} dAr, \tag{11.28}$$

and this is the phase of the Evans phase law (11.14). It has been shown elsewhere [3-10] that Eq. (11.28) quantitatively describes many phenomena which the MH phase (11.11) cannot describe qualitatively. In dynamics, the Foucault pendulum is another example of the Berry phase [15], now known to originate in the Evans phase of generally covariant unified field theory [3-10], and in electrodynamics the Pancharatnam phase [20] has similar characteristics and origin.

As discussed in reference [15] any vector parallel transported in a closed path produces the Berry phase. This procedure in the Evans unified field theory is an outcome of general relativity as argued already. As described in Ref. [15], the rotation angle of the Berry phase is related to the integral of the curvature of the surface bounded by the loops. This is precisely what is shown by the Evans phase in the forms (11.4), (11.14), or (11.28). The curvature of the helix is

$$R = \kappa^2, \tag{11.29}$$

and this realization gives rise [10,18] to the Sagnac effect in light traversing a circular path, a circle whose circumference is the same as the arc length of the helix drawn out from the circle as argued already.

So the Tomita-Chiao and Sagnac effects are essentially the same [18], and both are geometrical/topological phases originating in the Evans phase of general relativity, as does the Berry phase and Pancharatnam phase and all effects in physics in which these phases are observed [15]. All these effects therefore serve as further experimental verification of the Evans generally covariant unified field theory [3-10]. The Evans phase law is more fundamental than the Berry phase law because the former self-consistently describes both the dynamical phase of physical optics (observed for example in ordinarily reflection and interferometry) is an example of the Evans phase where the wave number κ in Eq. (11.14) is a property of the radiation itself (the electromagnetic field). The Berry phase is an example of the Evans phase where the wavenumber is an inverse distance or inverse periodic length or wavelength defined by the wavelength (or scalar curvature R) of spacetime itself as required by general relativity. In the Tomita-Chiao effect the set-up is an optical fiber wound into a helix. Similarly, the area in Eq. (11.14) is either a property of the radiation (dynamical phase from Evans phase) or of the set up (geometrical/topological phase from Evans phase). The Pancharatnam phase [20] can similarly be derived from the Evans phase by considering a closed loop in spacetime in given optical configurations. The Foucault pendulum is similarly an outcome of the Evans phase law applied to dynamics, as is the Sagnac and AB effects, and indeed, all effects in optics. The dynamical phase in optics always measures the $\boldsymbol{B}^{(3)}$, or Evans-Vigier, field [18,19].

The Evans phase law is intrinsically gauge invariant [10] because it always uses covariant derivatives. Self-consistently, the Berry phase is also gauge invariant for the same reason: It is a geometrical phase [15]. States in quantum mechanics are wavefunctions, and also acquire a Berry phase in general [15]. This well-known result is now understandable through the Evans phase and principle of least curvature [9], which unifies the Hamilton principle of least action and the Fermat principle of least time, giving rise to wave mechanics [9] and to the Schrödinger equation in the quantum weak field limit of the Evans unified field theory. The Berry phase in quantum mechanics becomes part of the Evans phase, a part that is observable experimentally in given configurations [1,2].

These inferences illustrate the fact that the generally covariant unified field theory [3-10] is a powerful and general theory of radiated and matter fields, and of geometrical and phases and related effects.

Acknowledgments

The Ted Annis Foundation, Craddock Inc., and ADAS are thanked for funding, and members of staff of AIAS and others for many interesting discussions.

References

1. A Tomita and R. Y. Chiao, *Phys. Rev. Lett.* **57**, 937 (1986).
2. T. W. Barrett in A. Lakhtakia, ed., *Essays on the Formal Aspects of Electromagnetic Theory* (World Scientific, Singapore, 1992).
3. M. W. Evans, *Found. Phys. Lett.* **16**, 367 (2003).
4. M. W. Evans, *Found. Phys. Lett.* **16**, 507 (2003).
5. M. W. Evans, *Found. Phys. Lett.* **17**, 25 (2004).
6. M. W. Evans, "Derivation of Dirac's equation from the Evans wave equation," *Found. Phys. Lett.* **17**(2) (2004), in press; preprint on www.aias.us.
7. M. W. Evans, "Unification of the gravitational and strong nuclear fields," *Found. Phys. Lett.* (2004), in press; preprint on www.aias.us
8. M. W. Evans, "The Evans lemma of differential geometry," *Found. Phys. Lett.*, submitted for publication; preprint on www.aias.us; "Derivation of the Evans wave equation from the Lagrangian and action, origin of the Planck constant in generally relativity," *Found. Phys. Lett.*, in press; preprint on www.aias.us; "Development of the Evans wave equation in the weak-field limit, the electrogravitic equation," *Found. Phys. Lett.*, in press; preprint on www.aias.us.
9. L. Felker, ed., *The Evans Equation* (World Scientific, Singapore, 2004, in preparation).
10. M. W. Evans, "Physical optics, the Sagnac effect and the Aharonov-Bohm effect in the Evans unified field theory," *Found. Phys. Lett.*, in press; preprint on www.aias.us.
11. M. W. Evans, "Generally covariant field and wave equations for gravitation and quantized gravitation in terms of the metric four vector, *Found. Phys. Lett.*, submitted for publication, preprint on www.aias.us.
12. S. M. Carroll, *Lecture Notes on General Relativity* (University of California, Santa Barbara, graduate course), arXiv:gr-qe/9712019 vl 3 Dec., 1997.
13. L. H. Ryder, *Quantum Field Theory*, 2nd edn. (University Press, Cambridge, 1996).
14. M. V. Berry, *Proc. Roy. Soc. A* **392**, 54 (1984).
15. www.mi.infn.it/manini/berryphase.
16. E. M. Milewski, ed., *Vector Analysis Problem Solver* (Research and Education Association, New York, 1987).
17. J. H. Hannay, *J. Phys. A* **18**, 221 (1985).

18. M. W. Evans, "O(3) electrodynamics," in M. W. Evans, ed., *Modern Nonlinear Optics*, a special topical issue of I. Prigogine and S. A. Rice, series eds., *Advances in Chemical Physics*, Vol. 119 (2), 2nd and e-book editions (Wiley Interscience, New York, 2001).
19. M. W. Evans, J.-P. Vigier, *et al.*, *The Enigmatic Photon* (Kluwer/Academic, Dordrecht, 1994 to 2003, hardback and paperback) in 10 volumes. M. W. Evans and L. B. Crowell, *Classical and Quantum Electrodynamics and the* $\boldsymbol{B}^{(3)}$ *Field* (World Scientific, Singapore, 2001).
20. T. W. Barrett and D. M. Grimes, *Advanced Electromagnetism* (World Scientific, Singapore, 1996).

12

Derivation Of The Lorentz Boost From The Evans Wave Equation

Summary. The Lorentz boost is derived from the Evans wave equation of generally covariant unified field theory by constructing the Dirac spinor from the tetrad in the SU(2) representation space of non-Euclidean spacetime. The Dirac equation in its wave formulation is then deduced as a well-defined limit of the Evans wave equation. By factorizing the d'Alembertian operator into Dirac matrices the Dirac equation in its original first differential form is obtained from the Evans wave equation. Finally the Lorentz boost is deduced from the Dirac equation using geometrical arguments. A self consistency check of the Evans wave equation is therefore forged by deducing therefrom the Lorentz boost in the appropriate limit. This procedure demonstrates that the Evans wave equation governs the properties of matter and anti-matter in general relativity and unified field theory, and leads both to Fermi-Dirac and Bose-Einstein statistics in general relativity.

Key words: Lorentz boost, Evans wave equation, generally covariant unified field theory.

12.1 Introduction

General relativity reduces to special relativity when one frame of reference moves at a constant velocity with respect to the other. This well defined limit is known as the Lorentz boost [1,2]. It follows that the recently derived Evans wave equation of generally covariant unified field theory [3-15] must self-consistently and non-trivially reduce to the Lorentz boost, while also suggesting experimentally measurable developments such as the Evans spin field, $\boldsymbol{B}^{(3)}$, observed in the magnetization of matter by circularly or elliptically polarized electromagnetic radiation - the inverse Faraday effect [16]. The Lorentz boost must be derivable analytically from the structure of Evans' generally covariant unified field theory, and therefore the derivation serves as one of many checks available 3-15on the self consistency of the Evans theory. The Lorentz boost or transformation was originally devised by Lorentz following a suggestion by Fitzgerald in response to the crisis in physics posed by the

Michelson-Morley experiment. It was developed by Lorentz and Poincaré and others in electrodynamics, and then by Einstein in 1905 in his famous theory of special relativity. Ten years later, in 1915, general relativity was proposed by Einstein in the form accepted today, and independently and slightly earlier by Hilbert. The theory of 1915 was as much of a revolutionary advance over its predecessor of 1905 as the latter was over earlier physics. The 1905 special relativity merged space and time, and the 1915 general relativity geometrized natural philosophy using the mathematical methods of Riemann devised in the nineteenth century. Relativity all depends on the Lorentz boost and the slightly earlier Lorentz-Fitzgerald contraction of about 1896.

Einstein was never satisfied that his theory was complete, because the 1915 theory of general relativity is confined to gravitation and spacetime with curvature but no torsion [2] - Riemannian spacetime described with the Christoffel connection. The first generally covariant unified field theory was devised by Evans in 2003 [5-15] using differential geometry to describe spacetime with both curvature and torsion [2]. The Evans theory has been tested experimentally and for analytical self consistency in many ways [5-15] and predicts from spacetime torsion the existence of the fundamental and experimentally well observed Evans spin field of electrodynamics, the $\boldsymbol{B}^{(3)}$ field. Tiny (nanotesla) $\boldsymbol{B}^{(3)}$ fields were first observed by Pershan *et al.* [17], shortly after the laser first became available, as magnetization of matter with circularly polarized electromagnetic radiation at visible frequencies. The existence of $\boldsymbol{B}^{(3)}$ was first inferred in 1992 by Evans [18] from these and other experimental data such as the magnetization of plasma observed by Deschamps *et al.* [19] with circularly polarized 3.0 GHz radiation. It is now possible to observe a $\boldsymbol{B}^{(3)}$ field of tens of tesla (i.e. megagauss) magnitude in under dense plasma [20]. It is also known now that in a correctly covariant electrodynamical theory (part of unified field theory) the $\boldsymbol{B}^{(3)}$ field is responsible for all physical optics, and so is a commonplace observable of everyday experience [3-15]. The revolutionary inference of $\boldsymbol{B}^{(3)}$ in electrodynamics gradually led to the broader and deeper inference [3-15] of the generally covariant unified field theory sought after by Einstein.

The fundamental concept in the Evans theory is the tetrad [2-15], which has a well defined meaning in differential geometry [2], a theory which shows that the two differential forms that define the gauge invariant fields of nature are the Riemann (curvature) and torsion forms. These are defined respectively by the second and first Maurer Cartan structure relations and in the Evans theory both are synthesized self-consistently and in an entirely original manner [3-15] from the tetrad form of differential geometry. The tetrad (a vector valued one form) is the potential field, the Riemann form (a tensor valued two form) is the gauge invariant gravitational field, and the torsion form (a vector valued two form) is the gauge invariant electromagnetic field, whose fundamental spin invariant is the well observed Evans spin field $\boldsymbol{B}^{(3)}$. The theory is generally covariant because it is a fully geometrized theory of general relativity, and it is unified because different types of radiated and matter fields all spring from one source, the tetrad.

In Section 12.2 the tetrad is defined in SU(2) representation space as the 2 x 2 transformation matrix between two Pauli spinors. One spinor (a two component column vector) is defined in the Euclidean orthonormal space [2] of the tetrad, and is labelled R and L, and the other in the non Euclidean base manifold, and is labelled 1 and 2. The tetrad defined in this way has four scalar components (arranged in a 2 x 2 matrix) which also define by simple transposition the Dirac spinor (a four component column vector) in the appropriate special relativistic limit, a limit in which we recover the Dirac wave equation from the Evans wave equation. The labels R and L in the orthonormal space indicate the existence of spin, torsion or handedness and the existence of anti-particles. They are therefore left and right handed spin labels, and originate in the fact that the tetrad must always be defined as a matrix linking two spaces in such as way as to create spin as well as mass in general relativity. The fundamental reason for this is that there are two Casimir invariants of the Einstein group (and also the Poincaré group): the mass and spin invariants [1,3-15] which in differential geometry are defined respectively by the Riemann and torsion forms, i.e by the second and first Maurer-Cartan structure relations [2-15] respectively. Only one of these appears in the original 1915 theory, the mass invariant, and then without recognition of the fact that the symmetric metric tensor used by Einstein is the dot product of two, more fundamental, tetrads. The Evans $\boldsymbol{B}^{(3)}$ field (not inferred until 1992) is now recognized as the fundamental spin invariant of generally covariant electrodynamics, missing entirely from both the Einstein theory of general relativity and the Maxwell-Heaviside theory of special relativity. The fundamental concept of spin, or torsion, is therefore missing entirely from the original 1915 theory of gravitation devised independently by Einstein and Hilbert. Recognition of spin and manifestations thereof such as $\boldsymbol{B}^{(3)}$ leads to the unified field theory of Evans [3-15]. The geometrical spin properties inherent in the tetrad are enough to lead to the existence of anti-particles in the Evans unified field theory, which is therefore a powerful predictive theory based on contemporary differential geometry [2-15]. The labels 1 and 2 in the base manifold are the labels of a Pauli spinor in the SU(2) representation space of the non-Euclidean base manifold. So the Dirac spinor is given meaning in general relativity as a causal property of spacetime, and not as a probability. This interpretation leads straightforwardly to an acceptable Klein-Gordon equation [3-15], the quantized version of the Einstein equation of free particle relativistic momentum. On the other hand the probabilistic interpretation of the Copenhagen School leads in the received opinion [1] to the "abandonment" of the Klein-Gordon equation and a great deal of obscurity and confusion. It is clear now from the Evans unified field theory that it is the obscure ontology of the Copenhagen School that should be abandoned, not the Klein Gordon equation. This is one of the broader philosophical consequences of the Evans theory in natural philosophy and clearly shows what Einstein and others meant when they rejected the Copenhagen quantum mechanics as incomplete. The Evans unified field theory shows that wave mechanics is a direct result of the tetrad postulate [2], the Evans Lemma and Evans wave and field equations [3-15]. The Evans theory (and copious experimental evidence [3-15] for the theory) shows clearly that wave mechanics is deducible from general relativity and is causal in nature.

In Section 12.3 the original, first differential, form of the Dirac equation is deduced analytically in the special relativistic limit by factorizing the d'Alembertian operator into Dirac matrices [1], making use of the metric tensor in the special relativistic limit.

Finally in Section 12.4 the Lorentz boost is deduced straightforwardly from the Dirac equation using hyperbolic half angles formulae, and the original Lorentz boost matrix recovered geometrically, as required, from the Evans wave equation. These procedures serve as a cross check on the Evans theory, give considerable insight into the meaning of the Dirac spinor in non-Euclidean spacetime and lead to many philosophical ramifications.

12.2 Derivation Of The Dirac Spinor And The Dirac Wave Equation From Evans' Theory

The Evans wave equation is

$$(\square + kT)q^a{}_\mu = 0 \tag{12.1}$$

and is the identity

$$D^\mu D_\mu q^a{}_\nu = 0 \tag{12.2}$$

based on the tetrad postulate

$$D_\mu q^a{}_\nu = 0. \tag{12.3}$$

Eq. (12.1) is based on the Evans Lemma:

$$\square q^a{}_\mu = R q^a{}_\mu, \tag{12.4}$$

which is the purely geometrical result:

$$D^\mu D_\mu = \square - R, \tag{12.5}$$

where R is scalar curvature. In everywhere flat spacetime

$$R \longrightarrow 0, \tag{12.6}$$

where

$$\square := \frac{1}{c^2}\frac{\partial^2}{\partial t^2} - \nabla^2, \tag{12.7}$$

is the d'Alembertian operator and where:

$$R = -kT. \tag{12.8}$$

Eq. (12.8) follows from the Evans field equation:

$$(R + kT)q^a{}_\mu = 0 \tag{12.9}$$

and applies for all radiated and matter fields, not only gravitation.

In order to derive the Dirac wave equation from Eq. (12.1) use an SU(2) representation space and define the tetrad by:

$$\begin{pmatrix} \zeta^R \\ \zeta^L \end{pmatrix} = \begin{pmatrix} \phi^R{}_1 & \phi^R{}_2 \\ \phi^L{}_1 & \phi^L{}_2 \end{pmatrix} \begin{pmatrix} \zeta^1 \\ \zeta^2 \end{pmatrix} \tag{12.10}$$

Eq. (12.10) may be written more concisely as:

$$\zeta^a = q^a{}_\mu \zeta^\mu, \tag{12.11}$$

where

$$\zeta^a = \begin{pmatrix} \zeta^R \\ \zeta^L \end{pmatrix}, \zeta^\mu = \begin{pmatrix} \zeta^1 \\ \zeta^2 \end{pmatrix}, q^a{}_\mu = \begin{pmatrix} \phi^R{}_1 & \phi^R{}_2 \\ \phi^L{}_1 & \phi^L{}_2 \end{pmatrix}. \tag{12.12}$$

Here ζ^μ is a Pauli spinor in the base manifold, and ζ^a is a Pauli spinor in the Euclidean orthonormal spacetime used to define the tetrad [2]. The latter is the 2 x 2 matrix defined by:

$$q^a{}_\mu = \begin{pmatrix} \phi^{RT} \\ \phi^{LT} \end{pmatrix}, \tag{12.13}$$

where

$$\phi^{RT} = (\phi^R{}_1 \phi^R{}_2), \tag{12.14}$$

$$\phi^{LT} = (\phi^L{}_1 \phi^L{}_2). \tag{12.15}$$

are row vectors transposed from column vectors. Therefore we may define the column four vector:

$$\psi = \begin{pmatrix} \phi^R \\ \phi^L \end{pmatrix}, \quad \phi^R = \begin{pmatrix} \phi^R{}_1 \\ \phi^L{}_2 \end{pmatrix}, \quad \phi^L = \begin{pmatrix} \phi^L{}_1 \\ \phi^R{}_2 \end{pmatrix}, \tag{12.16}$$

which obeys the Evans wave equation in the form:

$$(\Box + kT)\psi = 0. \tag{12.17}$$

Eq. (12.17) is equivalent to the Evans wave equation in the tetrad form:

$$(\Box + kT)q^a{}_\mu = 0. \tag{12.18}$$

Both equations (12.17) and (12.18) lead to the same set of simultaneous equations:

$$\begin{aligned} (\Box + kT^R{}_1)\phi^R{}_1 &= 0, \\ &\vdots \\ (\Box + kT^L{}_2)\phi^L{}_2 &= 0. \end{aligned} \tag{12.19}$$

Now use the principle that general relativity must reduce to special relativity when one frame moves at a constant velocity with respect to the other. In this limit the well known wave equations of special relativistic quantum mechanics must be recovered from the Evans wave equation. This principle implies [3-15]:

$$kT \longrightarrow 1/\lambda^2{}_C \tag{12.20}$$

where

$$\lambda_C := \hbar/mc \tag{12.21}$$

is the Compton wavelength of any matter or radiated field.

In this limit the Evans equation (12.1) becomes the Dirac wave equation [1]:

$$(\Box + m^2c^2/\hbar^2)\psi = 0. \tag{12.22}$$

The Dirac spinor is therefore recognized as the limiting form of a tetrad whose two rows have been transposed into column two vectors. The Dirac spinor is therefore a geometrical object, not a probability as in the Copenhagen interpretation. In the limit (12.20) there exists a non-zero least or minimum curvature

$$R_0 = -(mc/\hbar)^2 \tag{12.23}$$

that defines mass through the Evans principle of least curvature [3-15]. The total curvature:

$$R := \Box - R_0 = 0 \tag{12.24}$$

vanishes. Each of Eqns. (12.19) is a Klein Gordon equation, whose classical limit must be the Einstein equation of relativistic momentum:

$$p^\mu p_\mu = E^2/c^2 - p^2 = m^2c^2, \tag{12.25}$$

where E is the total relativistic kinetic energy, p is the relativistic momentum and E_0 the rest energy:

$$E_0 = mc^2. \tag{12.26}$$

The fact that Eq. (12.25) must be the classical limit of the Klein-Gordon equation implies the operator equivalence of quantum mechanics:

$$\Box = -(1/\hbar^2)p^\mu p_\mu, \tag{12.27}$$

which we have therefore deduced from the Evans unified field theory. Eq (12.25) is another form [21] of

$$p = \gamma m v. \tag{12.28}$$

In the non-relativistic limit

$$p = mv, \qquad T = \frac{1}{2} m v^2, \tag{12.29}$$

which are the Newton equations for momentum and kinetic energy of a free particle.

The famous equations of dynamics are therefore a consequence of the Evans principle of least curvature [3-15]. The total curvature vanishes in the Klein-Gordon equation and Dirac equation, but the individual components $\square$ and R_0 do not vanish. This result means that mass is a form of curvature and is defined for any elementary particle (including the neutrino and photon) by the least curvature (12.23). The Dirac spinor is defined in terms of the tetrad, and so the Dirac spinor introduces spin into the definition of a particle, as discussed in Section 12.1. Spin is introduced geometrically through the definition (12.10), a definition which implies that the famous half integral spin is spacetime torsion in SU(2) representation space. Eq. (12.10) means that each elementary particle is converted by the parity operator into its anti-particle, with the same mass but opposite handedness or helicity. The helicities originate in the Pauli spinor:

$$\zeta^a = \begin{pmatrix} \zeta^R \\ \zeta^L \end{pmatrix}, \tag{12.30}$$

where

$$a = L, R \tag{12.31}$$

are labels of the orthonormal space needed to define the tetrad. Analogously, in O(3) electrodynamics, these spin labels become:

$$a = (1), (2), (3) \tag{12.32}$$

of the complex circular basis[3-15], indicating three states of spin for the photon, (the transverse (1) and (2) and the longitudinal (3)), three sets of field equations in the appropriate limit, and giving the $\boldsymbol{B}^{(3)}$ field from general relativity.

These spin states are missing from Einstein's generally covariant theory of gravitation but are present in the Evans wave and field equations of generally covariant unified field theory.

The spin states are also missing from Einstein's theory of special relativity and Newtonian dynamics. They are observed experimentally however in numerous ways, for example the anomalous Zeeman effect, atomic and molecular spectra. ESR, NMR, MRI, the existence of anti-particles, in Fermi Dirac statistics, and so on. The Evans theory shows that these properties of nature are due to spacetime geometry in a particular limit, Eq. (12.20). The spin states are interconverted by the parity operator:

$$\begin{pmatrix} \zeta^R \\ \zeta^L \end{pmatrix} \longrightarrow \begin{pmatrix} \zeta^L \\ \zeta^R \end{pmatrix} . \tag{12.33}$$

Eq. (12.23) gives the important additional insight that mass vanishes in an everywhere flat spacetime. If mass vanishes, then so does rest energy and relativistic momentum, showing that the total relativistic kinetic energy E also vanishes in an everywhere flat spacetime. Therefore it makes no sense to assert the existence of a particle without mass, because such as particle would not have any kinetic energy and would not exist experimentally. Therefore there can be no massless neutrino and no massless photon. This is now known to be a consequence of the least curvature principle of Evans [3-15] and also a consequence of the Evans wave equation. In the appropriate limit the later reduces to the correctly covariant form of the Proca equation [1,3-15] for the photon with mass. Einstein's special relativistic theory is not therefore a theory in which spacetime is everywhere flat, because in such a spacetime there can be no mass, energy, spin, charge density and current density anywhere in the universe. For every physical theory therefore, the operator $D_\mu D_\mu$ never becomes $\Box$ identically. This statement is an expression of the Evans Lemma.

12.3 Derivation Of The Dirac Equation As A First-Order Differential Equation

Dirac originally inferred his famous equation as a first order differential equation. In this section we deduce this from the Dirac wave equation (12.22), which is a limit of the Evans wave equation as demonstrated already. The starting point is:

$$\partial^\mu = g^{\mu\nu} \partial_\nu, \tag{12.34}$$

where $g^{\mu\nu}$ is the Minkowskian metric tensor in the special relativistic limit [1]. Using Eq. (12.34) the d'Alembertian operator is:

$$\Box = \partial^\mu \partial_\mu = g^{\mu\nu} \partial_\nu \partial_\mu . \tag{12.35}$$

The metric is now factorized into the anticommutator of 4 x 4 γ^μ matrices [1]:

$$g^{\mu\nu} = \gamma^\mu \gamma^\nu = \frac{1}{2}\{\gamma^\mu, \gamma^\nu\} \tag{12.36}$$

so Eq. (12.22) becomes:

$$\left(\gamma^\mu\gamma^\nu\partial_\nu\partial_\mu + (mc/\hbar)^2\right)\psi = 0. \tag{12.37}$$

Now factorize the operator:

$$(i\gamma^\nu\partial_\nu + mc/\hbar)(i\gamma^\mu\partial_\mu - mc/\hbar) = -\gamma^\mu\gamma^\nu\partial_\mu\partial_\nu - (mc/\hbar)^2. \tag{12.38}$$

It follows that:

$$(i\gamma^\nu\partial_\nu + mc/\hbar)(i\gamma^\mu\partial_\mu - mc/\hbar)\psi = 0 \tag{12.39}$$

and that:

$$(i\gamma^\mu\partial_\mu - mc/\hbar)\psi = 0. \tag{12.40}$$

Finally use the operator equivalence:

$$p_\mu = i\hbar\partial_\mu \tag{12.41}$$

to obtain:

$$(\gamma^\mu p_\mu - mc)\psi = 0. \tag{12.42}$$

Eq. (12.40) is the Dirac equation in differential form, i.e. in representation space [1], and Eq. (12.42) is the Dirac equation in momentum space.

The γ^μ matrices are identified as the 4 x 4 Dirac matrices:

$$\gamma^0 = \begin{pmatrix} 0 & 1 \\ 1 & 0 \end{pmatrix}, \quad \gamma^i = \begin{pmatrix} 0 & -\sigma^i \\ \sigma^i & 0 \end{pmatrix}, \tag{12.43}$$

where:

$$\sigma^1 = \begin{pmatrix} 0 & 1 \\ 1 & 0 \end{pmatrix}, \quad \sigma^2 = \begin{pmatrix} 0 & -i \\ i & 0 \end{pmatrix}, \quad \sigma^3 = \begin{pmatrix} 1 & 0 \\ 0 & -1 \end{pmatrix} \tag{12.44}$$

are the Pauli matrices.

Recall that:

$$\psi = \begin{pmatrix} \phi^R \\ \phi^L \end{pmatrix} = \begin{pmatrix} {\phi^R}_1 \\ {\phi^R}_2 \\ {\phi^L}_1 \\ {\phi^L}_2 \end{pmatrix}. \tag{12.45}$$

Using Eq. (12.43) to (12.45) in Eq. (12.42) gives:

$$mc^2\phi^R = (E + \boldsymbol{\sigma}\cdot\boldsymbol{p}c)\phi^L, \tag{12.46}$$

$$mc^2\phi^L = (E - \boldsymbol{\sigma}\cdot\boldsymbol{p}c)\phi^R, \tag{12.47}$$

where Eq. (12.47) is the parity inverted Eq. (12.46). The Dirac equation is therefore a relation between the Pauli spinors ϕ^R and ϕ^L, which in turn are derived from the tetrad, the eigenfunction of the Evans wave equation and Evans Lemma

12.4 Derivation Of The Lorentz Boost

The Dirac equations (12.46) and (12.47) may be written in the simple form:

$$\phi^R = e^{\boldsymbol{\sigma}\cdot\boldsymbol{\theta}}\phi^L, \tag{12.48}$$

$$\phi^L = e^{-\boldsymbol{\sigma}\cdot\boldsymbol{\theta}}\phi^R, \tag{12.49}$$

where:

$$e^{\boldsymbol{\sigma}\cdot\boldsymbol{\theta}} = \cos h\theta + \boldsymbol{\sigma}\cdot\boldsymbol{n}\sin h\theta, \tag{12.50}$$

$$e^{-\boldsymbol{\sigma}\cdot\boldsymbol{\theta}} = \cos h\theta - \boldsymbol{\sigma}\cdot\boldsymbol{n}\sin h\theta. \tag{12.51}$$

Here:

$$\cos h\theta = E/mc^2 = \gamma \tag{12.52}$$

$$\sin h\theta = pc/mc^2 = \beta\gamma \tag{12.53}$$

The Einstein equation (12.25) follows with these definitions and the hyperbolic angle formula:

$$\cos h^2\theta - \sin h^2\theta = 1. \tag{12.54}$$

The parameters β and γ are defined by:

$$\gamma^2(1-\beta^2) = 1, \tag{12.55}$$

$$\beta = v/c, \qquad \gamma = (1 - v^2/c^2)^{-1/2}, \tag{12.56}$$

where v is the constant velocity of one frame with respect to another in special relativity.

Now write the Dirac equations as:

$$\phi^R = e^{(\boldsymbol{\sigma}\cdot\boldsymbol{\theta})/2}\left(e^{(\boldsymbol{\sigma}\cdot\boldsymbol{\theta})/2}\phi^L\right), \tag{12.57}$$

$$\phi^L = e^{-(\boldsymbol{\sigma}\cdot\boldsymbol{\theta})/2}\left(e^{-(\boldsymbol{\sigma}\cdot\boldsymbol{\theta})/2}\phi^R\right), \tag{12.58}$$

and define:

$$\phi^R(p) = e^{(\boldsymbol{\sigma}\cdot\boldsymbol{\theta})/2}\phi^R(0), \tag{12.59}$$

$$\phi^L(p) = e^{-(\boldsymbol{\sigma}\cdot\boldsymbol{\theta})/2}\phi^L(0), \tag{12.60}$$

to obtain the Lorentz boost [1]:

$$\psi(p) = \begin{pmatrix} e^{\boldsymbol{\sigma}\cdot\boldsymbol{\theta}/2} & 0 \\ 0 & e^{-\boldsymbol{\sigma}\cdot\boldsymbol{\theta}/2} \end{pmatrix}\psi(0). \tag{12.61}$$

The Lorentz boost is therefore a transformation between the Dirac spinors $\phi(p)$ and $\phi(0)$, and the Dirac equation is:

$$\psi(p) = \begin{pmatrix} e^{\boldsymbol{\sigma}\cdot\boldsymbol{\theta}} & 0 \\ 0 & e^{-\boldsymbol{\sigma}\cdot\boldsymbol{\theta}} \end{pmatrix} \psi^{**}(p). \tag{12.62}$$

where:

$$\psi^{**} = \hat{P}(\psi) \tag{12.63}$$

is the parity inverted Dirac spinor.

From Eqs. (12.61) and (12.62), it is seen that there is a simple geometrical relation between the Dirac equation and the Lorentz boost [1], one may be constructed from the other using hyperbolic half angle formulae. It follows that the Lorentz boost is a well-defined geometrical limit of the Evans wave equation, which is what we set out to prove.

Finally recognise that [1]:

$$\boldsymbol{K} = \pm i\boldsymbol{\sigma}/2, \tag{12.64}$$

where:

$$K_z = \frac{1}{i}\frac{\partial B}{\partial \theta}\bigg|_{\theta=0} \tag{12.65}$$

and where

$$B = \begin{pmatrix} \cos h\theta & \sin h\theta & 0 & 0 \\ \sin h\theta & \cos h\theta & 0 & 0 \\ 0 & 0 & 1 & 0 \\ 0 & 0 & 0 & 1 \end{pmatrix} \tag{12.66}$$

is the famous Lorentz boost matrix. By recognizing that the parameter θ of the Lorentz boost also appears in the Dirac equation, it can be seen that the Lorentz boost is derivable from the Evans wave equation. The unified field theory reduces to the most important and earliest inference of all relativity theory, the Lorentz-Fitzgerald contraction.

Acknowledgments

The Ted Annis Foundation, Applied Science Associates, Craddock Inc., and ADAS are thanked for generous funding. The staff of AIAS is thanked for many interesting discussions.

References

1. L. H. Ryder, *Quantum Field Theory*, (Cambridge Univ. Press, 1996, 2nd. Ed.).
2. S. M. Caroll, *Lecture Notes in General Relativity*, a graduate course in Univ. California, Santa Barbara, arXiv:gr-gq/9712019 v1 3 Dec 1997, available from author.
3. M. W. Evans, A Unified Field Equation for Gravitation and Electromagnetism, *Found. Phys. Lett.*, **16**, 367 (2003).
4. M. W. Evans, A Generally Covariant Wave Equation for Grand Unified Field Theory, *Found. Phys. Lett.*, **16**, 507 (2003).
5. M. W. Evans, The Equations of Grand Unified Field Theory in terms of the Maurer Cartan Structure Relations of Differential Geometry, *Found. Phys. Lett.*, **17**, 25 (2004).
6. M. W. Evans, Derivation of Diracs Equation from the Evans Wave Equation, *Found. Phys. Lett.*, **17**, 149 (2004).
7. M. W. Evans, Unification of the Gravitational and Strong Nuclear Fields, *Found. Phys. Lett.*, **17**, 267 (2004).
8. The Evans Lemma of Differential Geometry, *Found. Phys. Lett.*, **17**, 2004, in press, preprint on www.aias.us.
9. Derivation of the Evans Wave Equation from the Lagrangian and Action, Origin of the Planck Constant in General Relativity, *Found. Phys. Lett.*, **17**, (2004), in press, preprint on www.aias.us.
10. M. W. Evans and AIAS Author Group, Development of the Evans Equation in te Weak Field Limit, the Electrogravitic Equation, *Found. Phys. Lett.*, **17** (2004), in press.
11. M. W. Evans, Physical Optics, the Sagnac Effect, and the Aharonov Bohm Effect in the Evans Unified Field Theory, *Found. Phys. Lett.*, **17**, 301 (August 2004), preprint on www.aias.us.
12. M. W. Evans, Derivation of the Geometrical Phase from the Evans Phase Law of Generally Covariant Unified Field Theory, *Found. Phys. Lett.*, **17**, 393 (August 2004), preprint on www.aias.us.
13. M. W. Evans, New Concepts from the Evans Unified Field Theory, Part One: The Evolution of Curvature, Oscillatory Universe without Singularity, Causal Quantum Mechanics and General Force an Field Equations, *Found. Phys. Lett.*, submitted, preprint on www.aias.us.

14. Ibid. Part Two, "Derivation of the Heisenberg Equation and Replacement of the Heisenberg Uncertainty Principle", *Found. Phys. Lett.*, submitted, preprint in www.aias.us.
15. L. Felker (ed.), *The Evans Equations of Unified Field Theory*, (World Scientific, Singapore, in prep, 2004/2005); *The Collected Scientific Papers of Myron Wyn Evans*, (World Scientific, Singapore, from 2004), volumes one and two, the years 2000 to 2004; M. W. Evans (ed.), *Modern Nonlinear Optics*, a special topical issue in three parts of I. Prigogine and AS. A. Rice (series eds., *Advances in chemical Physics* (Wiley Interscience, New York, 2001, 2nd and e book editions), vol. 119(1) to 119(3); ibid. first edition, vol. 85(1) to 85(3) (1992, 1993 and 1997, hardbacj and softback editions); M. W. Evans and L. B; Crowell, *Classical and Quantum Electrodynamics and the* $\boldsymbol{B}^{(3)}$ *Field*, (World Scientific, 2001); M. W. Evans et al., *The Enigmatic Photon* (Kluwer Dordrecht, 1994 to 2003, hardback and softback editions), five volumes; M. W. Evans and A. A. Hasanein, *The Photomagneton in Quantum Field Theory* (World Scientific, Singapore, 1994).
16. J. P. van der Ziel, P. Pershan and L. D. Malmstrom, *Phys. Rev. Lett.*, **15**, 190 (1965)..
17. P. S. Pershan, J. P. van der Ziel, and L. D. Malmstrom, *Phys. Rev.*, bd 143, 574 (1966).
18. M. W. Evans, *Physica B*, **1**82, 227, 237 (1992).
19. J. Deschamps, M. Fitaire and M. Lagoutte, *Phys. Rev. Lett.*, **25**, 1330 (1970).
20. M. Tatarikis et al., "Measurement of the Inverse Faraday Effect in High Intensity Laser produced Plasmas" (CFL Annual Report Rutherford Appleton Laboratory, 1998./ 1999, m.tatarikis@ic.ac.uk).
21. J. B. Marion and S. T. Thornton, *Classical Dynamics of Particles and Systems*, (HBJ, New York, 1988, 3rd. Ed.).

13

The Electromagnetic Sector Of The Evans Field Theory

Summary. The equations of the electromagnetic sector of the Evans field theory are given in terms of differential geometry and are based on the well-known structure relations and Bianchi identities. The equations thus complete Einstein's basic axiom, that physics is derived from geometry, and extend the axiom to electrodynamics. Precise tests are suggested for the theory using the interaction of circularly polarized electromagnetic radiation with a non-relativistic electron beam. These tests include; the inverse Faraday effect (IFE), radiatively induced fermion resonance (RFR), and the electromagnetic Aharonov-Bohm (EMAB) effect.

Key words: The Evans field theory, electromagnetic sector, tests of the Evans theory, inverse Faraday effect, radiatively induced fermion resonance, electromagnetic Aharonov-Bohm effect.

13.1 Introduction

Recently the first successful generally covariant unified field theory has been developed and can be applied to all radiated and matter fields [1-20]. In Sec. 2 of this paper we give the equations of the electromagnetic sector of the theory and show that they are closely based on well known equations of differential geometry: the structure relations and Bianchi identities [21]. In Sec. 3, high precision tests of the theory are proposed in a non-relativistic electron beam interacting with circularly polarised electromagnetic radiation: the inverse Faraday effect (IFE), radiatively induced fermion resonance (RFR), and the electromagnetically induced Aharonov Bohm (EMAB) effect.

13.2 Equations Of The Electromagnetic Sector Of Evans' Field Theory

In this section we will give the equations of the electromagnetic sector of the Evans theory [1-20] in a form which is easily recognisable as standard differential geometry. The latter is transformed into Evans electromagnetic field theory using the *ansatz* [1-20]

$$A^a{}_\mu = A^{(0)} q^a{}_\mu . \tag{13.1}$$

Equation (13.1) defines the electromagnetic potential field $A^a{}_\mu$ as a vector-valued tetrad one-form within a C negative factor $A^{(0)}$. The tetrad $q^a{}_\mu$ [1-21] is the gravitational potential field, so the *ansatz* (13.1) illustrates field unification in the simplest way possible. In order to introduce the notation, we first give the older Maxwell-Heaviside (MH) field equations of special relativistic electrodynamics. In this way the advantages of, and the new information contained in, the new theory can be seen the most clearly.

The MH equations are expressed most eloquently in terms of differential forms [21,22] as follows:

$$F = d \wedge A, \tag{13.2}$$

$$d \wedge F = 0, \tag{13.3}$$

$$d \wedge \widetilde{F} = \mu_0 J. \tag{13.4}$$

Here A is the scalar-valued potential one-form, $d\wedge$ the exterior derivative of differential geometry, F the scalar-valued two-form that encapsulates the gauge-invariant electromagnetic field, $\widetilde{F}$ the dual of F [21,22], and J the scalar-valued three-form that encapsulates the charge current density of MH field theory. Here μ_0 is the permeability in vacuum, and SI units [1-20] have been used.

Equation (13.3) is a Jacobi identity [22], the most concise form of the homogeneous field equation MH field theory. The homogeneous field equation summarizes Gauss's law and Faraday's law of induction. These two laws are therefore geometrical identities within a C negative factor that turns the geometry into field equations of physics. This inference gives us a clue as to the geometrical origin of electrodynamics and leads us eventually to the Evans field equations given later in this section. Equation (13.4) is the inhomogeneous field equation of MH field theory but in this theory cannot be clearly recognised as an identity of geometry. The inhomogeneous field equation contains Coulomb's and the Ampère-Maxwell law. The gauge invariance of MH field theory follows from the Poincaré lemma [22]

$$d \wedge d = 0. \tag{13.5}$$

The electromagnetic two-form is gauge invariant because

$$d \wedge (d \wedge A) = 0, \tag{13.6}$$

and so F is known as the gauge invariant electromagnetic two-form. Therefore, in the concise and elegant notation of differential geometry, gauge invariance is simple to comprehend, it is the consequence of another geometrical identity, the Poincaré lemma. This suggests that the inhomogeneous field equation should also be a geometrical identity, and we will show later in this section that this is indeed the case in the Evans field theory of general relativity, but not in the MH field theory of special relativity. In the latter theory the inhomogeneous field equation is empirical, i.e., based on data and not inferred from geometry, as required by the Einsteinian principles of general relativity [23].

For clarity of exposition and as an introduction to the Evans field equations of electrodynamics, we give the well-known [1-22] tensor and vector forms Eqs. (13.2) to (13.4) as follows:

The tensor form of Eq. (13.2) is

$$F_{\mu\nu} = -F_{\nu\mu} = \partial_\mu A_\nu - \partial_\nu A_\mu \tag{13.7}$$

and defines the anti-symmetric field tensor $F_{\mu\nu}$ in terms of the potential field A_μ, a covariant four-vector. Equation (13.7) can also be written in contravariant form:

$$F^{\mu\nu} = -F^{\nu\mu} = \partial^\mu A^\nu - \partial^\nu A^\mu \tag{13.8}$$

The Jacobi identity (13.3) in tensor notation is

$$\partial^\lambda F^{\mu\nu} + \partial^\mu F^{\nu\lambda} + \partial^\nu F^{\lambda\nu} = 0. \tag{13.9}$$

On defining the dual tensor

$$\widetilde{F}^{\mu\nu} = \frac{1}{2}\epsilon^{\mu\nu\rho\sigma}F_{\rho\sigma}, \tag{13.10}$$

the Jacobi identity (13.9) becomes the well-known homogeneous field equation of MH field theory:

$$\partial_\mu \widetilde{F}^{\mu\nu} = 0. \tag{13.11}$$

Equation (13.11) is Eq. (13.3) in tensor notation.

In defining the dual tensor the totally antisymmetric unit tensor In four dimensions, $\epsilon^{\mu\nu\rho\sigma}$, has been used as usual. This tensor is the same for any non-Euclidean spacetime [21], but in the MH field theory is used in flat or Minkowski spacetime [1-23].

The inhomogeneous field equation in tensor notation is, in standard SI units,

$$\partial_\mu F^{\mu\nu} = \mu_0 j^\nu, \tag{13.12}$$

where j^ν is the charge current density four-vector. Equation (13.12) is the tensor equivalent of the form equation (13.4).

The homogeneous field equation (13.11) is equivalent to the following two vector equations:

$$\boldsymbol{\nabla} \cdot \boldsymbol{B} = 0, \tag{13.13}$$

$$\boldsymbol{\nabla} \times \boldsymbol{E} + \frac{\partial \boldsymbol{B}}{\partial t} = \boldsymbol{0}, \tag{13.14}$$

respectively Gauss's law and Faraday's law of induction, where $\boldsymbol{B}^{(3)}$ is magnetic flux density and $\boldsymbol{E}$ is electric field strength. Here c is the speed of light in vacuum. The inhomogeneous field equation (13.12) is equivalent to the following two vector equations:

$$\nabla \cdot \boldsymbol{D} = \rho, \tag{13.15}$$

$$\nabla \times \boldsymbol{H} = \boldsymbol{J} + \frac{\partial \boldsymbol{D}}{\partial t}, \tag{13.16}$$

respectively Coulomb's law and the Ampère-Maxwell law. Here $\boldsymbol{H}$ is the magnetic field strength, $\boldsymbol{D}$ the electric displacement, ρ the charge density, and $\boldsymbol{J}$ the current density.

The great MH field theory, which these well-known equations describe, has held sway in electrodynamics for over a hundred years, but is not a theory of general relativity. The theory is the archetypical theory of special relativity, in which spacetime is Minkowski spacetime, sometimes referred to as flat spacetime. It is well-known that special relativity evolved in 1905 from these same equations.

The Evans field theory infers electrodynamics from the twisting of spacetime and makes electrodynamics a theory of general relativity. Therefore, the Evans theory is the direct logical outcome of the Einsteinian principle that all theories of physics are theories of general relativity, i.e., all of physics is based on geometry. The most eloquent expression of geometry is differential geometry [1,22], which is valid for all spacetimes, and is more general [21] than the Riemannian geometry available to Einstein, a geometry in which the torsion tensor is assumed to be zero when constructing the relation between the Christoffel symbol and the symmetric metric tensor. In particular, differential geometry allows for the existence of the torsion tensor [21] as well as the Riemann or curvature tensor, and this fact has been utilized in the development of Evans field theory [1-20] to show that gravitation is the curving of spacetime, electromagnetism is the twisting of spacetime. The torsion tensor becomes the vector-valued torsion two-form $R^{a}{}_{b\mu\nu}$ in differential geometry. One of the great advantages of differential geometry is that it is valid for all possible spacetimes, so we can drop the indices $\mu\nu$ and simplify the notation [21], thus emphasising the most important principles of the geometry and thence the physics. In other words, we find all the laws of physics from the laws of geometry and take Einstein seriously. The torsion form becomes T^a and the Riemann form becomes $R^a{}_b$. The second major advantage is that differential geometry is developed in terms of the vector-valued tetrad one form $q^a{}_\mu$ which is more fundamental than the metric tensor $g_{\mu\nu}$ used by Einstein because

$$g_{\mu\nu} = q^{a}{}_{\mu} q^{b}{}_{\nu} \eta_{ab}. \tag{13.17}$$

In other words, the metric tensor is defined as the dot product of two tetrads, so the tetrad factorizes the metric tensor of the base manifold (non-Euclidean spacetime). Here η_{ab} is the diagonal metric tensor

$$\eta_{ab} = \eta^{ab} = \begin{pmatrix} 1 & 0 & 0 & 0 \\ 0 & -1 & 0 & 0 \\ 0 & 0 & -1 & 0 \\ 0 & 0 & 0 & -1 \end{pmatrix} \tag{13.18}$$

of the orthonormal spacetime of the tetrad, a flat or Minkowski spacetime. We may therefore express all the important equations in the flat, orthonormal, spacetime, remembering that each equation applies for the unwritten subscript $\mu\nu$ [21] of the non-Euclidean base manifold. In this way we find the equations of generally covariant unified field theory [1-20] by following the differential geometry. Similarly, Einstein discovered the equations of gravitation by following Riemannian geometry.

The field equations of electromagnetism in the Evans theory are found from the following fundamental and well-known equations of differential geometry. The first Maurer-Cartan structure relation [1-21] defines the vector-valued torsion two-form as the covariant exterior derivative (denoted by $D\wedge$) of the tetrad

$$T^a = D \wedge q^a = d \wedge q^a + \omega^a{}_b \wedge q^b, \tag{13.19}$$

where $\omega^a{}_b$ is the spin connection [21]. Using the *ansatz* (13.1) and multiplying both sides of Eq. (13.19) by $A^{(0)}$, we obtain

$$F^a = D \wedge A^a \tag{13.20}$$

$$= A^{(0)} T^a. \tag{13.21}$$

This is the equation defining the gauge-invariant electromagnetic field F^a in terms of the potential in the Evans theory and is the generally covariant form of Eq. (13.2) of the MH field theory of special relativity. By comparison of Eqs. (13.20) and (13.2), it is seen that $d\wedge$ has been replaced by $D\wedge$, A by A^a, and F by F^a. The potential and the field have each developed an *internal index* of the type normally associated with Yang-Mills gauge field theory [22]. We are now able to see, furthermore, that index is the direct result of differential geometry. Thus A^a is the potential field of the electromagnetic sector of Evans field theory, and F^a is the gauge-invariant field. In tensor notation, A_μ becomes $A^a{}_\mu$, a covariant four-vector with an internal index, and $F_{\mu\nu}$ becomes $F^a{}_{\mu\nu}$, an antisymmetric tensor with an internal index. In Eq. (13.20), the SI units of F^a, A^a, and $D\wedge$ are tesla $=$ JsC^{-1}m^{-2}, tesla m, and m^{-1}, respectively. If $a = (1), (2), (3)$ of the complex circular basis, then we obtain the field equations of O(3) electrodynamics [1-20]. The latter have been extensively developed theoretically and tested experimentally. In Sec. 3, we summarize three further tests so we have derived O(3) electrodynamics from differential geometry using the *ansatz* (13.1). In so doing, we have tested the *ansatz* experimentally, because O(3) electrodynamics has been tested experimentally [1-20].

The homogeneous and inhomogeneous field equations of the Evans theory are found from the first and second Bianchi identities of differential geometry [21]:

$$D \wedge R^a{}_b = 0, \tag{13.22}$$

$$D \wedge T^a = R^a{}_b \wedge q^b. \tag{13.23}$$

In terms of the spin connection, Eq. (13.22) becomes

$$d \wedge R^a{}_b + \omega^a{}_c \wedge R^c{}_b - R^a{}_c \wedge \omega^c{}_b = 0 \tag{13.24}$$

and is a generalization of the well-known Bianchi identity of Riemannian geometry:

$$D_\rho R^\kappa{}_{\lambda\mu\nu} + D_\mu R^\kappa{}_{\lambda\nu\rho} + D_\nu R^\kappa{}_{\lambda\rho\mu} = 0. \tag{13.25}$$

In terms of the spin connection, Eq. (13.23) reads

$$d \wedge T^a + \omega^a{}_b \wedge T^b = R^a{}_b \wedge q^b \tag{13.26}$$

and is a generalization of the following identity in Riemannian geometry:

$$R_{\rho\sigma\mu\nu} + R_{\rho\mu\nu\sigma} + R_{\rho\nu\sigma\mu} = 0. \tag{13.27}$$

Equation (13.27) is equivalent to the following fundamental symmetries of the Riemann tensor:

$$R_{\rho\sigma\mu\nu} = -R_{\sigma\rho\mu\nu}, \tag{13.28}$$

$$R_{\rho\sigma\mu\nu} = -R_{\rho\sigma\nu\mu}. \tag{13.29}$$

The structure of the electromagnetic sector of the Evans theory is therefore summarized as follows:

$$\boxed{\begin{aligned} F &= d \wedge A, \\ d \wedge F &= 0, \\ d \wedge \widetilde{F} &= \mu_0 J. \end{aligned}} \longrightarrow \boxed{\begin{aligned} T^a &= D \wedge q^a, \\ D \wedge T^a &= R^a{}_b \wedge q^b, \\ D \wedge R^a{}_b &= 0. \end{aligned}}$$

The homogeneous field equation of the electromagnetic sector of the Evans unified field theory is

$$D \wedge F^a = R^a{}_b \wedge A^b = A^{(0)} R^a{}_b \wedge q^b = \mu_0 j^a. \tag{13.30}$$

When the electromagnetic and gravitational fields decouple:

$$D \wedge F^a = 0, \tag{13.31}$$

$$R^a{}_b \wedge q^b = 0; \tag{13.32}$$

and, in the MH limit,

$$D \wedge F^a \rightarrow d \wedge F = 0. \tag{13.33}$$

When the gravitational field influences the electromagnetic field or vice versa, the unified field is governed by the equation

$$D \wedge F^a = A^{(0)} R^a{}_b \wedge q^b \neq 0, \tag{13.34}$$

and the homogeneous field equation of the MH theory, Eq. (13.33), becomes the homogeneous field equation of the Evans unified field theory, Eq. (13.34).

Equation (13.31) is the intermediate step towards the completed unified field theory; and, when $a(1),(2),(3)$, Eq. (13.31) becomes the homogeneous field equation of O(3) electrodynamics [17]. In vector notation, Eq. (13.31) becomes the following six equations:

$$\boldsymbol{\nabla} \cdot \boldsymbol{B}^{(1)} = 0, \tag{13.35}$$

$$\boldsymbol{\nabla} \cdot \boldsymbol{B}^{(2)} = 0, \tag{13.36}$$

$$\boldsymbol{\nabla} \cdot \boldsymbol{B}^{(3)} = 0, \tag{13.37}$$

$$\boldsymbol{\nabla} \cdot \boldsymbol{E}^{(1)} + \partial \boldsymbol{B}^{(1)} / \partial t = 0, \tag{13.38}$$

$$\boldsymbol{\nabla} \cdot \boldsymbol{E}^{(2)} + \partial \boldsymbol{B}^{(2)} / \partial t = 0, \tag{13.39}$$

$$\partial \boldsymbol{B}^{(3)} / \partial t = 0. \tag{13.40}$$

In order to search experimentally for the influence of the gravitational field on the electromagnetic field, it is necessary to search for departures from Gauss's law and Faraday's law of induction. Equation (13.34) is obtained from Eq. (13.33) by symmetry building ("inverse symmetry breaking"), as summarized in the following diagram: Similarly the inhomogeneous field equation of

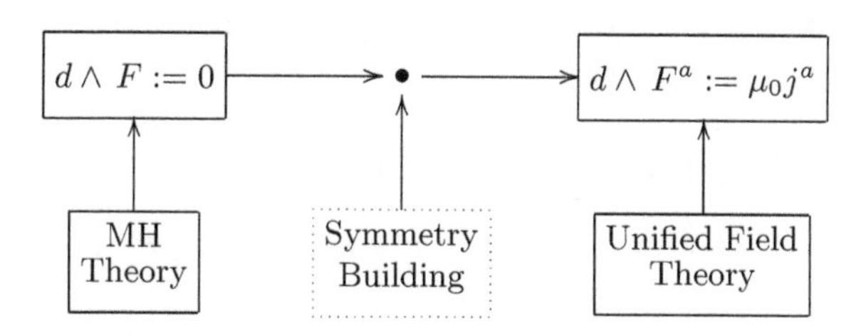

Fig. 13.1. Homogeneous Unified Field Theory

the unified field theory is obtained by symmetry building as follows: The MH field equations in this case are the Coulomb and the Ampère-Maxwell laws. So, to look for the effects of gravitation on electromagnetism, it is necessary to search experimentally for departures from these laws.

In tensor notation, the electromagnetic field in the Evans unified field theory is

$$F^a{}_\mu = A^{(0)} T^a{}_\mu = A^{(0)} T^\kappa{}_{\mu\nu} \, q^a{}_\kappa, \tag{13.41}$$

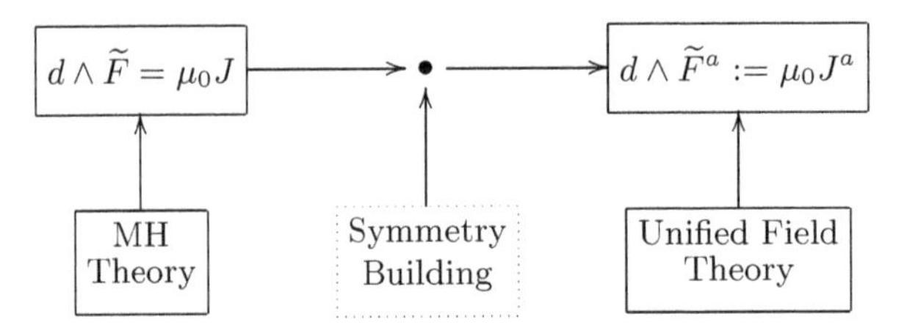

Fig. 13.2. Inhomogeneous Unified Field Theory

where $T^{\kappa}{}_{\mu\nu}$ is the torsion tensor. Therefore, the inhomogeneous field equation of the unified field theory is

$$\partial^{\mu} F^{a}{}_{\mu\nu} = \mu_0 J^{a}{}_{\nu}, \tag{13.42}$$

where the charge current density is defined by

$$J^{a}{}_{\nu} := \frac{A^{(0)}}{\mu_0} q^{a}{}_{\kappa} D^{\mu} \left(\Gamma^{\kappa}{}_{\mu\nu} - \Gamma^{\kappa}{}_{\nu\mu} \right). \tag{13.43}$$

In differential-form notation, Eq. (13.42) reads

$$d \wedge \widetilde{F}^a = \mu_0 J^a, \tag{13.44}$$

and so the electromagnetic sector of the Evans unified field theory is described by

$$d \wedge F^a = \mu_0 j^a, \tag{13.45}$$

$$d \wedge \widetilde{F}^a = \mu_0 J^a. \tag{13.46}$$

The unified field theory evolves from the original MH field theory by building symmetry as follows:

$$\boxed{\begin{aligned} d \wedge F &= 0, \\ d \wedge \widetilde{F} &= \mu_0 J. \end{aligned}} \longrightarrow \boxed{\begin{aligned} d \wedge F^a &= \mu_0 j^a, \\ d \wedge \widetilde{F}^a &= \mu_0 J^a. \end{aligned}}$$

The symmetry of the unified field equation is that of the Einstein group with sixteen irreducible representations. This is a higher symmetry than that of O(3) electrodynamics, the intermediate theory in which the electromagnetic and gravitational fields are still decoupled, and O(3) electrodynamics has in turn a higher symmetry than U(1) electrodynamics, the MH symmetry. The experimental evidence for O(3) electrodynamics is summarized in Ref. [17].

In tensor notation, Eq. (13.23) reads

$$\begin{aligned} &\partial_{\mu} T^{a}{}_{\nu\rho} + \omega^{a}{}_{\mu b} T^{b}{}_{\nu\rho} + \partial_{\nu} T^{a}{}_{\rho\mu} + \omega^{a}{}_{\nu b} T^{b}{}_{\rho\mu} \\ &+ \partial_{\rho} T^{a}{}_{\mu\nu} + \omega^{a}{}_{\rho b} T^{b}{}_{\mu\nu} = R^{a}{}_{\mu\nu\rho} + R^{a}{}_{\nu\rho\mu} + R^{a}{}_{\rho\mu\nu} \end{aligned} \tag{13.47}$$

and is a balance of identities of differential geometry [21]. On the right-hand side is the Bianchi identity used in gravitational general relativity [27], and the left-hand side is the covariant exterior derivative of the torsion form. In tensor notation, the homogeneous field equation of the unified field theory is obtained directly from Eqs. (13.47) and (13.1) and reads

$$\begin{aligned} &\partial_\mu F^a{}_{\nu\rho} + \omega^a{}_{\mu b} F^b{}_{\nu\rho} + \partial_\nu F^a{}_{\rho\mu} + \omega^a{}_{\nu b} F^b{}_{\rho\mu} \\ &+ \partial_\rho F^a{}_{\mu\nu} + \omega^a{}_{\rho b} F^b{}_{\mu\nu} = A^{(0)} \left(R^a{}_{\mu\nu\rho} + R^a{}_{\nu\rho\mu} + R^a{}_{\rho\mu\nu} \right) \end{aligned} \tag{13.48}$$

Equation (13.48) governs the unified gravitational and electromagnetic fields. When the two fields are unified they interact and mutually influence each other. This process is governed by

$$R^a{}_{\mu\nu\rho} + R^a{}_{\nu\rho\mu} + R^a{}_{\rho\mu\nu} \neq 0. \tag{13.49}$$

When there is no influence of gravitation on electromagnetism both the left-hand and right-hand sides of Eq. (13.48) vanish identically

$$d \wedge F^a = 0. \tag{13.50}$$

$$R^a{}_b \wedge q^b = \omega^a{}_b \wedge T^b \tag{13.51}$$

Eq. (13.50) is the familiar Bianchi identity of the gravitational field uninfluenced by the electromagnetic field, and Eq. (13.51) is the homogeneous field equation of the electromagnetic field uninfluenced by the gravitational field.

The electromagnetic field is therefore the spinning of spacetime described by the torsion tensor and the gravitational field is the curving of spacetime described by the curvature tensor. Differential geometry is needed to build the symmetry of the unified field theory and in differential geometry spinning is described by the torsion form and curvature by the curvature form. The homogeneous field equation of the unified field is an identity, Eq. (13.48), of differential geometry, which is valid for a spacetime in which there is both spin and curvature. Such an identity is not available in Riemann geometry when the Christoffel connection is symmetric in its lower two indices. This is the type of spacetime used in the Einstein theory of gravitation, and in this spacetime Eq. (13.50) always holds. So such a spacetime cannot be used to unify gravitation and electromagnetism. The Maxwell-Heaviside field theory is philosophically different from the Einstein field theory in that the electromagnetic field is superimposed on a flat (Minkowski) spacetime in which the spin connection vanishes. In the Evans unified field theory this fundamental philosophical difference (which prevents field unification in the standard model and string theory) is bridged by recognizing that the electromagnetic field is the torsion form of differential geometry, so electromagnetism is the spinning of spacetime *itself.* This spinning is in general influenced by the curving of spacetime through Eqs. (13.47) or (13.48). We therefore obtain field unification by symmetry building. The intermediate stage in this process, O(3) electrodynamics, has been tested extensively against experimental data [17].

13.3 Experimental Tests Of The Evans Theory

It has been established and well accepted [1-20] that the Evans field theory infers the inverse Faraday effect through the Evans spin field

$$B^{(3)} = -igA^{(1)} \wedge A^{(2)}, \tag{13.52}$$

which is dual to torsion two-form of spacetime. The inverse Faraday effect therefore originates in the torsion or twisting motion of spacetime. In order to test this hypothesis accurately it is suggested that a high precision inverse Faraday effect calibrating experiment first be carried out as follows in a non-relativistic electron beam.

The inverse Faraday effect is the orbital angular momentum imparted to the electron beam by the Evans spin field. For a non-relativistic electron beam, the energy exchanged during the process is [1-20]

$$En_{IFE} = \frac{e\hbar}{2m}B^{(3)} = \frac{e^2A^{(0)2}}{2m} = \frac{p^2}{2m}. \tag{13.53}$$

Here m is the mass of the electron,

$$m = 9.10953 \times 10^{-31}kg, \tag{13.54}$$

and $p = eA^{(0)}$ its linear momentum. Therefore, the energy imparted to the electron beam by the circularly polarized electromagnetic beam is the Newtonian kinetic energy $p^2/2m$. The energy in Eq. (13.53) may be written as [1-20]

$$En_{IFE} = \frac{e^2A^{(0)2}}{2m} = \left(\frac{e^2\mu_0 c}{2m}\right)\frac{I}{\omega^2}, \tag{13.55}$$

where I is the beam power density (Wm^{-2}) and ω is the beam angular frequency (rads^{-1}). The energy exchanged is characteristically dependent on the power density and the inverse square of the angular frequency.

The orbital angular momentum J imparted to the electron beam can be calculated from the fundamental relation between energy and angular momentum,

$$En = \frac{1}{2}\omega J, \tag{13.56}$$

and so

$$J = \left(\frac{e^2\mu_0 c}{2m}\right)\frac{I}{\omega^3}. \tag{13.57}$$

From Eq. (13.57) the magnetic flux density is tesla (1.0 T = 10^4G) imparted to the non-relativistic electron beam may be calculated [1-20] to be

$$B^{(3)}_{\text{sample}} = \mu_0 M^{(3)}_{\text{sample}} = \frac{N}{V}\mu_0\frac{e}{2m}J = \frac{N}{V}\left(\frac{\mu_0^2 e^3 c}{2m^2}\right)\frac{I}{\omega^3}. \tag{13.58}$$

Here $M^{(3)}$ is the magnetization produced in the inverse Faraday effect for N electrons in a sample volume V; see Eq. (F5), p. 208, *The Enigmatic Photon, Vol. 3* (Ref. 15). For a pulsed Nd YaG laser with $I = 5.5 \times 10^{12} Wm^{-2}, \omega = 1.77 \times 10^{16}$ rads^{-1} and $N/V = 10^{26}$ electrons m^{-3} (Avogadro's number), we find

$$B^{(3)}_{\text{sample}} \sim 10^{-9} T = 10^{-5} G, \tag{13.59}$$

an order of magnitude observed experimentally in liquids, solids and gases, for example. The inverse Faraday effect has been observed many times [1-20] in a variety of materials, and solidly indicates the existence of the Evans spin field $\boldsymbol{B}^{(3)}$. However a high precision, baseline experiment in a non-relativistic electron beam has yet to be carried out.

Having carried out this baseline experiment, it is then necessary to develop radiatively induced fermion resonance (RFR), which originates in the spin angular momentum imparted to the electron beam by the Evans spin field [1-20]:

$$\sigma_0 En_{RFR} = \frac{e\hbar}{2m} B^{(3)} \sigma_Z = \frac{e^2 A^{(0)2}}{2m} \sigma_Z = \frac{p^2 \sigma_Z}{2m}. \tag{13.60}$$

Here σ_Z is the third Pauli matrix and σ_0 the zero-order Pauli matrix. It is seen from Eq. (13.60) that RFR originates in the Newtonian kinetic energy multiplied by σ_Z. Electron spin resonance between the states of σ_Z gives rise to RFR as follows:

$$En = 2\pi\hbar f_{\text{res}} = \frac{e^2 A^{(0)}}{2m} [1 - (-1)] = \left(\frac{e^2 \mu_0 c}{m} \right) \frac{I}{\omega^2}, \tag{13.61}$$

so the resonance frequency in hertz is

$$f_{\text{res}} = \left(\frac{e^2 \mu_0 c}{2\pi\hbar m} \right) \frac{I}{\omega^2} = 1.007 \times 10^{28} \frac{I}{\omega^2}. \tag{13.62}$$

It displays the characteristic I/ω^2 dependence of the IFE (Eq. (13.55)). This property of RFR means that it can produce fermion resonance at high resolution without magnets, for example electron spin resonance, nuclear magnetic resonance and magnetic resonance imaging without permanent magnets [1-20]. The magnet is replaced by a circularly polarized radio frequency beam.

Thirdly, it is necessary to carry out an accurate test of EMAB by replacing the magnetized iron whisker of the original Chambers experiment by a circularly polarized radio frequency beam. The EMAB is caused by the $\boldsymbol{B}^{(3)}$ of the radio frequency field and detected as a shift in the fringe pattern of two interfering electron beams:

$$\Delta x = \frac{L}{d} \frac{\lambda}{2\pi} \Delta\delta \tag{13.63}$$

in a Young interferometer. Here L is the distance between the plates of the interferometer, d, the distance between the openings in the first plate, and λ

the electron wavelength. In the original Chambers experiment used to detect the Aharonov-Bohm effect:

$$\Delta\delta = \frac{e}{\hbar} BAr = \frac{e}{\hbar}\Phi, \tag{13.64}$$

where B is the magnetic flux density produced by the iron whisker and Ar the area enclosed by the electron beams. This shift was first observed by Chambers in 1959 following the original theory by Aharanov and Bohm. This has recently been understood in terms of the Evans phase law [1-20]:

$$\Delta\phi = \exp\left(i\frac{e}{\hbar} \int \boldsymbol{B}^{(3)} \cdot \boldsymbol{k} dAr \right), \tag{13.65}$$

which follows from the fact that any magnetic field must be defined by

$$B = D \wedge A = d \wedge A + gA \wedge A, \tag{13.66}$$

where $g = e/\hbar$. In the complex circular basis, Eq. (13.66) becomes

$$B^{(3)} = -igA^{(1)} \wedge A^{(2)}. \tag{13.67}$$

The conventional magnetic field is described by Eq. (13.66), when B is the magnetic flux density of the iron whisker [1-20]. So the Evans field theory gives a consistent description both of the conventional and electromagnetic Aharonov-Bohm effects in terms of the covariant derivative (13.66).

In vector notation,

$$\boldsymbol{B}^{(3)} = -ig\boldsymbol{A}^{(1)} \times \boldsymbol{A}^{(2)}, \tag{13.68}$$

and the magnitude of the field is

$$B^{(3)} = |\boldsymbol{B}^{(3)}| = \frac{e}{\hbar} A^{(0)2}, \tag{13.69}$$

where

$$A^{(0)2} = \mu_0 cI/\omega^2, \tag{13.70}$$

showing again the characteristic I/ω^2 dependence. In SI units:

$$\begin{aligned} \mu_0 &= 4\pi \times 10^{-7}\ Js^2\ C^{-2}m^{-1}, c = 2.997925 \times 10^8\ ms^{-1}, \\ e &= 1.60219 \times 10^{-19}C, \hbar = 1.05459 \times 10^{-34}\ Js, \end{aligned} \tag{13.71}$$

and

$$B^{(3)} = \left(\frac{e\mu_0 c}{\hbar}\right) \frac{I}{\omega^2}. \tag{13.72}$$

From Eqs. (13.64) and (13.72):

$$\frac{\Delta\delta}{Ar} = \left(\frac{e^2\mu_0 c}{\hbar^2}\right) \frac{I}{\omega^2} = 8.6954 \times 10^{34} \frac{I}{\omega^2}, \tag{13.73}$$

which is the phase shift per unit area enclosed by the electron beams due to a circularly polarized radio frequency field replacing the iron whisker in the Chambers experiment. This is the electromagnetic Aharanov-Bohm (EMAB) effect.

The equivalent of Eq. (13.74) for the magnetic Aharonov-Bohm effect is

$$\frac{\Delta\delta}{Ar} = \frac{e}{\hbar}B \tag{13.74}$$

for a magnetic field 1G ($= 10^{-4}$T), and the phase shift per unit area from Eq. (13.74) is $1.519 \times 10^{11}\, m^{-2}$.

For an electromagnetic field of 1.0 G Hz angular frequency and 10^{-4} Wm^{-2} power density (1 watt per cm^2) the phase shift per unit area from Eq. (13.73) is $8.695 \times 10^{12} m^{-2}$. Under such conditions, the EMAB effect should be observable using a modification of the Chambers experiment. However, if a Nd YaG laser is used at the same power density but at angular frequency $1.77 \times 10^{16}\mathrm{rads}^{-1}$, the phase shift per unit area becomes $0.013\, m^{-2}$ and unobservable. An EMAB effect might however be observable with a Nd YaG laser if the latter is pulsed to very high power densities.

The observation of the IFE in this way would be important for fundamental physics, the observation of the RFR would open up a new industry, and the observation of the EMAB effect would lead to several new radar technologies.

Acknowledgments

Craddock Inc., Applied Science Associates, and the Ted Annis Foundation are thanked for funding.

References

1. M. W. Evans, "Derivation of O(3) electrodynamics from generally covariant unified field theory," *Found. Phys. Lett.*submitted; preprint on www.aias.us.
2. M. W. Evans, *Found. Phys. Lett.***16**, 367 (2003).
3. M. W. Evans, *Found. Phys. Lett.***16**, 507 (2003).
4. M. W. Evans, *Found. Phys. Lett.***17**, 25 (2004).
5. M. W. Evans, *Found. Phys. Lett.***17**, 149 (2004).
6. M. W. Evans, *Found. Phys. Lett.***17**, 267 (2004).
7. M. W. Evans, *Found. Phys. Lett.***17**, 301 (2004).
8. M. W. Evans, *Found. Phys. Lett.***17**, 393 (2004).
9. M. W. Evans, "The Evans lemma of differential geometry, *Found. Phys. Lett.***17**, in press (2004); preprint on www.aias.us.
10. M. W. Evans, "Derivation of the Evans wave equation from the Lagrangian and action. Origin of the Planck constant in general relativity," *Found. Phys. Lett.* **17**, 535 (2004); preprint on www.aias.us.
11. M. W. Evans and the AIAS Author Group, "Development of the Evans wave equations in the weak-field limit," *Found. Phys. Lett.***17**, 497 (2004); preprint on www.aias.us.
12. M. W. Evans, "New concepts from the Evans unified field theory. Part One: The evolution of curvature, oscillatory universe without singularity, causal quantum mechanics, and general force and field equations," *Found. Phys. Lett.*, submitted (2004); preprint on www.aias.us.
13. M. W. Evans, "New concepts from the Evans unified field theory. Part Two: Derivation of the Heisenberg equation and replacement of the Heisenberg uncertainty principle," *Found. Phys. Lett.*, submitted (2004); preprint on www.aias.us.
14. M. W. Evans, "Derivation of the Lorentz-boost from the Evans wave equation," *Found. Phys. Lett.*, in press (2004).
15. M. W. Evans *et al.*, *The Enigmatic Photon, Vols. 1-5* (Kluwer Academic, Dordrecht, 1994 to 2002, hardback and softback).
16. M. W. Evans, *Generally Covariant Unified Field Theory: The Geometrization of Physics* (in press, 2005). M. W. Evans and L. Felker, *The Evans Equations of Unified Field Theory* (World Scientific, Singapore, 2004, in preparation).
17. M. W. Evans, ed., *Modern Non-Linear Optics*, a special topics issue of I. Prigogine and S. A. Rice, series eds., *Advances in Chemical Physics* (Wiley-Interscience, New York, 2001, 2nd and e-book edns.), Vols. 119(1) to 119(3); *ibid.*, 1st edn., Vol. 85.

18. M. W. Evans and L. B. Crowell, *Classical and Quantum Electrodynamics and the $\boldsymbol{B}^{(3)}$ Field* (World Scientific, Singapore, 2001).
19. M. W. Evans and A. A. Hasanein, *The Photomagneton in Quantum Field Theory* (World Scientific, Singapore, 1994).
20. M. W. Evans, *Physica B* **182**, 227, 237 (1992).
21. S. P. Carroll, *Lecture Notes in General Relativity* (University of California, Santa Barbara, graduate course), arXiv:gr-gq/9712019 vl 3 Dec 1997).
22. L. H. Ryder, *Quantum Field Theory*, 2nd edn. (University Press, Cambridge, 1996).
23. A. Einstein, *The Meaning of Relativity* (Princeton University Press, 1921).

14

New Concepts From The Evans Unified Field Theory. Part One: The Evolution Of Curvature, Ocillatory Universe Without Singularity, Causal Quantum Mechanics, And General Force And Field Equations

Summary. The Evans field equation is solved to give the equations governing the evolution of scalar curvature R and contracted energy-momentum T. These equations show that R and T are always analytical, oscillatory, functions without singularity and apply to all radiated and matter fields from the sub-atomic to the cosmological level. One of the implications is that all radiated and matter fields are both causal and quantized, contrary to the Heisenberg uncertainty principle. The wave equations governing this quantization are deduced from the Evans field equation. Another is that the universe is oscillatory without singularity, contrary to contemporary opinion based on singularity theorems. The Evans field equation is more fundamental than, and leads to, the Einstein field equation as a particular example, and so modifies and generalizes the contemporary Big Bang model. The general force and conservation equations of radiated and matter fields are deduced systematically from the Evans field equation. These include the field equations of electrodynamics, dark matter, and the unified or hybrid field.

Key words: Evans field equation, equations of R, oscillatory universe, general field and force equations, causal quantization.

14.1 Introduction

The Evans unified field theory [1-10] is based on the well-known geometrical concept of the tetrad [11], but uses and develops the tetrad, in several novel ways. All physics is reduced essentially to the tetrads, and thus to geometry. Unification of the radiated and matter fields of nature is achieved by the synthesis of new equations which are all based on the properties of the tetrad in differential geometry. The basic field equation of nature in this theory is the Evans field equation [1-10]

$$Rq^a{}_\mu = -kTq^a{}_\mu, \tag{14.1}$$

which can be developed systematically in many directions. In Eq. (1), R is scalar curvature [1-12], T the contracted energy-momentum tensor, and k the Einstein constant. The tetrad is denoted by $q^a{}_\mu$ and in differential geometry is a vector-valued one-form [11]. In the Evans unified field theory the tetrad is the potential field and is also governed by the Evans wave equation [1-10]

$$\Box q^a{}_\mu = -kTq^a{}_\mu, \tag{14.2}$$

derived from the well-known tetrad postulate [11]:

$$D^\nu q^a{}_\mu = 0. \tag{14.3}$$

Figure (1) is a schematic of how the Evans field equation reduces to known equations of physics, and Fig. (2) is a similar schematic for the Evans wave equation.

In this paper, the first of a series dealing with new concepts from the Evans unified field theory, it is shown that the Evans field equation, when combined with the tetrad postulate, produces the following equations for the evoluation of R and T:

$$\frac{1}{R}\partial^\mu R = \pm R\partial^\mu\left(\frac{1}{R}\right), \tag{14.4}$$

$$\frac{1}{T}\partial^\mu T = \pm T\partial^\mu\left(\frac{1}{T}\right). \tag{14.5}$$

This is shown in Sec. 2. In Sec. 3 the structure of the most general gauge field and force equations of nature is deduced systematically from Eq. (1). The well-known Einstein field equation for gravitation is one example, out of several new possible structures or classes, of equations to emerge from the Evans field equation. If we accept general relativity, it is therefore likely that these new equations contain a great deal of hitherto undiscovered and unexplored physics. In other words, we proceed rigorouslly on the basic assumption that all physics is causal and generally covariant, as originally proposed by Einstein [12]. These papers [1-10] can therefore be viewed as completing the theory of general relativity, and as extending it to all radiated and matter fields in nature. This is what is meant by "unified field theory."

14.2 The Evolution Equations Of R And T

The field equation (1) is a balance of the identity

$$D^{\mu}(Rq^{a}{}_{\mu}) = 0 \tag{14.6}$$

of differential geometry and the conservation equation

$$D^{\mu}(Tq^{a}{}_{\mu}) = 0 \tag{14.7}$$

and is therefore a geometrization of physics in terms of the tetrad. The usual form of the Bianchi identity in the general-relativistic theory of gravitation

$$D^{\mu}G_{\mu\nu} = 0, \tag{14.8}$$

$$G_{\mu\nu} = R_{\mu\nu} - \frac{1}{2}Rg_{\mu\nu} \tag{14.9}$$

is a special case of Eq. (6), and the well-known Noether theorem

$$D^{\mu}T_{\mu\nu} = 0 \tag{14.10}$$

is a special case of Eq. (7). The Einstein field equation balances Eqs. (8) and (10) to give

$$R_{\mu\nu} - \frac{1}{2}Rg_{\mu\nu} = kT_{\mu\nu} \tag{14.11}$$

and can be deduced [1-10] as a special case of the Evans field equaton. Here

$$G_{\mu\nu} = R_{\mu\nu} - \frac{1}{2}Rg_{\mu\nu} \tag{14.12}$$

is the well-known Einstein field, where $R_{\mu\nu}$ is the Ricci tensor and $g_{\mu\nu}$ is the symmetric metric tensor of Einstein's original theory [12]. Finally, $T_{\mu\nu}$ is the well-known symmetric canonical energy-momentum tensor.

These familiar Einsteinian tensors are now known to be special cases of more general tetrad matrices of the Evans field theory and are defined by dot products of tetrads [1-11] as follows:

$$R_{\mu\nu} = R^{a}{}_{\mu}q^{b}{}_{\nu}\eta_{ab}, \tag{14.13}$$

$$T_{\mu\nu} = T^{a}{}_{\mu}q^{b}{}_{\nu}\eta_{ab}, \tag{14.14}$$

$$g_{\mu\nu} = q^{a}{}_{\mu}q^{b}{}_{\nu}\eta_{ab}. \tag{14.15}$$

Here η_{ab} is the (diagonal) metric of the Euclidean orthonormal space labelled by the a index of the tetrad [1-11]. The more general field and force equations of nature, introduced systematically in Sec. 3, follow from the fundamental fact that one can define wedge (or cross) and outer products of tetrads as well as dot products. The wedge products give rise to antisymmetric torsion fields such as the electromagnetic field dual to antisymmetric curvature fields such as the Riemann curvature field [1-11], and the outer products gives rise to fields in nature which are hitherto ill-understood or unexplored and must be classified theoretically on the basis of symmetry. This is one purpose of this series of papers. One of these fields may be that of dark matter [13]. Another type of field is the unified or hybrid field which may become observable in the tiny and hitherto ill-understood effects of electromagnetism on gravitation [14,15]. Most generally there exist in the Evans unified field theory symmetric, antisymmetric, and asymmetric fields which originate in the well-known fact [16] that any square (in general asymmetric) matrix can be resolved into the sum of symmetric and antisymmetric components. The geometrization of physics inherent in the basic Evans field equation means that each component has a physical significance, i.e., each component is a type of radiated or matter field in nature, some are known, others are unexplored, but all are understood consistently and are thus unified philosophically.

To understand nature, study geometry: the essence of general relativity.

In order to derive Eqs. (4) and (5), start from the Evans field Eq. (1) and use the following relations:

$$G^a{}_\mu = -\frac{1}{4} R q^a{}_\mu, \tag{14.16}$$

$$T^a{}_\mu = \frac{1}{4} T q^a{}_\mu. \tag{14.17}$$

Equations (16) and (17) are derived from the definitions [1-12] of R and T introduced originally by Einstein [12]:

$$R = g^{\mu\nu} R_{\mu\nu}, \quad T = g^{\mu\nu} T_{\mu\nu}. \tag{14.18}$$

Using the Einstein convention [12]

$$g^{\mu\nu} g_{\mu\nu} = 4 \tag{14.19}$$

and the Cartan convention [11]

$$q^a{}_\mu q_\mu{}^a = 1, \tag{14.20}$$

together with the definitions (13) and (14), we obtain:

$$\begin{aligned} R &= g^{\mu\nu} R_{\mu\nu} \\ &= q_\mu{}^a q^\nu{}_b \eta^{ab} R^a{}_\mu q^b{}_\nu \eta_{ab} \\ &= (\eta^{ab}\eta_{ab})(q^\nu{}_b q^b{}_\nu)(q_\mu{}^a R^a{}_\mu) = 4 q_\mu{}^a R^a{}_\mu. \end{aligned} \tag{14.21}$$

Multiply either side of Eq. (21) by $q^a{}_\mu$ to obtain

$$R^a{}_\mu = \frac{1}{4} R q^a{}_\mu, \tag{14.22}$$

$$G^a{}_\mu = R^a{}_\mu - \frac{1}{2} R q^a{}_\mu = -\frac{1}{4} R q^a{}_\mu, \tag{14.23}$$

which is Eq. (16). Similarly, we obtain Eq. (17).

Substitution of Eqs. (16) and (17) in the Evans field Eq. (1) gives

$$G^a{}_\mu = kT^a{}_\mu. \tag{14.24}$$

We first show as follows that Eq. (24) leads to the Einstein field equation as a particular case by writing Eq. (24) in the form

$$\frac{1}{4} R q^a{}_\mu - \frac{1}{2} R q^a{}_\mu = \frac{1}{4} kT q^a{}_\mu. \tag{14.25}$$

Multiply both sides of Eq. (25) by $q^b{}_\nu \eta_{ab}$ to get the Einstein field Eq. (11). The latter is therefore a structure that is derivable straightforwardly from the more general Evans field Eqs. (1) or (24) by forming dot products of tetrads. The latter reveal the inner or deeper structure of the well-known Einstein field equation. It follows that the most general form of the Bianchi identity of geometry is

$$D^\mu G^a{}_\mu = 0, \tag{14.26}$$

and the most general conservation law of physics is consequently

$$D^\mu T^a{}_\mu = 0. \tag{14.27}$$

Consider now a special case of the tetrad postulate (3):

$$D^\mu q^a{}_\mu = 0, \tag{14.28}$$

a special case which follows from Eq. (3) on using

$$D^\mu q^a{}_\mu = D^0 q^a{}_0 + D^1 q^a{}_1 + D^2 q^a{}_2 + D^3 q^a{}_3 = 0. \tag{14.29}$$

The Evans identity (26) can therefore be developed using Eq. (28) as:

$$D^\mu (R q^a{}_\mu) = R D^\mu q^a{}_\mu + q^a{}_\mu D^\mu R = q^a{}_\mu D^\mu R = 0. \tag{14.30}$$

Similarly the Evans conservation law (27) can be developed as

$$D^\mu (T q^a{}_\mu) = q^a{}_\mu D^\mu T = 0, \tag{14.31}$$

where $R = -kT$ for all radiated and matter fields.

On using the well-known geometrical result [11] that the covariant derivative acting on any scalar quantity is the ordinary derivative:

$$D^\mu R = \partial^\mu R, \quad D^\mu T = \partial^\mu T, \tag{14.32}$$

Equations (30) and (31 become

$$q^a{}_\mu \partial^\mu R = 0, \quad q^a{}_\mu \partial^\mu T = 0, \tag{14.33}$$

and the Evans field equation becomes the identity

$$q^a{}_\mu \partial^\mu (R + kT) = 0, \tag{14.34}$$

where

$$q^a{}_\mu \partial^\mu R = q^a{}_\mu \partial^\mu T = 0. \tag{14.35}$$

Equation (34) is similar in structure to the Evans wave equation (2), but Eq. (34) is an identity because R is $-kT$. The Evans identity (30) shows that $\partial^\mu R$ is orthogonal to the tetrad $q^a{}_\mu$ in the non-Minkowski base manifold indexed μ. Similarly, the Evans conservation law (31) shows that $\partial^\mu T$ is orthogonal to the tetrad. These results give deeper insight into the meaning of the Bianchi identity and the Noether theorem.

From Eq. (22),

$$q^a{}_\mu = \frac{4}{R} R^a{}_\mu, \tag{14.36}$$

and, using Eq. (28),

$$D^\mu \left(\frac{4}{R} R^a{}_\mu \right) = 0, \tag{14.37}$$

i.e.,

$$R^a{}_\mu D^\mu \left(\frac{4}{R} \right) = 0. \tag{14.38}$$

Therefore, Eq. (38) shows that

$$R q^a{}_\mu D^\mu \left(\frac{1}{R} \right) = 0. \tag{14.39}$$

Equations (30) and (39) show that

$$\partial^\mu R \propto \pm R \partial^\mu \left(\frac{1}{R} \right), \tag{14.40}$$

where the coefficient of proportionality must in general be scalar curvature R. Thus we arrive at Eq. (4). Replacing R by $-kT$ gives Eq. (5), and differentiating Eq. (4) leads to

$$\square R = \pm R^2 \square \left(\frac{1}{R} \right). \tag{14.41}$$

Equations (4) and (41) must have analytical solutions in general, i.e., R must be continuously differentiable. If we consider

$$\frac{1}{R} \partial^\mu R = -R \partial^\mu \left(\frac{1}{R} \right) \tag{14.42}$$

or, specifically, the time component

$$\frac{1}{R}\frac{\partial R}{\partial t} = -R\frac{\partial}{\partial t}\left(\frac{1}{R}\right), \tag{14.43}$$

then a solution of Eq. (43) is seen to be

$$R = R_0 e^{i\omega t}, \tag{14.44}$$

with a real part

$$\mathrm{Re}(R) = R_0 \cos \omega t. \tag{14.45}$$

The cosine function is bounded by plus or minus unity and never goes to infinity. Therefore there can no singularity in the scalar curvature R. It follows from Eq. (18) that there is never a singularity in the metric $g_{\mu\nu}$ or Ricci tensor $R_{\mu\nu}$. In other words, the universe evolves without a singularity, and it follows that the well-known singularity theorems built around the Einstein field equation do not have any physical meaning. These singularity theorems are complicated misinterpretations. In other words, general relativity must always be a field theory that is everywhere analytical [17]. Similarly, the older Newton theory must be everywhere analytical. There are no singularities in nature. Equation (45) shows that the universe can contract to a dense state, but then re-expands and re-contracts. Apparently we are currently in a state of evolution where the universe is on the whole expanding. This does not mean that every individual part of the universe is expanding. Some parts may be contracting or may be stable with respect to the laboratory observer.

Equations (4) and (41) suggest that all radiated and matter fields evolve through a wave equation. This inference leads to causal wave mechanics. The Evans wave equation (2) is an equation of wave mechanics in which the eigenfunction is the tetrad. The evolution of the tetrad is causal, in the sense that there is no Heisenberg uncertainty principle in general relativity, and there is no need for such a principle to describe nature. The uncertainty principle is subjective; it essentially asserts that nature is unknowable and that measurement is subjective. Any assertion about the unknowable, however, is inevitably subjective in itself, is unmeasurable by definition, and is therefore outside the domain of natural philosophy. General relativity is fundamentally incompatible with this principle, because general relativity asserts that nature is knowable and objectively measurable, given the equations that govern it. These are now known to be the equations of the Evans unified field theory. Figures (1) and (2) summarize how well-known and tested equations of classical and quantum mechanics emerge [1-10] from the Evans unified field theory. The latter unifies all radiated and matter fields and also unifies general relativity and quantum mechanics. Wave equations may also be constructed in which the eigenfunction is R or T, proving in another way that nature is knowable, because R and T are governed by general relativity. In the rest of this section we illustrate the construction of this class of wave equations from the original Einstein field theory itself. This exercise can be repeated to give a more general class of such wave equations based on the Evans unified field theory.

The starting point for this class of wave equation are the Einsteinian definitions [12]

$$R = R_{\mu\nu} g^{\mu\nu}, \tag{14.46}$$

$$T = T_{\mu\nu} g^{\mu\nu}. \tag{14.47}$$

Multiplication on both sides by $g_{\mu\nu}$ gives

$$R g_{\mu\nu} = 4 R_{\mu\nu}, \quad T g_{\mu\nu} = 4 T_{\mu\nu}, \tag{14.48}$$

i.e.,

$$R_{\mu\nu} = \frac{1}{4} R g_{\mu\nu}, \quad T_{\mu\nu} = \frac{1}{4} T g_{\mu\nu}. \tag{14.49}$$

The restricted or conventional Noether theorem [11] therefore reads

$$D_\mu (T g^{\mu\nu}) = g^{\mu\nu} D_\mu T + T D_\mu g^{\mu\nu} = 0. \tag{14.50}$$

Multiplying this equation by $g_{\mu\nu}$, we get

$$D_\rho R = \alpha_\rho R, \tag{14.51}$$

where

$$\alpha_\rho = -\frac{1}{4} g_{\rho\nu} D_\mu g^{\mu\nu}. \tag{14.52}$$

Equation (51) is a first-order differential equation. Similarly,

$$D_\rho T = \alpha_\rho T. \tag{14.53}$$

Use of Eqs. (32) gives

$$\partial_\rho R = \alpha_\rho R, \quad \partial_\rho T = \alpha_\rho T \tag{14.54}$$

which are first-order equations in the ordinary rather than the covariant derivative.

The wave equations follow straightforwardly by differentiation:

$$\Box R = \partial^\rho (\alpha_\rho R), \quad \Box T = \partial^\rho (\alpha_\rho T). \tag{14.55}$$

On using

$$\partial^\rho (\alpha_\rho R) = \alpha^\rho \partial^\rho R + R \partial^\rho \alpha_\rho, \tag{14.56}$$

Equation (55) becomes the second-order differential, or wave, equation

$$\Box R = (\alpha_\rho \alpha^\rho + \partial^\rho \alpha_\rho) R, \tag{14.57}$$

where

$$\alpha_\rho \alpha^\rho = \frac{1}{16} \left(g_{\rho\nu} D_\mu g^{\mu\nu} \right) \left(g^{\rho\nu} D^\mu g_{\mu\nu} \right). \tag{14.58}$$

The wave equation can be written as

$$(\Box + \beta)R = 0, \tag{14.59}$$

where

$$\beta = -(\partial^\rho \alpha_\rho + \alpha^\rho \alpha_\rho). \tag{14.60}$$

Similarly,

$$(\Box + \beta)T = 0. \tag{14.61}$$

Equations (59) and (61) have the structure (see Fig. (2)) of the main wave equations of physics; but, along with the Evans wave equation [1-10], they are also equations of general relativity and therefore causal. They show that R and T are quantized for all radiated and matter fields of nature. We describe this procedure as "causal quantization," to distinguish it from Heisenberg's quantization, which is subjective as argued and should have no place within objective, and objectively measurable, natural philosophy. The Evans unified field theory imparts a deterministic structure to nature and resolves the twentieth century debate between the Copenhagen school and the deterministic school in physics, coming down firmly on the latter's side. The Evans unified field theory also suggests the existence of unexplored areas of physics and develops the standard model into a generally covariant field theory of all radiated and matter fields, while recovering (see Figs. (1) and (2)) the previously known and tested equations of physics. Some of them, for example the Maxwell-Heaviside equations, are developed into structures such as O(3) electrodynamics [18], which has been fully tested experimentally and is compatible with general relativity.

14.3 General Wave, Field And Force Equations Of The Evans Theory

The wave and field equations of this section are generalizations to unified field theory of the well-known wave and gauge field equations of electrodynamics [19]. Consider the Evans field equation in the form

$$G q^a{}_\mu = k T q^a{}_\mu. \tag{14.62}$$

The unified potential field is the tetrad or vector-valued one-form $q^a{}_\mu$, which is in general an asymmetric square matrix. The latter can always be written as the sum of symmetric and antisymmetric component square matrices, components that are physically meaningful potential fields of nature:

$$q^a{}_\mu = q^a{}_\mu{}^{(S)} + q^a{}_\mu{}^{(A)}. \tag{14.63}$$

In the Evans unified field theory the gravitational potential field is identified [1-10] as the tetrad $q^a{}_\mu$ and the electromagnetic potential field as $A^{(0)} q^a{}_\mu$, where $A^{(0)}$ is measured in $V{\cdot}s/m$. The unit of magnetic flux, i.e., the weber (or $V{\cdot}s$) belongs to $\hbar/e$, the magnetic fluxon, and both $\hbar$ and e are manifestations of the principle of least curvature [1-10] of the Evans unified field theory. Both the gravitational and the electromagnetic potential fields can in general have symmetric and antisymmetric components:

$$A^a{}_\mu = A^{(0)} q^a{}_\mu = A^a{}_\mu{}^{(S)} + A^a{}_\mu{}^{(A)}, \tag{14.64}$$

and all four components appearing in Eqs. (63) and (64) are objectively measurable fields of nature. Geometry shows that there can be two types of gravitational potential fields: $q^a{}_\mu{}^{(S)}$ and $q^a{}_\mu{}^{(A)}$, and two types of electromagnetic potential field, $A^a{}_\mu{}^{(S)}$ and $A^a{}_\mu{}^{(A)}$. These four types of field are governed by four Evans field equations:

$$R_1 q^a{}_\mu{}^{(S)} = -kT_1 q^a{}_\mu{}^{(S)}, \tag{14.65}$$

$$R_2 q^a{}_\mu{}^{(A)} = -kT_2 q^a{}_\mu{}^{(A)}, \tag{14.66}$$

$$R_3 A^a{}_\mu{}^{(S)} = -kT_3 A^a{}_\mu{}^{(S)}, \tag{14.67}$$

$$R_4 A^a{}_\mu{}^{(A)} = -kT_4 A^a{}_\mu{}^{(A)}, \tag{14.68}$$

in which appear four types of canonical energy momentum tensor: symmetric gravitational, antisymmetric gravitational, symmetric electromagnetic, and antisymmetric electromagnetic. In Einstein's generally covariant theory of gravitation [12], only one type of canonical energy-momentum tensor appears, the symmetric gravitational. In the weak-field limit, the latter gives Newtonian dynamics, in which the force field is centrally directed along the line between two point masses in Newton's inverse square law. The latter emerges from Einstein's theory of 1915 [12] as a flat spacetime limit of Riemannian geometry with curvature but no torsion. There is no sense of torsion or spin in Newtonian dynamics and no sense of torsion or spin in Einstein's generally covariant theory of gravitation [12]. In electrostatics, the force field corresponding to the Coulomb inverse square law is also central, and there is no torsion present. In electrodynamics however, there exists the magnetic field, signifying spin, and electrodynamics in the Evans unified field theory [1-10] is a generally covariant theory with torsion as well as curvature. We conclude that the symmetric part of the tetrad $q^a{}_\mu{}^{(S)}$ represents the central, gravitational potential field, and the symmetric $A^a{}_\mu{}^{(S)}$ represents the central, electrostatic potential field. The antisymmetric $A^a{}_\mu{}^{(A)}$ represents the rotating and translating electrodynamic potential field.

The antisymmetric $q^a{}_\mu{}^{(A)}$ represents a type of spinning potential field which is C positive, where C is charge conjugation symmetry [20]. This fundamental potential field of nature is not present in the Einsteinian or Newtonian theories of gravitation as argued and is a field that is governed by the

Evans equation (66). It may be the potential field of dark matter [13], which is observed to constitute the great majority of mass in the vicinity of spiral galaxies. Significantly, the latter are thought to be formed by spinning motion, responsible for their characteristic spiral shape. The field $q^{a}{}_{\mu}{}^{(A)}$ is not centrally directed and so does not manifest itself in the Newtonian inverse square law in the weak-field limit. (Similarly, the antisymmetric electrodynamic $A^{a}{}_{\mu}{}^{(A)}$ does not reduce to the Coulomb inverse square law, which must be obtained [1-10] from the symmetric electrostatic $A^{a}{}_{\mu}{}^{(S)}$.) The antisymmetric $q^{a}{}_{\mu}{}^{(A)}$ is also the root cause of the well-known Coriolis and centripetal accelerations, which conventionally require a rotating frame not present in Newtonian dynamics. The rotating frame is built into the Evans unified field theory as spacetime torsion.

All of these fields emerge systematically from the tetrad $q^{a}{}_{\mu}$ by splitting it into its symmetric and antisymmetric components and by multiplying them by a C negative coefficient whose unit is the volt. The original asymmetric tetrad is the unified potential field of nature. The C negative manifestation of the unified field is $\zeta^{(0)}q^{a}{}_{\mu}$, where $\zeta^{(0)}$ must be determined experimentally. The coefficient $\zeta^{(0)}$ determines for example the way in which an electrostatic field affects the gravitational field. All forms of energy-momentum are interconvertible, implying that

$$T^{a}{}_{\mu} = T^{a}{}_{\mu}{}^{(S)} + T^{a}{}_{\mu}{}^{(A)}. \tag{14.69}$$

The interaction field $\zeta^{(0)}q^{a}{}_{\mu}$ and its concomitant $T^{a}{}_{\mu}$ may for example be measurable in the influence of an electrostatic field on the gravitational field [14,15]. If so, the total interaction between two charged particles would be the sum of the Newton and Coulomb inverse square laws and a hitherto unknown interaction component which must be looked for with high precision balance experiments [14]. For example, there may be a tiny effect of an electrostatic field on a perfect insulator in one arm of a high precision (e.g., picogram resolution) balance. There may also be a tiny effect of mass (perfect insulators) used to unbalance a high precision device such as a Wheatstone bridge. In other words, the fundamental term $\zeta^{(0)}$ is not present in Einstein's original theory of general relativity, because that deals only with gravitation. So $\zeta^{(0)}$ is an example of a new concept of the Evans theory, the subject of this series of papers. The concept of $\zeta^{(0)}$ has its fundamental origin in thermodynamics: All types of energy-momentum are interconvertible, and so all types of potential field are interconvertible. The mechanism of the interconversion must be found by experiment. Reproducible and repeatable effects of an electric field on gravitation, for example, can be understood within the Evans unified field theory, but not within the standard model.

We may always define the unified field by

$$G^{a}{}_{\mu} = G^{(0)}\left(R^{a}{}_{\mu} - \frac{1}{2}Rq^{a}{}_{\mu}\right) \tag{14.70}$$

and the unified energy-momentum by

$$G^a{}_\mu = G^{(0)} k T^a{}_\mu . \tag{14.71}$$

The unified (i.e., most general form of the) Bianchi identity is Eq. (26), and the unified conservation theorem is Eq. (27). By covariant differentiation, we obtain

$$D_\mu (D^\mu G^a{}_\mu) = (D_\rho D^\rho) G^a{}_\mu = 0, \tag{14.72}$$

and therefore [1-10] arrive at the unified wave equations

$$(\Box + kT) G^a{}_\mu = (\Box + kT) T^a{}_\mu = 0, \tag{14.73}$$

whose eigenfunctions are the unified field $G^a{}_\mu$ and the unified $T^a{}_\mu$. Self-consistently, these wave equations can be obtained straightforwardly from the Evans wave Eq. (2) on using Eqs. (16) or (17). If $\zeta^{(0)}$ exists, it is also governed by a wave equation

$$(\Box + kT)(\zeta^{(0)} q^a{}_\mu) = 0. \tag{14.74}$$

The Evans lemma [1-10]

$$\Box q^a{}_\mu = R q^a{}_\mu \tag{14.75}$$

gives rise to a class of identities:

$$\Box G^a{}_\mu = R G^a{}_\mu, \tag{14.76}$$

$$\Box T^a{}_\mu = R T^a{}_\mu, \tag{14.77}$$

so that the Bianchi identity is generalized to the wave equation

$$\Box G^a{}_\mu = R G^a{}_\mu = -\frac{1}{4} R^2 q^a{}_\mu \tag{14.78}$$

and the Noether theorem to the wave equation

$$\Box T^a{}_\mu = R T^a{}_\mu = \frac{1}{4} R T q^a{}_\mu . \tag{14.79}$$

The most general asymmetric gauge field is

$$G^{ab}{}_{\mu\nu} = \frac{G}{4} q^a{}_\mu q^b{}_\nu \tag{14.80}$$

and most general $T^{ab}{}_{\mu\nu}$ tensor is

$$T^{ab}{}_{\mu\nu} = \frac{T}{4} q^a{}_\mu q^b{}_\nu . \tag{14.81}$$

These can also be written as sums of symmetric and antisymmetric components:

$$G^{ab}{}_{\mu\nu} = G^{ab}{}_{\mu\nu}{}^{(S)} + G^{ab}{}_{\mu\nu}{}^{(A)}, \tag{14.82}$$

$$T^{ab}{}_{\mu\nu} = T^{ab}{}_{\mu\nu}{}^{(S)} + T^{ab}{}_{\mu\nu}{}^{(A)}. \tag{14.83}$$

The homogeneous field equation is then defined by the Jacobi identity for any antisymmetric matrix,

$$D^{\mu} G^{ab}{}_{\mu\nu}{}^{(A)} = 0, \tag{14.84}$$

and the general inhomogeneous field equation is defined by

$$D^{\mu} \widetilde{G}^{ab}{}_{\mu\nu}{}^{(A)} = J^{ab}{}_{\nu}, \tag{14.85}$$

where $\widetilde{G}^{ab}{}_{\mu}{}^{(A)}$ is the dual of $G^{ab}{}_{\mu\nu}$:

$$\widetilde{G}^{ab}{}_{\mu\nu} = \frac{1}{2}\varepsilon_{\mu\nu\rho\sigma} G^{\rho\sigma ab(A)} \tag{14.86}$$

and where $J^{ab}{}_{\nu}$ is a general charge-current density. It is also possible to define a general force equation from the asymmetric matrix $T^{ab}{}_{\mu\nu}$:

$$D^{\mu} T^{ab}{}_{\mu\nu} = -f^{ab}{}_{\nu}. \tag{14.87}$$

Note that Eq. (87) is valid only for a subsystem [21]. For a closed system the net force of Eq. (87) may be zero. Finally, the most general form of the Lorentz force equation may be written as

$$f^{ab}{}_{\nu}\,(\text{Lorentz}) = G^{ab}{}_{\mu\nu} J^{\nu ab}, \tag{14.88}$$

showing that the Lorentz force equation is also an equation of general relativity.

Acknowledgments

The Ted Annis Foundation, ADAS, and Craddock Inc. are thanked for funding, and the staff of AIAS for many interesting discussions.

References

1. M. W. Evans, *Found. Phys. Lett.* **16**, 367 (2003).
2. M. W. Evans, *Found. Phys. Lett.* **16**, 507 (2003).
3. M. W. Evans, *Found. Phys. Lett.* **17**, 25 (2004).
4. M. W. Evans, "Derivation of Dirac's equation from the Evans wave equation," *Found. Phys. Lett.* **17**, 149 (2004).
5. M. W. Evans, "Unification of the gravitational and strong nuclear fields," *Found. Phys. Lett.* **17**, 267 (2004).
6. M. W. Evans, "Physical optics, the Sagnac effect and the Aharonov-Bohm effect in the Evans unified field theory," *Found. Phys. Lett.* **17**, 301 (2004).
7. M. W. Evans, "Derivation of the geometrical phase from the Evans phase law of generally covariant unified field theory," *Found. Phys. Lett.* **17**, 393 (2004).
8. M. W. Evans, "The Evans lemma of differential geometry," *Found. Phys. Lett.* **17**, 443 (2004).
9. M. W. Evans, "Derivation of the Evans wave equation from the Lagrangian and action," *Found. Phys. Lett.* **17**, 535 (2004).
10. *The Collected Scientific Papers of Myron Wyn Evans*, National Library of Wales and Artspeed California, collection and website of circa 600 collected papers in preparation, to be cross-linked to www.aias.us.
11. S. M. Carroll, *Lecture Notes in General Relativity* (University of California, Santa Barbara, graduate course, arXiv:gr-qe/9712019 vl 3 Dec., 1997), complete course available from author on request.
12. A. Einstein, *The Meaning of Relativity* (Princeton University Press, 1921).
13. L. H. Ryder, *Quantum Field Theory*, 2nd edn. (Cambridge University Press, 1996).
14. M. W. Evans and AIAS Author Group, Development of the Evans Wave equation in weak-field: the electrogravitic equation," *Found. Phys. Lett.* **17**, 497 (2004).
15. L. Felker, ed., *The Evans Equations*, in preparation.
16. G. Stephenson, *Mathematical Methods for Science Students* (Longmans, London, 1968).
17. M. Sachs, in M. W. Evans, ed., *Modern Non-linear Optics*, a special topical issue in three parts of I. Prigogine and S. A. Rice, eds., *Advances in Chemical Physics*, Vol. 119(1) (Wiley-Interscience, New York, 2001), 2nd edn. and 3-book end.

18. M. W. Evans, J.-P. Vigier, *et al.*, *The Enigmantic Photon* (Kluwer Academic, Dordrecht, 1994-2002), in five volumes hardback and paperback. M. W. Evans and L. B. Crowell, *Classical and Quantum Electrodynamics and the* $\boldsymbol{B}^{(3)}$ *Field* (World Scientific, Singapore, 2001).
19. M. W. Evans, reviews in Vols. 119(2) and 119(3) of Ref. (17), and reviews in the 1st edn., Vol. 85 of *Advances in Chemical Physics* (Wiley Interscience, New York, 1992, 1993, and 1997), paperback edn.
20. J. L. Jimenez and I. Campos, in T. W. Barrett and D. M. Grimes, *Advanced Electromagnetism* (World Scientific, Singapore, 1995), foreword by Sir Roger Penrose.

15

New Concepts From The Evans Unified Field Theory, Part Two: Derivation Of The Heisenberg Equation and Replacement Of The Heisenberg Uncertainty Principle

Summary. The Heisenbergequation of motion is derived from the Evans wave equation of motion of generally covariant unified field/matter theory as the non-relativistic quantum limit of general relativity. The method used is first to derive the free particle Klein Gordon wave equation in the special relativistic limit of the Evans wave equation. The free particle Einstein equation for special relativistic motion (the equation of special relativistic momentum) is found in the classical limit of the Klein Gordon equation, together with the fundamental operator equivalence of quantum mechanics. The free particle Newton equation is found as the non-relativistic limit of the Einstein equation of special relativity (the relativistic momentum equation) by using the definition of relativistic kinetic energy. The time-independent free particle Schrödinger equation is found from the already derived operator equivalence applied to Newton's equation. Finally the Heisenberg equation is found as a form of the Schrödinger equation. The self consistency of the method used is checked with the rotational form of the Evans field equation. The well known commutator relation which is the basis of the acausal and therefore subjective "uncertainty" principle is derived from causal and therefore objective general relativity using the geometrical concepts of tetrad and torsion. The Planck constant is defined as the minimum quantity of action, or angular momentum, allowed by the Evans Principle of Least Curvature, which together with the Evans Lemma and wave equation, give all the known principles of quantum mechanics from causal general relativity, and also suggest several new laws of nature from which new engineering can evolve and in some sectors, has been implemented already.

Key words: Evans unified field theory, generally covariant unified field theory, Evans wave equation, Evans field equation, Evans Lemma, Heisenberg equation, the Evans principle of least curvature.

15.1 Introduction

The Heisenberg equation [1] is the basis of much of quantum mechanics and matrix mechanics in the non-relativistic limit. In this paper, the second of a series [2] on new concepts from Evans' generally covariant unified field/matter theory [3]–[12], the Heisenberg equation is derived self consistently both from the Evans wave equation and the Evans field equation in rotational form. The acausal and subjective uncertainty principle of Heisenberg is replaced by a causal and objective law from general relativity. This new law has the same mathematical structure as the uncertainty principle but is interpreted using the well known tetrad and torsion forms of differential geometry [2]–[13]. The complicated twentieth century dialogue between general relativity and quantum mechanics is therefore resolved straightforwardly by realizing that the wave-function originates in non-Euclidean spacetime which is in general can produce both curvature and torsion. The key to this philosophical resolution is the Evans Lemma [2]–[12], which is the subsidiary proposition of the Evans wave equation, and which asserts that:

$$\Box q^a{}_\mu = R q^a{}_\mu \tag{15.1}$$

where $q^a{}_\mu$ is the tetrad, and R a scalar curvature. Scalar curvature R is therefore the a eigenvalue of the d'Alembertian $\Box$, operating on the tetrad $q^a{}_\mu$, as eigenfunction. This quantization of scalar curvature leads directly to the Evans wave equation:

$$(\Box + kT)\, q^a{}_\mu = 0 \tag{15.2}$$

using the fundamental Einsteinian relation:

$$R = -kT \tag{15.3}$$

where k is the Einstein constant of general relativity and where T is the index contracted canonical energy-momentum tensor [14].

Quantum mechanics is therefore obtained straightforwardly from general relativity by recognising that the wavefunction is the tetrad. The latter is defined for any vector V^μ by:

$$V^a = q^a{}_\mu V^\mu \tag{15.4}$$

where a is the index of the orthonormal spacetime [13] of the base manifold indexed μ. The Lemma (15.1)and wave equation (15.2) are then direct mathematical consequences [2]–[12] of the well known and fundamental [13] tetrad postulate of differential geometry:

$$D_\nu q^a{}_\mu = 0 \tag{15.5}$$

where D_ν the covariant derivative. The tetrad is essentially the matrix linking two frames of reference, one of which [13] is orthogonal and normalized, i.e. is a Euclidean or flat spacetime, and the other is the non-Euclidean base manifold. The flat spacetime is labelled a and the base manifold is labelled μ. The vector V^μ can be defined using a representation space of any dimension, and the basis set a can be defined in any appropriate way [13] . The tetrad postulate, Lemma and wave equation are true for any dimension of representation space and any basis representation [2]–[12] (e.g. unit vectors or Pauli matrices). Therefore the Dirac equation and chromodynamics can be derived from the Evans equation by appropriate definition of the tetrad. There are only four physically significant dimensions however, time and three space dimensions. This means that the Evans unified field theory has a great advantage of simplicity over string theory (with its many spurious, unphysical "dimensions"). The Evans theory has been tested extensively against data [2]–[12] and has produced new engineering in several sectors [14]. String theory has produced nothing because it is an unphysical theory. Similarly the Heisenberg uncertainty principle can be of no interest to engineers, or in fact, to physicists. There are many other examples of spurious concepts in the history of science, for example the Aristotelian epicycles which preceded the Kepler laws and the great Newtonian synthesis of 1665.

In Section 15.2 the Heisenberg equation is derived as a quantum nonrelativistic limit of the Evans wave equation (15.2). The derivation is checked using the rotational form of the Evans field equation. In Section 15.3 the Heisenberg uncertainty principle is discarded in favor of a new commutator law built up straightforwardly from differential geometry and the Evans principle of least curvature [2]–[12]. The latter identifies the Planck constant as the least unit of action or angular momentum in general relativity. The new law has the same mathematical structure as the older uncertainty principle, but is causal and is derived from general relativity. It is therefore objective in the sense that all physical laws must be objective, must predict events using mathematics, and must be objectively measurable against data. Heisenberg and the Copenhagen School introduced spurious concepts which are not useful in natural philosophy. The most obscure of these is the assertion that certain events are "unknowable" [1]. As soon as this assertion is made, however, the boundaries of natural philosophy are breached, because any assertion on the unknowable is subjective or theistic. Natural philosophy measures data objectively. The Copenhagen School asserts that the measuring process affects data in an acausal or unknowable way. General relativity is diametrically at odds with this assertion, and within general relativity measurement is objective: any event has a measurable cause. Recently [15] copious experimental evidence has been published which shows that the uncertainty principle is, unsurprisingly, untenable.

The Heisenberg EQUATION on the other hand is simply a restatement of the slightly earlier Schrödinger equation, and hardly deserves an appellation. It is given one because of the spurious uncertainty principle whose structure is derived from the form of the Heisenberg equation, or more properly, from the Schrödinger equation and de Broglie wave particle dualism. So in this paper we are at pains to separate the notion of uncertainty principle from our unified field theory. We CAN derive the mathematical form of the uncertainty principle, but do not interpret it in the manner of the Copenhagen School.

15.2 Derivation of the Heisenberg Equation

In its simplest form [1] the Heisenberg equation is:

$$(xp - px)\,\psi = i\hbar\psi \tag{15.6}$$

where:

$$p = -i\hbar\frac{\partial}{\partial x} \tag{15.7}$$

is a differential operator representing momentum, and where x represents position. Here ψ is the wavefunction or eigenfunction and $\hbar$ is the reduced Planck constant $(h/(2\pi))$. Eq. (15.6) is often stated as:

$$xp - px = i\hbar \tag{15.8}$$

and was spuriously elevated by Heisenberg into a principle [1] which has come to be known as the Heisenberg uncertainty principle, rejected by Einstein and the deterministic School. The operators in Eq. (15.7) are defined as follows:

$$xp\psi = -ix\hbar\frac{\partial\psi}{\partial x} \tag{15.9}$$

$$px\psi = -i\hbar\frac{\partial}{\partial x}(x\psi) = -i\hbar\left(x\frac{\partial\psi}{\partial x} + \psi\right). \tag{15.10}$$

Note that p operates on the product $x\psi$ using the Leibnitz Theorem. From Eqs. (15.9) and (15.10)

$$(xp - px)\,\psi = -i\hbar x\frac{\partial\psi}{\partial x} + i\hbar x\frac{\partial\psi}{\partial x} + i\hbar\psi = i\hbar\psi. \tag{15.11}$$

The method used to derive Eq. (15.7) from Eq. (15.2) is to first derive the time independent Schrödinger equation [1] in the appropriate limit, then recognise that the Heisenberg equation is a restatement of the Schrödinger equation using the operator equivalence of quantum mechanics, an example of which is Eq. (15.7). The relativistic form of the operator equivalence is DERIVED from the Evans wave equation in the limit of special relativity. In this limit we obtain the Klein-Gordon equation [2]–[12, 16] from the Evans wave equation. The classical form of the Klein Gordon equation is Einstein's original definition of relativistic momentum in the special relativistic limit. We then implement standard methods [16] to find the Newton equation as the limit of special relativity when velocities are small compared to the speed of light, c. The time independent Schrödinger equation is the Newton equation after implementation of the operator equivalence already derived from the equivalence of the Klein Gordon and Einstein equations. The Heisenberg equation (15.7) is the time independent Schrödinger equation written in a different way but using the same operator definition.

The Klein Gordon equation is the limit in which the Evans wave equation becomes:

$$\left(\Box + \frac{m^2c^2}{\hbar^2}\right) q^a{}_\mu = 0 \tag{15.12}$$

i.e. is the limit represented by:

$$kT = (mc/\hbar)^2 = 1/\lambda^2{}_c \tag{15.13}$$

where λ_c is the Compton wavelength and where m is mass. The limit (15.13) is an example of the principle that equations of general relativity must reduce to equations of special relativity when accelerations (forces) are small in magnitude (free particle limit). This Einsteinian principle dictates that the component of kT originating in rest energy must become the inverse square of the Compton wavelength in the limit of special relativity. The Evans wave equation of motion then describes the motion of a free particle upon which no external forces or accelerations are acting. The modulus of R defined in this limit is the least curvature, and this an example of the Evans principle of least curvature [2]–[12]. The least, or minimum amount of, curvature needed to define the rest energy of any particle is:

$$R_0 = -\frac{m^2c^2}{\hbar^2} \tag{15.14}$$

and defines the first Casimir invariant [17] of a particle, its characteristic mass. So spacetime is never everywhere flat. (In an everywhere flat spacetime the universe is empty, there are no particles and no fields, and all energy vanishes identically.) The least curvature (15.14) characterises the mass and Compton wavelength of all elementary particles, including the photon, which must therefore have mass in the Evans unified field theory. More precisely, the artificial distinction between particle and field is rejected in favor of radiated and matter fields [18], and there are no point particles, because there are no singularities. Similarly the singularities of quantum electrodynamics and chromodynamics are eliminated in favor of general relativity because there are no point charges. These illustrate some of the major advantages of the new unified field theory [2]–[12].

The eigenfunction in the limit represented by (15.12) is the tetrad..The wave function of the Klein-Gordon equation is a scalar component of the tetrad, a component which is conventionally [17] written as the scalar field:

$$\phi = q^0{}_0, q^0{}_1, \cdots, q^4{}_4. \tag{15.15}$$

The Evans unified field theory therefore shows that the older concept of scalar field in special relativity must be extended to allow for the fact that ϕ must always be a tetrad component.

Having found the Klein Gordon equation, the time independent Schrödinger equation can be deduced by first deriving the definition of special relativistic momentum in the form:

$$E^2 = p^2c^2 + m^2c^4. \tag{15.16}$$

Here E is conventionally [16] referred to as the total energy, but is more accurately the total kinetic energy of a free particle, which in special relativity has a component:

$$E_0 = mc^2 \tag{15.17}$$

known as the rest energy. Eq. (15.16) follows from the Klein-Gordon equation by using the fundamental operator equivalence of quantum mechanics [17]:

$$p^\mu = i\hbar\partial^\mu, \tag{15.18}$$

i.e.

$$E = i\hbar\frac{\partial}{\partial x}, \qquad \boldsymbol{p} = -i\hbar\boldsymbol{\nabla}. \tag{15.19}$$

We have therefore derived the operator equivalence from general relativity using the Evans wave equation, the reason being that Eq. (15.16) reduces to the Newton equation in the appropriate limit. In order to derive the Einstein equation (15.16) and the Newton equation the operator equivalence (15.18) is needed. In other words if these well known equations are to be deduced from general relativity UNIFIED with quantum mechanics we are led to the operator equivalence (15.18). Note carefully that in the received opinion the operator equivalence has no justification in general relativity.

Einstein's original definition of relativistic momentum is:

$$p = \gamma mv = \left(1 - \frac{v^2}{c^2}\right)^{-1/2} mv \tag{15.20}$$

and is equivalent to Eq.(15.18) To see this develop Eq. (15.20) as follows:

$$p^2c^2 = \gamma^2m^2v^2c^2 = \gamma^2m^2c^4\left(\frac{v^2}{c^2}\right) \tag{15.21}$$

with:

$$\frac{v^2}{c^2} = 1 - \frac{1}{\gamma^2} \tag{15.22}$$

where

$$\frac{1}{\gamma^2} = 1 - \frac{v^2}{c^2}. \tag{15.23}$$

So Eq. (15.20) is:

$$\begin{aligned} p^2c^2 = \gamma^2m^2c^4\left(1 - \frac{1}{\gamma^2}\right) &= \gamma^2m^2c^4 - m^2c^4 \\ &:= E^2 - E^2_0 \end{aligned} \tag{15.24}$$

which is Eq. (15.16).

This means that the Newton law in special relativity is also modified to:

$$\boldsymbol{F} = \frac{d\boldsymbol{p}}{dt} = \frac{d}{dt}\left(\gamma m \boldsymbol{v}\right). \tag{15.25}$$

The kinetic energy in special relativity is calculated from the definition (15.25) using the work:

$$W_{12} = \int_1^2 \boldsymbol{F} \cdot d\boldsymbol{r} = T_2 - T_1. \tag{15.26}$$

If we start from rest:

$$W = T = \int \frac{d}{dt}\left(\gamma m \boldsymbol{v}\right) \cdot \boldsymbol{v} dt \tag{15.27}$$

and

$$T = m \int_0^v v d(\gamma v). \tag{15.28}$$

Integrating by parts [16]

$$\begin{aligned} T &= \gamma m v^2 - m \int_0^v \gamma v dv \\ &= \gamma m v^2 + mc^2 \left(1 - v^2/c^2\right)^{1/2} |_0^v \\ &= \gamma m v^2 + mc^2 \left(1 - v^2/c^2\right)^{1/2} - mc^2. \end{aligned} \tag{15.29}$$

The special relativistic kinetic energy is therefore:

$$T = mc^2 \left(\gamma - 1\right). \tag{15.30}$$

Therefore the rest energy is this kinetic energy multiplied $(\gamma - 1)^{-1}$, showing that the rest energy is kinetic energy and not potential energy.

The Newtonian kinetic energy is now deduced from the equation:

$$T = mc^2 \left(1 - \frac{v^2}{c^2}\right)^{-1/2} - mc^2 \tag{15.31}$$

using the Maclaurin expansion:

$$\left(1 - \frac{v^2}{c^2}\right)^{-1/2} = 1 + \frac{1}{2}\frac{v^2}{c^2} + \cdots. \tag{15.32}$$

when $v << c$ in Eq. (15.31) we obtain Newton's kinetic energy for a free particle:

$$T = \frac{1}{2} m v^2 = \frac{p^2}{2m}. \tag{15.33}$$

The Schrödinger equation for a free particle is, in consequence:

$$\frac{p^2}{2m}\psi = T\psi \tag{15.34}$$

where p is the operator defined by Eq. (15.19). Using Eq. (15.19) in Eq. (15.34) produces the usual form of the time independent Schrödinger equation [1]:

$$-\frac{\hbar^2}{2m}\nabla^2\psi = T\psi. \tag{15.35}$$

If there is potential energy V present (as in the harmonic oscillator or Coulomb interaction between proton and electron in an atom) the kinetic energy is written as:

$$T = E - V \tag{15.36}$$

where E is the total energy (sum of the kinetic and potential energy). In this case the time independent Schrödinger equation is written as:

$$-\frac{\hbar^2}{2m}\nabla^2\psi = (E - V)\,\psi. \tag{15.37}$$

Eq. (15.37) is most well known as:

$$\hat{H}\psi = E\psi \tag{15.38}$$

where

$$\hat{H} := -\frac{\hbar^2}{2m}\nabla^2 + V \tag{15.39}$$

is the hamiltonian operator [1].

Therefore we have derived the time independent Schrödinger equation from a well defined limit of the generally covariant Evans wave equation without using the spurious and unphysical Heisenberg uncertainty principle. The wave function in the Klein Gordon equation has been identified as a scalar component of the tetrad, and not as a probability. It is well known that the probabilistic interpretation of the wave function of the Klein Gordon equation led to its abandonment in favour of the Dirac equation, and therefore the probabilistic interpretation is untenable. This problem with the Klein Gordon equation is remedied in the Evans unified field theory, which shows that its wave function is a tetrad component (a scalar).A tetrad matrix may have negative scalar components in general, whereas a probability cannot be negative. The Dirac equation may also be derived from the Evans wave equation [2]–[12] and in the wave function for the Dirac equation is a four component column vector, the Dirac spinor [17]:

$$\psi = \begin{bmatrix} \psi^R \\ \psi^L \end{bmatrix} = \begin{bmatrix} {\psi^R}_1 \\ {\psi^R}_2 \\ {\psi^L}_1 \\ {\psi^L}_2 \end{bmatrix}. \tag{15.40}$$

The four components of the Dirac spinor are four components of a tetrad which links two Pauli spinors as follows:

$$\begin{bmatrix} \psi^R \\ \psi^L \end{bmatrix} = \begin{bmatrix} \psi^R{}_1 & \psi^R{}_2 \\ \psi^L{}_1 & \psi^L{}_2 \end{bmatrix} \begin{bmatrix} \psi^1 \\ \psi^2 \end{bmatrix} . \tag{15.41}$$

One Pauli spinor (two component column vector) is defined in the orthonormal space:

$$\psi^a = \begin{bmatrix} \psi^R \\ \psi^L \end{bmatrix} \tag{15.42}$$

and the other in the base manifold:

$$\psi^\mu = \begin{bmatrix} \psi^1 \\ \psi^2 \end{bmatrix} . \tag{15.43}$$

So the Dirac spinor contains the four tetrad components arranged in a column vector instead of in the 2 x 2 tetrad matrix defined in Eq. (15.41). (Strictly speaking the term "tetrad" should be reserved for a 4 x 4 matrix, and the term "diad" used for a 2 x 2 matrix, but the term "tetrad" is conventionally used generically [13]. The Dirac equation is therefore derived from the Evans wave equation in the free particle limit:

$$\left(\Box + \frac{m^2c^2}{\hbar^2} \right) \psi = 0 \tag{15.44}$$

and can be considered as four Klein Gordon equations:

$$\begin{gathered} \left(\Box + \frac{m^2c^2}{\hbar^2} \right) \psi^R{}_1 = 0 \\ \vdots \\ \left(\Box + \frac{m^2c^2}{\hbar^2} \right) \psi^L{}_2 = 0. \end{gathered} \tag{15.45}$$

The Dirac equation in the form (15.44) is equivalent to:

$$\left(\gamma^\mu p_\mu - \frac{mc}{\hbar} \right) \psi = 0 \tag{15.46}$$

which is the form originally derived by Dirac [17]. It is well known that the Dirac equation predicts the existence of observable anti-particles in nature, and also predicts the existence of half integral spin, Fermi Dirac statistics, and useful spectral techniques such as ESR, NMR and MRI.

The Evans unified field theory allows the Dirac equation to be identified as a limiting form of the generally covariant Evans wave equation. In the received opinion (the "standard model") this inference is not made, and the standard model is not generally covariant, a major shortcoming remedied by the Evans wave equation.

The Heisenberg equation (15.6)is clearly seen now as a restatement of the Schrödinger equation (15.35) using the same operator equivalence (15.18). Note carefully that we have DERIVED this operator definition from general relativity developed into our unified field theory. The reason why the operator equivalence has been derived is that we have started from the tetrad postulate and Evans Lemma (15.1), which gives a fundamental wave equation of motion, the Evans wave equation, from the geometrical first principles of general relativity. Historically the operator equivalence was used empirically (i.e. because of the need to explain experimental data [1]) to convert the Newton equation into a wave equation, the Schrödinger equation. So the Evans unified field theory is a major advance in understanding because it derives quantum mechanics from general relativity, an aim of physics throughout the twentieth century.

The Schrödinger equation (15.35) may also be derived self-consistently from the a weak field limit of the Evans wave equation [2]–[12]. The tetrad $q^a{}_\mu$ is interpreted classically as a gravitational potential or interaction field analogous to the original metric used by Einstein [14]. In the weak field limit the field is assumed to be time-independent and velocities are low compared with c. This means that the relevant component of the tetrad to consider is $q^0{}_0$ and that the d'Alembertian becomes the negative of the Laplacian. Therefore:

$$\left(-\nabla^2 + kT\right) q^0{}_0 = 0, \qquad T = \frac{m}{V}. \tag{15.47}$$

In the weak field limit the tetrad component $q^0{}_0$ is well approximated by unity, so we recover the Poisson equation for gravitation:

$$\nabla^2 q^0{}_0 = \frac{km}{V} \tag{15.48}$$

which may be expressed [6] as the inverse square law of Newton:

$$\boldsymbol{F} = m\boldsymbol{g} = \frac{mM}{r^2}\boldsymbol{k}. \tag{15.49}$$

The principle of equivalence of gravitational and inertial mass means that Eq. (15.49) also gives the Newtonian kinetic energy (15.33) for a free particle (upon which no external forces act). In the Newtonian limit the relativistic momentum (15.20) reduces to mv, where m is the inertial mass and the kinetic energy of the free particle in the Newtonian limit is calculated from the work integral (15.28) as:

$$T = m\int_0^v v dv = \frac{1}{2}mv^2. \tag{15.50}$$

The work integral is in turn calculated from the Newton force equation (second law):

$$\boldsymbol{F} = d\boldsymbol{p}/dt \tag{15.51}$$

which is derivable from the Poisson equation (15.48), thus checking for self consistency. Having derived the Newtonian kinetic energy self consistently from the Evans equation in two complementary ways, the Schrödinger equation follows by applying the operator equivalence (15.18) to the Newtonian kinetic energy as shown already.

Before proceeding to the derivation of the Heisenberg equation (15.1) from the rotational form of the Evans field equation [2]–[12] a few comments are offered as follows on the correct interpretation of the quantities appearing in the wave equation (15.2), in order to identify and define precisely the kinetic energy, rest energy and potential energy and their associated scalar curvatures.

The total scalar curvature appearing in the Lemma (15.1) is:

$$R_{\text{total}} = -\Box + R \tag{15.52}$$

where the d'Alembertian operator is:

$$\Box = -\frac{1}{\hbar^2} p^\mu p_\mu \tag{15.53}$$

through the operator equivalence (15.18). So this operator contains the information on the kinetic energy momentum p^μ of a free particle. The rest curvature (15.14) contains the information on the rest energy of the free particle, whose TOTAL curvature vanishes in the limit of special relativity:

$$R_{\text{total}} = \frac{1}{\hbar^2}\left(p^\mu p_\mu - m^2 c^2\right) = 0. \tag{15.54}$$

This is what the Einstein equation (15.16) and its quantized version, the Klein Gordon equation, tells us. In general relativity on the other hand, frames can move arbitrarily with respect to each other, and there are accelerations present, represented by curvatures which are not present for a free particle moving with a constant velocity. These curvatures generalize [13, 14] the concept of force in Newtonian dynamics. The characteristics of field/matter are now represented by the Evans wave equation (15.2) and its subsidiary geometrical proposition, the Evans Lemma (15.2). The scalar curvature of the Lemma and wave equation is now made up in general of the rest curvature and curvature due to potential energy:

$$R = -\frac{m^2 c^2}{\hbar^2} + R_{\text{potential}}. \tag{15.55}$$

The potential curvature arises from interaction between particles and fields, for example the Coulomb interaction within an atom, or the Hooke's law interaction of the harmonic oscillator [1], or in charge/field interaction in electrodynamics. The generally covariant d'Alembert equation for example, is obtained from the Evans wave equation (15.2) by incorporating a current term whose origin is potential curvature. The generally covariant Proca equation is the Klein Gordon equation with the generally covariant tetrad potential $A^a{}_\mu$ [2]–[12] as eigenfunction or wave function. The Proca equation is for the free photon or field, and contains no potential curvature or energy. The latter is introduced through charge current density when there is field/charge interaction. Consideration of the harmonic oscillator potential energy gives rise to zero point energy in the Evans wave equation, both in a dynamical and electrodynamical context. So in general the contracted energy momentum tensor T contains information on potential energy, depending on the type of problem being considered in areas such as dynamics or electrodynamics. For the free particle the rest curvature is defined by:

$$R_0 = -kT_0 \tag{15.56}$$

where To is the relevant part of T for the free particle, i.e. its rest energy density mc^2/V_0 We therefore arrive at the concept of rest volume of a particle [2]–[12], a volume defined by:

$$m_0 V_0 = \frac{\hbar^2 k}{c^2}. \tag{15.57}$$

The existence of rest volume shows that there are no point particles (mathematical singularities) in our unified field theory, and no point particles in general relativity. This leads to a major advantage over the re-normalization procedures inherent in quantum electrodynamics and quantum chromodynamics, while retaining undiminished and perhaps improving on the accuracy of quantum electrodynamics. Re-normalization was rejected by Dirac and other leading thinkers, despite its accuracy. The latter is an unsurprising consequence of the fact that quantum electrodynamics is a perturbation theory. It would be surprising if it were not accurate. It has been shown [19] that O(3) electrodynamics improves on the accuracy of quantum electrodynamics, and it is now known that O(3) electrodynamics is a consequence of our generally covariant unified field theory [2]–[12]. TOTAL energy in our theory is always conserved, and this is the Noether Theorem. Total charge I current density in our theory is also conserved. These conservation laws are recognized as being equivalent to identities of geometry [2]–[12]. In Einstein's original theory [14] the Noether Theorem is:

$$D^\mu T_{\mu\nu} = 0 \tag{15.58}$$

and the Bianchi identity is:

$$D^\mu G_{\mu\nu} := 0 \tag{15.59}$$

where:

$$G_{\mu\nu} = kT_{\mu\nu} \tag{15.60}$$

is the Einstein field tensor [13] for classical gravitation. The generally covariant unified field theory of Evans [2]–[12] extends these considerations self consistently to all radiated and matter fields, and also to quantum mechanics, self consistently including quantum electrodynamics as a generally covariant theory WITHOUT singularities.

We may summarize these considerations through a new concept, total curvature is conserved.

Returning now to our main theme of deriving the Heisenberg equation it is now shown that it is the result of the rotational form of the Evans field equation [2]–[12], which was proposed in March 2003 slightly earlier than the wave equation (April 2003). The Heisenberg equation in the form (15.8) is well known [1] to define angular momentum:

$$\hat{J}_Z = xp_y - yp_x = \pm m_J \hbar \tag{15.61}$$

where $\hat{J}_Z$ is the angular momentum operator and m_J its quantum number, observed for example in atomic and molecular spectra. In our generally covariant theory these rotational quantities are introduced through the torsion form of differential geometry [2]–[13], missing from Einstein's original theory of classical gravitation, and angular motion or spin is the spinning of spacetime itself, not the motion of an entity imposed on the reference frame of a flat spacetime. Angular momentum components (orbital or intrinsic) are scalar elements of torsion within a factor $\hbar$. The latter is the minimum quantity of angular momentum in the universe, and so appears for example in the definition (15.14) of minimum particulate mass (the first Casimir invariant of our theory). The second Casimir invariant is the spin invariant, $\boldsymbol{B}^{(3)}$ examples being $\hbar$ itself or the fundamental $\boldsymbol{B}^{(3)}$ spin field introduced by Evans [20] in 1992 to describe the inverse Faraday effect. The concept of Casimir invariants of a particle (its mass and spin) were first introduced by Wigner in 1939 [17] for the Poincare group of special relativity, but in our theory they apply for the Einstein group of general relativity [18].

In cylindrical polar coordinates [1]:

$$\hat{J}_Z = -i\hbar \left(x\frac{\partial}{\partial y} - y\frac{\partial}{\partial x} \right) = -i\hbar \frac{\partial}{\partial \phi} \tag{15.62}$$

and for a particle on a ring:

$$\hat{J}_Z \psi = \pm m_e \hbar \psi \tag{15.63}$$

which may be interpreted either as a Heisenberg equation or a Schrödinger equation. The torsion form in our theory is defined by the wedge product of tetrads [2]–[12]:

$$\tau^c{}_{\mu\nu} = q^a{}_\mu \wedge q^b{}_\nu \tag{15.64}$$

and is a vector valued two-form anti-symmetric in μ and ν. The definition (15.64) is true for any base manifold, (i.e. for any type of spacetime with both curvature and torsion present) and so may be written as:

$$\tau^c := q^a \wedge q^b . \tag{15.65}$$

Eq. (15.65) means that the tetrad is a vector valued one-form that may represent angular momentum within a proportionality factor with the correct units. (The tetrad itself is unitless.) It is therefore possible to define a generally covariant angular momentum operator:

$$\hat{J}q^a{}_\mu = \pm m_J \hbar q^a{}_\mu \tag{15.66}$$

for any base manifold. Eq. (15.66) is a rotational form of the Evans field equation [2]–[12]:

$$\hat{R}q^a{}_\mu = -kTq^a{}_\mu \tag{15.67}$$

where:

$$\hat{R}q^a{}_\mu = Rq^a{}_\mu \tag{15.68}$$

and

$$\hat{R} = \square. \tag{15.69}$$

Eq. (15.69) is the operator equivalence which allows the Evans wave equation to be obtained from the Evans field equation, thus unifying general relativity with wave (or quantum) mechanics. Eq. (15.69) therefore gives more insight to the well known operator equivalence (15.18), which is at the root of the Heisenberg equation.

Consider the solution:

$$\hat{J}q^a{}_\mu = -m_J \hbar q^a{}_\mu \tag{15.70}$$

and multiply each side of Eq. (15.70) by angular frequency ω:

$$\omega \hat{J}q^a{}_\mu = -m_J \hbar \omega q^a{}_\mu. \tag{15.71}$$

The quantum $\hbar\omega$ is the quantum of kinetic energy, and so Eq. (15.71) may be written as:

$$\hat{T}q^a{}_\mu = -Tq^a{}_\mu \tag{15.72}$$

where

$$T = m_J \frac{\hbar\omega}{c^2 V}. \tag{15.73}$$

Eq. (15.72) can be identified now as the Evans field equation:

$$Rq^a{}_\mu = -kTq^a{}_\mu \tag{15.74}$$

or Evans wave equation:

$$\hat{R}q^a{}_\mu = \square q^a{}_\mu = -kTq^a{}_\mu \tag{15.75}$$

where

$$\hat{R} = \frac{k\omega\hat{J}}{c^2 V}. \tag{15.76}$$

Eq. (15.76)implies that the quantum of energy $\hbar\omega$ for any particle (including of course the photon, for which it was originally defined by Planck) originates in scalar curvature:

$$R = \pm k m_J \frac{\hbar\omega}{c^2 V}. \tag{15.77}$$

The Heisenberg equation has therefore been derived in a second way from our generally covariant theory [2]–[12] as the non-relativistic quantum limit of Eq. (15.66)

15.3 Replacement of the Heisenberg Uncertainty Principle

In classical general relativity it is always possible to define tetrad matrices with equations such as:

$$x^a = q^a{}_\mu x^\mu, \qquad p^a = q^a{}_\nu p^\nu \tag{15.78}$$

which correspond [2]–[12] to:

$$x = 4q^a{}_\mu x^\mu{}_a, \qquad p = 4q^a{}_\nu p^\nu{}_a. \tag{15.79}$$

Using the Cartan convention [13]

$$q^a{}_\mu q^\mu{}_a = 1 \tag{15.80}$$

we obtain:

$$x^\mu{}_a = \frac{1}{4} x q^\mu{}_a, \qquad p^\nu{}_b = \frac{1}{4} p q^\nu{}_b. \tag{15.81}$$

So it is always possible to define:

$$J_c = \frac{1}{16} x p q_a \wedge q_b. \tag{15.82}$$

In the limit of special relativity, the least curvature principle applies and:

$$J_c \rightarrow \hbar q_c \tag{15.83}$$

so the uncertainty principle is replaced by the classical expression:

$$x_a \wedge p_b \rightarrow \hbar \tau_c. \tag{15.84}$$

Eqn. (15.82) shows that an angular momentum in general relativity is always the wedge product of tetrads. Eqn. (15.84)can be written as:

$$x^a \wedge p^b \geq \hbar \tau^c \tag{15.85}$$

and is the law governing the least amount of angular momentum in the limit of special relativity. The essence of our argument is to replace the classical and Euclidean:

$$xp - px = 0 \tag{15.86}$$

by a commutator or wedge product of tetrads (Eq. (15.85)) in non-Euclidean spacetime with torsion. Such a spacetime was not considered by Einstein [14] in his original theory of gravitation because he restricted attention to Riemannian spacetime with a zero torsion tensor [13]. The tetrads which make up Eq. (15.85) each obey the tetrad postulate and Evans Lemma and wave equation, so are independently determined in a causal manner.

Heisenberg asserted that complementary observables such as x and p cannot be determined simultaneously to an arbitrarily high precision [1]. In other words complete knowledge of both x and p is impossible. Atkins [1] describes this obscure assertion as follows: "Some pairs of properties are not just simultaneously unknown, they are unknowable." This type of subjective (and therefore arbitrary) assertion was rejected by Einstein, de Broglie, Schrödinger and followers such as Bohm, Vigier and Sachs throughout the twentieth century. Therefore the originators of quantum mechanics and wave particle dualism steadfastly rejected Heisenberg's interpretation.

Using the wave particle dualism of de Broglie in its simplest form [1] we may replace $\hbar\kappa$ for ANY particle, not only the photon. This is the essence of Louis de Broglie's great theorem. The wave-number κ also becomes a tetrad in the Evans unified field theory and Eq. (15.85) means that the commutator of the wave-number and position in general relativity with torsion has a minimum value, the Planck constant$\hbar$. It can now be seen that Eq. (15.85) is an extension to angular motion of the phase of a wave, which is conventionally written as the relativistically invariant product $\kappa^\mu x_\mu$ In the new unified field theory [2]–[12] the phase is governed by the Evans phase law, essentially a generally covariant Stokes law.

We therefore arrive at the equation:

$$\tau^c = q^a \wedge q^b = \frac{1}{\hbar} x^a \wedge p^b \tag{15.87}$$

for the torsion form, which is missing from Einstein's original theory of general relativity applied to gravitation [14]. The wave particle dualism is expressed in differential geometry by:

$$p^b = \hbar\kappa^b \tag{15.88}$$

and so the torsion form is the wedge product:

$$\tau^c = x^a \wedge \kappa^b . \tag{15.89}$$

Similarly, the symmetric metric is the scalar product:

$$g_{\mu\nu} = q^a{}_\mu q^b{}_\nu \eta_{ab} = x^a{}_\mu \kappa^b{}_\nu \eta_{ab} \tag{15.90}$$

and is therefore identified as a phase tensor. The conventional phase of a wave is defined as the generally covariant scalar $x^\mu \kappa_\mu$ and is generalized to the Evans phase law [2]–[12] in the new unified field theory. The Berry phase and other topological phases are examples of the Evans phase law.

The action is in general the tensor valued two form:

$$S^{ab}{}_{\mu\nu} = \hbar q^{a}{}_{\mu} q^{b}{}_{\nu} = \hbar x^{a}{}_{\mu} \kappa^{b}{}_{\nu} = x^{a}{}_{\mu} p^{b}{}_{\nu} = \frac{1}{c} \int \mathcal{L} dx^{a}{}_{\mu} dx^{b}{}_{\nu} \tag{15.91}$$

where $\mathcal{L}$ is the scalar lagrangian density. The Evans least curvature principle [2]–[12] unifies the Hamilton least action and Fermat least time principles [1] and implies that the tensor action (15.91) is minimized in generalized Euler Lagrange equations of motion. Experimentally [1], the least observable action is $\hbar$ the Planck constant, and this experimental result is expressible in differential geometrical equations such as:

$$\left|x^{a} p^{b}\right| \geq \hbar \delta^{ab}, \qquad \left|x^{a} \kappa^{b}\right| \geq \delta^{ab}, \tag{15.92}$$

$$\left|q^{a} q^{b}\right| \geq \delta^{ab}. \tag{15.93}$$

Eqn.(15.92) for example means that if the wave-number form κ^{b} becomes very large (high a frequency wave) then the position form x^{a} becomes very small in to keep $\hbar$ a universal constant as observed. Eq.(15.93) means that the minimum value of the tensor product of tetrads is unity, the everywhere flat-spacetime limit. The Evans wave equation shows that a wave of any kind is a spacetime perturbation, not an entity superimposed on a frame in the flat spacetime limit. The reason for this is that the tetrad is the eigenfunction and that scalar curvature is in consequence itself quantized through the Evans Lemma.

Eq. (15.92) is the generally covariant expression that replaces the Heisenberg uncertainty principle, and Eq. (15.93) replaces Born's idea that the product of a wavefunction with its complex conjugate is a probability density [1]. These two well known but misguided principles are therefore replaced by straightforward geometrical constraints based on the experimental fact that $\hbar$ is the least observable unit of action and is a universal constant. Every consideration is therefore reduced to geometry as required by general relativity. In the course of this analysis it becomes clear that Einstein and the deterministic School were correct in thinking that quantum mechanics is an incomplete theory. The weaknesses in conventional quantum mechanics are what are conventionally considered its strengths: the uncertainty principle and the Born interpretation of the wave-function. The Evans unified field theory removes these weaknesses and replaces them with equations of differential geometry. The evolution of the wave-function in quantum mechanics is thus recognised as the causal evolution of the tetrad, which is the eigenfunction of a wave equation and therefore characterized by eigenvalues. These eigenvalues are the quanta. The action is in general the tensor valued product of tetrads multiplied by $\hbar$, and therefore the most general tensor valued action form is an object of differential geometry. There is nothing in general relativity which is unknowable, because there is nothing in geometry that is unknowable.

Acknowledgements

The Ted Annis Foundation, ADAS and Craddock Inc. and individual contributors are thanked for generous funding. The Fellows and Emeriti of AIAS are thanked for many interesting discussions.

References

1. P. W. Atkins, *Molecular Quantum Mechanics*, (Oxford Univ. Press, 1983, 2"a ed.).
2. M. W. Evans, *Found. Phys. Lett.*, **16**, 367 (2003).
3. M. W. Evans, *Found. Phys. Lett.*, **16**, 507 (2003).
4. M. W. Evans, *Found. Phys. Lett.*, **17**, 25 (2004).
5. M. W. Evans, Derivation of Dirac's Equation from the Evans Wave Equation, *Found. Phys. Lett.*, **17**, (2) (2004), in press, preprint on www.aias.us.
6. M. W. Evans, Unification of the Gravitational and Strong Nuclear Fields, *Found. Phys. Lett.*, **17** (2004), in press, preprint ion www.aias.us.
7. M. W. Evans, Physical Optics, the Sagnac Effect and the Aharonov Bohm Effect in the Evans Unified Field Theory, *Found. Phys. Lett.*, in press, **17** (2004), preprint on www.aias.us.
8. M. W. Evans, Derivation of the Geometrical Phase from the Evans Phase Law of Generally Covariant Unified Field Theory, *Found. Phys. Lett.*, **17** (2004), preprint on www.aias.us.
9. M. W. Evans, Derivation of the Evans Wave Equation from the Lagrangian and Action, *Found. Phys. Lett.*, **17** (2004), in press, preprint on www.aias.us.
10. M. W. Evans, The Evans Lemma of Differential Geometry, *Found. Phys. Lett.*, submitted for publication, preprint on www.aias.us.
11. M. W. Evans, New Concepts from the Evans Unified Field Theory, Part One: The Evolution of Curvature, Oscillatory Universe without Singularities, Causal Quantum Mechanics and General Force and Field Equations, *Found. Phys. Lett.*, in press, preprint on www.aias.us; M. W. Evans and AIAS Author Group, Development of the Evans Wave Equation in the Weak Field Limit, The Electrogravitic Equation, Found. Phys, Lett., **17**, 497 (2004) preprint on www.aias.us.
12. The Collected Scientific Papers of Myron Wyn Evans, Volumes One and Two, The Years 2000 to 2004 (www.myronevanscollectedworks.com by Artspeed of Los Angeles, California); L. Felker, ed., The Evans Equations, (World Scientific, in prep).
13. S. M. Carroll, *Lecture Notes on General Relativity* (Univ. California, Santa Barbara, graduate course, arXiv:gr-qe / 9712019 vI 3 Dec., 1997), complete course available from author on request.
14. A. Einstein, *The Meaning of Relativity* (Princeton Univ. Press, 1921).
15. J. R. Croca, *Towards a Non-Linear Quantum Physics*, (World Scientific, 2003).

16. J. B. Marion and S. T. Thornton, Classical Dynamics (HBJ, New York, 1988, 3'd ed.).
17. L. H. Ryder, *Quantum Field Theory*, (Cambridge, 1996, 2d ed.).
18. M. W. Evans (ed.), *Modern Nonlinear Optics*, a special topical issue of I. Prigogine and S. A. Rice (series eds.), *Advances in Chemical Physics*, (Wiley-Interscience, New York, 2001), vol. 119.
19. M. W. Evans and L. B. Crowell, *Classical and Quantum Electrodynamics and the* $\boldsymbol{B}^{(3)}$ *Field*, (World Scientific, 2001); M. W. Evans, J-P. Vigier et al., *The Enigmatic Photon* (Kluwer, Dordrecht 1994 to 2002, hardback and softback), five volumes.
20. M. W. Evans, Physica B, **182**, 227, 237 (1992).

16

Development Of The Evans Wave Equations In The Weak-Field Limit: The Electrogravitic Equation

Summary. The Evans wave equation [1]–[3] is developed in the weak-field limit to give the Poisson equation and an electrogravitic equation expressing the electric field strength $\boldsymbol{E}$ in terms of the acceleration g due to gravity and a fundamental scalar potential $\phi^{(0)}$ with the units or volts (joules per coulomb). The electrogravitic equation shows that an electric field strength can be obtained from the acceleration due to gravity, which in general relativity is non-Euclidean spacetime. Therefore an electric field strength can be obtained, in theory, from scalar curvature R.

Key words: Evans wave equation, unified field theory, generally covariant electrodynamics, weak field limit, electrogravitic equation.

16.1 Introduction

Recently, field and wave equations [1]–[3] for grand united field theory (GUFT) have been inferred on the basis that the electromagnetic sector must be generally covariant and that the electromagnetic potential is a tetrad. The tetrad is the one form that is the eigenfunction of the generally covariant Evans wave equation [2], which describes all four fields in GUFT. The gauge invariant electromagnetic field is the torsion form, a wedge product of two tetrads, and is defined by the first Maurer- Cartan structure relation [4]. The homogeneous and inhomogeneous field equations of generally covariant electrodynamics are identities of differential geometry [3, 4] that follow from the fact that the generally covariant potential is a tetrad one form. These inferences follow from the identification of the tangent space of general relativity with the fiber bundle space of gauge theory.

In this Letter, the Evans wave equation is developed in the weak- field limit (16.2) to give the Poisson equations of Newtonian dynamics and of electrostatics. The two Poisson equations are then compared to derive a simple but fundamental electrogravitic equation which shows that electric field strength $\boldsymbol{E}$ between two charged particles originates in the acceleration due to gravity

g generated by the classes of the two particles. The field strength $\boldsymbol{E}$ is proportional to the acceleration g through the fundamental scalar potential $\phi^{(0)}$ with the units of volts (joules per coulomb). Therefore, it may be inferred from the electrogravitic equation that the electric field strength $\boldsymbol{E}$ originates in non-Euclidean spacetime in the weak-field limit [1]–[4] and therefore from scalar curvature R. This theoretical result is supported qualitatively by reproducible and repeatable results from devices such as the motionless electromagnetic generator (MEG) [5]. Quantitative experimental tests of the electrogravitic equation will require measurements of the effect of changing mass on electric field strength, in the simplest case the electric field strength generated between two charged particles.

16.2 Derivation of the Electrogravitic Equation

The derivation starts from the Evans wave equation for gravitation [2]

$$(\square + kT)q^a{}_\mu = 0, \tag{16.1}$$

where $q^a{}_\mu$ is the tetrad one-form that describes the gravitational potential, k is Einstein's constant, T is the contracted energy momentum tensor [6] of Einstein, and $\square$ is the d'Alembertian operator for flat, or Euclidean, spacetime. The Evans wave equation for electromagnetism is then

$$(\square + kT)A^a{}_\mu = 0, \tag{16.2}$$

where the electromagnetic potential is the tetrad one-form

$$A^a{}_\mu = A^{(0)}q^a{}_\mu = \frac{\phi^{(O)}}{c}q^a{}_\mu. \tag{16.3}$$

Here $\phi^{(0)}$ is a fundamental scalar potential and c is the speed of light in vacuum. The gravitational field is given by the second Maurer Cartan structure equation

$$R^a{}_b = D \wedge \omega^a{}_b, \tag{16.4}$$

where $\omega^a{}_b$ is the spin connection, and within a factor $A^{(0)}$ the electromagnetic field is given by the first Maurer Cartan structure equation:

$$T^a = D \wedge q^a \tag{16.5}$$

where q^a is the tetrad.

In the weak-field limit [2]–[4], Eq. (16.1) reduces to the Poisson equation for Newtonian gravitation:

$$\nabla^2 \Phi = 4\pi G\rho, \tag{16.6}$$

where Φ is the gravitational potential in units of $(ms^{-1})^2$ and p is the mass density in units of kgm^{-3}. Here G is the Newton gravitational constant. In the same weak-field limit, Eq. (16.2) becomes the Poisson equation for electrostatics [3]:

$$\nabla^2 \left(\phi^{(0)}\Phi\right) = 4\pi G \left(\phi^{(0)}\rho\right); \tag{16.7}$$

so, in order to unify the theory of electrostatics with that of Newtonian gravitation, replace Φ in the equations of gravitation by $\phi^{(0)}\Phi$ to generate the equations of electrostatics.

For example, the acceleration due to gravity is

$$\boldsymbol{g} = -\boldsymbol{\nabla}\Phi, \tag{16.8}$$

and thus the electric field strength is

$$\boldsymbol{E} = -\frac{1}{c^2}\nabla\left(\Phi^{(0)}\Phi\right) \tag{16.9}$$

in S. I. units. Comparison of Eqs. (16.8) and (16.9) gives the electrogravitic equation

$$\boldsymbol{E} = -\frac{\phi^{(0)}}{c^2}\boldsymbol{g}, \tag{16.10}$$

which shows that the electric field strength between two charged particles originates in the acceleration due to gravity between the two particles. The electric field strength and the acceleration due to gravity therefore become two parts of one field, the electrogravitic field.

16.3 Discussion

The motionless electromagnetic generator [5] (MEG) may provide qualitative evidence for the fact that electric field strength in a circuit can be obtained from non-Euclidean spacetime, and the electrogravitic equation is a simple example of how this process occurs. The electric field in a circuit is generated from the product of the fundamental potential $\phi^{(0)}$ and the acceleration due to gravity, which in general relativity is non-Euclidean spacetime. The fundamental potential in volts is the scaling factor that links the electromagnetic potential to the scalar curvature [3]. The MEG has been precisely replicated [5] and thus is an example of how electric field strength and electromagnetic energy can be obtained from spacetime. However, the MEG is a complicated device; and, in order to test the electrogravitic equation quantitatively, experiments are needed on the simplest level, the interaction of two charged particles. For a given potential $\phi^{(0)}$, the equation shows that changing the mass of one particle, keeping the other mass and two charges constant, should result in a change in the electric field generated between the two particles. If the fundamental potential is known, this effect can be predicted quantitatively. Similarly changing the charge on one particle, keeping the other charge and both masses constant, should result in a small change in the acceleration due to gravity between the two particles. Again, if we know the fundamental potential precisely, this effect can be calculated quantitatively for comparison with experimental data.

Therefore these are proposed experimental tests of the hypothesis leading to the Evans equation [1]–[3]. A considerable amount of data in optics and other effects are available which prove beyond reasonable doubt the existence of the Evans-Vigier field $\boldsymbol{B}^{(3)}$ [7], the fundamental field of generally covariant electrodynamics.

Acknowledgments

The Ted Annis Foundation and Craddock Inc. are thanked for funding, and the staff of AIAS for many interesting discussions.

References

1. M. W. Evans, *Found. Phys. Lett.*, **16**, 367 (2003).
2. M. W. Evans, "A generally covariant wave equation for grand unified field theory," *Found. Phys. Lett.*, **16**, 507 (2003).
3. M. W. Evans, "The equations of grand unified field theory in terms of the Maurer-Cartan structure relations," *Found. Phys. Lett.*, **17**, 25 (2004).
4. S. P. Carroll, *Lecture Notes in General Relativity* (Univ California Santa Barbara, graduate course) arXiv:gr-gq/9712019 v1 3 Dec 1997.
5. T. E. Bearden in M. W. Evans,ed., *Modern Non-linear Optics*, in I. Prigogine and S.A. Rice eds. *Advances in Chemical Physics* (Wiley Interscience, New York, 2001) vol 119(2).
6. A. Einstein, *The Meaning of Relativity* (Princeton Univ. Press, 1921 and subsequent editions).
7. M. W. Evans, Physica B, **182**, 227 (1992). M. W. Evans, J.-P. Vigier et al., *The Enigmatic Photon* (Kluwer Academic, Dordrecht, 1994 to 2002 hardback and softback), Vols. 1 to 5; reviews in Ref. (5)

17

The Spinning and Curving of Spacetime: The Electromagnetic and Gravitational Fields In The Evans Unified Field Theory

Summary. The unification of the gravitational and electromagnetic fields achieved geometrically in the generally covariant unified field theory of Evans implies that electromagnetism is the spinning of spacetime, gravitation is the curving of spacetime. The homogeneous unified field equation of Evans is a balance of spacetime spin and curvature and governs the influence of electromagnetism on gravitation using the first Bianchi identity of differential geometry. The second Bianchi identity of differential geometry is shown to lead to the conservation law of the Evans unified field, and also to a generalization of the Einstein field equation for the unified field. Rigorous mathematical proofs are given in appendices of the four equations of differential geometry which are the cornerstones of the Evans unified field theory: the first and second Maurer Cartan structure relations and the first and second Bianchi identities. As an example of the theory the origin of wavenumber and frequency is traced to elements of the torsion tensor of spinning spacetime.

Key words: Evans unified field theory, spinning and curving spacetime, origin of the wavenumber and frequency.

17.1 Introduction

From 1925 to 1955 Einstein made various attempts to unify the gravitational and electromagnetic fields within general relativity. These attempts are summarized in updated appendices of various editions of ref. [1] and are all based on geometry. The gravitational sector of the unified field was developed by Einstein and others in terms of Riemann geometry with a symmetric Christoffel connection, $\Gamma^{\kappa}{}_{\mu\nu}$, which implies the first Bianchi identity:

$$R_{\sigma\mu\nu\rho} + R_{\sigma\nu\rho\mu} + R_{\sigma\rho\mu\nu} = 0 \tag{17.1}$$

by symmetry [2]. In Eq (17.1) $R_{\sigma\mu\nu\rho}$ is the Riemann or curvature tensor with lowered indices, defined by:

$$R_{\sigma\mu\nu\rho} = g_{\sigma\kappa} R^{\kappa}{}_{\mu\nu\rho} \tag{17.2}$$

where $g_{\sigma\kappa}$ is the symmetric metric tensor [1]. The Riemann curvature tensor is defined in terms of the gamma connection $\Gamma^{\kappa}{}_{\mu\nu}$ by:

$$R^{\rho}{}_{\sigma\mu\nu} = \partial_{\mu}\Gamma^{\rho}{}_{\nu\sigma} - \partial_{\nu}\Gamma^{\rho}{}_{\mu\sigma} + \Gamma^{\rho}{}_{\mu\lambda}\Gamma^{\lambda}{}_{\nu\sigma} - \Gamma^{\rho}{}_{\nu\lambda}\Gamma^{\lambda}{}_{\mu\sigma} \,. \tag{17.3}$$

Eq (17.3) is true for any kind of gamma connection, as is the second Bianchi identity:

$$D_{\lambda}R^{\rho}{}_{\sigma\mu\nu} + D_{\rho}R^{\sigma}{}_{\lambda\mu\nu} + D_{\sigma}R^{\lambda}{}_{\rho\mu\nu} := 0 \tag{17.4}$$

where $D\wedge$ is the covariant derivative [2] defined with the general gamma connection of any symmetry. The symmetric Christoffel connection is the special case where the gamma connection is symmetric and defined by:

$$\Gamma^{\kappa}{}_{\mu\nu} = \Gamma^{\kappa}{}_{\nu\mu} \,. \tag{17.5}$$

Using the metric compatibility postulate [2]:

$$D_{\rho}g^{\mu\nu} = 0 \tag{17.6}$$

the symmetric Christoffel connection can be expressed in terms of the symmetric metric:

$$\Gamma^{\sigma}{}_{\mu\nu} = \frac{1}{2}g^{\sigma\rho}\left(\partial_{\mu}g_{\nu\rho} + \partial_{\nu}g_{\rho\mu} - \partial_{\rho}g_{\mu\nu}\right) . \tag{17.7}$$

The use of Eq. (17.7) automatically implies that the torsion tensor $T^{\kappa}{}_{\mu\nu}$ vanishes:

$$T^{\kappa}{}_{\mu\nu} = \Gamma^{\kappa}{}_{\mu\nu} - \Gamma^{\kappa}{}_{\nu\mu} \tag{17.8}$$

So Einstein's famous gravitational theory is one in which there is no spacetime torsion or spinning. The first Bianchi identity (1) is also a special case therefore, defined by Eq. (17.5). More generally the cyclic sum in Eq. (17.1) is NOT zero if the gamma connection is not symmetric, and this turns out to be of fundamental importance for unified field theory: any mutual influence of gravitation upon electromagnetism depends on the fact that Eq. (17.1) does not hold in general. In contrast, note carefully that the second Bianchi identity (4) is ALWAYS true for any type of connection, because it is fundamentally the cyclic sum of commutators of covariant derivatives [2]:

$$[[D_{\lambda}, D_{\rho}], D_{\sigma}] + [[D_{\rho}, D_{\sigma}], D_{\lambda}] + [[D_{\sigma}, D_{\lambda}], D_{\rho}] := 0. \tag{17.9}$$

The above are the well known geometrical equations that are the cornerstones of Einstein's generally covariant theory of gravitation [1, 2].

The type of Riemann geometry almost always used by Einstein and others [2] for generally covariant gravitational field theory is a special case of the more general Cartan differential geometry [2] in which the connection is no longer symmetric and which the metric is in general the outer or tensor product of two more fundamental tetrads$q^{a}{}_{\mu}$. Thus, in differential geometry the metric tensor is in general an asymmetric matrix. Any asymmetric matrix is always the sum of a symmetric matrix and an antisymmetric matrix [3], so it is possible to construct an antisymmetric metric tensor. The symmetric metric used by Einstein to describe gravitation is therefore the special case defined by the inner or dot product of two tetrads [2]:

$$g_{\mu\nu} = q^a{}_\mu q^b{}_\nu \eta_{ab} \tag{17.10}$$

where η_{ab} is the diagonal, constant metric in the orthonormal or flat spacetime of the tetrad, indexed a:

$$\eta_{ab} = diag\,(-1, 1, 1, 1)\,. \tag{17.11}$$

Any attempt to construct a generally covariant unified field theory of all radiated and matter fields must therefore be based on differential geometry and must be based on the tetrad rather than the metric. The fundamental reason for this is that electromagnetism is known experimentally to be a spin phenomenon, and spin does not enter into Einstein's theory of gravitation because the torsion tensor vanishes as we have argued. Therefore a unified field theory must be based on geometry (such as differential geometry) that considers a non-zero torsion tensor as well as a non-zero curvature tensor. The conclusive advantage of a geometrical theory of fields over a gauge theory of fields is that the tangent bundle spacetime indexed a in the former theory is geometrical and therefore physical from the outset, whereas the fiber bundle spacetime of gauge theory is abstract and is a mathematical construct imposed for convenience on the base manifold without any reference to geometry. Thus gauge theory can never be a valid theory of general relativity because the latter is fundamentally based on geometry and must always be developed logically therefrom. Proceeding on this fundamental geometrical hypothesis of general relativity, therefore, the Evans unified field theory follows straightforwardly by tracing its origins to the fundamental equations that define differential geometry. These fundamental equations of differential geometry become the fundamental equations of the unified field theory through the Evans Ansatz [4]:

$$A^a{}_\mu = A^{(0)} q^a{}_\mu . \tag{17.12}$$

In Eq. (17.4) $A^a{}_\mu$ is the potential field of the electromagnetic sector of the unified field and the tetrad $q^a{}_\mu$ is the fundamental building block of the gravitational sector. Here $A^{(0)}$ denotes a $\hat{C}$ negative scalar originating in the magnetic fluxon $\hbar/e$, a primordial and universal constant of physics. Here $\hbar$ is the reduced Planck constant $h/(2\pi)$ and e the charge on the proton (the negative of the charge on the electron):

$$\hbar = 1.05459 \times 10^{-34} Js, \tag{17.13}$$

$$e = 1.60219 \times 10^{-19} C. \tag{17.14}$$

In Section 17.2 we give the four fundamental equations of differential geometry: the first and second Maurer Cartan structure relations and the first and second Bianchi identities and transform them into the equations of the Evans unified field [5]–[20]using Eq. (17.12). The rigorous mathematical proofs of all four equations are given in Appendices A to D.

In Section 17.3 the Maxwell Heaviside and Einstein limits of the Evans unified field theory are derived and discussed and in Section 17.4 a discussion is given of the implications of the unified field theory in evolution and various new technologies based on the ability of the gravitational and electromagnetic fields to be mutually influential.

17.2 The Fundamental Equations

The fundamental equations of the unified field theory are the fundamental equations of differential geometry [2], namely the two Maurer Cartan structure relations and the two Bianchi identities. There are two fundamental differential forms [2] that together describe any spacetime, the torsion or spin form and Riemann or curvature form. Any radiated or matter field in general relativity is therefore defined in terms of these forms. The structure relations of differential geometry define the spin and curvature forms respectively as the covariant exterior derivatives of the tetrad form and spin connection one-form:

$$T^a = D \wedge q^a = d \wedge q^a + \omega^a{}_b \wedge q^b, \tag{17.15}$$

$$R^a{}_b = D \wedge \omega^a{}_b = d \wedge \omega^a{}_b + \omega^a{}_c \wedge \omega^c{}_b \tag{17.16}$$

It is shown rigorously in the Appendices that these definitions are equivalent to the definitions of the spin and curvature tensors in terms of the gamma connection of any symmetry. Differential geometry is valid for any spacetime and any type of connection, and this realization is a key step towards the evolution of the Evans unified field theory, the first successful unified theory of the gravitational and electromagnetic fields.

The other two fundamental equations of differential geometry are the first and second Bianchi identities [2]:

$$D \wedge T^a := R^a{}_b \wedge q^b, \tag{17.17}$$

$$D \wedge R^a{}_b := 0. \tag{17.18}$$

These are written out in tensor notation and rigorously proven in the Appendices. The first Bianchi identity (17) generalizes Eq. (17.1), and the second Bianchi identity, Eq. (17.18), is Eq. (17.4) defined for any type of connection.

Eqs. (17.15) to (17.18) are the four cornerstones of any unified field theory based on geometry, i.e. of any generally covariant unified field theory. They are transformed into equations of the unified field using the Ansatz (17.12), and so Eqs. (17.15) and (17.17) become:

$$F^a = D \wedge A^a = d \wedge A^a + \omega^a{}_b \wedge A^b, \tag{17.19}$$

$$D \wedge F^a := R^a{}_b \wedge A^b. \tag{17.20}$$

Eq. (17.19) defines the field in terms of the potential, Eq. (17.20) is the homogeneous field equation of the electromagnetic sector. Eqs. (17.16) and (17.18) define the gravitational sector for any connection. In general both the spin and curvature forms are non-zero and so Eq. (17.20) demonstrates the way in which the gravitational field may influence the electromagnetic field and vice-versa. The extent to which this occurs must be found experimentally but Eq (17.20) shows that it is possible through a balance of spin and curvature in differential geometry. When both sides of Eq. (17.20) are non-zero the electromagnetic field can be influenced by the gravitational field and the gravitational field can be influenced by the electromagnetic field. In the first instance it then becomes possible to build electric power stations from spacetime curved by mass, and in the second instance it becomes possible to build counter gravitational devices built from electromagnetic technology. The possibility of such technologies must be tested by high precision experiments [21]. It seems likely that the chances of success are maximised by using high intensity femtosecond laser pulses incident on a high precision gravimeter in a high vacuum. The latter is used to remove "ion wind" artifact, i.e. extraneous effects due to atmospheric charging.

If it is found within experimental uncertainty that there is no effect of the gravitational field on the electromagnetic field and vice versa then the primordial Evans field has split entirely during the course of billions of years of evolution into what we term "pure electromagnetism" and "pure gravitation". These independent fields are described by the unified field equation:

$$d \wedge F^a = 0. \tag{17.21}$$

This is evidently an equation of differential geometry in the limit:

$$d \wedge F^a = 0, \tag{17.22}$$

$$R^a{}_b \wedge A^b = \omega^a{}_b \wedge F^b, \tag{17.23}$$

and so is a generally covariant unified field equation. In the following section we discuss this equation further in order to define precisely the Einsteinian and Maxwell - Heaviside limits of the Evans field theory.

17.3 Limiting Forms of the Evans Field

The Einsteinian limit is defined by:

$$T^a = 0, \tag{17.24}$$

so the torsion or spin form vanishes and we recover the equations of the introduction. In the language of differential geometry the Einstein field theory is therefore:

$$D \wedge q^a = 0, \tag{17.25}$$

$$R^a{}_b = D \wedge \omega^a{}_b, \tag{17.26}$$

$$R^a{}_b \wedge q^b = 0, \tag{17.27}$$

$$D \wedge R^a{}_b = 0. \tag{17.28}$$

It is defined by the two Bianchi identities with a symmetric Christoffel symbol, and by the structure relations for zero torsion or spin form. The first structure relation in the Einstein theory gives a differential equation for the tetrad in terms of the spin connection:

$$d \wedge q^a = -\omega^a{}_b \wedge q^b \tag{17.29}$$

which is equivalent to Eq. (17.7) of the introduction. The second Bianchi identity of the Einstein field theory, Eq. (17.28), leads directly [2] to the well known Einstein field equation. In tensor notation this is:

$$G_{\mu\nu} = kT_{\mu\nu} \tag{17.30}$$

where:

$$G_{\mu\nu} = R_{\mu\nu} - \frac{1}{2} R g_{\mu\nu} \tag{17.31}$$

is the Einstein field tensor. The Ricci tensor $R_{\mu\nu}$ and the metric $g_{\mu\nu}$ are symmetric in the Einstein field theory because the Christoffel connection is symmetric in its lower two indices (Eq. (17.5)). In the more general Evans unified field theory the Ricci tensor and the metric tensor are asymmetric matrices with anti-symmetric components representing spin. In the Einstein theory the tetrad postulate [2] of differential geometry:

$$D_\nu q^a{}_\mu = 0 \tag{17.32}$$

is specialized to the metric compatibility condition [2] for a symmetric metric:

$$D_\nu g^{\mu\rho} = D_\nu g_{\mu\rho} = 0. \tag{17.33}$$

Finally the canonical energy-momentum tensor is symmetric in the Einstein limit and is used in the well known Noether Theorem [2]:

$$D^\mu T_{\mu\nu} = 0. \tag{17.34}$$

More generally $T_{\mu\nu}$ is asymmetric in the Evans theory and therefore has an anti-symmetric component representing canonical angular energy/angular momentum.

So the Einstein limit of the Evans unified field theory is a special case in which the spinning of spacetime is not considered.

The Maxwell Heaviside theory of the electromagnetic field is older than general relativity and is conceptually a different theory in which the field is an abstract, mathematical entity superimposed on flat or Minkowski spacetime. In general relativity on the other hand the field is always non-Euclidean geometry itself and so must be the frame of reference itself. General relativity is simpler (one concept, the frame, instead of two concepts, field and frame) and is therefore the preferred theory by Ockham's Razor. The simplest way of thinking about this conceptual jump is to think of a helix. In general relativity the helix is the spinning and translating baseline, while in Maxwell Heaviside theory the helix is the abstract field superimposed on a static frame. If we restrict attention to three space dimensions the flat frame is the static Cartesian frame. In differential geometry the spinning and translating frame is the base manifold for the electromagnetic field, labelled, μ, and the tangent bundle is described by a Minkowski spacetime labelled a. The tetrad is defined for the electromagnetic field by:

$$V^a = q^a{}_\mu V^\mu \tag{17.35}$$

where V^μ is a vector in the base manifold, and where V^a is a vector in the tangent bundle. The tetrad is therefore the four by four invertible transformation matrix [2] between base manifold and tangent bundle. The tetrad is therefore a geometrical construct as required for general relativity. Circular polarization, discovered experimentally by Arago in 1811, is described geometrically by elements of $A^a{}_\mu$ from Eq. (17.12), i.e. by the following complex valued tetrad elements:

$$A^{(1)}{}_x = \left(A^{(0)}/\sqrt{2}\right) e^{i\phi}, \tag{17.36}$$

$$A^{(1)}{}_y = -i\left(A^{(0)}/\sqrt{2}\right) e^{i\phi}, \tag{17.37}$$

where ϕ is the electromagnetic phase. The complex conjugates of these elements are:

$$A^{(2)}{}_x = \left(A^{(0)}/\sqrt{2}\right) e^{-i\phi}, \tag{17.38}$$

$$A^{(2)}{}_y = i\left(A^{(0)}/\sqrt{2}\right) e^{-i\phi}. \tag{17.39}$$

Therefore these tetrad elements are individual components of the following vectors:

$$\boldsymbol{A}^{(1)} = \left(A^{(0)}/\sqrt{2}\right)(\boldsymbol{i} - i\boldsymbol{j})\, e^{i\phi}, \tag{17.40}$$

$$\boldsymbol{A}^{(2)} = \left(A^{(0)}/\sqrt{2}\right)(\boldsymbol{i} + i\boldsymbol{j})\, e^{-i\phi}, \tag{17.41}$$

representing a spinning and forward moving frame. This frame is multiplied by $A^{(0)}$ to give the generally covariant electromagnetic potential field. In 1992 it was inferred by Evans [22] that these vectors define the Evans spin field, $\boldsymbol{B}^{(3)}$, of electromagnetism:

$$\boldsymbol{B}^{(3)*} = -ig\boldsymbol{A}^{(1)} \times \boldsymbol{A}^{(2)}, \tag{17.42}$$

where g is defined by the wavenumber:

$$g = \frac{\kappa}{A^{(0)}}. \tag{17.43}$$

The Evans spin field is a fundamental spin invariant of general relativity and is observed through the fact [23] that circularly polarized electromagnetic radiation of any frequency magnetizes any material. This reproducible and repeatable phenomenon is known as the inverse Faraday effect.

In order to reach the Maxwell Heaviside limit from the Evans field theory we must make the above important conceptual adjustments and consider the limit:

$$R^a{}_b \wedge q^b = \omega^a{}_b \wedge T^b \tag{17.44}$$

i.e. the limit reached when there is no gravitation curvature, but when there is spacetime spin defined by the following differential geometry:

$$T^a = D \wedge q^a, \tag{17.45}$$

$$d \wedge T^a = 0, \tag{17.46}$$

$$D \wedge \omega^a{}_b = 0. \tag{17.47}$$

In this geometry the Riemann or curvature form is zero so the second Bianchi identity becomes a differential equation for the gravitational spin connection:

$$d \wedge \omega^a{}_b = -\omega^a{}_c \wedge \omega^c{}_b. \tag{17.48}$$

This is therefore the underlying differential geometry that defines the pure, generally covariant, electromagnetic field if it has split away completely from the gravitational field during the course of billions of years of evolution. If there is any residual influence of gravitation upon electromagnetism and vice versa to be found experimentally then BOTH the curvature and spin forms are experimentally non-zero, and the Evans unified field is described by the differential geometry of Section 17.2 - the geometry of the primordial Evans field.

In general relativity, once we have found the geometry we understand the physics.

For pure electromagnetism the geometry translates into its field equations using the Ansatz (17.12) to give:

$$F^a = D \wedge A^a, \tag{17.49}$$

$$d \wedge F^a = 0, \tag{17.50}$$

and the gravitational equation

$$d \wedge \omega^a{}_b = -\omega^a{}_c \wedge \omega^c{}_b. \tag{17.51}$$

The electromagnetic field is thus always defined through the spin connection by:

$$F^a = D \wedge A^a = d \wedge A^a + \omega^a{}_b \wedge A^b. \tag{17.52}$$

Generally covariant non-linear optics is therefore generated by expanding the spin connection in terms of the tetrad or potential as follows:

$$\begin{aligned}\omega^a{}_b \wedge A^b = -gA^b \wedge \big(&A^c + g\epsilon^{cde} A^d A^e \\ &+g^2 \epsilon^{cfgh} A^f A^g A^h + \cdots\big).\end{aligned} \tag{17.53}$$

All observable non-linear optical phenomena are therefore phenomena of spinning spacetime always describable by a well defined spin connection. The Maxwell Heaviside theory is the weak field limit or linear limit described by:

$$F^a = d \wedge A^a. \tag{17.54}$$

The Maxwell Heaviside theory is further restricted by the fact that it implicitly suppresses the index a, meaning that only one unwritten scalar component of the tangent bundle spacetime is considered, and then only implicitly. In other words the tangent bundle is ill defined in the Maxwell Heaviside field theory. More generally, in the Evans unified field theory, there are four physical components of the Minkowski spacetime of the tangent bundle: (ct, X, Y, Z), although a can represent any suitable set of basis elements such as unit vectors or Pauli matrices [2] in any well defined mathematical representation space of the physical tangent bundle spacetime). Suppressing the index a gives the familiar equations:

$$F = d \wedge A, \tag{17.55}$$

$$d \wedge F = 0. \tag{17.56}$$

These are the Maxwell Heaviside equations in differential geometric notation. The second equation is a combination of the Gauss Law and the Faraday Law of induction. Therefore in order to construct a unified field theory with differential geometry it is necessary to recognise that the Maxwell Heaviside structure is incomplete.

The inference of the Evans spin field in 1992 [21] was the first step towards this recognition of the incompleteness of the Maxwell Heaviside theory, and therefore towards the the unified field theory long sought after by Einstein[1] and others. The Evans spin field is the torsion or spin form component defined by the wedge product:

$$B^{(3)*} = -igA^{(1)} \wedge A^{(2)}. \tag{17.57}$$

The inference of $\boldsymbol{B}^{(3)}$ led to the development [5]–[20] of O(3) electrodynamics, in which the field is defined by:

$$\begin{gathered}F^{(3)*} = d \wedge A^{(3)*} - igA^{(1)} \wedge A^{(2)} \\ et\ \ cyclicum\end{gathered} \tag{17.58}$$

and so O(3) electrodynamics is a special case of Eqns. (17.49) to (17.53). The Gauss and Faraday Laws in O(3) electrodynamics are given by:

$$d \wedge F^a = R^a{}_b \wedge A^b - \omega^a{}_b \wedge F^b = 0. \tag{17.59}$$

Experimentally, it is found [5]–[20] that Eq. (17.59) must split into the particular solution:

$$d \wedge F^a = 0, \tag{17.60}$$

$$\omega^a{}_b \wedge F^b = R^a{}_b \wedge A^b. \tag{17.61}$$

These equations are obeyed by the circularly polarized electromagnetic potential field described in Eqs (17.40) and (17.41). Eqs. (17.60)and (17.61)also imply:

$$\boldsymbol{\nabla} \cdot \boldsymbol{B}^{(3)} = 0, \tag{17.62}$$

$$\frac{\partial \boldsymbol{B}^{(3)}}{\partial t} = \boldsymbol{\nabla} \times \boldsymbol{B}^{(3)} = \boldsymbol{0}. \tag{17.63}$$

for the Evans spin field. Therefore the latter does not give rise to Faraday induction, but gives rise to the magnetization observed in the inverse Faraday effect. Once we recognise the existence of the index a the inverse Faraday effect and all of non-linear optics follows logically. In the linear Maxwell Heaviside theory these non-linear optical effects have to be described [5]–[20] with additional ad hoc and non-linear constitutive relations which are obviously extraneous to the original linear Maxwell Heaviside structure. This original linear structure was inferred in the nineteenth century, long before the advent of non-linear optics. The latter never became available to Einstein, who never realized its significance to unified field theory. All sectors of a generally covariant unified field theory must be non-linear, because geometry is non-linear. Self consistently therefore, the linear Maxwell Heaviside structure can be obtained only if the spin connection vanishes, in which case spacetime becomes the Minkowski spacetime of special relativity, and Maxwell Heaviside theory was the first theory of special relativity. The frame covariance of the latter was first inferred (circa 1900 - 1904) by Poincare and Lorentz. Only later, in 1905, did Einstein finally extend the concept of special relativity to all of physics from electromagnetism.

The Evans unified field theory is therefore much more powerful than the earlier Maxwell Heaviside field theory, being a generally covariant theory of all radiated and matter fields. One example out of many possible examples is given to end this section, the description of the class of all Aharonov Bohm effects [5]–[20] for all fields. This class of phenomena can be defined within the context of Evans' field when F^a is zero but the potential A^a is non-zero. For the electromagnetic sector this means that:

$$d \wedge A^a = 0, \tag{17.64}$$

$$F^a = d \wedge A^a + \omega^a{}_b \wedge A^b \neq 0, \tag{17.65}$$

and for the gravitational sector it means that:

$$d \wedge q^a = -\omega^a{}_b \wedge q^b, \tag{17.66}$$

$$R^a{}_b = 0, \tag{17.67}$$

The potential field $A^a$of electromagnetism for example can interact with matter fields such as electrons when F^a is zero. The first experimental evidence for this inference was given by Chambers using a static magnetic field, but the Evans unified field theory shows that there is also an optical or electromagnetic Aharonov Bohm effect [5]–[20] and also a gravitational Aharonov Bohm effect. The latter has been observed precisely but the electromagnetic Aharonov Bohm effect has not been observed experimentally yet. However the theory of the latter effect has been given in considerable detail [5]–[20] and has major technological consequences if observed. The class of Aharonov Bohm effects is therefore explained straightforwardly as spacetime phenomena in experimental situations when F^a or T^a is zero but when q^a is non-zero. It is also possible to explain them when $R^a{}_b$ is zero and when the spin connection is non-zero. This will be the subject of a future paper. They are simply the consequence of geometry in general relativity

17.4 Consequences for Evolutionary Theory and New Technology

Before proceeding to a discussion of the implications of the Evans unified field theory we note the generally covariant wave equation:

$$D \wedge (D \wedge F^a) = R^a{}_b \wedge F^b \tag{17.68}$$

which can be derived from the two Bianchi identities written as:

$$D \wedge (D \wedge q^a) = \left(D \wedge \omega^a{}_b\right) \wedge q^b, \tag{17.69}$$

$$D \wedge \left(D \wedge \omega^a{}_b\right) := 0. \tag{17.70}$$

Using the Ansatz (17.12) Eqs. (17.69) and (17.70) become:

$$D \wedge (D \wedge A^a) := \left(D \wedge \omega^a{}_b\right) \wedge A^b, \tag{17.71}$$

$$D \wedge \left(D \wedge \omega^a{}_b\right) := 0. \tag{17.72}$$

Differentiating the right hand side of Eq (17.68) and using the Leibnitz Theorem [2]:

$$D \wedge \left(\left(D \wedge \omega^a{}_b\right) \wedge q^b\right) := \left(D \wedge \omega^a{}_b\right) \wedge \left(D \wedge q^b\right) \tag{17.73}$$

Using Eq (17.16), Eq (17.73) becomes:

$$D \wedge \left(\left(D \wedge \omega^a{}_b\right) \wedge q^b\right) := R^a{}_b \wedge T^b. \tag{17.74}$$

Use Eq (17.17) in Eq (17.74) to give:

$$D \wedge (D \wedge T^a) := R^a{}_b \wedge T^b, \tag{17.75}$$

which using the Ansatz (17.12) translates into Eq (17.68). Using the latter the condition for independent fields (no mutual interaction of gravitation and electromagnetism) becomes:

$$D \wedge (D \wedge F^a) := R^a{}_b \wedge F^b = 0 \tag{17.76}$$

which means that the fields are independent when the wedge product of the Riemann tensor and electromagnetic field tensor vanishes. Conversely if this wedge product is non-zero the fields can influence each other. This influence, if found to be non-zero experimentally with precise, well designed experiments, implies major new technology as discussed briefly already. This type of technology is governed in general by Eq. (17.68) which is a wave equation or quantum equation with the field F^a as eigenfunction or wave-function and the field $R^a{}_b$ as eigenvalues or quantum values. These new technologies would therefore depend on the fact that the quantum values of F^a are $R^a{}_b$ within a factor $A^{(0)}$, in other words it would depend on a generally covariant quantum mechanics of the unified Evans field.

Using the tetrad postulate [2]:

$$\partial_\mu q^a{}_\nu + \omega^a{}_{\mu b} q^b{}_\nu = \Gamma^\lambda{}_{\mu\nu} q^a{}_\lambda \tag{17.77}$$

the torsion form becomes the equivalent torsion tensor (see Appendix One):

$$T^a{}_{\mu\nu} = \left(\Gamma^\lambda{}_{\mu\nu} - \Gamma^\lambda{}_{\nu\mu}\right) q^a{}_\lambda, \tag{17.78}$$

$$T^\lambda{}_{\mu\nu} = q^a{}_\lambda T^a{}_{\mu\nu} = \Gamma^\lambda{}_{\mu\nu} - \Gamma^\lambda{}_{\nu\mu}. \tag{17.79}$$

Therefore, using the Ansatz (17.12) the magnetic field is defined by:

$$B^a{}_{\mu\nu} = T^\lambda{}_{\mu\nu} A^a{}_\lambda \tag{17.80}$$

and it becomes particularly clear from Eq (17.80) that spacetime spinning gives rise to electromagnetism.

The Maxwell Heaviside limit has the mathematical structure:

$$\boldsymbol{B}^{(1)} = \boldsymbol{\nabla} \times \boldsymbol{A}^{(1)} \tag{17.81}$$

where $A^{(1)}$ is given by Eq (17.40). So the magnetic field from Eqs (17.40) and (17.81) is:

$$\boldsymbol{B}^{(1)} = \frac{B^{(0)}}{\sqrt{2}} \left(i\boldsymbol{i} + \boldsymbol{j}\right) e^{i\phi}, \tag{17.82}$$

and is the limit of Eq. (17.80) when the spin connection vanishes. Taking components of Eqs. (17.39) and (17.82)

$$B^{(1)}{}_x = -B^{(1)}{}_1 = -B^{(1)}{}_{23} = B^{(1)}{}_{32} = i\frac{B^{(0)}}{\sqrt{2}}e^{i\phi}, \tag{17.83}$$

$$B^{(1)}{}_y = -B^{(1)}{}_2 = -B^{(1)}{}_{31} = B^{(1)}{}_{13} = \frac{B^{(0)}}{\sqrt{2}}e^{i\phi}, \tag{17.84}$$

$$A^{(1)}{}_0 = A^{(1)}{}_3 = B^{(1)}{}_0 = B^{(1)}{}_3 = 0. \tag{17.85}$$

Therefore simple algebra gives:

$$iB^{(0)} = A^{(0)}\left(T^1{}_{23} - iT^2{}_{23}\right). \tag{17.86}$$

As in any paradigm shift Eq (17.86) gives new insight into known things. Eq (17.86) shows that the Maxwell Heaviside field theory defines the magnetic field as a complex sum of torsion tensor components. From Eq (17.86):

$$B^{(0)} = A^{(0)}\left(T^2{}_{32} + iT^1{}_{32}\right). \tag{17.87}$$

In the Maxwell Heaviside limit we know however that [5]–[20]:

$$B^{(0)} = \kappa A^{(0)} \tag{17.88}$$

where

$$\kappa = \frac{\omega}{c}. \tag{17.89}$$

Here ω is the angular frequency and c the speed of light. From Eqs. (17.87) and (17.88) therefore

$$\kappa = T^2{}_{32} + iT^1{}_{32}, \tag{17.90}$$

a result which traces the origin of wave-number in the Maxwell Heaviside limit to a complex sum of scalar valued torsion tensor components.

It is then possible to define the scalar curvature in the Maxwell Heaviside limit:

$$R = \kappa\kappa^* = \left(T^2{}_{32}\right)^2 - \left(T^1{}_{32}\right)^2, \tag{17.91}$$

a result which illustrates the fact that we are now thinking of the electromagnetic field as a spinning of spacetime, and not as an abstract mathematical field superimposed on a static frame of reference in Minkowski (flat) spacetime. In flat spacetime the scalar curvature is zero, in spinning spacetime it is non-zero. In flat spacetime initially parallel lines remain parallel, in spinning spacetime they become geodesics [2]of the electromagnetic field. The dielectric permittivity and the absorption coefficient [24] are defined in terms of a complex wavenumber, so these fundamental spectroscopic properties are traced to a geometrical origin. Photon mass is defined by the Evans Principle of Least Curvature [5]–[20] in which is subsumed the Hamilton Principle of Least Action in dynamics and the Fermat Principle of Least Time in optics. The Evans Principle asserts that in the limit of Minkowski spacetime:

$$\kappa \rightarrow \frac{2\pi}{\lambda_0} = 2\pi\frac{mc}{\hbar}, \tag{17.92}$$

where λ_0 is the Compton wavelength of any particle of mass m, including the photon. Therefore mass is also identified has having a geometrical origin, a least or minimized curvature in the Minkowski limit. Indeed, we see from Eqs. (17.90) and (17.92) that within a factor $2\pi c/\hbar$, mass is a torsion element for any matter field or radiated field:

$$T^2_{\ 32} \rightarrow 2\pi \frac{mc}{\hbar}, \tag{17.93}$$

Therefore the origin of all fundamental physical concepts in the Evans unified field theory is differential geometry. Recently Pinter [25], in a remarkable multi-disciplinary work, has extended this basic insight to infer in a rigorously logical sequence of arguments that the origin of life is also differential geometry. Live organisms are extensions of general relativity itself and have evolved to their present condition through a sequence of transitions brought about by the theory of general relativity. The laws of physics, chemistry, biology, geology and genetics for example, are the laws of differential geometry. Within the context of the Evans unified field theory it is now recognized that electromagnetism as well as gravitation, originates in differential geometry, i.e. in the primordial or unified Evans field. Electromagnetism and gravitation are two parts of the same thing, and both are essential for the evolution of life. This inference justifies the fundamental and closely argued hypothesis used by Pinter [25], that gravitational effects evolve into effects driven by electrodynamics, for example photosynthesis in the early planet Earth. In the older Maxwell Heaviside field theory electrodynamics appears "out of the blue" but in the Evans unified field theory its origin is the same as that of gravitation. Similarly the origin of the weak and strong fields is also differential geometry in the Evans unified field theory. There are no longer any abstract internal gauge spaces or strings in nature.

The technological implications of the Evans field theory depend on its inference that one type of field may affect another, as briefly discussed already. If it were possible to obtain electromagnetic power from spacetime curved, for example, by the Earth's mass or by the mass of an electron in a circuit, the problem of burning fossil fuel would be obviated. Conversely, if it were possible for electromagnetic devices to counter or enhance the gravitational force, great technological strides would be made in any future aerospace industry. The Evans unified field theory shows that this is indeed possible but very careful, very precise, experiments are needed to measure the extent of the interaction (if any) between the sectors of the Evans field.

Acknowledgments

The Ted Annis Foundation, Craddock Inc., and Applied Science Associates are thanked for funding, and the AIAS environment for many interesting discussions. Paul Pinter is thanked for a copy of his important multi-disciplinary work. Bob Gray, Jan Abas and Michael Anderson are thanked for the construction of websites.

A

The First Maurer-Cartan Structure Relation

The first structure relation [2] defines the torsion or spin form as the exterior covariant derivative of the tetrad form:

$$T^a{}_{\mu\nu} = (D \wedge q^a)_{\mu\nu} = (d \wedge q^a)_{\mu\nu} + \omega^a{}_{\mu b} q^b{}_\nu - \omega^a{}_{\nu b} q^b{}_\mu \tag{A.1}$$

where $\omega^a{}_{\mu b}$ is the spin connection. The torsion tensor is therefore:

$$T^\lambda{}_{\mu\nu} = q^\lambda{}_a T^a{}_{\mu\nu} \tag{A.2}$$

and using the tetrad postulate:

$$T^\lambda{}_{\mu\nu} = q^\lambda{}_a \left(\partial_\mu q^a{}_\nu - \partial_\nu q^a{}_\mu + \omega^a{}_{\mu b} q^b{}_\nu - \omega^a{}_{\nu b} q^b{}_\mu \right) \tag{A.3}$$

we obtain:

$$T^\lambda{}_{\mu\nu} = \Gamma^\lambda{}_{\mu\nu} - \Gamma^\lambda{}_{\nu\mu} . \tag{A.4}$$

This is an expression for the torsion tensor in terms of the gamma connection of any symmetry. If the gamma connection is the symmetric Christoffel symbol:

$$\Gamma^\lambda{}_{\mu\nu} = \Gamma^\lambda{}_{\nu\mu} \tag{A.5}$$

then the torsion tensor vanishes.

B

The Second Maurer-Cartan Structure Relation

The second structure relation defines the Riemann or curvature form as the exterior covariant derivative of the spin connection, regarded as a one-form:

$$R^a{}_b = D \wedge \omega^a{}_b \tag{B.1}$$

i.e

$$R^a{}_{b\nu\mu} = \partial_\nu \omega^a{}_{\mu b} - \partial_\mu \omega^a{}_{\nu b} + \omega^a{}_{\nu c} \omega^c{}_{\mu b} - \omega^a{}_{\mu c} \omega^c{}_{\nu b}. \tag{B.2}$$

It is proven in this appendix that the second structure relation is equivalent to the definition of the Riemann tensor for a gamma connection of any symmetry.

The proof starts with the tetrad postulate expressed as:

$$\omega^a{}_{\mu b} = q^a{}_\nu q^\lambda{}_b \Gamma^\nu{}_{\mu\lambda} - q^\lambda{}_b \partial_\mu q^a{}_\lambda. \tag{B.3}$$

Multiplying both sides of Eq (B.3) by $q^b{}_\lambda$ and using:

$$q^b{}_\lambda q^\lambda{}_b = 1 \tag{B.4}$$

the tetrad postulate can be expressed as:

$$\partial_\mu q^a{}_\lambda = q^a{}_\nu \Gamma^\nu{}_{\mu\lambda} - q^b{}_\lambda \omega^a{}_{\mu b}. \tag{B.5}$$

Differentiating Eq (B.3) and using the Leibnitz Theorem:

$$\begin{aligned} \partial_\nu \omega^a{}_{\mu b} =& \partial_\nu \left(q^a{}_\sigma q^\lambda{}_b \Gamma^\sigma{}_{\mu\lambda} \right) - \partial_\nu \left(q^\lambda{}_b \partial_\mu q^a{}_\lambda \right) \\ =& \partial_\nu \left(q^a{}_\sigma q^\lambda{}_b \right) \Gamma^\sigma{}_{\mu\lambda} + q^a{}_\sigma q^\lambda{}_b \partial_\nu \Gamma^\sigma{}_{\mu\lambda} \\ & - \left(\partial_\nu q^\lambda{}_b \right) \left(\partial_\mu q^a{}_\lambda \right) - q^\lambda{}_b \left(\partial_\nu \partial_\mu \left(q^a{}_\lambda \right) \right). \end{aligned} \tag{B.6}$$

Now use the Leibnitz Theorem again:

$$\partial_\nu \left(q^\lambda{}_b q^a{}_\sigma \right) = q^a{}_\sigma \partial_\nu q^\lambda{}_b + q^\lambda{}_b \partial_\nu q^a{}_\sigma \tag{B.7}$$

to obtain:

$$\begin{aligned}\partial_\nu \omega^a{}_{\mu b} = &\left(q^a{}_\sigma \Gamma^\sigma{}_{\mu\lambda} - \partial_\mu q^a{}_\lambda\right) \partial_\nu q^\lambda{}_b \\ &+ q^\lambda{}_b \Gamma^\sigma{}_{\mu\lambda}\, \partial_\nu q^a{}_\sigma + q^a{}_\sigma q^\lambda{}_b \partial_\nu \Gamma^\sigma{}_{\mu\lambda} \\ &- q^\lambda{}_b \left(\partial_\nu \partial_\mu \left(q^a{}_\lambda\right)\right)\end{aligned} \tag{B.8}$$

Now use Eq. (B.5)in Eq. (B.8):

$$\begin{aligned}\partial_\nu \omega^a{}_{\mu b} =& q^b{}_\lambda \omega^a{}_{\mu b} \partial_\nu q^\lambda{}_b + q^\lambda{}_b \Gamma^\sigma{}_{\mu\lambda}\, \partial_\nu q^a{}_\sigma \\ &+ q^\lambda{}_b q^a{}_\sigma \partial_\nu \Gamma^\sigma{}_{\mu\lambda} - q^\lambda{}_b \left(\partial_\nu \partial_\mu \left(q^a{}_\lambda\right)\right).\end{aligned} \tag{B.9}$$

Switching the μ and ν indices gives:

$$\begin{aligned}\partial_\mu \omega^a{}_{\nu b} =& q^b{}_\lambda \omega^a{}_{\nu b} \partial_\mu q^\lambda{}_b + q^\lambda{}_b \Gamma^\sigma{}_{\nu\lambda}\, \partial_\mu q^a{}_\sigma \\ &+ q^\lambda{}_b q^a{}_\sigma \partial_\mu \Gamma^\sigma{}_{\nu\lambda} - q^\lambda{}_b \left(\partial_\mu \partial_\nu \left(q^a{}_\lambda\right)\right)\end{aligned} \tag{B.10}$$

which implies:

$$\begin{aligned}\partial_\nu \omega^a{}_{\mu b} - \partial_\mu \omega^a{}_{\nu b} =& q^b{}_\lambda \left(\omega^a{}_{\mu b} \partial_\nu q^\lambda{}_b - \omega^a{}_{\nu b} \partial_\mu q^\lambda{}_b\right) \\ &+ q^b{}_\lambda \left(\Gamma^\sigma{}_{\mu\lambda}\, \partial_\nu q^a{}_\sigma - \Gamma^\sigma{}_{\nu\lambda}\, \partial_\mu q^a{}_\sigma\right) \\ &+ q^\lambda{}_b q^a{}_\sigma \left(\partial_\nu \Gamma^\sigma{}_{\mu\lambda} - \partial_\mu \Gamma^\sigma{}_{\nu\lambda}\right)\end{aligned} \tag{B.11}$$

because

$$\left(\partial_\nu \partial_\mu - \partial_\nu \partial_\mu\right) q^a{}_\lambda = 0. \tag{B.12}$$

In order to evaluate the Riemann form:

$$R^a{}_{b\nu\mu} = \partial_\nu \omega^a{}_{\mu b} - \partial_\mu \omega^a{}_{\nu b} + \omega^a{}_{\nu c} \omega^c{}_{\mu b} - \omega^a{}_{\mu c} \omega^c{}_{\nu b} \tag{B.13}$$

we need:

$$\omega^a{}_{\nu c} = q^a{}_\mu q^\lambda{}_c \Gamma^\mu{}_{\nu\lambda} - q^\lambda{}_c \partial_\nu q^a{}_\lambda \tag{B.14}$$

$$\omega^a{}_{\mu b} = q^c{}_\nu q^\lambda{}_b \Gamma^\nu{}_{\mu\lambda} - q^\lambda{}_b \partial_\mu q^c{}_\lambda \tag{B.15}$$

$$\omega^a{}_{\mu c} = q^a{}_\nu q^\lambda{}_c \Gamma^\nu{}_{\mu\lambda} - q^\lambda{}_c \partial_\mu q^a{}_\lambda \tag{B.16}$$

$$\omega^c{}_{\nu b} = q^c{}_\mu q^\lambda{}_b \Gamma^\mu{}_{\nu\lambda} - q^\lambda{}_b \partial_\nu q^c{}_\lambda. \tag{B.17}$$

It is then possible to evaluate products such as:

$$\omega^a{}_{\nu c} \omega^c{}_{\mu b} = \left(q^a{}_\mu q^\lambda{}_c \Gamma^\mu{}_{\nu\lambda} - q^\lambda{}_c \partial_\nu q^a{}_\lambda\right)\left(q^c{}_\nu q^\lambda{}_b \Gamma^\nu{}_{\mu\lambda} - q^\lambda{}_b \partial_\mu q^c{}_\lambda\right). \tag{B.18}$$

The Riemann tensor can then be evaluated using:

$$R^\sigma{}_{\lambda\nu\mu} = q^\sigma{}_a q^b{}_\lambda R^a{}_{b\nu\mu}. \tag{B.19}$$

In order to evaluate Eq (B.19) first rearrange dummy indices in Eq. (B.18) as follows:

$$\begin{aligned}&q^{\lambda}{}_{c}q^{a}{}_{\mu}q^{\lambda}{}_{b}q^{c}{}_{\nu}\Gamma^{\mu}{}_{\nu\lambda}\Gamma^{\nu}{}_{\mu\lambda}\\&\qquad\downarrow(\mu\rightarrow\sigma)\\&q^{\lambda}{}_{c}q^{a}{}_{\sigma}q^{\lambda}{}_{b}q^{c}{}_{\nu}\Gamma^{\sigma}{}_{\nu\lambda}\Gamma^{\nu}{}_{\mu\lambda}\\&\qquad\downarrow(\lambda\rightarrow\rho,\nu\rightarrow\rho)\\&q^{\rho}{}_{c}q^{a}{}_{\sigma}q^{\lambda}{}_{b}q^{c}{}_{\rho}\Gamma^{\sigma}{}_{\nu\rho}\Gamma^{\rho}{}_{\mu\lambda}=q^{a}{}_{\sigma}q^{\lambda}{}_{b}\Gamma^{\sigma}{}_{\nu\rho}\Gamma^{\rho}{}_{\mu\lambda}\end{aligned}\tag{B.20}$$

Secondly cancel the term $q^{\lambda}{}_{b}\Gamma^{\sigma}{}_{\nu\rho}\partial_{\nu}q^{a}{}_{\sigma}$ in Eq. (B.11) with the term $-(q^{\lambda}{}_{c}\partial_{\nu}q^{a}{}_{\lambda})$ $(q^{\lambda}{}_{b}q^{c}{}_{\nu}\Gamma^{\nu}{}_{\mu\lambda})$ In Eq. (B.18) by rearranging dummy indices as follows:

$$\begin{aligned}&-q^{\lambda}{}_{c}q^{\lambda}{}_{b}q^{c}{}_{\nu}\Gamma^{\nu}{}_{\mu\lambda}\partial_{\nu}q^{a}{}_{\lambda}\\&\qquad\downarrow(\lambda\rightarrow\sigma)\\&-q^{\sigma}{}_{c}q^{\lambda}{}_{b}q^{c}{}_{\nu}\Gamma^{\nu}{}_{\mu\lambda}\partial_{\nu}q^{a}{}_{\sigma}\\&\qquad\downarrow(\nu\rightarrow\sigma)\\&-q^{\sigma}{}_{c}q^{\lambda}{}_{b}q^{c}{}_{\sigma}\Gamma^{\sigma}{}_{\mu\lambda}\partial_{\nu}q^{a}{}_{\sigma}=-q^{\lambda}{}_{b}\Gamma^{\sigma}{}_{\mu\lambda}\partial_{\nu}q^{a}{}_{\sigma}.\end{aligned}\tag{B.21}$$

Finally cancel the term $-q^{b}{}_{\lambda}\omega^{a}{}_{\nu b}\partial_{\mu}q^{\lambda}{}_{b}$ in Eq. (B.11) with the term $q^{\lambda}{}_{c}q^{\lambda}{}_{b}$ $(\partial_{\nu}q^{a}{}_{\lambda})(\partial_{\mu}q^{c}{}_{\lambda})-q^{a}{}_{\mu}q^{\lambda}{}_{c}q^{\lambda}{}_{b}\Gamma^{\mu}{}_{\nu\lambda}\partial_{\mu}q^{c}{}_{\lambda}$ in Eq. (B.18). To do this rewrite the Eq (B.18) term as $q^{\lambda}{}_{c}q^{\lambda}{}_{b}\partial_{\mu}q^{c}{}_{\lambda}(\partial_{\nu}q^{a}{}_{\lambda}-q^{a}{}_{\mu}\Gamma^{\mu}{}_{\nu\lambda})$ and use the tetrad postulate:

$$\partial_{\nu}q^{a}{}_{\lambda}=q^{a}{}_{\mu}\Gamma^{\mu}{}_{\nu\lambda}-q^{b}{}_{\lambda}\omega^{a}{}_{\nu b}\tag{B.22}$$

to obtain:

$$-q^{\lambda}{}_{c}q^{\lambda}{}_{b}q^{b}{}_{\lambda}\omega^{a}{}_{\nu b}\partial_{\mu}q^{c}{}_{\lambda}=-q^{c}{}_{\lambda}\omega^{a}{}_{\nu b}\partial_{\mu}q^{c}{}_{\lambda}.\tag{B.23}$$

We therefore obtain:

$$-q^{b}{}_{\lambda}\omega^{a}{}_{\nu b}\partial_{\mu}q^{\lambda}{}_{b}-\left(-q^{\lambda}{}_{c}\omega^{a}{}_{\nu b}\partial_{\mu}q^{c}{}_{\lambda}\right)=-\omega^{a}{}_{\nu b}\left(q^{c}{}_{\lambda}\partial_{\mu}q^{\lambda}{}_{c}+q^{\lambda}{}_{c}\partial_{\mu}q^{c}{}_{\lambda}\right).\tag{B.24}$$

In order to show that this is zero use:

$$q^{\lambda}{}_{c}q^{c}{}_{\lambda}=1\tag{B.25}$$

and differentiate:

$$\partial_{\mu}(q^{\lambda}{}_{c}q^{c}{}_{\lambda})=0.\tag{B.26}$$

Finally use the Leibnitz Theorem to obtain:

$$q^{\lambda}{}_{c}\partial_{\mu}q^{c}{}_{\lambda}+q^{c}{}_{\lambda}\partial_{\mu}q^{\lambda}{}_{c}=0.\tag{B.27}$$

The remaining terms give the Riemann tensor for any gamma connection:

$$R^{\lambda}{}_{\sigma\nu\mu}=\partial_{\nu}\Gamma^{\sigma}{}_{\mu\lambda}-\partial_{\mu}\Gamma^{\sigma}{}_{\nu\lambda}+\Gamma^{\sigma}{}_{\nu\rho}\Gamma^{\rho}{}_{\mu\lambda}-\Gamma^{\sigma}{}_{\mu\rho}\Gamma^{\rho}{}_{\nu\lambda}\tag{B.28}$$

quod erat demonstrandum.

C

The First Bianchi Identity

The first Bianchi identity of differential geometry is a balance of spin and curvature

$$D \wedge T^a := R^a{}_b \wedge q^b \tag{C.1}$$

and becomes the homogeneous field equation of the Evans unified field theory:

$$D \wedge F^a := R^a{}_b \wedge A^b \tag{C.2}$$

using the Evans Ansatz:

$$A^a = A^{(0)} q^a \tag{C.3}$$

So it is important to thoroughly understand the structure and meaning of the first Bianchi identity as in this Appendix. In order to proceed we need the following general definitions [2] of the exterior derivative and wedge product for any differential form:

$$(d \wedge A)_{\mu_1 \cdots \mu_{p+1}} = (p+1)\, \partial_{[\mu_1} A_{\mu_2 \cdots \mu_{p+1}]} \tag{C.4}$$

$$(A \wedge B)_{\mu_1 \cdots \mu_{p+q}} = \frac{(p+q)!}{p!q!} (p+1)\, A_{[\mu_1 \cdots \mu_p} B_{\mu_{p+1} \cdots \mu_{p+q}]}. \tag{C.5}$$

Eq (C.4) defines the exterior derivative of a p-form and Eq. (C.5) defines the wedge product of a p-form and a q-form. We also use the fact that the spin connection is a one-form [2]. The exterior covariant derivative of a one-form $X^a{}_\mu$, for example, then follows as:

$$(D \wedge X)^a{}_{\mu\nu} = (d \wedge X)^a{}_{\mu\nu} + (\omega \wedge X)^a{}_{\mu\nu} \tag{C.6}$$

where:

$$(d \wedge X)^a{}_{\mu\nu} = \partial_\mu X^a{}_\nu - \partial_\nu X^a{}_\mu \tag{C.7}$$

$$(\omega \wedge X)^a{}_{\mu\nu} = \omega^a{}_{\mu b} X^b{}_\nu - \omega^a{}_{\nu b} X^b{}_\mu . \tag{C.8}$$

Eqs. (C.7) and (C.8) follow using:

$$p = 1, q = 1, \mu_1 = \mu, \mu_2 = \nu \tag{C.9}$$

and

$$(d \wedge A)_{\mu_1 \mu_2} = (d \wedge A)_{\mu\nu} = 2\partial_{[\mu} A_{\nu]} = \partial_\mu A_\nu - \partial_\nu A_\mu \tag{C.10}$$

$$(A \wedge B)_{\mu_1 \cdots \mu_{p+q}} = (A \wedge B)_{\mu\nu} = \frac{2!}{1!1!} A_{[\mu} B_{\nu]} = A_\mu B_\nu - A_\nu B_\mu \tag{C.11}$$

Now extend this method to the exterior covariant derivative of a two-form, using:

$$\begin{aligned} (d \wedge A)_{\mu_1 \mu_2 \mu_3} &= 3\partial_{[\mu_1} A_{\mu_2 \mu_3]} \\ &= \partial_\mu A_{\nu\rho} + \partial_\nu A_{\rho\mu} + \partial_\rho A_{\mu\nu} \end{aligned} \tag{C.12}$$

and

$$\begin{aligned} (A \wedge B)_{\mu_1 \mu_2 \mu_3} &= \frac{3!}{2!1!} A_{[\mu_1} B_{\mu_2 \mu_3]} = 3A_{[\mu} B_{\nu\rho]} \\ &= A_\mu B_{\nu\rho} + A_\nu B_{\rho\mu} + A_\rho B_{\mu\nu} \end{aligned} \tag{C.13}$$

Therefore the exterior covariant derivative of the torsion or spin form used in the first Bianchi identity is:

$$(D \wedge T)^a{}_{\mu\nu\rho} = (d \wedge T)^a{}_{\mu\nu\rho} + (\omega \wedge T)^a{}_{\mu\nu\rho} \tag{C.14}$$

where:

$$(d \wedge T)^a{}_{\mu\nu\rho} = \partial_\mu T^a{}_{\nu\rho} + \partial_\nu T^a{}_{\rho\mu} + \partial_\rho T^a{}_{\mu\nu} \tag{C.15}$$

$$(\omega \wedge T)^a{}_{\mu\nu\rho} = \omega^a{}_{\mu b} T^b{}_{\nu\rho} + \omega^a{}_{\nu b} T^b{}_{\rho\mu} + \omega^a{}_{\rho b} T^b{}_{\mu\nu} \tag{C.16}$$

and where:

$$T^a{}_{\mu\nu} = \left(\Gamma^\lambda{}_{\mu\nu} - \Gamma^\lambda{}_{\nu\mu} \right) q^a{}_\lambda \tag{C.17}$$

Similarly:

$$\begin{aligned} R^a{}_b \wedge q^b &= R^a{}_{b\mu\nu} q^b{}_\rho + R^a{}_{b\nu\rho} q^b{}_\mu + R^a{}_{b\rho\mu} q^b{}_\nu \\ &= R^a{}_{\mu\nu\rho} + R^a{}_{\nu\rho\mu} + R^a{}_{\rho\mu\nu} \\ &= \left(R^\sigma{}_{\mu\nu\rho} + R^\sigma{}_{\nu\rho\mu} + R^\sigma{}_{\rho\mu\nu} \right) q^a{}_\sigma . \end{aligned} \tag{C.18}$$

So the first Bianchi identity becomes:

$$\partial_\mu T^a{}_{\nu\rho} + \omega^a{}_{\mu b} T^b{}_{\nu\rho} + \cdots = R^\sigma{}_{\mu\nu\rho} q^a{}_\sigma + \cdots . \tag{C.19}$$

Using Eq. (C.17), Eq. (C.19) becomes:

$$\begin{aligned} \partial_\mu \left(\left(\Gamma^\lambda{}_{\nu\rho} - \Gamma^\lambda{}_{\rho\nu} \right) q^a{}_\lambda \right) + \omega^\lambda{}_{\mu b} \left(\Gamma^\lambda{}_{\nu\rho} - \Gamma^\lambda{}_{\rho\nu} \right) q^b{}_\lambda + \cdots \\ = R^\lambda{}_{\mu\nu\rho} q^a{}_\lambda + \cdots \end{aligned} \tag{C.20}$$

Using the Leibnitz Theorem Eq. (C.20) becomes:

$$\begin{aligned}\left(\partial_\mu \Gamma^\lambda_{\ \nu\rho} - \partial_\mu \Gamma^\lambda_{\ \rho\nu}\right) q^a_{\ \lambda} + \left(\partial_\mu q^a_{\ \lambda} + \omega^a_{\ \mu b} q^b_{\ \lambda}\right)\left(\Gamma^\lambda_{\ \nu\rho} - \Gamma^\lambda_{\ \rho\nu}\right) \\ + \cdots \qquad = R^\lambda_{\ \mu\nu\rho} q^a_{\ \lambda} + \cdots\end{aligned} \tag{C.21}$$

Now use the tetrad postulate:

$$\partial_\mu q^a_{\ \rho} + \omega^a_{\ \mu b} q^b_{\ \sigma} = \Gamma^\lambda_{\ \mu\sigma} q^a_{\ \lambda} \tag{C.22}$$

in Eq. (C.21) to obtain:

$$\begin{aligned}&\partial_\mu \Gamma^\lambda_{\ \nu\rho} - \partial_\nu \Gamma^\lambda_{\ \mu\rho} + \Gamma^\lambda_{\ \mu\sigma}\Gamma^\sigma_{\ \nu\rho} - \Gamma^\lambda_{\ \nu\sigma}\Gamma^\sigma_{\ \mu\rho} \\ &+ \partial_\nu \Gamma^\lambda_{\ \rho\mu} - \partial_\rho \Gamma^\lambda_{\ \nu\mu} + \Gamma^\lambda_{\ \nu\sigma}\Gamma^\sigma_{\ \rho\mu} - \Gamma^\lambda_{\ \rho\sigma}\Gamma^\sigma_{\ \nu\mu} \\ &+ \partial_\rho \Gamma^\lambda_{\ \mu\nu} - \partial_\mu \Gamma^\lambda_{\ \rho\nu} + \Gamma^\lambda_{\ \rho\sigma}\Gamma^\sigma_{\ \mu\nu} - \Gamma^\lambda_{\ \mu\sigma}\Gamma^\sigma_{\ \rho\nu} \\ &:= R^\lambda_{\ \rho\mu\nu} + R^\lambda_{\ \mu\nu\rho} + R^\lambda_{\ \nu\rho\mu}.\end{aligned} \tag{C.23}$$

The Riemann tensor for any connection (Appendix two) is:

$$R^\lambda_{\ \rho\mu\nu} = \partial_\mu \Gamma^\lambda_{\ \nu\rho} - \partial_\nu \Gamma^\lambda_{\ \mu\rho} + \Gamma^\lambda_{\ \mu\sigma}\Gamma^\sigma_{\ \nu\rho} - \Gamma^\lambda_{\ \nu\sigma}\Gamma^\sigma_{\ \mu\rho}, \tag{C.24}$$

and so Eq. (C.23) is an identity made up of the cyclic sum of three Riemann tensors on either side. The familiar Bianchi identity of the famous Einstein gravitational theory is the SPECIAL CASE when the cyclic sum vanishes:

$$R^\lambda_{\ \rho\mu\nu} + R^\lambda_{\ \mu\nu\rho} + R^\lambda_{\ \nu\rho\mu} = 0. \tag{C.25}$$

Eq. (17.25) is true if and only if the gamma connection is the symmetric Christoffel symbol:

$$\Gamma^\lambda_{\ \mu\nu} = \Gamma^\lambda_{\ \nu\mu}. \tag{C.26}$$

It is not at all clear using tensor notation (Eq. (17.23)) that the first Bianchi identity is a balance of spin and curvature. In order to see this we need the differential form notation of Eq. (C.1) and this is of key importance for the development of the Evans unified field theory.

D

The Second Bianchi Identity

The second Bianchi identity is:

$$\begin{aligned} D \wedge R^a{}_b &= d \wedge R^a{}_b + \omega^a{}_c \wedge R^c{}_b + \omega^c{}_b \wedge R^a{}_c \\ &:= 0. \end{aligned} \tag{D.1}$$

using the results of Appendix Three we may write out Eq. (D.1) in tensor notation:

$$D_\rho R^a{}_{b\mu\nu} + D_\mu R^a{}_{b\nu\rho} + D_\nu R^a{}_{b\rho\nu} := 0 \tag{D.2}$$

where:

$$\begin{aligned} D_\rho R^a{}_{b\mu\nu} &= \partial_\rho R^a{}_{b\mu\nu} + \omega^a{}_{\rho c} R^c{}_{b\mu\nu} + \omega^c{}_{\rho b} R^a{}_{c\mu\nu} \\ & \textit{et cyclicum.} \end{aligned} \tag{D.3}$$

Now use:

$$R^a{}_{b\mu\nu} = q^\sigma{}_b R^a{}_{\sigma\mu\nu}. \tag{D.4}$$

The Leibnitz Theorem and tetrad postulate give the result:

$$D_\rho R^a{}_{b\mu\nu} = D_\rho \left(q^\sigma{}_b R^a{}_{\sigma\mu\nu} \right) = q^\sigma{}_b D_\rho R^a{}_{\sigma\mu\nu} \tag{D.5}$$

which implies:

$$D_\rho R^a{}_{\sigma\mu\nu} + D_\mu R^a{}_{\sigma\nu\rho} + D_\nu R^a{}_{\sigma\rho\mu} := 0 \tag{D.6}$$

Now use:

$$R^a{}_{\sigma\mu\nu} = q^a{}_\kappa R^\kappa{}_{\sigma\mu\nu} \tag{D.7}$$

The Leibnitz Theorem and tetrad postulate are used again to find:

$$D_\rho R^\kappa{}_{\sigma\mu\nu} + D_\mu R^\kappa{}_{\sigma\nu\rho} + D_\nu R^\kappa{}_{\sigma\rho\mu} := 0 \tag{D.8}$$

which is the second Bianchi identity in tensor notation for any gamma connection, quod erat demonstrandum.

The second Bianchi identity is true for ANY gamma connection because it is equivalent to:

$$[D_\rho, [D_\mu, D_\nu]] + [D_\mu, [D_\nu, D_\rho]] + [D_\nu, [D_\rho, D_\mu]] := 0. \tag{D.9}$$

and Eq. (D.9) can be summarized symbolically as a round trip with covariant derivatives around a cube: The second Bianchi identity is the geometrical

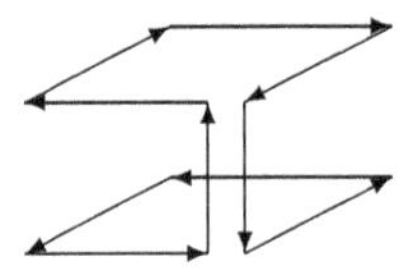

Fig. D.1. Covariant Derivatives Around a Cube

foundation for the conservation law of the Evans unified field theory:

$$D \wedge T^a{}_b := 0. \tag{D.10}$$

Eq. (D.10) is the required generalization of the Noether Theorem for the unified field theory. The second Bianchi identity is also the foundation for the generalization of the Einstein field equation in the Evans unified field theory:

$$R^a{}_b = kT^a{}_b. \tag{D.11}$$

References

1. A. Einstein, *The Meaning of Relativity* (Princeton Univ. Press, 1921 and subsequent editions).
2. S. P. Carroll, *Lecture Notes in General Relativity* (Univ California Santa Barbara, graduate course) arXiv:gr-gq/9712019 v1 3 Dec 1997.
3. G. Stephenson, *Mathematical Methods for Science Students* (Longmans, London, 1968).
4. M. W. Evans, *Generally Covariant Unified Field Theory: the Geometrization of Physics*, (in press, 2005).
5. M. W. Evans, *Found. Phys. Lett.*, **16**, 367 (2003).
6. M. W. Evans, *Found. Phys. Lett.*, **16**, 507 (2003).
7. M. W. Evans, *Found. Phys. Lett.*, **17**, 25 (2004).
8. M. W. Evans, *Found. Phys. Lett.*, **17**, 149 (2004).
9. M. W. Evans, *Found. Phys. Lett.*, **17**, 267 (2004).
10. M. W. Evans, *Found. Phys. Lett.*, **17**, 301 (2004).
11. M. W. Evans, *Found. Phys. Lett.*, **17**, 393 (2004).
12. M. W. Evans, The Evans Lemma of Differential Geometry, *Found. Phys. Lett.*, **17**, 433 (2004)
13. M. W. Evans, Derivation of the Evans Wave Equation from the Lagrangian and Action, Origin of the Planck Constant in General Relativity, *Found. Phys. Lett.*, **17**, 535 (2004)
14. M. W. Evans and the AIAS Author Group, Development of the Evans Wave Equation in the Weak Field Limit, *Found. Phys. Lett.*, **17**, 497 (2004)
15. M. W. Evans, New Concepts from the Evans Unified Field Theory: Part One, The Evolution of Curvature, Oscillatory Universe without Singularity, Causal Quantum Mechanics, and General Force and Field Equations, *Found. Phys. Lett.*, in press (preprint on www.aias.us).
16. M. W. Evans, ibid., Part two: Derivation of the Heisenberg Equation and Replacement of the Heisenberg Uncertainty Principle, *Found. Phys. Lett.*, in press (preprint on www.aias.us).
17. M. W. Evans, Derivation of the Lorentz Boost from the Evans Wave Equation, *Found. Phys. Lett.*, **17**, 663 (2004)
18. M. W. Evans, The Electromagnetic Sector of the Evans Unified Field Theory, *Found. Phys. Lett.*, in press.

19. M. W. Evans (ed.), *Modern Non-Linear Optics*, special topical issues of I. Prigogine an S. A. Rice, *Advances in Chemical Physics* (Wiley Interscience, New York, 1992 to 2001, first and second eds.), vol. 85 and 119; M. W. Evans, J.-P. Vigier et al., *The Enigmatic Photon* (Kluwer, Dordrecht, 1994 to 2002 hardback and softback), in five volumes; M. W. Evans and L. B. Crowell, *Classical and Quantum Electrodynamics and the* $\boldsymbol{B}^{(3)}$ *Field*, (World Scientific, Singapore, 2001); M. W. Evans and A. A. Hasanein, The Photomagneton in Quantum field Theory, (World Scientific, Singapore, 1994).
20. L. Felker, *The Evans Equations of Unified Field Theory* preprint on www.aias.us.
21. M. W. Evans, STAIF 2004 papers on www.aias.us.
22. M. W. Evans, Physica B, **182**, 227, 237 (1992).
23. M. Tatarikis et al., *Measurement of the Inverse Faraday Effect in High Intensity Laser produced Plasmas*, CFDL Annual report, 1998 / 1999, Rutherford Appleton Laboratory.
24. M. W. Evans, G. J. Evans, W. T. Coffey and P. Grigolini, *Molecular Dynamics and the Theory of Broad Band Spectroscopy* (Wiley-Interscience, New York, 1980).
25. P. Pinter, private communication.

18

Derivation of O(3) Electrodynamics from Generally Covariant Unified Field Theory

Summary. The equations of O(3) electrodynamics are derived as an example of Evans' generally covariant unified theory of radiated and matter fields, a theory which is based on the equations of differential geometry and which extends Einstein's theory of general relativity to electrodynamics. The latter is therefore developed into a correctly and generally covariant sector of unified field theory, one of the basic outcomes of which is the Evans spin field $\boldsymbol{B}^{(3)}$ observed in the inverse Faraday effect (magnetization by circularly or elliptically polarized electromagnetic radiation) and in many other ways now known. Another direct result of making electrodynamics a generally covariant field theory is that it is described in terms of the covariant derivatives appropriate to spacetime with torsion. The electromagnetic field is thereby recognized as the spinning of spacetime itself rather than an entity superimposed on a frame in flat spacetime. Consequently Maxwell Heaviside field theory is developed into a generally covariant form, one example of which is O(3) electrodynamics. The latter can be thought of as a gauge field theory with O(3) orthonormal space symmetry. The equations of O(3) electrodynamics are derived in detail from the generally covariant unified field theory.

Key words: Generally covariant unified field theory, O(3) electrodynamics, Evans spin field $\boldsymbol{B}^{(3)}$inverse Faraday effect.

18.1 Introduction

General relativity is in essence the geometrization of physics. The laws of natural philosophy are derived from the theorems of geometry, a giant leap forward in thought brought about by Hilbert and independently by Einstein in 1915, to whom the theory of general relativity is attributed [1]–[2]. Einstein's original theory was developed for the gravitational field and resulted in the Einstein field equation. In its most condensed form the latter is:

$$R = -kT \tag{18.1}$$

where R is scalar curvature, k the Einstein constant and T the index contracted canonical energy-momentum tensor. The left hand side is geometry, the right hand side is physics. In logic therefore, such an equation applies for all radiated and matter fields, not only gravitation. This logical outcome of the 1915 Eq. (18.1) was finally achieved in 2003 by Evans [3]–[16] and is a generally covariant unified field theory based on differential geometry. In Section 18.2 of this paper the relevant equations of differential geometry are summarized, equations from which the whole of theoretical physics can be derived logically, given Eq. (18.1). In Section 18.3 the theory is used to derive in detail the equations of O(3) electrodynamics, and to infer the existence of the Evans spin field [17] $\boldsymbol{B}^{(3)}$, which has turned out to be the key to field unification sought after by Einstein for forty years (1915 to the mid fifties). The Evans spin field is observed experimentally in numerous ways now known [16], for example in the inverse Faraday effect [18], the magnetization of matter by circularly polarized electromagnetic radiation, and in the whole of physical optics via the Evans phase law. The latter encompasses both the dynamical and the geometrical phase and produces the correctly and generally covariant Berry phase [3]–[16]in unified field matter theory

18.2 The Fundamental Geometrical Equations of the Unified Field Matter Theory

The theory has been developed in comprehensive detail elsewhere [3]–[18]. It is the purpose of this Section to conveniently summarize the equations of differential geometry upon which the field matter theory is built. These equations therefore define the fundamental structure of the theory and provide the guidelines for the development of any generally covariant theory of physics, i.e. any field matter theory, classical or quantum, provided one accepts the axioms of general relativity encapsulated in Eq. (18.1).

Adopting the condensed but well known notation [2] of contemporary differential geometry the four equations of differential geometry from which the unified field theory is developed are as follows:

$$D \wedge V^a := d \wedge V^a + \omega^a{}_b \wedge V^b \tag{18.2}$$

$$Dq^a = 0 \tag{18.3}$$

$$\tau^c = D \wedge q^c \tag{18.4}$$

$$R^a{}_b = D \wedge q^a{}_b. \tag{18.5}$$

These equations define the fundamental properties of any non-Euclidean spacetime with both curvature and torsion in terms of differential forms. Eq. (18.2) defines the covariant exterior derivative of any vector V^a, where $\omega^a{}_b$ is the spin connection and where $d\wedge$ denotes the ordinary exterior derivative [2] of differential geometry. Eq. (18.3) is the tetrad postulate, which asserts that the ordinary (as distinct for the exterior) covariant derivative of the vector valued tetrad one form vanishes for any spacetime. The covariant ordinary derivative is therefore denoted by D and the covariant exterior derivative by $D\wedge$ where $\wedge$ is the wedge operator [2]. Eqs. (18.4) and (18.5) are the first and second Maurer Cartan structure relations [2], defining respectively the torsion (τ^c) and Riemann or curvature ($R^a{}_b$) forms in terms of the tetrad and spin connection respectively.

These four equations are inter-connected in free space by the recently inferred and fundamental first and second Evans duality equations of differential geometry [3]–[17]:

$$\omega^a{}_b = -\kappa \epsilon^a{}_{bc} q^c \tag{18.6}$$

$$R^a{}_b = -\kappa \epsilon^a{}_{bc} \tau^c \tag{18.7}$$

where κ is wave-number. The novel Evans duality equations were inferred in free space from the fact that the torsion and Riemann forms are both anti-symmetric in their base manifold indices μ and ν, so that one must be the dual of the other in the orthonormal space of the tetrad [2]:

$$R^a{}_{b\mu\nu} = -\kappa \epsilon^a{}_{bc} \tau^c{}_{\mu\nu} \tag{18.8}$$

where:

$$\epsilon^{abc} = \eta_{da} \epsilon^d{}_{bc}. \tag{18.9}$$

Here ϵ_{abc} is the Levi Civita symbol (totally anti-symmetric third rank unit tensor) and where η_{da} is the metric in this orthonormal (orthogonal and normalized [2]) space. The torsion form, a vector valued two form with index c, is therefore dual to the curvature or Riemann form, a tensor valued two form anti-symmetric in its a, b indices [2]. Using the Maurer Cartan structure relations it is therefore inferred that the tetrad is dual to the spin connection, the second Evans duality equation (18.6).

The well known tetrad postulate has been developed [3]–[17] into the fundamentally important Evans Lemma (or subsidiary proposition) of differential geometry:

$$\Box q^a = R q^a \tag{18.10}$$

which gives the Evans wave equation using Eq. (18.1):

$$(\Box + kT)\, q^a = 0 \tag{18.11}$$

Eq. (18.11) unifies general relativity and quantum mechanics, making the latter a causal theory of physics and rendering the Copenhagen interpretation of the wave function unnecessary. The Evans wave equation is the generally covariant development of all the well known wave equations of physics, for example the Dirac equation. In well defined limits the Evans wave equation reduces to the generally covariant form of the Proca equation, indicating conclusively that the photon must have a non-zero mass.

The scalar curvature appearing in Eq. (18.10) may always be defined from dimensionality as the square of a wave-number:

$$R := \kappa^2 \tag{18.12}$$

and in the limit of special relativity (a free particle translating with constant velocity), this wave-number becomes the Compton wave-number of any particle (including the neutrino and photon, which must both have finite mass m):

$$\kappa_c = \frac{2\pi}{\lambda_c} = 2\pi\frac{mc}{\hbar}. \tag{18.13}$$

Here $\hbar$ is the Planck constant, and c the speed of light in vacuo. The well known and observed Compton wave-number (or wavelength) of any particle is therefore recognized for the first time to be the Evans least curvature [3]–[17] that defines mass.

These are therefore the equations of geometry, specifically the equations of differential geometry, from which all the known equations of physics may be derived, and new equations and fundamental properties such as the Evans spin field $\boldsymbol{B}^{(3)}$, inferred.

18.3 The Equations of O(3) Electrodynamics

The field theory of O(3) electrodynamics [3]–[18] is a special case of Evans' generally covariant unified field theory outlined in Section 18.2 and has been extensively tested against experimental data [19]. It has numerous known advantages [3]–[18]over the older Maxwell Heaviside field theory, and produces novel properties of physics such as the Evans spin field $\boldsymbol{B}^{(3)}$, now known to be a fundamental spin invariant of the Einstein group missing from the original 1915 theory because the latter is confined to spacetimes with zero torsion form [2] (Christoffel symbols symmetric in the lower two indices and Riemann geometry). The electrodynamical sector of Evans' unified field theory recognizes the potential field of electrodynamics to be [3]–[18]:

$$A^a = A^{(0)}q^a \tag{18.14}$$

where $A^{(0)}$ is a fundamental scalar, negative under charge conjugation symmetry ($\hat{C}$) and with the units of tesla m. The origin of the scalar $A^{(0)}$ is the universal constant $\hbar/e$, the magnetic fluxon, with units of magnetic flux (weber = volts). Here $\hbar$ is the reduced Planck constant $(h/(2\pi))$ and e the charge on the proton. (The charge on the electron is $-e$).

The generally covariant magnetic field in Evans' generally covariant unified field theory [3]–[18] must always be defined by:

$$F^a = D \wedge A^a \tag{18.15}$$

and has the units of tesla = weber m^{-2}. The reason for this is that the covariant exterior derivative is always needed [2] in differential geometry for arbitrary spacetimes with in general non-zero torsion form and Riemann form.

So it is seen that differential geometry guides us towards the generally covariant definition of the magnetic field, (and also the electric field), and as we shall see, gives us the correct and generally covariant field equations of electrodynamics. The Maxwell Heaviside field theory, although well known, is not a correct theory of general relativity because it uses ordinary derivatives in a flat spacetime (zero Riemann and torsion forms).We have seen in Section 18.2 that the torsion form is sometimes the dual of the Riemann form for all spacetimes, (the second Evans duality equation (18.7) of differential geometry), so the existence of the Riemann form sometimes implies the existence of the torsion form. In other words curvature implies torsion. It is therefore seen that Einstein's omission of the torsion form is geometrically incorrect, and this explains why he was never able to develop a unified field theory. The contemporary "standard model" is not a theory of general relativity, and is therefore not correctly covariant, because within the standard model, a flat spacetime is used for three sectors out of four (electrodynamical, weak and strong fields). Unsurprisingly therefore , the standard model is unable to account for the fundamental and generally covariant Evans spin field $\boldsymbol{B}^{(3)}$, now known to be observable in many ways [3]–[18]. Contemporary string theory is a mathematical construct (i.e. string theory is not a theory of physics, it is a construct of mathematics that uses several unphysical "dimensions") and for this reason can make no predictions about nature. In other words string theory is an obscure and elaborate mathematical way of trying to describe things that are already known in physics, and already describable more simply with already known theories of physics. For this reason string theory is not capable of predicting anything new in physics, and is not a unified field theory. String theory may be interesting for pure mathematics, but the correct geometrical basis for physics is now well known to be the Evans field matter theory [3]–[18], whose origins in differential geometry are summarized briefly in Section 18.2. The Evans theory uses only the four physical dimensions: time and three space dimensions. These are used as in standard relativity to construct four dimensional spacetime. For this reason the Evans theory is a powerful and predictive theory of nature which also reduces to the known equations of both classical and quantum physics [3]–[18].

From Eqs. (18.4) and (18.6):

$$F^a = d \wedge A^a + \frac{\kappa}{A^{(0)}} A^b \wedge A^c. \tag{18.16}$$

Defining:

$$g := \frac{\kappa}{A^{(0)}} \tag{18.17}$$

we obtain the magnetic field in general relativity:

$$F^a = D \wedge A^a = d \wedge A^a + g A^b \wedge A^c. \tag{18.18}$$

The electric field in general relativity is similarly defined with appropriate indices. The precise way of doing this in O(3) electrodynamics is developed later in this Section. The field theory of O(3) electrodynamics [3]–[18] is defined by:

$$a, b, c = (1), (2), (3) \tag{18.19}$$

so that:

$$\begin{gathered} \boldsymbol{B}^{(3)*} = \boldsymbol{\nabla} \times \boldsymbol{A}^{(3)*} - ig\boldsymbol{A}^{(1)} \times \boldsymbol{A}^{(2)} \\ et\ cyclicum. \end{gathered} \tag{18.20}$$

Here (1), (2) and (3) are the indices of the complex circular basis with O(3) group symmetry, whose three complex unit vectors, $\boldsymbol{e}^{(1)}$, $\boldsymbol{e}^{(2)}$ and $\boldsymbol{e}^{(3)}$, are cyclically inter-related:

$$\begin{gathered} \boldsymbol{e}^{(1)} \times \boldsymbol{e}^{(2)} = i\boldsymbol{e}^{(3)*} \\ et\ cyclicum \end{gathered} \tag{18.21}$$

and related to the Cartesian $\boldsymbol{i}$, $\boldsymbol{j}$ and $\boldsymbol{k}$ by:

$$\boldsymbol{e}^{(1)} = \boldsymbol{e}^{(2)*} = \frac{1}{\sqrt{2}}(\boldsymbol{i} - i\boldsymbol{j}) \tag{18.22}$$

$$\boldsymbol{e}^{(3)} = \boldsymbol{k}. \tag{18.23}$$

The inverse Faraday effect is then defined in general relativity by the magnetization:

$$\boldsymbol{M}^{(3)*} = \frac{1}{\mu_0}\frac{g'}{g}\boldsymbol{B}^{(3)*} = \frac{-i}{\mu_0}g'\boldsymbol{A}^{(1)} \times \boldsymbol{A}^{(2)} \tag{18.24}$$

where μ_0 is the permeability in vacuo and g' a coefficient in units of $e/\hbar$, the inverse fluxon. Therefore the inverse Faraday effect (magnetization by circularly or elliptically polarized electromagnetic radiation) observes $\boldsymbol{B}^{(3)}$ directly and is the magnetization of matter due to $\boldsymbol{B}^{(3)}$, as originally inferred by Evans in Dec. 1991 [17].

The correct homogeneous and inhomogeneous field equations of electrodynamics are found from Evans' generally covariant unified field theory by using the guidelines of differential geometry. The inhomogeneous field equation of generally covariant electrodynamics follows from the first Maurer Cartan structure relation (18.4) and is the fundamental Bianchi identity of differential geometry for spacetimes with both torsion and curvature [2]:

$$D \wedge \tau^a := R^a{}_b \wedge q^b. \tag{18.25}$$

Using the second Evans duality equation, Eq. (18.25) becomes:

$$D \wedge \tau^a := -\kappa\epsilon^a{}_{bc}\tau^c \wedge q^b. \tag{18.26}$$

Therefore the inhomogeneous field equation is a differential equation in the torsion form and can be solved analytically or numerically for any given situation in electrical and electronic engineering. Using Eq. (18.15), Eq. (18.26) becomes:

$$D \wedge F^a = \frac{-\kappa}{A^{(0)}} \epsilon^a{}_{bc} B^c \wedge A^b \tag{18.27}$$

i.e.

$$D \wedge F^a = g \epsilon^a{}_{bc} A^b \wedge B^c. \tag{18.28}$$

Eq. (18.28) replaces the familiar inhomogeneous field equation of Maxwell Heaviside field theory [19], and so replaces the Coulomb Law and the Ampere Maxwell Law.

The correct homogeneous field equation of electrodynamics is found from the following identity of differential geometry:

$$d \wedge \tau^a := R^a{}_b \wedge q^b - \omega^a{}_b \wedge \tau^b = 0 \tag{18.29}$$

where:

$$\widetilde{\tau}^a{}_{\rho\sigma} = \frac{1}{2} \epsilon_{\rho\sigma\mu\nu} \tau^{a\mu\nu} \tag{18.30}$$

is the dual of the torsion form in the base manifold. Using Eq. (18.15) the homogeneous field equation of generally covariant electrodynamics is therefore:

$$d \wedge F^a \sim 0 \tag{18.31}$$

and can again be solved analytically or numerically for any spacetime. The homogeneous equation (18.28) and the inhomogeneous equation (18.31) must be solved simultaneously for quantities of interest in practical electrical and electronic engineering, but this should be easily possible with contemporary software and hardware. Eq. (18.31) replaces the familiar homogeneous field equation of Maxwell Heaviside field theory and therefore replaces the Gauss Law and the Faraday Law of induction. Eq. (18.31) is an identity obeyed by any anti-symmetric tensor such as the torsion tensor, which in differential geometry becomes the vector valued torsion two form (vector valued because of the single index c, two form because of the two indices μ and ν, indicating an anti-symmetric tensor for each c [2]). In the older Maxwell Heaviside field theory the c index is missing and the electromagnetic field tensor is not recognized as a torsion form dual to the curvature or Riemann form. The reason for this is that the Maxwell Heaviside field theory is the archetypical theory of special relativity (flat spacetime) and historically preceded (circa 1900 to1905) the theory of general relativity (1915) (Riemann or curved spacetime but torsion form incorrectly omitted). It was first shown by Evans in 2003 [3]–[18] that field unification occurs correctly only in a spacetime with non-zero curvature correctly dual to non-zero torsion through the fundamental Evans duality equations (18.6) and (18.7) of differential geometry.

It follows that the correct equation of charge current density (J^a)in generally covariant electrodynamics is found from the right hand side of Eq. (18.28):

$$d \wedge F^a = \mu_0 j^a \tag{18.32}$$

From Eq. (18.31) it is seen that there are no physical magnetic monopoles in Evans' generally covariant unified field theory. The basic reason for this is that the covariant derivative of the dual of the torsion form is identically zero - a geometrical theorem obeyed in all spacetimes. There are, however, observable topological magnetic monopoles given by the covariant derivative [3]–[18]. These originate again in the fact that spacetime in general has curvature and torsion from differential geometry.

Having summarized the basic concepts the rest of this Section illustrates in detail the derivation of O(3) electrodynamics [3]–[18], which is now known to be an example of the more general Evans unified field theory.

From Eq. (18.15) it is seen that the following generally covariant gauge field is part of the general definition of gauge field:

$$G^c{}_{\mu\nu} = G^{(0)} \left(q^a{}_\mu q^b{}_\nu - q^a{}_\nu q^b{}_\mu \right). \tag{18.33}$$

Here $G^{(0)}$ is a scaling factor and $q^a{}_\mu$ and $q^b{}_\nu$ are tetrads. The magnetic field components from Eq (18.33) are:

$$\begin{gathered} B^c{}_{ij} = B^{(0)} \left(q^a{}_i q^b{}_j - q^a{}_j q^b{}_i \right), \\ i, j, k = 1, 2, 3 \end{gathered} \tag{18.34}$$

and the electric field components are:

$$E^c{}_{0i} = E^{(0)} \left(q^a{}_0 q^b{}_i - q^a{}_i q^b{}_0 \right). \tag{18.35}$$

In the complex circular basis these equations become:

$$B^{(1)*}{}_{ij} = -iB^{(0)} \left(q^{(2)}{}_i q^{(3)}{}_j - q^{(2)}{}_j q^{(3)}{}_i \right) \tag{18.36}$$

$$B^{(2)*}{}_{ij} = -iB^{(0)} \left(q^{(3)}{}_i q^{(1)}{}_j - q^{(3)}{}_j q^{(1)}{}_i \right) \tag{18.37}$$

$$B^{(3)*}{}_{ij} = -iB^{(0)} \left(q^{(1)}{}_i q^{(2)}{}_j - q^{(1)}{}_j q^{(2)}{}_i \right) \tag{18.38}$$

$$E^{(1)*}{}_{0i} = -iE^{(0)} \left(q^{(0)}{}_0 q^{(2)}{}_i - q^{(0)}{}_i q^{(2)}{}_0 \right) \tag{18.39}$$

$$E^{(2)*}{}_{0i} = iE^{(0)} \left(q^{(0)}{}_0 q^{(1)}{}_i - q^{(0)}{}_i q^{(1)}{}_0 \right) \tag{18.40}$$

$$E^{(3)*}{}_{0i} = -iE^{(0)} \left(q^{(0)}{}_0 q^{(3)}{}_i - q^{(0)}{}_i q^{(3)}{}_0 \right). \tag{18.41}$$

The tetrad is defined by the Evans wave equation for the electromagnetic potential field:

$$(\square + kT)\, A^a{}_\mu = 0. \tag{18.42}$$

The tetrad components appropriate to circularly polarized electromagnetic radiation uninfluenced by gravitation are as follows:

$$q^{(1)}{}_1 = -q^{(1)}{}_x = -ie^{i\phi}/\sqrt{2}, \tag{18.43}$$

$$q^{(1)}{}_2 = -q^{(1)}{}_y = -ie^{i\phi}/\sqrt{2}, \tag{18.44}$$

$$q^{(2)}{}_1 = -q^{(2)}{}_x = ie^{i\phi}/\sqrt{2}, \tag{18.45}$$

$$q^{(2)}{}_2 = -q^{(2)}{}_y = -ie^{i\phi}/\sqrt{2}, \tag{18.46}$$

$$q^{(0)}{}_0 = -q^{(3)}{}_z = 1. \tag{18.47}$$

The electric field components are therefore defined by:

$$\begin{aligned}
E^{(2)}{}_{01} &= E^{(1)*}{}_{01} = -iE^{(0)}q^{(0)}{}_0 q^{(2)}{}_1 \\
&= -E^{(2)}{}_1 = E^{(2)}{}_x = E^{(0)}e^{-i\phi}/\sqrt{2}, \\
E^{(2)}{}_{02} &= E^{(1)*}{}_{02} = -iE^{(0)}q^{(0)}{}_0 q^{(2)}{}_2 \\
&= -E^{(2)}{}_2 = E^{(2)}{}_y = iE^{(0)}e^{-i\phi}/\sqrt{2}, \\
E^{(1)}{}_{01} &= E^{(2)*}{}_{01} = iE^{(0)}q^{(0)}{}_0 q^{(1)}{}_1 \\
&= -E^{(1)}{}_1 = E^{(1)}{}_x = E^{(0)}e^{i\phi}/\sqrt{2}, \\
E^{(1)}{}_{02} &= E^{(2)*}{}_{02} = iE^{(0)}q^{(0)}{}_0 q^{(1)}{}_2 \\
&= -E^{(1)}{}_2 = E^{(1)}{}_y = -iE^{(0)}e^{i\phi}/\sqrt{2}, \\
E^{(3)}{}_{03} &= -iE^{(0)} = -E^{(3)}{}_3 = E^{(3)}{}_z,
\end{aligned} \tag{18.48}$$

i.e

$$\begin{aligned}
E^{(2)}{}_{01} &= -E^{(2)}{}_1 \\
E^{(2)}{}_{02} &= -E^{(2)}{}_2 \\
E^{(1)}{}_{01} &= -E^{(1)}{}_1 \\
E^{(1)}{}_{02} &= -E^{(1)}{}_2 \\
E^{(3)}{}_{03} &= -E^{(3)}{}_3.
\end{aligned} \tag{18.49}$$

and the magnetic field components by:

$$
\begin{aligned}
B^{(3)*}{}_{12} &= B^{(3)}{}_{12} = B^{(3)}{}_{z} = -B^{(3)}{}_{3} \\
&= -iB^{(0)}\left(q^{(1)}{}_{1}q^{(2)}{}_{2} - q^{(1)}{}_{2}q^{(2)}{}_{1}\right) \\
&= B^{(0)}, \\
B^{(1)*}{}_{23} &= B^{(2)}{}_{23} = B^{(2)}{}_{x} = -B^{(2)}{}_{1} \\
&= -iB^{(0)}\left(q^{(2)}{}_{2}q^{(3)}{}_{3} - q^{(2)}{}_{3}q^{(3)}{}_{2}\right) \\
&= -iB^{(0)}e^{-i\phi}/\sqrt{2}, \\
B^{(1)*}{}_{31} &= B^{(2)}{}_{31} = -B^{(2)}{}_{y} = B^{(2)}{}_{2} \\
&= -iB^{(0)}\left(q^{(2)}{}_{3}q^{(3)}{}_{1} - q^{(2)}{}_{1}q^{(3)}{}_{3}\right) \\
&= -B^{(0)}e^{-i\phi}/\sqrt{2}, \\
B^{(1)*}{}_{13} &= B^{(2)}{}_{13} = B^{(2)}{}_{y} = -B^{(2)}{}_{2} \\
&= -iB^{(0)}\left(q^{(2)}{}_{1}q^{(3)}{}_{3} - q^{(2)}{}_{3}q^{(3)}{}_{1}\right) \\
&= B^{(0)}e^{-i\phi}/\sqrt{2}, \\
B^{(1)*}{}_{32} &= B^{(2)}{}_{32} = -B^{(2)}{}_{x} = B^{(2)}{}_{1} \\
&= -iB^{(0)}\left(q^{(2)}{}_{3}q^{(3)}{}_{2} - q^{(2)}{}_{2}q^{(3)}{}_{3}\right) \\
&= iB^{(0)}e^{-i\phi}/\sqrt{2},
\end{aligned}
\tag{18.50}
$$

i.e.

$$
\begin{aligned}
B^{(3)}{}_{12} &= -B^{(3)}{}_{21} = -B^{(3)}{}_{3}, \\
B^{(2)}{}_{23} &= -B^{(2)}{}_{32} = -B^{(2)}{}_{1}, \\
B^{(2)}{}_{13} &= -B^{(2)}{}_{31} = -B^{(2)}{}_{2},
\end{aligned}
\tag{18.51}
$$

and

$$
\begin{aligned}
B^{(1)}{}_{i} &= -\frac{1}{2}\epsilon_{ijk}B^{(1)}{}_{jk}, \\
B^{(2)}{}_{i} &= -\frac{1}{2}\epsilon_{ijk}B^{(2)}{}_{jk}, \\
B^{(3)}{}_{i} &= -\frac{1}{2}\epsilon_{ijk}B^{(3)}{}_{jk}.
\end{aligned}
\tag{18.52}
$$

There are three sets of equations which give the correctly covariant form of the familiar Maxwell Heaviside field equations, to which we refer now only because of historical context. In other words the Maxwell Heaviside field equations must be regarded now as particular special cases of the more general Evans field equations defined as follows. The equations of index (1) are deduced from the particular geometrical relations implied by using the complex circular basis:

$$\begin{aligned} \boldsymbol{B}^{(1)} &= i\frac{\boldsymbol{E}^{(1)}}{c}, \\ \boldsymbol{E}^{(1)} &= -ic\boldsymbol{B}^{(1)}. \end{aligned} \tag{18.53}$$

The duality relations (18.53) are obeyed by the following set of index (1) equations:

$$\begin{aligned} \boldsymbol{\nabla}\cdot\boldsymbol{B}^{(1)} = 0,\ \ \boldsymbol{\nabla}\cdot\boldsymbol{E}^{(1)} &= 0, \\ \frac{\partial \boldsymbol{B}^{(1)}}{\partial t} + \boldsymbol{\nabla}\times\boldsymbol{E}^{(1)} &= 0, \\ \boldsymbol{\nabla}\times\boldsymbol{B}^{(1)} - \frac{1}{c^2}\frac{\partial \boldsymbol{E}^{(1)}}{\partial t} &= 0. \end{aligned} \tag{18.54}$$

These are the O(3) electrodynamical field equations of index (1) [3]–[18]. The electric and magnetic fields for index (1) can be expressed as the well known transverse plane waves

$$\begin{aligned} \boldsymbol{E}^{(1)} &= E^{(0)}\left(\boldsymbol{i} - i\boldsymbol{j}\right)e^{i\phi}/\sqrt{2}, \\ \boldsymbol{B}^{(1)} &= B^{(0)}\left(i\boldsymbol{i} + \boldsymbol{j}\right)e^{i\phi}/\sqrt{2}. \end{aligned} \tag{18.55}$$

From elementary vector analysis:

$$\frac{\partial \boldsymbol{B}^{(1)}}{\partial t} = -\omega B^{(0)}\left(\boldsymbol{i} - i\boldsymbol{j}\right)e^{i\phi}/\sqrt{2}, \tag{18.56}$$

and the curl is defined as:

$$\begin{aligned} \boldsymbol{\nabla}\times\boldsymbol{E}^{(1)} &= \frac{E^{(0)}}{\sqrt{2}}\begin{vmatrix} \boldsymbol{i} & \boldsymbol{j} & \boldsymbol{k} \\ \frac{\partial}{\partial X} & \frac{\partial}{\partial Y} & \frac{\partial}{\partial Z} \\ e^{i\phi} & -ie^{i\phi} & 0 \end{vmatrix} \\ &= \kappa E^{(0)}\left(\boldsymbol{i} - i\boldsymbol{j}\right)e^{i\phi}/\sqrt{2}. \end{aligned} \tag{18.57}$$

This verifies eq. (18.54).

Similarly:

$$\frac{\partial \boldsymbol{E}^{(1)}}{\partial t} = \omega E^{(0)}\left(i\boldsymbol{i} + \boldsymbol{j}\right)e^{i\phi}/\sqrt{2} \tag{18.58}$$

and

$$\begin{aligned} \boldsymbol{\nabla}\times\boldsymbol{B}^{(1)} &= \frac{B^{(0)}}{\sqrt{2}}\begin{vmatrix} \boldsymbol{i} & \boldsymbol{j} & \boldsymbol{k} \\ \frac{\partial}{\partial X} & \frac{\partial}{\partial Y} & \frac{\partial}{\partial Z} \\ ie^{i\phi} & e^{i\phi} & 0 \end{vmatrix} \\ &= \omega E^{(0)}\left(i\boldsymbol{i} + \boldsymbol{j}\right)e^{i\phi}/\left(\sqrt{2}c^2\right) \end{aligned} \tag{18.59}$$

thus verifying Eq. (18.54).

The O(3) field equations of index (2) are built up from the geometrical duality:

$$\begin{aligned} \boldsymbol{B}^{(2)} &= -i\frac{\boldsymbol{E}^{(2)}}{c}, \\ \boldsymbol{E}^{(2)} &= ic\boldsymbol{B}^{(2)}, \end{aligned} \tag{18.60}$$

which obey the O(3) electrodynamical field equations of index (2):

$$\begin{aligned} \boldsymbol{\nabla}\cdot\boldsymbol{B}^{(2)} = 0,\ \boldsymbol{\nabla}\cdot\boldsymbol{E}^{(2)} &= 0, \\ \frac{\partial\boldsymbol{B}^{(2)}}{\partial t} + \boldsymbol{\nabla}\times\boldsymbol{E}^{(2)} &= \boldsymbol{0}, \\ \boldsymbol{\nabla}\times\boldsymbol{B}^{(2)} - \frac{1}{c^2}\frac{\partial\boldsymbol{E}^{(2)}}{\partial t} &= \boldsymbol{0}. \end{aligned} \tag{18.61}$$

The electric and magnetic fields form Eqs. (18.61) are found to be the plane waves:

$$\begin{aligned} \boldsymbol{E}^{(2)} &= E^{(0)}\left(\boldsymbol{i}+i\boldsymbol{j}\right)e^{-i\phi}/\sqrt{2}, \\ \boldsymbol{B}^{(2)} &= B^{(0)}\left(-i\boldsymbol{i}+\boldsymbol{j}\right)e^{-i\phi}/\sqrt{2}. \end{aligned} \tag{18.62}$$

These are complex conjugates of the plane waves (18.55).

From elementary vector analysis:

$$\frac{\partial\boldsymbol{B}^{(2)}}{\partial t} = \omega B^{(0)}\left(-\boldsymbol{i}-i\boldsymbol{j}\right)e^{-i\phi}/\sqrt{2} \tag{18.63}$$

and the curl is:

$$\begin{aligned} \boldsymbol{\nabla}\times\boldsymbol{E}^{(2)} &= \frac{E^{(0)}}{\sqrt{2}}\begin{vmatrix} \boldsymbol{i} & \boldsymbol{j} & \boldsymbol{k} \\ \frac{\partial}{\partial X} & \frac{\partial}{\partial Y} & \frac{\partial}{\partial Z} \\ e^{-i\phi} & ie^{-i\phi} & 0 \end{vmatrix} \\ &= \kappa E^{(0)}\left(\boldsymbol{i}+i\boldsymbol{j}\right)e^{-i\phi}/\sqrt{2} \end{aligned} \tag{18.64}$$

thus verifying Eq. (18.61). Similarly

$$\frac{\partial\boldsymbol{E}^{(2)}}{\partial t} = -i\omega E^{(0)}\left(\boldsymbol{i}+i\boldsymbol{j}\right)e^{-i\phi}/\sqrt{2} \tag{18.65}$$

and

$$\begin{aligned} \boldsymbol{\nabla}\times\boldsymbol{B}^{(2)} &= \frac{B^{(0)}}{\sqrt{2}}\begin{vmatrix} \boldsymbol{i} & \boldsymbol{j} & \boldsymbol{k} \\ \frac{\partial}{\partial X} & \frac{\partial}{\partial Y} & \frac{\partial}{\partial Z} \\ -ie^{-i\phi} & e^{-i\phi} & 0 \end{vmatrix} \\ &= \kappa B^{(0)}\left(-i\boldsymbol{i}+\boldsymbol{j}\right)e^{-i\phi}/\sqrt{2} \end{aligned} \tag{18.66}$$

thus verifying Eq. (18.61).

The field equations of index (3) are fond from the geometrical duality:

$$\begin{aligned} \boldsymbol{B}^{(3)} &= i\frac{\boldsymbol{E}^{(3)}}{c}, \\ \boldsymbol{E}^{(3)} &= -ic\boldsymbol{B}^{(3)}, \end{aligned} \tag{18.67}$$

and are:

$$\begin{aligned} \boldsymbol{\nabla}\cdot\boldsymbol{B}^{(3)} = 0,\ \boldsymbol{\nabla}\cdot\boldsymbol{E}^{(3)} &= 0, \\ \frac{\partial\boldsymbol{B}^{(3)}}{\partial t} + \boldsymbol{\nabla}\times\boldsymbol{E}^{(3)} &= \boldsymbol{0}, \\ \boldsymbol{\nabla}\times\boldsymbol{B}^{(3)} - \frac{1}{c^2}\frac{\partial\boldsymbol{E}^{(3)}}{\partial t} &= \boldsymbol{0}. \end{aligned} \tag{18.68}$$

The fields for index (3) are missing entirely from Maxwell Heaviside field theory (spacetime with no torsion) and are the fundamental spin fields of general relativity (spacetime with torsion):

$$\begin{aligned} \boldsymbol{B}^{(3)} &= B^{(0)}\boldsymbol{k}, \\ \boldsymbol{E}^{(3)} &= -ic\boldsymbol{B}^{(3)}, \\ Re\left(\boldsymbol{E}^{(3)}\right) &= \boldsymbol{0}. \end{aligned} \tag{18.69}$$

These duality relations, field equations and fields of O(3) electrodynamics all follow from the fundamental definition (18.32), which is part of the more general definition (18.18).

The older Maxwell Heaviside field equations have the structure of Eq. (18.54), (18.61) and (18.68) but there are no indices (1), (2) and (3), and no spin field $\boldsymbol{B}^{(3)}$. The fundamental reason for this is now known to be the fact that the Maxwell Heaviside field theory is not correctly (i.e. generally) covariant. The Evans field theory is correctly covariant and is also a unified field theory which contains much more information about for example electricity and magnetism or electronics, computing and communications devices than the older Maxwell Heaviside field theory. From the Bianchi identity (18.25) and the Evans duality equations, it is clear that the Evans theory also contains information about the way in which electromagnetism influences gravitation, and this information is of importance in space propulsion engineering for example. Another example of the practical usefulness of the Evans theory stems from the fact that the Evans theory shwons that the electromagnetic field is a property of spacetime torsion. So in theory, it is possible to obtain energy from spacetime with torsion, an exceedingly important goal of energy engineering [20]. Historically, the Evans field theory was gradually inferred from the Evans spin field $\boldsymbol{B}^{(3)}$, which was in turn inferred from the experimental inverse Faraday effect.

Acknowledgments

Craddock Inc., the Ted Annis Foundation, Applied Science Associates and ADAS are thanked for generous funding of this work, and the staff of AIAS for many interesting discussions.

References

1. A. Einstein, *The Meaning of Relativity*, (Princeton Univ. Press, 1921).
2. S. M. Carroll, *Lecture Notes in General Relativity*, (Univ. California Santa Barbara, graduate course), arXiv:gr-gq/9712019 v1 3 Dec, 1997.
3. M. W. Evans, *Found. Phys. Lett.*, **16**, 367 (2003).
4. M. W. Evans, *Found. Phys. Lett.*, **16**, 507 (2003).
5. M. W. Evans, *Found. Phys. Lett.*, **17**, 25 (2004).
6. M. W. Evans, *Found. Phys. Lett.*, **17**, 149 (2004).
7. M. W. Evans, *Found. Phys. Lett.*, **17**, 267 (2004).
8. M. W. Evans, *Found. Phys. Lett.*, **17**, 301 (2004).
9. M. W. Evans, *Found. Phys. Lett.*, **17**, 393 (2004).
10. M. W. Evans, The Evans Lemma of Differential Geometry, *Found. Phys. Lett.*, **17** (2004), in press, preprint on www.aias.us.
11. M. W. Evans, Derivation of the Evans Wave Equation form the Lagrangian and Action, Origin of the Planck Constant in General Relativity, Found., Phys. Lett., **17** (2004), in press, preprint on www.aias.us.
12. M. W. Evans and AIAS Author Group, Development of the Evans Wave Equation in the Weak Field Limit, the Electrogravitic Equation, *Found. Phys. Lett.*, **17** (2004), in press, preprint on www.aias.us.
13. M. W. Evans, New Concepts form the Evans Unified Field Theory, Part One, The Evolution of Curvature, Oscillatory Universe without Singularity, Causal Quantum Mechanics and General Force and Field Equations, *Found. Phys,. Lett*, **17** (2004), submitted, preprint on www.aias.us.
14. ibid., part two: Derivation of the Heisenberg Equation and Replacement of the Heisenberg Uncertainty Principle, *Found. Phys. Lett.*, **17** (2004), submitted, preprint on www.aias.us.
15. M. W. Evans, Derivation of the Lorentz Boost from the Evans Wave Equation, *Found. Phys. Lett.*, **17** (2004), submitted, preprint on www.aias.us.
16. M. W. Evans, *Generally Covariant Unified Field Theory, the Geometrization of Physics*, (in press, 2005); L. Felker and M. W. Evans, *The Evans Equations of Unified Field Theory* (World Scientific, Singapore, in prep., 2004 / 2005); *The Collected Scientific Papers of Myron Wyn Evans*, (World Scientific , Singapore, from 2004, circa fifteen to twenty volumes), vols. 1 and 2, 2000-2004; M. W. Evans (ed.), *Modern Nonlinear optics*, a special topical issue of I. Prigogine and AS. A Rice, (series eds.), *Advances in Chemical Physics*, (Wiley Interscience,

New York, 2001, 2nd and e book eds.), vols.119(1) to 119(3), ibid., vol. 85(1) to 85(3), first ed. (Wiley Interscience, New York, 1992, 1993, 1997, hardback and softback); M. W. Evans and L. B. Crowell, *Classical and Quantum Electrodynamics and the* $\boldsymbol{B}^{(3)}$ *Field* (World Scientific, Singapore, 2001); M. W. Evans, J,.-P, Vigier et al., *The Enigmatic Photon* (Kluwer Dordrecht, 1994 to 2002, hardback and softback), in five volumes; M. W. Evans and A. A Hasanein, *The Photomagneton in Quantum Field Theory* (World Scientific, Singapore, 1994).

17. M. W. Evans, Physica B, **182**, 227, 237 (1992).
18. P. Pershan, J. P. van der Ziel, and L. D. Malmstrom, Phys. Rev., **143**, 574 (1966); J .Deschamps, M. Fitaire and M. Lagoutte, *Phys. Rev. Lett.*, **25**, 1330 (1970); M. Tatarikis et al., Measurements of the Inverse Faraday Effect in High Intensity Laser Induced Plasma, (Ann. Rep., Rutherford Appleton Lab., 1998/1999).
19. M. W. Evans, reviews in vols. 119(2) and 119(3) of Modern Nonlinear Optics (2nd edition), see these and other entries of .ref. (16).
20. T. E. Bearden et al, ref (19), vol 119(2) and www.cheniere org.

19

The Derivation of O(3) Electrodynamics from The Evans Unified Field Theory

Summary. The theory of O(3) electrodynamics is derived from the generally covariant Evans unified field theory as a well defined particular case. The latter is described by spin connection components of the electromagnetic field after it has split from the gravitational field during the course of billions of years of evolution.

Key words: O(3) electrodynamics; generally covariant unified field theory; spin connection elements.

19.1 Introduction

Recently a generally covariant unified field theory of gravitation and electromagnetism has been developed [1]–[25] from a precursor gauge field theory - O(3) electrodynamics. The unified field theory is based on differential geometry, and completes Einstein's search [26] of 1925 to 1955 for a structure that is a logical extension of his earlier 1915 theory of gravitation. The latter is based on a Riemann geometry in which the torsion tensor is zero by construction. Einstein subsequently made attempts to develop a unified field theory in which electromagnetism, as well as gravitation, is thought of as a generally covariant property of spacetime. In so doing, spin or torsion has to be incorporated within general relativity. To obtain a unified field theory it is therefore insufficient to replace derivatives by covariant derivatives with Christoffel connections, as in the Einstein Maxwell equations. The reason is that the Christoffel connection is symmetric in its lower two indices, so that the torsion tensor is subsequently zero by construction:

$$T^{\kappa}{}_{\mu\nu} = \Gamma^{\kappa}{}_{\mu\nu} - \Gamma^{\kappa}{}_{\nu\mu} = 0. \tag{19.1}$$

This means that spin is not self-consistently included in the Einstein Maxwell equations. Einstein realized this and pursued the correct geometry for thirty years. Evidently this has to be a geometry in which the torsion tensor is non-zero and this is differential geometry [27]. It has been shown recently [1]–[25]

that this geometry can be translated directly into a generally covariant unified field theory in which the basic building block is the vector-valued tetrad one-form $q^a{}_\mu$. The potential field of the electromagnetic sector is then defined by:

$$A^a{}_\mu = A^{(0)} q^a{}_\mu \tag{19.2}$$

where $A^{(0)}$ is a C negative coefficient originating in the primordial fluxon $\hbar/e$. The electromagnetic sector of the unified field is defined by the covariant exterior derivative:

$$F^a = D \wedge A^a = d \wedge A^a + \omega^a{}_b \wedge A^b \tag{19.3}$$

where $d\wedge$ denotes the exterior derivative and $\omega^a{}_b$ the spin-connection.

In Section 19.2 the structure of O(3) electrodynamics is obtained from Eq. (19.3), thus proving that O(3) electrodynamics is a well defined limit of the Evans unified field theory. A brief discussion of this result is given in Section 19.3.

19.2 The Limit of O(3) Electrodynamics

The spin connection in the unified field theory [1]–[25] must also obey the following cyclic equation:

$$\begin{aligned} d \wedge F^a &= A^{(0)} \left(R^a{}_b \wedge A^b - \omega^a{}_b \wedge F^b \right) = \mu_0 j^a \\ &\sim 0. \end{aligned} \tag{19.4}$$

In vector notation Eq. (19.4) is:

$$\boldsymbol{\nabla} \cdot \boldsymbol{B}^a = 0 \tag{19.5a}$$

$$\boldsymbol{\nabla} \times \boldsymbol{E}^a + \frac{\partial \boldsymbol{B}^a}{\partial t} = 0 \tag{19.5b}$$

and for each index a Eq. (19.5a) is the Gauss law of magnetism and Eq.(19.5b) is the Faraday law of induction Eq.(19.4) is the experimental constraint:

$$R^a{}_b \wedge q^b = \omega^a{}_b \wedge T^b \tag{19.6}$$

or free space condition. Using the Maurer-Cartan structure equations:

$$T^b = D \wedge q^b \tag{19.7}$$

$$R^a{}_b = D \wedge \omega^a{}_b \tag{19.8}$$

it is seen that a particular solution of Eq. (19.6) is:

$$\omega^a{}_b = -\frac{1}{2} \kappa \epsilon^a{}_{bc} q^c \tag{19.9}$$

where κ is a wavenumber and

$$\epsilon^a{}_{bc} = g^{ad}\epsilon_{dbc}. \tag{19.10}$$

Here

$$g^{ad} = \text{diag}(1, -1, -1, -1) \tag{19.11}$$

is the metric of the orthonormal tangent spacetime and ϵ_{dbc} is the Levi-Civita symbol.

It follows that:

$$F^1 = d \wedge A^1 + gA^2 \wedge A^3 \tag{19.12}$$
$$F^2 = d \wedge A^2 + gA^3 \wedge A^1 \tag{19.13}$$
$$F^3 = d \wedge A^3 + gA^1 \wedge A^2 \tag{19.14}$$

where

$$g = \frac{\kappa}{A^{(0)}}. \tag{19.15}$$

In the complex circular basis [1]–[25], Eqs.(19.12)–(19.14) become:

$$F^{(1)*} = d \wedge A^{(1)*} - igA^{(2)} \wedge A^{(3)} \tag{19.16}$$
$$F^{(2)*} = d \wedge A^{(2)*} - igA^{(3)} \wedge A^{(1)} \tag{19.17}$$
$$F^{(3)*} = d \wedge A^{(3)*} - igA^{(1)} \wedge A^{(2)}, \tag{19.18}$$

and these are the fundamental definitions of the field tensors of O(3) electrodynamics [1]–[25].

The Evans spin field is:

$$B^{(3)*} = -igA^{(1)} \wedge A^{(2)} \tag{19.19}$$

and is observed experimentally in the magnetization of all materials by circular polarized electromagnetic radiation at any frequency, the inverse Faraday effec [1]–[25].

It is seen that the $\boldsymbol{B}^{(3)}$ spin field is the result of , and a fundamental property of, objective physics, i.e. the result of general relativity. Specially, $\boldsymbol{B}^{(3)}$ arises from the spinning frame which represents electromagnetism in objective physics, it is defined by the spin connection. The $\boldsymbol{B}^{(3)}$ spin field is not defined in the Maxwell-Heaviside field theory of special relativity because in that nineteenth century theory the electromagnetic field is an entity superimposed on a static frame in Minkowski spacetime("flat spacetime"). In the Minkowski spacetime there is no spin connection, and no $\boldsymbol{B}^{(3)}$ spin field. Therefore the Evans unified field theory of objective physics is preferred experimentally and on the fundamental basis of objectivity in physics.

There is no $\boldsymbol{E}^{(3)}$ field because:

$$\begin{aligned} F_{03}^{(3)*} &= \left(d \wedge A^{(3)}\right)^*_{03} - igA_0^{(1)} \wedge A_3^{(2)} \\ &= 0 \end{aligned} \tag{19.20}$$

as again observed experimentally. There is no electric analogue of the inverse Faraday effect. Analogously, the plane of polarization of light is rotated only by a static magnetic field (the Faraday effect), and is not rotated by a static electric field.

19.3 Discussion

The O(3) electrodynamic structure derived in Section 19.2 is a geometric structure in which the tangent bundle index a is a well defined property of differential geometry and therefore of generally covariant physics. The original O(3) electrodynamics [1]–[25] was developed as a gauge field theory in which the internal index a is an index of the fiber bundle imposed on a flat spacetime. The mathematical structure is the same in both cases, but the geometrical interpretation is preferred because it is generally covariant. All laws of physics must be generally covariant and objective in any frame of reference. The latter interpretation allows field unification through the first Bianchi identity of differential geometry, for example, producing the homogeneous field equation of the unified field:

$$D \wedge F^a = R^a{}_b \wedge A^b. \tag{19.21}$$

When the electromagnetic and gravitational fields split into separate entities, equation (19.1) splits into:

$$d \wedge F^a = 0 \tag{19.22}$$

$$R^a{}_b \wedge A^b = \omega^a{}_b \wedge F^b. \tag{19.23}$$

The Bianchi identity used in Einstein's gravitational theory of 1915 is also obeyed, in tensor notation it becomes the familiar:

$$R_{\sigma\mu\nu\rho} + R_{\sigma\nu\rho\mu} + R_{\sigma\rho\mu\nu} = 0. \tag{19.24}$$

Eq. (19.24) is true if and only if Eq. (19.1) is true, i.e. if and only if the Christoffel connection is symmetric, in which case the torsion tensor vanishes, and with it the electromagnetic field. This is therefore a self-consistent argument, the electromagnetic field being the torsion tensor.

More generally the Bianchi identity (19.24) is not true, this means physically that there can be a tiny influence of gravitation on electromagnetism and vice versa. This influence must be tiny because it is known that the Gauss Law and Faraday Law of induction hold to great precision., i.e. it is known experimentally that Eq. (19.22) must be true to great precision. The familiar Maxwell-Heaviside structure is:

$$F = d \wedge A \tag{19.25}$$

$$d \wedge F = 0 \tag{19.26}$$

and is well known to be the archetypical theory of special relativity, invariant only upon Lorentz transformation, and not under general coordinate transformation [27]. The Evans unified field theory in contrast is generally covariant, and thus objective under any type of coordinate transformation. It is therefore preferred to the Maxwell-Heaviside field theory. Experimentally, the Evans field theory is preferred because it is capable of self-consistently generating non-linear optical phenomena such as magnetization by circularly polarized electromagnetic radiation (the inverse Faraday effect [1]–[25]. The latter is magnetization in any material and at any frequency by the Evans spin field:

$$\boldsymbol{B}^{(3)*} = -igA^{(1)} \wedge A^{(2)}, \tag{19.27}$$

a fundamental property of electromagnetic radiation at any frequency and in any state of polarization. The Evans spin field is a generally covariant property of nature, i.e a property and prediction of Einstein's theory of general relativity recently extended by Evans [1]–[25] to a unified field theory of all radiated and matter fields. The Evans spin field is a non-linear property of electromagnetic radiation and is an element of the torsion form. For these reasons it is not defined in the Maxwell-Heaviside field theory of electromagnetism, because the latter theory is linear and Lorentz covariant only. The Maxwell-Heaviside field theory does not consider the torsion form or spin connection of differential geometry because it is a flat spacetime theory in which the spin connection, torsion and Riemann forms all vanish. Thus, the homogeneous field equation of the Maxwell-Heaviside field theory is Eq. (19.26), in which the covariant exterior derivative $D\wedge$ is replaced by the exterior derivative $d\wedge$ and in which the spin connection vanishes. There being no spin connection, the internal index a loses meaning, so Eq. (19.22) becomes Eq. (19.26). In this way we may recover the Maxwell-Heaviside structure from the Evans unified field theory in the limit of flat spacetime where the spin connection approaches zero asymptotically. In this limit gravitation is obviously absent, because the spacetime is flat and the Riemann form has vanished.

However, this procedure loses a great deal of information, all of non-linear optics is thrown away, and the basic axiom of Einsteinian natural philosophy is thrown away: general covariance. So we reject this procedure despite the fact that it gives the Maxwell-Heaviside structure straightforwardly as an asymptotic limit of the Evans field theory. Much preferable is the O(3) electrodynamics procedure represented by:

$$F^a = D \wedge A^a \tag{19.28}$$

$$d \wedge F^a = 0 \tag{19.29}$$

and well tested experimentally as summarized in the literature [1]–[25]. Eq. (19.29) allows us to recover the well tested Gauss Law and Faraday Law of induction without losing general covariance and without losing the inverse Faraday effect and the Evans spin field. More generally, O(3) electrodynamics can be developed into a theory capable of describing any type of non-linear optical effect by choosing the spin connection to satisfy the experimental requirements of non-linear optics while retaining general covariance.

The Evans unified field theory also unifies causal general relativity with wave mechanics by developing the well known tetrad postulate of differential geometry:

$$D_\nu q^a{}_\mu = 0 \tag{19.30}$$

into the Evans Lemma:

$$\Box q^a{}_\mu = R q^a{}_\mu . \tag{19.31}$$

The lemma (19.31) is a subsidiary geometrical proposition which shows that spacetime is quantized. The wave-function is the tetrad $q^a{}_\mu$. In the electrodynamic sector the Lemma becomes:

$$\Box A^a{}_\mu = R A^a{}_\mu , \tag{19.32}$$

an equation which shows that the electromagnetic field is quantized in a causal manner. We therefore reject the acausal assertions of the Copenhagen School in favour of the causal axioms of the Determinist School. Finally a general form of the famous Einstein field equation:

$$R = -kT \tag{19.33}$$

is used to translate the Lemma into the Evans wave equation [1]–[25]:

$$(\Box + kT)\, q^a{}_\mu = 0. \tag{19.34}$$

Here k is the Einstein constant, T a contracted energy-momentum tensor and R a scalar curvature. Note carefully however that the quantities R and T are no longer confined to gravitation, they are properties of the Evans unified field. In the latter the Ricci tensor $R_{\mu\nu}$ and the canonical energy-momentum tensor $T_{\mu\nu}$ are asymmetric for the unified Evans field, and so both tensors must always be a sum of symmetric and anti-symmetric components from a basic theorem of matrices [28]. The metric $g_{\mu\nu}$ is also asymmetric in the Evans unified field, i.e. is a tensor product of tetrads:

$$g^{ab}{}_{\mu\nu} = q^a{}_\mu q^b{}_\nu . \tag{19.35}$$

However, we may still contract in the same way as Einstein:

$$R = g^{ab}{}_{\mu\nu} R^{\mu\nu}{}_{ab} ; T = g^{ab}{}_{\mu\nu} T^{\mu\nu}{}_{ab} ; g^{ab}{}_{\mu\nu} g^{\mu\nu}{}_{ab} = 4, \tag{19.36}$$

and thus recover Eq. (19.35) from the famous Einstein wave equation:

$$R^{ab}{}_{\mu\nu} - \frac{1}{2} R g^{ab}{}_{\mu\nu} = k T^{ab}{}_{\mu\nu} . \tag{19.37}$$

Evidently, the latter also becomes an equation in asymmetric tensors in the Evans unified field theory and not just symmetric tensors as in the original 1915 gravitational theory of Einstein.

The Evans wave equation has been tested [1]–[25] in well known limits and shown to produce all the main equations of physics such as Dirac's equation. It is possible to recover the latter equation because the tangent bundle index a of the tetrad can be used in any representation space [27]. Having obtained Dirac's equation we have inferred fermions and elementary anti-particles from general relativity. Similarly the structure of nuclear weak and strong field theory can be inferred using appropriate representation spaces for the tangent bundle of the tetrad. A lagrangian can always be defined from which the field equation is recovered by variation and Euler Lagrange equation. The covariant derivative can always be chosen to build any type of generally covariant unified field theory of elementary particles. Having already obtained the Maxwell-Heaviside limit we can if desired automatically obtain the electro-weak field of Glashow, Weinberg and Salaam (GWS) as a limit of the Evans unified field theory. However we reject this procedure because it has the same inherent difficulties as described already: the Maxwell-Heaviside field throws away non-linear optics and general covariance, and therefore so does GWS theory. Far preferable is to develop O(3) electrodynamics into an electro-weak field theory and some attempts have already been made in this direction [1]–[25]. More generally the Evans unified field theory should be developed rigorously into a generally covariant electro-weak theory by suitable choice of representation space and geometry. The overall structure of quantum strong field theory (quantum chromodynamics) has also been recovered from the Evans unified field theory [1]–[25] using the required SU(3) representation space. Therefore if quarks exist as the most elementary particles they can if desired be described in this way from the Evans unified field theory. However quarks are postulated to be unobservable (confined) and in natural philosophy that which is unobservable is not physical. Finally the Evans spin field is the archetypical string [1]–[25] of the electromagnetic field and the Evans field theory could be developed with string theory. However the latter is often regarded as a mathematical construct rather than a theory of physics.

Acknowledgments

The Ted Annis Foundation, Craddock Inc., and Applied Science Associates are thanked for generous funding, and the AIAS group for many interesting discussions.

References

1. M. W. Evans, *Found. Phys. Lett.*, **16**, 367 (2003).
2. M. W. Evans, *Found. Phys. Lett.*, **16**, 507 (2003).
3. M. W. Evans, *Found. Phys. Lett.*, **17**, 25 (2004).
4. M. W. Evans, *Found. Phys. Lett.*, **17**, 149 (2004).
5. M. W. Evans, *Found. Phys. Lett.*, **17**, 267 (2004).
6. M. W. Evans, *Found. Phys. Lett.*, **17**, 301 (2004).
7. M. W. Evans, *Found. Phys. Lett.*, **17**, 393 (2004).
8. M. W. Evans, *Found. Phys. Lett.*, **17**, 433 (2004).
9. M. W. Evans and the AIAS Author Group, *Found. Phys. Lett.*, 17, 497 (2004).
10. M. W. Evans, New Concepts from the Evans Unified Field Theory: Part One, The Evolution of Curvature, Oscillatory Universe Without Singularity, Causal Quantum Mechanics and General Force and Field Equations, *Found. Phys. Lett.*, in press, preprint on www.aias.us.
11. Ibid., part 2, Derivation of the Heisenberg Equation and Replacement of the Heisenberg Uncertainty Principle, *Found. Phys. Lett.*, in press, preprint on www.aias.us.
12. M. W. Evans, "Derivation of the Lorentz Boost from the Evans Wave Equation", *Found. Phys. Lett.*, in press 2004 / 2005 (preprint on www.aias.us).
13. M. W. Evans, "The Electromagnetic Sector of the Evans Unified Field Theory.", *Found. Phys. Lett.*, in press 2004 / 2005 (preprint on www.aias.us).
14. M. W. Evans, "The Spinning and Curving of Spacetime: The Electromagnetic and Gravitational Fields in the Evans Unified Field Theory", *Found. Phys. Lett.*, in press, (preprint on www.aias.us).
15. M. W. Evans, *Generally Covariant Unified Field Theory: The Geometrization of Physics* (in press, 2005).
16. M. W. Evans and S. Kielich (eds.), *Modern Non-Linear Optics*, a special topical issue in three parts of I. Prigogine and S. A. Rice (Series Editors.), *Advances in Chemical Physics* (Wiley-Interscience, New York , 1992, 1993, 1997, first edition, hardback and softback), vol. 85.
17. M. W. Evans and A. A. Hasanein, *The Photomagneton in Quantum Field Theory* (World Scientific, Singapore, 1994).
18. M. W. Evans, J.-P. Vigier et alii, *The Enigmatic Photon* (Kluwer, Dordrecht, 1994 to 2002, hardback and softback), in five volumes.

19. M. W. Evans (ed.), *Modern Non-Linear Optics*, a special topical issue in three parts of I. Prigogine and S. A. Rice (Series Editors), *Advances in Chemical Physics* (Wiley-Interscience, New York, 2001, hardback and e-book, second edition) vol. 119.
20. M. W. Evans and L. B. Crowell, *Classical and Quantum Electrodynamics and the* $\boldsymbol{B}^{(3)}$ *Field* (World Scientific, Singapore, 2001).
21. L. Felker, *The Evans Equations* (World Scientific in prep.) preprint on www.aias.us.
22. M. W. Evans, Physica B., **182**, 227, 237 (1992).
23. M. W. Evans, *The Photon's Magnetic Field: Optical NMR Spectroscopy* (World Scientific, Singapore, 1992).
24. The Collected Scientific Papers of M. W. Evans circa seven hundred papers in twenty volumes, in prep. These give all details of O(3) Electrodynamics. Papers to become fully available on a website www.myronevanscollectedworks.com.
25. M. W. Evans, formal manuscripts and informal notes on www.aias.us and www.atomicprecision.com.
26. A Einstein, The Meaning of Relativity (Princeton Univ Press, 1921 and subsequent editions).
27. S. P. Carroll, *Lecture Notes in General Relativity* (a graduate course at Harvard Univ., Univ. California Santa Barbara and Univ Chicago, arXiv: gr-gq/9712019 v1 3 Dec 1997).
28. G. Stephenson, Mathematical Methods for Science Students (Longmans, London, 1968).

20

Calculation of the Anomalous Magnetic Moment of the Electron from the Evans-Unified Field Theory

Summary. The Evans Lemma is used to calculate the g factor of the electron to ten decimal places, in exact agreement with experimental data within contemporary experimental uncertainty. The calculation proves that the Evans unified field theory is a fully quantized and generally covariant field theory of all known radiated and matter fields. It therefore has clear advantages over quantum electrodynamics, which is a theory only of special relativity and only of photons and electrons without reference to gravitation. The Evans unified field theory has none of the well known weaknesses of quantum electrodynamics, for example unphysical infinities, renormalization and dimensional regularization. Quantum electrodynamics is also fundamentally self-contradictory, being at once acausal (sum over histories construction of the wave-function at a fundamental level, "the electron can do anything it likes") and causal (use of the Huygens Principle at a fundamental level, "the electron cannot do anything it likes"). The Evans theory is rigorously causal and based on general relativity.

Key words: Evans unified field theory; anomalous electron magnetic moment; anomalous electron g factor; criticism of quantum electrodynamics.

20.1 Introduction

The anomalous g factor of the electron can be explained only by a fully quantized field theory. Semi-classical field theories produce the result:

$$g = 2 \tag{20.1}$$

from the Dirac equation and minimal prescription with a classical electromagnetic potential which is zero in the vacuum. In semi-classical theories therefore the matter field (electron) is quantized with Dirac's equation (a limiting form of the Evans wave equation) but the electromagnetic field is still classical In Section 20.2 the recently developed and generally covariant unified field theory of Evans [1]–[25] is used to calculate the observed g factor of the electron [26]:

$$g = 2.0023193048. \tag{20.2}$$

The theoretical result from the Evans theory agrees with the above experimental value to ten decimal places and within contemporary experimental uncertainty [26]. In Section 20.2 it is shown how this precise agreement between the Evans theory and experiment is obtained by quantizing both the matter field (the electron) and the radiated field (electromagnetic field) with the fundamental Evans Lemma. In Section 20.3 some criticisms of quantum electrodynamics are made, and it is argued that the Evans theory is to be preferred for several fundamental reasons.

20.2 Vacuum or Zero-point Energy in the Evans Theory

The famous anomaly in the g factor and magnetic moment of the electron originates in the existence of zero-point energy or vacuum energy in a photon ensemble regarded as a collection of harmonic oscillators. This ensemble of photons is described by the Evans Lemma [1]–[25]:

$$\square A^a{}_\mu = RA^a{}_\mu \tag{20.3}$$

where the potential field is the tetrad one-form within a $\hat{C}$ negative coefficient $A^{(0)}$. The latter originates in the primordial magnetic fluxon $\hbar/e$ (weber) where $\hbar$ is the reduced Planck constant and where $-e$ is the charge on the electron. The well known transverse plane wave [27] is a solution of the Evans Lemma (20.3) with the scalar tetrad components:

$$\begin{aligned} A^{(1)}{}_X &= \tfrac{A^{(0)}}{\sqrt{2}} e^{i\phi} \;, \; A^{(1)}{}_Y = -i \tfrac{A^{(0)}}{\sqrt{2}} e^{i\phi} \\ A^{(2)}{}_X &= \tfrac{A^{(0)}}{\sqrt{2}} e^{-i\phi} \;, \; A^{(1)}{}_Y = i \tfrac{A^{(0)}}{\sqrt{2}} e^{-i\phi} . \end{aligned} \tag{20.4}$$

The following constant valued tetrad components:

$$A^{(0)}{}_0 = A^{(3)}{}_Z = A^{(0)} \tag{20.5}$$

are timelike and longitudinal solutions of the Evans Lemma. Here ϕ is the electromagnetic phase:

$$\phi = \omega t - \boldsymbol{\kappa r} \tag{20.6}$$

a special case of the Evans phase law[1]–[25] of generally covariant unified field theory. Here ω is the angular frequency

$$\omega = \kappa c \tag{20.7}$$

at instant t and $\boldsymbol{\kappa}$ is the wavenumber at coordinate $\boldsymbol{r}$. In these equations c is the speed of light in vacuo. The transverse tetrad elements (20.4) are solutions of the Evans Lemma in the form:

$$\left(\Box + \kappa^2\right) A^a{}_\mu = 0. \tag{20.8}$$

The photon ensemble is obtained by expanding the wavefunction:

$$\psi = e^{i\phi} \tag{20.9}$$

in a Fourier series [26, 27]. The energy levels of the photon ensemble are well known to be the energy levels of the harmonic oscillator problem:

$$En = \left(n + \frac{1}{2}\right) \hbar\omega \tag{20.10}$$

when there are no photons present (n = 0), there is a non-zero vacuum energy or zero-point energy:

$$En_0 = \frac{1}{2}\hbar\omega. \tag{20.11}$$

The photon ensemble is therefore regarded in the way that was originally inferred by Planck [26, 27], as a collection of harmonic oscillators. Atkins [27] describes the vacuum energy as being due to fluctuating electric and magnetic fields when there are no photons present. Therefore in the absence of photons the electron records the influence of these vacuum fields in its zitterbewegung or jitterbug motion. The result is that the g factor is changed from 2 of the Dirac equation to the experimental value in Eq. (20.2). This means that the electron wobbles about its equatorial axis as it spins [27].

There therefore exists a QUANTIZED VACUUM POTENTIAL:

$$A^a{}_\mu{}^{(vac)} = A^{(0)} q^a{}_\mu{}^{(vac)} \tag{20.12}$$

which is the eigenfunction of the Evans Lemma corresponding to its minimum (zero-point) eigenvalue R_0 in the harmonic oscillator problem:

$$\Box A^a{}_\mu{}^{(vac)} = R_0 A^a{}_\mu{}^{(vac)}. \tag{20.13}$$

The minimum eigenvalue is a minimum or least curvature as required by the Evans Principle of Least Curvature (a generally covariant synthesis [1]–[25] of the Hamilton Principle of Least Action and the Fermat Principle of Least Time [27]). The vacuum or zero point tetrad $q^a{}_\mu{}^{(vac)}$ is this eigenfunction within a scalar factor $A^{(0)}$, indicating self-consistently that the vacuum is the non-Minkowskian spacetime of Einstein's general relativity extended by Evans [1]–[25] into a generally covariant and rigorously causal unified field theory.

The Dirac wave equation has been inferred [1]–[25] as a limiting form of the Evans Lemma

$$\Box q^a{}_\mu = R q^a{}_\mu \tag{20.14}$$

where the minimum eigenvalue or least curvature is defined by:

$$R \to R_0 = -\left(mc/\hbar\right)^2. \tag{20.15}$$

Here

$$\lambda = \frac{\hbar}{mc} \tag{20.16}$$

is the Compton wavelength of the electron where m is its mass. The Dirac spinor is defined by transposition of four tetrad elements in SU(2) representation space into a column vector:

$$q^a{}_\mu = \begin{bmatrix} q^R{}_1 & q^R{}_2 \\ q^L{}_1 & q^L{}_2 \end{bmatrix} \rightarrow \begin{bmatrix} q^R{}_1 \\ q^R{}_2 \\ q^L{}_1 \\ q^L{}_2 \end{bmatrix} \tag{20.17}$$

So the Dirac spinor consists of right and left Pauli spinors:

$$\psi = \begin{bmatrix} \phi^R \\ \phi^L \end{bmatrix}, \phi^R = \begin{bmatrix} q^R{}_1 \\ q^R{}_2 \end{bmatrix}, \phi^L = \begin{bmatrix} q^L{}_1 \\ q^L{}_2 \end{bmatrix} \tag{20.18}$$

which are identified as elements of [1]–[25], of a two component column vector (i.e. spinor) in SU(2) representation space. The two Pauli spinors are interconverted by parity inversion:

$$\hat{\mathbf{P}}\left(\phi^R\right) = \phi^L. \tag{20.19}$$

The electron therefore has right and left handed half integral spin as well as mass. In this description it is a fermion with g factor of exactly 2. This is a description of causal general relativity without yet taking note of the all important effect of the quantized vacuum potential, and in this description the Dirac spinor is a solidly defined geometrical object, not an abstract and unknowable probability as in the special relativistic and acausal Copenhagen interpretation. The introduction of an unknowable (i.e. acausal) probability into causal natural philosophy is one of many reasons why Einstein, Schrodinger, de Broglie and the Determinist School reject Copenhagen quantum mechanics as an incomplete theory. The electron CANNOT do anything it likes [26], it is governed by a causal wave equation of motion (20.14), a limiting case of the Evans wave equation [1]–[25]. Analogously a dynamical object cannot do anything it likes, it is governed by Newton's laws of motion.

The wave equation (20.14) can be factorized [1]–[26] into a first order differential equation of motion:

$$\left(\gamma^a p_a - mc\right) q^b{}_\mu = 0 \tag{20.20}$$

where γ^a is the Dirac matrix. The equation (20.20) is generally covariant, and can be written in the tangent bundle spacetime of Evans' theory as:

$$\left(\gamma^a p_a - mc\right) q^b = 0 \tag{20.21}$$

for all indices μ of the base manifold. The latter is a non-Minkowskian or curved spacetime In general all equations of generally covariant physics can always be written in this way as equations of the orthonormal tangent bundle spacetime, with indices such as μ of the base manifold suppressed or implied. This is standard practice [28] in differential geometry and the Evans theory is the geometrization of all physics, not only gravitation. The great advantage of writing the equations of generally covariant physics in the tangent bundle spacetime is that the latter is orthonormal by construction, with Minkowskian metric η^{ab}:

$$\eta^{ab} = \gamma^a \gamma^b = \begin{bmatrix} 1 & 0 & 0 & 0 \\ 0 & -1 & 0 & 0 \\ 0 & 0 & -1 & 0 \\ 0 & 0 & 0 & -1 \end{bmatrix} . \tag{20.22}$$

Here γ^a are standard Dirac matrices [1]–[26]. Any suitable orthonormal basis set such as the Pauli matrices can be used in the tangent bundle spacetime and so a complex valued spinor or two-vector can also always be defined in the tangent bundle spacetime. The tetrad is the invertible matrix that links vector components and basis vectors [28] in the tangent bundle and base manifold, and in the Evans unified field theory [1]–[25] the tetrad is the gravitational potential field. Within a factor $A^{(0)}$ the tetrad is also the electromagnetic potential field.

In this way therefore it is always possible to introduce gravitation into any equation of physics through the tetrad. The symmetric metric of the base manifold, the object used by Einstein and Hilbert in 1915 to define gravitation, is the dot or scalar product of two tetrads:

$$g_{\mu\nu} = q^a{}_\mu q^b{}_\nu \eta_{ab} \tag{20.23}$$

and so it is always possible to express the symmetric metric of the base manifold in terms of Dirac matrices of the tangent bundle spacetime:

$$g_{\mu\nu} = q^a{}_\mu q^b{}_\nu \gamma_a \gamma_b . \tag{20.24}$$

Therefore it is possible in theory to measure the effect of gravitation on the Dirac equation, both in the absence and presence of the electromagnetic field. This is possible in the Evans unified field theory but not possible in the contemporary "standard model" or in quantum electrodynamics.

In the presence of the vacuum potential the Evans equation (20.21) becomes:

$$\left(\gamma^a \left(p_a + eA^{(vac)}{}_a \right) - mc \right) q^b = 0. \tag{20.25}$$

The magnitude of the vacuum potential is defined by the ratio:

$$\alpha := 4\pi \frac{eA^{(vac)}}{\hbar\kappa} = 4\pi c \frac{eA^{(vac)}}{\hbar\omega} \tag{20.26}$$

where α is the fine structure constant [1]–[27]:

$$\alpha = \frac{e^2}{4\pi\epsilon_0 \hbar c} \tag{20.27}$$

in S.I. units. Here ϵ_0 is the vacuum permittivity. In reduced units of

$$4\pi\epsilon_0 \hbar c = 1 \tag{20.28}$$

it is seen that:

$$eA^{(vac)} = \frac{e^2}{4\pi} En^{(vac)}. \tag{20.29}$$

This is the most fundamental way of making $A^{(vac)}$ proportional to $E^{(vac)}$ through the fundamental proportionality constant e, and this is the fundamental meaning of the vacuum potential and the reason why the fine structure constant appears in our calculation. The covariant version of Eq. (20.29) is:

$$eA^{(vac)} = \frac{e^2}{4\pi} p^{a(vac)} = \frac{\alpha}{4\pi} p^{a(vac)} \tag{20.30}$$

and is written in the tangent bundle spacetime following our general rule that all equations of generally covariant physics may be written in the tangent bundle with indices of the base manifold suppressed or implied. Thereby we arrive at the inference that Eq (20.30) is the minimal prescription [1]–[27] or covariant derivative for the vacuum potential.

The Evans equation (20.25) now becomes:

$$\left(\gamma^a p_a \left(1 + \alpha^{'}\right) - mc\right) q^b = 0 \tag{20.31}$$

where:

$$\alpha^{'} = \frac{\alpha}{4\pi}. \tag{20.32}$$

Eqn. (20.31) can be interpreted as an increase in the Dirac gamma matrix for constant p_a:

$$\gamma^a \rightarrow \gamma^a \left(1 + \alpha^{'}\right), \tag{20.33}$$

so the existence of $\alpha^{'}$ is seen clearly through Eq. (20.34) to be a property of spacetime itself, and thus self-consistently as a vacuum property. It is always possible to express the Minkowski metric as the anti-commutator of Dirac matrices:

$$\left[\gamma^a, \gamma^b\right] = \gamma^a \gamma^b + \gamma^b \gamma^a = 2\eta^{ab} \tag{20.34}$$

and the factor 2 of Eq. (20.34) can be interpreted as the Dirac g factor of the electron. From Eq (20.33) the g factor is therefore increased by the vacuum potential to:

$$g = \left(1 + \alpha^{'}\right)^2 + \left(1 + \alpha^{'}\right)^2 = 2\left(1 + \frac{\alpha}{2\pi} + \frac{\alpha^2}{16\pi^2}\right). \tag{20.35}$$

This theoretical result is in agreement with the experimental result of Eq (20.2) to ten decimal places, the greatest known precision of contemporary physics.

This agreement is therefore a very precise test of the Evans unified field theory and is due fundamentally to the least or minimum curvature produced by the vacuum energy:

$$En^{(vac)} = \frac{1}{2}\hbar\omega = \frac{1}{2}c\hbar\kappa = \frac{1}{2}c\hbar\left|R_0\right|^{1/2}. \tag{20.36}$$

20.3 Criticisms of Quantum Electrodynamics

In the received opinion at present the calculation of g is carried out with quantum electrodynamics (qed) by increasing the value of the Dirac matrix with the convergent vertex

$$\gamma_\mu \rightarrow \gamma_\mu + {\Lambda^{(2)}}_\mu. \tag{20.37}$$

This calculation is carried out in Minkowski or flat spacetime and so is not generally covariant as required by Einsteinian natural philosophy. The convergent vertex is defined by

$${\Lambda^{(2)}}_\mu = \alpha' \left(p_\mu + p'_\mu\right) / (mc) \tag{20.38}$$

[1]–[26]. In qed dimensional regularization is used to remove primitive divergences of the path integral formalism [1]–[26], which has the effect:

$$e \rightarrow \mu^{2-d/2}e \tag{20.39}$$

in the lagrangian. Here μ is an arbitrary mass and d is the mass dimension of the lagrangian. This procedure leads to ${\Lambda^{(2)}}_\mu$ from the vertex graph:

$$-ie\Lambda_\mu\left(p, q, p+q\right). \tag{20.40}$$

The removal of infinities of this graph results in a change in the physical properties of the electron. The convergent ${\Lambda^{(2)}}_\mu$ is that part of the overall vertex with no k in the nominator of the integrand.

Quantum electrodynamics is not therefore a foundational theory of natural philosophy because it obtains the right result by arbitrary means: dimensional regularization, which changes e, and renormalization, which artificially removes infinities of the path integral method. Quantum electrodynamics is Lorentz covariant only (it is a theory of special relativity). Quantum electrodynamics uses the sum over histories description of the wavefunction. This is an acausal description in which the electron can do anything it likes [26], go backwards or forwards in time for example. This acausality or unknowability is contradicted fundamentally and diametrically in qed by use [26] of the Huygens Principle, which expresses causality or knowability - the wavefunction is built up by superposition in causal historical sequence - an event is always preceded by a cause, and nothing goes backwards in time. For these and other reasons qed was rejected by Einstein, Schrodinger, de Broglie, Dirac and many others from its inception in the late forties.

As we have seen the much vaunted precision of qed is obtained much more simply in the generally covariant and causal unified field theory of Evans. By Okham's Razor and for several other fundamental reasons discussed already, the Evans theory is preferred to quantum electrodynamics. Similarly the Evans theory is preferred to quantum chromodynamics, which has the same fundamental flaws as quantum electrodynamics with the added problem of quark confinement. Quarks are postulated to exist but to be unobservable [26]. Something that is both unknowable and unobservable springs from theism and has no place in natural philosophy - the causal description of the natural world. If nature were unknowable and unobservable we could never know anything about nature and this is what the Copenhagen School would have us swallow. Add to this the many "dimensions" of string theory and whither physics?

20.4 Discussion

The interaction of the electron and photon ensemble may always be described in the Evans unified field theory by:

$$\left(\gamma^a\left(i\hbar\partial_a - eA_a\right) - mc\right)q^b = 0. \tag{20.41}$$

This equation can be developed into a wave equation as follows:

$$\left(\gamma^a\left(i\hbar\partial_a - eA_a\right) - mc\right)\left(\gamma^b\left(-i\hbar\partial_b - eA_b^*\right) - mc\right)q^b = 0. \tag{20.42}$$

Here the whole of the operator acts on the tetrad as follows:

$$\left(\Box + \frac{m^2c^2}{\hbar^2} + \frac{emc}{\hbar^2}\left(A_a + A_a^*\right) + \frac{e^2}{\hbar^2}A^aA_a^*\right)q^b = 0 \tag{20.43}$$

where we have used the result [26]:

$$\Box = \partial^a\partial_a = g^{ab}\partial_a\partial_b = \gamma^a\gamma^b\partial_a\partial_b. \tag{20.44}$$

In Eq. (20.43) it becomes clear that the overall effect of the electromagnetic field on the electron is to always to add a scalar curvature [1]–[25]:

$$R_{em} = -\frac{emc}{\hbar^2}\left(A_a + A_a^*\right) - \frac{e^2}{\hbar^2}A^aA_a^*. \tag{20.45}$$

In the presence of gravitation (curvature of the base manifold):

$$\frac{m^2c^2}{\hbar^2} \rightarrow R \tag{20.46}$$

so the Evans equation (20.43) can be used to calculate the effect of the gravitational field on the electromagnetic field interacting with an electron. More generally the calculation and computation can be extended to atoms and molecules in a gravitational field. This is twenty first century physics, i.e. not possible in the contemporary standard model, but very important to new technologies. Finally, the Evans theory is more precise than quantum electrodynamics because it is able to calculate the g factor of the electron to any order in the fine structure constant. In quantum electrodynamics such calculations involve tremendous complexity and are deeply flawed philosophically. All said and done, physics is natural philosophy, and if the philosophy is flawed fundamentally it must, as always in the development of human thought, be replaced by something better.

Acknowledgements

Craddock Inc., the Ted Annis Foundation and Applied Science Associates are thanked for funding and AIAS Fellows for many interesting discussions.

References

1. M. W. Evans, A Unified Field Theory for Gravitation and Electromagnetism, *Found. Phys. Lett.*, **16**, 367 (2003).
2. M. W. Evans, A Generally Covariant Wave Equation for Grand Unified Field Theory, *Found. Phys. Lett.*, **16**, 507 (2003).
3. M. W. Evans, The Equations of Grand Unified Field Theory in terms of the Maurer Cartan Structure Relations of Differential Geometry, *Found. Phys. Lett.*, **17**, 25 (2004).
4. M. W. Evans, Derivation of Dirac's Equation from the Evans Wave Equation, *Found. Phys. Lett.*, **17**, 149 (2004).
5. M. W. Evans, Unification of the Gravitational and Strong Nuclear Fields, *Found. Phys. Lett.*, **17**, 267 (2004).
6. M. W. Evans, The Evans Lemma of Differential Geometry, *Found. Phys. Lett.*, **17**, 433 (2004).
7. M. W. Evans, Derivation of the Evans Wave Equation from the Lagrangian and Action: Origin of the Planck Constant in General Relativity, *Found. Phys. Lett.*, **17**, 535 (2004).
8. M. W. Evans and the AIAS Author Group, Development of the Evans Wave Equation in the Weak Field Limit: the Electrogravitic Equation, *Found. Phys. Lett.*, **17**, 497 (2004).
9. M. W. Evans, Physical Optics, the Sagnac Effect and the Aharonov Bohm Effect in the Evans Unified Field Theory, *Found. Phys. Lett.*, **17**, 301 (2004).
10. M. W. Evans, Derivation of the Geometrical Phase from the Evans Phase Law of Generally Covariant Unified Field Theory, *Found. Phys. Lett.*, **17**, 393 (2004).
11. M. W. Evans, New Concepts from the Evans Unified Field Theory, Part One: The Evolution of Curvature, Oscillatory Universe without Singularity and General Force and Field Equations, *Found. Phys. Lett.*, in press, preprint on www.aias.us.
12. M. W. Evans, New Concepts from the Evans Unified Field Theory, Part Two: Derivation of the Heisenberg Equation and Replacement of the Heisenberg Uncertainty Principle, *Found. Phys. Lett.*, in press, preprint on www.aias.us.
13. M. W. Evans, Derivation of the Lorentz Boost from the Evans Wave Equation, *Found. Phys. Lett.*, in press (2004 / 2005), preprint on www.aias.us
14. M. W. Evans, Derivation of O(3) Electrodynamics from Generally Covariant Unified Field Theory, *Found. Phys. Lett.*, in press, preprint on www.aias.us.

15. M. W. Evans, The Electromagnetic Sector of the Evans Field Theory, *Found. Phys. Lett.*, in press (2004 / 2005), preprint on www.aias.us.
16. M. W. Evans, The Spinning and Curving of Spacetime, the Electromagnetic and Gravitational Fields in the Evans Unified Field Theory, *Found. Phys. Lett.*, in press, preprint on www.aias.us.
17. M. W. Evans, The Derivation of O(3) Electrodynamics from the Evans Unified Field Theory, *Found. Phys. Lett.*, in press, preprint on www.aias.us.
18. M. W. Evans, Generally Covariant Unified Field Theory: the Geometrization of Physics (in press, 2005)
19. M. W. Evans, *The Photon's Magnetic Field*, Optical NMR Spectroscopy (World Scientific, Singapore, 1992).
20. M. W. Evans and S. Kielich (eds.), *Modern Nonlinear Optics*, a special topical issue of I. Prigogine and S. A Rice (eds.), *Advances in Chemical Physics* (Wiley-Interscience, New York, 1992, 1993, 1997, first edition hardback and softback), vols. 85(1), 85(2), 85(3).
21. M. W. Evans and A. A. Hasanein, *The Photomagneton in Quantum Field Theory*, (World Scientific, Singapore, 1994).
22. M. W. Evans, J.-P. Vigier et alii, *The Enigmatic Photon* (Kluwer, Dordrecht, 1994 to 2002, hardback and softback) in five volumes.
23. M. W. Evans (ed.), *Modern Nonlinear Optics*, a special topical issue of I. Prigogine and S. A. Rice (series eds.), *Advances in Chemical Physics* (Wiley Interscience, New York, 2001, second edition, hardback and e book), vols. 119(1), 119(2), and 119(3).
24. M. W. Evans and L. B. Crowell, *Classical and Quantum Electrodynamics and the B Field* (World Scientific, Singapore, 2001).
25. M. W. Evans and L. Felker, *The Evans Equations* (World Scientific, 2005 in prep.)
26. L. H. Ryder, *Quantum Field Theory* (Cambridge, 1996, second edition softback).
27. P. W. Atkins, *Molecular Quantum Mechanics*, (Oxford, 1983, second edition softback).
28. S. P. Carroll, *Lecture Notes in General Relativity* , (a graduate course at Harvard, UC Santa Barbara and Univ Chicago, arXiv: gr-gq/9712019 v1 3 Dec 1997).

21

Generally Covariant Electro–Weak Theory

Summary. A generally covariant electro-weak theory is developed by factorizing the Evans Lemma into first order differential equations and using the appropriate minimal prescription. The differential equations are written in the tangent bundle spacetime for all base manifolds so are generally covariant. The masses of the weak field bosons are understood in terms of scalar curvature. Therefore the electro-weak theory is developed without having to use the concept of spontaneous symmetry breaking and the Higgs mechanism. The latter does not occur in Einsteins general relativity and is not generally covariant nor is it a foundational concept. The electro-weak theory of Glashow, Weinberg and Salaam (GWS) is a theory of special relativity and for this reason is not generally covariant. The Evans unified field theory is foundational because it is a theory of general relativity, and so is preferred to the GWS/Higgs theory when used to describe the electro-weak field. The Evans theory has the advantage of being able to incorporate the gravitational and strong fields into electro-weak field theory. Boson masses in the Evans theory are spacetime scalar curvatures, well defined in the special relativistic limit by the Evans Principle of Least Curvature.

Key words: Generally covariant electro-weak theory, Evans unified field theory. Evans Lemma

21.1 Introduction

The electro-weak field theory is a theory of special relativity based directly on the concept of spontaneous symmetry breaking [1] and the Higgs mechanism. The masses of the weak field bosons are introduced through the latter mechanism, which uses an adjustable parameter to force agreement between theory and experimental data, such as data from neutrino electron scattering (weak neutral current) . This type of electro-weak theory was developed independently by Glashow, Weinberg and Salaam and is known as GWS theory. The theory is not foundational because it is not generally covariant, and because the Higgs mechanism is an ad hoc method of introducing masses in such as way that pre-conceived ideas about the neutrino and the photon are maintained intact. The GWS theory, because it is not generally covariant, can never be used to explore the effect of gravitation on electro-weak phenomena and is not a true unified field theory. The masses are introduced in GWS theory in a carefully contrived manner: it is assumed at the outset that the photon and neutrino masses are zero and must be KEPT zero by juggling parameters in the minimal prescription. These assumptions are basically in contravention of Einsteins general relativity, in which zero mass means zero energy and identically flat spacetime in which no fields or particles can exist. An everywhere Minkowski spacetime in Einsteinian general relativity means an empty universe devoid of all fields and all particles. It is now generally accepted [2] on experimental evidence that the neutrino is not a massless particle, so the basic assumption of GWS theory collapses. The weak field boson masses are introduced in GWS theory through the Higgs mechanism, in which a preconceived vacuum symmetry of special relativity is assumed to be spontaneously broken. The theory is delicately glued together in such a way that the photon and neutrino masses remain zero. So the GWS theory is a circular argument. It makes sure that the initial assumption is artificially proven. The experimentally observed weak field boson masses are not predicted foundationally in terms of the basic constants of physics, the data are FITTED with the adjustable parameter of the Higgs mechanism, the basic data in this case being scattering peaks from particle colliders, a type of spectrum of energies. The Higgs boson is postulated to exist but has not been observed experimentally in forty years of very expensive searching. The Higgs boson furthermore cannot be a foundational feature of natural philosophy because the Higgs mechanism, as we have argued, distils down to an adjustable parameter in special relativity. The Higgs boson mass must always be ill defined in general relativity, and any claim to have observed this non-existent boson will be a costly and elaborate curve fitting exercise. There will remain no experimental evidence whatsoever for a vacuum whose symmetry must be spontaneously broken - and then only in special relativity. This much serves to illustrate the mixture of ill-defined concepts known as the standard model. Added to these basic problems of GWS theory is the use of the path integral method and renormalization of the unphysical infinities introduced thereby. The GWS theory is renormalizable only if the Higgs mechanism is used [1]. Without the Higgs mechanism the path integral formalism cannot be used, so the former mechanism is a means of circumventing the fatal flaws of the path integral method by the introduction of Higgs unprovable ideas about the special relativistic vacuum. In general relativity however there is always a special relativistic vacuum by definition - the vacuum IS Minkowski (or flat) spacetime. General relativity tells us no less than this and certainly no more. So the assumed symmetric vacuum of Higgs is extraneous to the general relativity of Einstein. GWS/Higgs starts and ends by telling us that the vacuum of special relativity must have a symmetry which must be broken. GWS /Higgs, then, must always be a statement about an universe devoid of all matter and fields (Minkowski spacetme) and therefore a statement about the nature of nothing at all - primordial theism.

In Section 2 a generally covariant electro-weak theory is introduced based on differential geometry and the recently developed Evans unified field theory [3]–[27]. In sharp contrast to GWS/Higgs the Evans field theory is rigorously a theory of general relativity, and is a straightforward geometrization of all particle and field theory as fundamentally required by general relativity. The Higgs mechanism is not used, and no pre-conceptions or initial assumptions are made about the photon mass and neutrino mass. The path integral method is rigorously avoided,, and the fundamental wave equation of physics, the Evans Lemma, is derived directly from the fundamental tetrad postulate of differential geometry itself [3]–[27]. The Lemma is the subsidiary proposition which, together with Einsteins field equation in index contracted form, leads to the Evans wave equation. In so doing the Einstein field equation is interpreted in the manner originally intended [28] by Einstein, i.e. is interpreted as applying to ALL fields in nature and not only the gravitational field. A generally covariant electro-weak theory is then developed in Section 2 by factorizing the Lemma into first order differential equations in which the interaction between particles (particle scattering) is described with the appropriate covariant derivative. The latter is essentially a change in the tetrad, i.e. a change in the nature of spacetime itself, brought about by particle particle scattering, collisions, or interaction processes. The ad hoc isospinor of GWS Higgs theory [1] is given a physical interpretation in general relativity as a two component vector made up of tetrads, one for the left handed muon neutrino or left handed electron neutrino and one for the left handed electron. This procedure automatically defines a representation space of a particular, SU(2), symmetry in analogy to the SU(2) symmetry used in the original Dirac theory. (In the latter however, one component of the two vector (the Dirac spinor) is a right handed Pauli spinor and the other is a left handed Pauli spinor, both components applying to the electron.) The electro-weak two component vector is governed by its appropriate Evans Lemma, whose eigenvalue matrix is one of scalar curvatures - particle masses or energies within coefficients made up of fundamental constants. The two component vector is an object of differential geometry and not of gauge theory, and the magnitude of this two-vector is invariant under an SU(2) transformation. It is assumed on the basis of experimental data that there is no right handed neutrino in nature, but there is a right handed electron. The latter is also governed by its appropriate Evans Lemma. In the tangent bundle spacetime of the Evans field theory the Lemma can always be factorized into a Dirac equation for all base manifold geometries. Any observed elementary particle spectrum (particle particle scattering process) may be built up directly by appropriate choice of covariant derivative in the factorized Evans Lemma. The peaks in the spectrum correspond to particular terms built up from the minimal prescription, and so these peaks (or masses) are, self-consistently, manifestations of particular scalar curvatures and general relativity, not of the Higgs mechanism. There are no loose parameters in the Evans field theory, and each peak in the elementary particle spectrum is defined by a field intensity. In the case of neutrino electron scattering processes these are field intensities corresponding to the vector boson components. These intensities can be interpreted in terms of mass or energy. No assumptions are made in the Evans theory about the nature of the vacuum apart from the fact that the vacuum is Minkowski spacetime, and the path integral method is not used under any circumstances. The Evans field theory consists analytically of second order differential (wave) equations or first order differential equations to be solved simultaneously for matter fields (particles) and radiated fields. These equations are solved numerically, rigorously avoiding the path integral formalism, if they happen to be analytically intractable. This process completely removes the problem of infinities and renormalization thereof. Finally, the radiated fields in the Evans theory are the agents of particle interaction, as in any field theory, and the concepts of general relativity are unified causally with those of wave mechanics by virtue of geometry in the Evans Lemma [3]–[27].

In Section 21.3 a discussion is given of the advantages of the Evans electro-weak theory over its GWS/Higgs predecessor: the former theory is generally covariant, simpler in structure, and easily applied to experimental data; the latter theory has the fatal weaknesses already described in this introduction.

21.2 The Evans Electro-Weak Theory

The generally covariant electro-weak theory originates in the Evans Lemma [3]–[27]:

$$\square q^a{}_\mu = R q^a{}_\mu \tag{21.1}$$

whose eigenfunction is the tetrad $q^a{}_\mu$ and whose eigenvalues are scalar curvatures R (in units of inverse square metres). The Evans Lemma is the subsidiary geometrical proposition leading to the Evans wave equation:

$$\left(\square + kT\right) q^a{}_\mu = 0, \tag{21.2}$$

where k is Einsteins constant and T the index contracted energy - momentum tensor. Eq. (21.2) follows from Eq. (21.1) using the Einstein field equation in index contracted form [28]:

$$R = -kT. \tag{21.3}$$

As discussed originally by Einstein [28] Eq. (21.3) must be interpreted as applying to all fields and particles, not only the gravitational field.

As is customary in differential geometry [29] Eq. (21.1) may be written simply as:

$$\square q^a = R q^a, \tag{21.4}$$

i.e. as an equation of the orthonormal tangent bundle spacetime for all indices μ of the base manifold. The tetrad is the invertible matrix defined for any vector $\boldsymbol{V}$ by:

$$V^a = q^a{}_\mu V^\mu \tag{21.5}$$

where $\boldsymbol{V}^a$ is defined in the tangent bundle and $\boldsymbol{V}^\mu$ in the base manifold [29]. Any suitable basis set and representation space can be used to describe the orthonormal tangent bundle spacetime for any type of base manifold (non-Minkowski spacetime). It follows that all the equations of physics are equations of the tangent bundle spacetime for all base manifolds. In the limit of an empty universe (the vacuum) devoid of all matter fields and radiated fields the base manifold asymptotically approaches Minkowski spacetime everywhere, there is no mass, spin or helicity, and the tangent and base manifold spacetimes become static and indistinguishable. All tetrad components become constants and R vanishes. The Evans Principle of Least Curvature [3]–[27] states that the minimum R of the Evans Lemma is:

$$R_0 = -\left(\frac{mc}{\hbar}\right)^2 \tag{21.6}$$

where m is the mass of the particle, $\hbar$ is the reduced Planck constant and c the velocity of light in vacuo. Here:

$$\lambda_0 = \frac{\hbar}{mc} \tag{21.7}$$

is the Compton wavelength of any particle. Eq. (21.6) is the special relativistic limit of the Evans field theory. Note carefully that the special relativistic limit is not defined as the limit of everywhere flat (Minkowski) spacetime without spin. In the latter limit there is no R and no mass anywhere in the universe. Evidently, mass m must be non-zero in special relativity, mass is the first Casimir invariant of the Poincare group of special relativity [3]–[27]. The other Casimir invariant of special relativity is spin, and the two Casimir invariants define any particle. In general relativity however the appropriate Lie group is the Einstein group, so in the Evans field theory mass m is the first Casimir invariant of the Einstein group, and spin is the second Casimir invariant of the Einstein group. The Evans field theory is generally covariant for all matter and radiated fields, i.e objective in any frame of reference moving with respect to any other frame of reference in any way. This is a fundamental requirement of everywhere objective physics which is missing entirely in the standard model and GWS/Higgs electro-weak theory. General covariance is of course the fundamental axiom of general relativity, and without it physics is not an objective subject, physics to one observer would be different from physics to another observer. In the Evans field theory the absence of gravitational interaction between particles is defined by Eq. (21.6). In this asymptotic limit of no gravitation, the tetrad of Eq. (21.6) defines the spinning of the base manifold with respect to the tangent bundle spacetime. This spinning motion defines the electromagnetic, weak and strong fields in the absence of gravitation, and so defines the electro-weak field. In the presence of gravitation the base manifold is both spinning and curving with respect to the orthonormal (Minkowski) spacetime of the tangent bundle. The Evans Principle of Least Curvature embodied in Eq. (21.6) is so called because the least possible total curvature in the universe occurs when there is no gravitational attraction between particles of mass m, in which limit Eq. (21.6) applies. The Evans Principle of Least Curvature is a unification of the Hamilton Principle of Least Action and the Fermat Principle of Least Time [30]. Therefore all the equations of the electromagnetic, weak and strong fields in the absence of gravitation are equations of spin in either the tangent bundle spacetime or base manifold. One frame is spinning with respect to the other and both are Minkowski spacetimes when there is no gravitation present. The equations of physics have the same form in both frames and so are generally covariant as required. In geometrical terms this statement means that the equations are valid both for the tetrad and inverse tetrad. In the rest of this paper we develop a generally covariant electro-weak theory in the absence of gravitation. The effect of gravitation on this theory can always be considered by curving the base manifold. In GWS/Higgs theory and the standard model, the effect of gravitation on the other three sectors cannot be analyzed and the concept of tetrad is confined to gravitation only. Spin in GWS/Higgs and the standard model is something extraneous to general relativity, as is the Higgs mechanism itself. This means that the concept of spin is not objective in the standard model, a fatal flaw.

The Evans Lemma in the asymptotic limit of no gravitation, Eq. (21.4), can be factorized into a Dirac equation by expanding the dAlembertian operator in terms of the Dirac matrices :

$$\square = \partial^b \partial_b = \eta_{ab} \partial^a \partial^b = \eta^{ab} \partial_a \partial_b = \gamma^a \gamma^b \partial_a \partial_b, \tag{21.8}$$

$$g_{\mu\nu} = q^a{}_\mu q^b{}_\nu \eta_{ab} \tag{21.9}$$

where $g_{\mu\nu}$ and η_{ab} are the manifold and tangent space metrics respectively. It follows that Eq. (21.4) is:

$$\left(-i\gamma^b \partial_b - m_e c/\hbar\right) \left(i\gamma^a \partial_a - m_e c/\hbar\right) q^c{}_\mu = 0 \tag{21.10}$$

and that there exist two Dirac equations, one the complex conjugate of the other:

$$\left(i\gamma^a \partial_a - m_e c/\hbar\right) q^c{}_\mu = 0 \tag{21.11}$$

$$\left(-i\gamma^b \partial_b - m_e c/\hbar\right) q^c{}_\mu{}^* = 0. \tag{21.12}$$

It also follows that the Dirac spinor originates in a tetrad of differential geometry [3]–[27]. Therefore the effect of gravitation on the Dirac equation can be analyzed by curving the base manifold. In the absence of gravitation the Dirac equation is an equation of spin, in this case the half integral spin of the Dirac electron with right and left handed components. It follows that the right and left handed electrons are each described by tetrad components - the right and left Pauli spinors. The tetrad appearing in Eqs. (21.10) is the four by four matrix defined by:

$$\sigma^a = q^a{}_\mu \sigma^\mu \tag{21.13}$$

where σ are Pauli matrices (basis elements) respectively in the tangent bundle spacetime and base manifold. The Dirac spinor is obtained by transposing the row vectors of the tetrad $(q^a{}_\mu)$ into column vectors, giving a column four-vector:

$$q^a{}_\mu = \begin{bmatrix} q^1{}_1 \; q^1{}_2 \\ q^2{}_1 \; q^2{}_2 \end{bmatrix} \rightarrow \begin{bmatrix} q^1{}_1 \\ q^1{}_2 \\ q^2{}_1 \\ q^2{}_2 \end{bmatrix} \tag{21.14}$$

and the two Pauli spinors (for the right and left handed electron) are:

$$\xi^1 = \begin{bmatrix} q^1{}_1 \\ q^1{}_2 \end{bmatrix}, \xi^2 = \begin{bmatrix} q^2{}_1 \\ q^2{}_2 \end{bmatrix}, \tag{21.15}$$

and are therefore column two-vectors made up of two tetrad elements. The Dirac spinor is therefore a column vector made up of two Pauli spinors, one right handed, the other left handed. All the elements in these column vectors are tetrad elements defined by geometry as required in general relativity.

In the presence of gravitation the Dirac equations (21.10) and (21.11) become the first order differential Evans equations:

$$\left(i\gamma^a \partial_a - |R|^{1/2}\right) q^c{}_\mu = 0 \tag{21.16}$$

$$\left(-i\gamma^b \partial_b - |R|^{1/2}\right) q^c{}_\mu{}^* = 0. \tag{21.17}$$

The generally covariant electro-weak theory of this paper is built up from these first order differential equations in the absence of gravitation. To illustrate the method used first consider the interaction of two electrons mediated by the electromagnetic potential field. The latter is defined by a tetrad within a factor $A^{(0)}$:

$$A^a{}_\mu = A^{(0)} q^a{}_\mu \tag{21.18}$$

In the absence of gravitation this tetrad is also governed by the Evans Lemma:

$$\left(\square + (m_p c/\hbar)^2\right) A^a{}_\mu = 0 \tag{21.19}$$

where m_p is the exceedingly small but non-zero mass of the photon [3]–[27]. The interaction of the photon and electron is accordingly described by the covariant derivative or minimal prescription written in the tangent bundle spacetime:

$$\left(i\hbar\gamma^{a}\left(\partial_{a}-ieA_{a}\right)m_{e}c\right)q^{c}=0. \tag{21.20}$$

This equation is a form of the fundamental tetrad postulate of differential geometry [3]–[29]:

$$D^{\mu}\left(D_{\mu}q^{a}{}_{\nu}\right)=0 \tag{21.21}$$

i.e. the covariant derivative of a tetrad is always zero. Therefore the interaction of an electron and a photon is analyzed by solving Eqs. (21.20) simultaneously with:

$$\left(i\hbar\gamma^{a}\left(\partial_{a}-ieA_{a}\right)m_{e}c/\right)A^{c}=0. \tag{21.22}$$

This can be done numerically on contemporary computers without use of the path integral method. In so doing the problems of infinities and renormalization are completely avoided. Eq (21.22) describes the momentum lost by the photon, Eq. (21.20) describes the momentum gained by the electron, Eq. (21.11) governs the motion of the free electron and Eq. (21.19) that of the free photon. Here m_e and m_p are the electron and photon masses respectively.

It is clear from the complex conjugate Eqs. (21.11) and (21.12) that there exist:

$$\left(\gamma^{a}\left(i\hbar\partial_{a}-eA_{a}\right)-m_{e}c/\right)q^{c}=0. \tag{21.23}$$

$$\left(\gamma^{a}\left(-i\hbar\partial_{a}-eA_{a}^{*}\right)-m_{e}c/\right)q^{c*}=0. \tag{21.24}$$

Therefore a wave equation can be constructed as follows:

$$\left(\gamma^{a}\left(-i\hbar\partial_{a}-eA_{a}^{*}\right)-m_{e}c/\right)\left(\gamma_{b}\left(i\hbar\partial^{b}-eA^{b}\right)-m_{e}c/\right)q^{c}=0. \tag{21.25}$$

This wave equation is:

$$\left(\Box+\frac{em_{e}c\gamma^{a}}{\hbar^{2}}\left(A_{a}+A_{a}^{*}\right)+\left(\frac{e}{\hbar}\right)^{2}A_{a}^{*}A^{a}+\left(\frac{mc}{\hbar}\right)^{2}\right)q^{c}=0 \tag{21.26}$$

and is the Evans Lemma describing the interaction of a photon and an electron. The interaction is described through extra scalar curvatures:

$$|R_{1}|=e^{2}A_{a}^{*}A^{a}/\hbar^{2}, \tag{21.27a}$$

$$|R_{1}|=em_{e}c\gamma^{a}\left(A_{a}+A_{a}^{*}\right)/\hbar^{2} \tag{21.27b}$$

which appear when the photon and electron interact, collide or scatter. Depending on the preferred terminology . The scalar curvatures (21.27) are characteristic of the scattering process and do not exist in Eq (21.11) for the free electron or in Eq. (21.19) for the free photon.

It is these scalar curvatures that describe the electro-weak interactions in the generally covariant Evans theory. Having set the scene in this way it is now possible to develop a simple type of electro-weak theory for the scattering of the neutrino and electron. We proceed by setting up the appropriate tetrad postulate and finding the interaction scalar curvatures for neutrino electron scattering analogous to (21.27a) for photon electron scattering. Similar procedures can be used for any type of particle scattering, but in this paper the theory is illustrated by neutrino electron scattering. The fundamental principle is that all scattering processes are governed by the tetrad postulate, i.e. by the first order Evans equations and the Evans Lemma.

These equations are straightforwardly generalized to any type of fermion boson scattering (or any type of particle scattering) by generalizing the electron to the fermion and the photon to the boson. In so doing no preconceived ideas concerning a hypothetical massless photon or neutrino are used, and no preconceived ideas about hyper-charge and vacuum symmetry breaking. These ideas are all extraneous to general relativity and thus to fundamental physics. Furthermore the abstract fiber-bundle index of gauge theory is replaced by the physical (i.e geometrical) tangent-bundle index, an index which is rigorously defined and governed (or constrained) by fundamental differential geometry. Again, the abstract fiber-bundle index of gauge field theory is extraneous to general relativity, and is not needed in our geometrical development. We therefore reject almost all of the ideas of the standard model, retaining general relativity. Only in this way can a truly foundational theory of particle scattering ever evolve, and only in this way can we ever hope to evolve a theory in which the effect of gravitation on radio-activity (electro-weak field) can be analyzed foundationally. It is known experimentally [31] that radio-activity evolved from gravitational events and the standard model is unable to analyze these data even at a qualitative level. The reason is that in the standard model there is no mechanism with which the effect of gravitation on the electro-weak field can be analyzed. It should come as no surprise therefore that the SU(2) internal (gauge) space used in GWS/Higgs must also be discarded. The much vaunted internal gauge space of GWS/Higgs (used to define the iso-spinor) is no more than a useful summary of a particular mathematical structure that can be understood much more simply and more clearly in the Evans unified field theory. The latter is generally covariant and can be used to analyze the effect of gravitation on radio-activity, and to prove conclusively that radio-activity evolved from gravitation [31]. Gauge theory must therefore itself be rejected in favor of a theory such as the Evans field theory, a generally covariant unified field theory developed [3]–[27] with the geometrical guidelines drawn up by Einstein. If we adhere to these guidelines, particle scattering theory becomes much far clearer and much simpler than that offered by GWS/Higgs. The latter is a mixture of concepts based on Minkowski spacetime. The use in GWS/Higgs of ideas extraneous to general relativity is effectively the use of loose parameters which are adjusted to force agreement with experimental data from particle colliders. In the Evans electro-weak theory there appear only the fundamental constants of physics and the foundational boson intensities analogous to electromagnetic field intensity which must be determined experimentally. These intensities indicate the observed boson masses as energy peaks in particle collider data of many different varieties. When the so called standard model attempts field unification all that really happens is the introduction of more loose parameters. This contemporary situation is strikingly reminiscent of the use of epicycles (many loose parameters) before Kepler discovered the laws of planetary motion, and before Newton rationalized these laws into powerful, simple equations using ONLY the Newtonian gravitational constant G, a fundamental constant of physics proportional to the Einsteinian constant k of general relativity. It becomes painfully clear therefore that the standard model (like Aristotelian epicycles) is historically another example of pathological or pseudo physics. Even worse is the infinity plagued complexity of Feynman calculus, and the meaningless and multidimensional mathematical process known as string theory: any rational scientist must surely know that an overhaul and drastic simplification of academic physics is long overdue before the subject loses all credibility and predictive ability.

For example, if we wish to consider the collision of a weak neutral Z boson with a neutrino, then we solve numerically the following simultaneous Evans equations (generally covariant Dirac equations):

$$(i\hbar\gamma^a\left(\partial_a - igZ_a\right)m_\nu c/)\,\nu^b = 0 \tag{21.28}$$

$$(i\hbar\gamma^a\left(\partial_a + igZ_a\right)m_z c/)\,Z^b = 0 \tag{21.29}$$

where ν^b is the neutrino wave-function and where Z^b is the weak neutral boson wave-function, a tetrad defined by:

$$Z^a{}_\mu = Z^{(0)}q^a{}_\mu. \tag{21.30}$$

The neutrino is a fermion and its wave-function is a tetrad of the type defined in Eq. (21.12). The non-zero neutrino mass appears in Eq. (21.28). The fact that the neutrino has a mass has now been established experimentally [2]. This one experimental fact is enough for the rejection of GWS/Higgs, because the latter is built entirely around the supposition that there exist two massless particles, the photon and neutrino. This supposition is again extraneous to general relativity (in which there can be no massless particles, as in Newtonian physics) and the supposition is thus extraneous to objective physics. Unsurprisingly the supposition has been found experimentally to be a false one. As in the case of epicycles, it might take some time for the standard model to be rejected, but if physics is to remain a generally covariant and thus objective study of nature, rejected it must be.

In Eqs. (21.28) and (21.29) m_ν and m_Z are therefore the non-zero neutrino and Z boson masses respectively, and g is the appropriate coupling constant which is C negative and so is proportional to the charge on the electron $-e$. There is no need for the obscure notion of modified hypercharge, introduced by Weinberg and accepted uncritically in GWS/Higgs. The use of hypercharge originated in gauge theory of the strong nuclear field, shortly after the introduction of the abstract internal gauge index by Yang and Mills. From the point of view of Einsteinian (i.e. objective or generally covariant) natural philosophy, this was another false turn in the development of physics. The index a of the tetrad, in contrast, is geometrical in origin, and thus physical according to general relativity. The abstract fiber-bundle index is just that, an abstract or loose parameter arbitrarily superimposed on flat or Minkowski spacetime without any regard to base manifold geometry and thus without any regard to gravitation, and worse, to objective physics. The abstract fiber-bundle index may be used to define internal gauge symmetries, but these must always remain extraneous to general relativity. Worse still is the use in the standard model of approximate internal gauge symmetries in nuclear strong field and quark theory. The obvious truth in mathematics is that a symmetry is exact and can never be approximate. Quarks cannot exist approximately, yet this is what we are told, i.e. what must follow logically from the use of approximate symmetry as a foundational idea. The rational mind would conclude that quarks do not exist, they have merely been postulated to exist.

In the Evans field theory the abstract index of gauge theory is replaced by the geometrical index a of the tetrad, and that index is of course governed rigorously by the rules of differential geometry itself. There is no room for approximate geometry in human thought, and no room for subjective thought-entities such as quarks which exist approximately and are confined so as to be unobservable. Natural philosophy is the objective study of the observable in nature. Having rid ourselves of this cupboard full of skeletons known as the standard model it becomes much easier to see that the interaction of a Z boson and a neutrino is a matter of solving the Evans equations (21.28) and (21.29) on a desktop computer, avoiding the floating point overflow inevitably caused by infinities, i.e. avoiding the path integral method by using robust integrating software. Nature abhors a Feynman infinity as much as it abhors a broken Higgs vacuum. Both the infinity and the broken vacuum are untested products of the human mind (i.e. of subjective thought untested by data) and cannot exist in nature. The latter can be defined only by objective measurement.

In the diagrammatic form of the type familiar in particle scattering theory textbooks the Evans equations (21.20) and (21.22) are summarized by: This diagram summarizes the interaction of two electrons through the

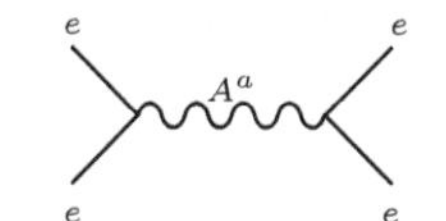

Fig. 21.1. Feynman Diagram

photon. Eqs. (21.28) and (21.29) are summarized by the diagram: illustrating the weak neutral current. An interaction between a Z boson and an electron is defined by the following two simultaneous Evans equations:

$$\left(i\hbar\gamma^{a}\left(\partial_{a}-ig_{1}Z_{a}\right)-m_{e}c\right)q^{a}=0 \qquad (21.31)$$

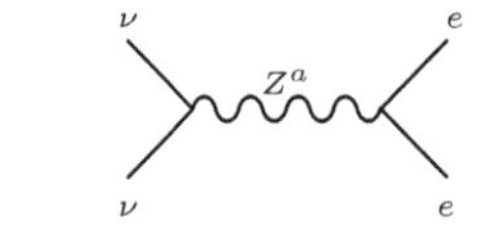

Fig. 21.2. Feynman Diagram

$$\left(i\hbar\gamma^a\left(\partial_a + ig_1 Z_a\right) - m_Z c\right) Z^a = 0 \tag{21.32}$$

where g_1 is the appropriate coupling constant again proportional to e.

In general scattering theory it is customary to use the momentum exchange diagram: which indicates the following processes:

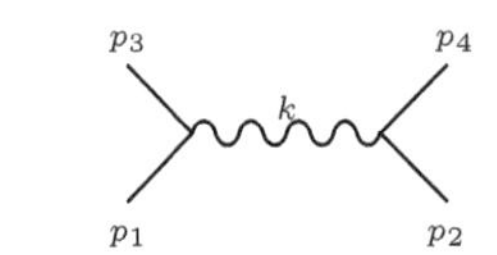

Fig. 21.3. Feynman Diagram

$$p_1 + p_2 = p_3 + p_4 \tag{21.33}$$

$$p_1 + k = p_3 \tag{21.34}$$

$$p_2 - k = p_4 \tag{21.35}$$

By adding Eqs. (21.34) and (21.35) it becomes clear that a boson momentum k is gained and lost simultaneously as follows:

$$(p_1 + k) + (p_2 - k) = p_3 + p_4 \tag{21.36}$$

This is what is known with traditional obscurity of language as a virtual boson. This general process is also describable by the appropriate simultaneous Evans equations. In order to describe the transmutation processes that occur in radio activity more than two Evans equations must solved simultaneously using powerful enough contemporary hardware and software. This fact is illustrated by the scattering process: mediated by the charged weak field boson W^-. In

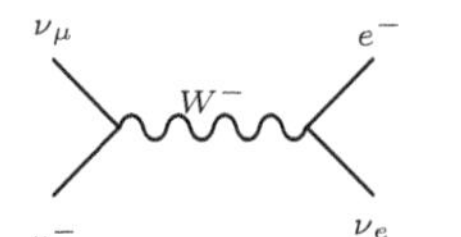

Fig. 21.4. Feynman Diagram

the above diagram the customary notation of particle scattering theory has been followed. Here $\mu^-$is the muon, a fermion with a mass about 207 times greater than the electron and a lifetime of 2.2×10^{-8} sec, ν_μ is the muon–neutrino, e^- is the electron, and ν_e is the electron-neutrino. By reference to diagram (21.4) the process in diagram (21.3) is the following conservation of momentum:

$$p\left(\mu^{-}\right)+p\left(\nu_{e}\right)=p\left(\nu_{\mu}\right)+p\left(e^{-}\right) \tag{21.37}$$

and Eq. (21.4) is denoted by the nuclear reaction, transmutation or radioactive process:

$$\mu^{-}+\nu_{e}=\nu_{\mu}+e^{-} \tag{21.38}$$

observed in particle colliders. By reference to Eq. (21.34) diagram (21.4) means that the muon momentum plus W^{-} momentum gives the muon-neutrino momentum:

$$p\left(\mu^{-}\right)+p\left(W^{-}\right)=p\left(\nu_{\mu}\right). \tag{21.39}$$

By reference to Eq. (21.35) diagram (21.4) also means that the electron-neutrino momentum minus the W^{-} momentum gives the electron momentum:

$$p\left(\nu_{e}\right)-p\left(W^{-}\right)=p\left(e^{-}\right). \tag{21.40}$$

The total momentum of all four particles is conserved as follows:

$$\left(p\left(\mu^{-}\right)+p\left(W^{-}\right)\right)+\left(p\left(\nu_{e}\right)-p\left(W^{-}\right)\right)=p\left(\nu_{\mu}\right)+p\left(e^{-}\right). \tag{21.41}$$

The general momentum exchange or particle scattering process illustrated in diagram (21.3) can now be seen to be described by the Evans equations:

$$\left(i\hbar\gamma^{a}\left(\partial_{a}+igk_{a}\right)-m_{1}c\right)p^{b}{}_{1}=\left(i\hbar\gamma^{a}\partial_{a}-m_{3}c\right)p^{b}{}_{3} \tag{21.42}$$

$$\left(i\hbar\gamma^{a}\left(\partial_{a}-igk_{a}\right)-m_{2}c\right)p^{b}{}_{2}=\left(i\hbar\gamma^{a}\partial_{a}-m_{4}c\right)p^{b}{}_{4}. \tag{21.43}$$

In this notation $p^{b}{}_{1},\cdots,p^{b}{}_{4}$ are the four wavefunctions of the interacting matter waves (particles). If this collision process results in transmutation (radio activity) then the particles emerging after collision are two different particles. The wavefunctions after collision are those of the two different particles. Each wave function is a tetrad and so carries the label b of the tangent bundle spacetime. If there is no collision or scattering, then:

$$k_{a}=0, p_{1}=p_{3}, p_{2}=p_{4} \tag{21.44}$$

and the following equations describing the two free particles are obtained self-consistently:

$$\left(i\hbar\gamma^{a}\partial_{a}-m_{1}c\right)p^{b}{}_{1}=0 \tag{21.45}$$

$$\left(i\hbar\gamma^{a}\partial_{a}-m_{2}c\right)p^{b}{}_{2}=0. \tag{21.46}$$

Eq. (21.42) is the description of Eq. (21.34) in the Evans unified field theory, and Eq. (21.43) is the description of Eq. (21.34). These are the basic equations which describe any particle scattering process in the Evans unified field theory. Evidently, the scattering process may or may not involve transmutation, and is mediated by the boson k_{a}. The free boson is itself governed by an Evans equation:

$$\left(i\hbar\gamma^a\partial_a - m_k c\right) k^b{}_i = 0 \tag{21.47}$$

where m_k is the mass of the boson. In Eq. (21.47), $k^b{}_a$ is the wavefunction of the boson before colliding with the particle. Eq. (21.47) is found using Eq. (21.9) by factorizing the Evans wave equation:

$$\left(\square + \left(\frac{m_k c}{\hbar}\right)^2\right) k^b{}_i = 0. \tag{21.48}$$

The collision of the boson with the particle then reduces the boson momentum as follows:

$$\left(i\hbar\gamma^a\left(\partial_a - igk_a\right) - m_k c\right) k^b = \left(i\hbar\gamma^a\partial_a - m_k c\right) k^b{}_f = 0 \tag{21.49}$$

where $k^b{}_f$ is the final wavefunction of the boson after collision. Eq. (21.49) is balanced through conservation of momentum by:

$$\left(i\hbar\gamma^a\left(\partial_a + igk_a\right) - m_k c\right) k^b = \left(i\hbar\gamma^a\partial_a - m_k c\right) k^b{}_i = 0. \tag{21.50}$$

Eqs (21.42), (21.43), (21.49) and (21.50) must be solved simultaneously, and are generally covariant unified field equations describing any type of collision between two particles mediated by any type of boson (field quantum). They describe the virtual boson exchange process summarized in:

$$(p_1 + k) - (p_2 - k) = p_3 + p_4. \tag{21.51}$$

In integrating these equations robust contemporary software should be used, and not the sixty year old path integral method, which produces well known pathological infinities. The criterion for acceptability of any theory must be general covariance (objectivity) and not renormalizability as in the standard model. The vastly complicated process of renormalization is merely a response to a flawed theory of special relativity (the standard model) and string theory merely compounds the problem with more loose parameters and meaningless concepts.

We are now ready to describe a transmutation process such as that in diagram (21.4) with the appropriate Evans equations of objective and unified field theory. An objective theory of physics is by definition a theory of general relativity, so the standard model fails this first and fundamentally important test of natural philosophy. In diagram (21.4), a muon μ^- of momentum p_1 collides with an electron-neutrino ν_e of momentum p_2. The collision is mediated or buffered by the weak charged boson W^-. The two particles which emerge from the collision are different from the two particles that were present before collision. The emerging or transmuted particles are the muon-neutrino ν_μ with momentum p_3 and the electron e^- with momentum p_4. Eq. ((21.42) and (21.34) means that the final momentum p_3 of the muon neutrino is the sum of the initial momentum of the muon and the momentum of the boson. This is represented by the Evans equation (21.42):

$$\left(i\hbar\gamma^a\left(\partial_a + igW_a^-\right) - m_\mu c\right)\mu^b = \left(i\hbar\gamma^a\partial_a - m_{\nu_\mu}c\right)\nu^b{}_\mu = 0 \tag{21.52a}$$

$$\left(i\hbar\gamma^a\left(\partial_a - igW_a^-\right) - m_W c\right)W^b = 0 \tag{21.52b}$$

where in accord with contemporary practice in particle physics we have denoted the wavefunction of the muon by μ^b and that of the muon-neutrino by $\mu^b{}_\mu$. The coupling constant g in equation (21.52) measures the strength of the collision, during the course of which the boson $W^-{}_a$ must lose the momentum it has transferred to the muon. The left hand side of Eq (21.52) describes the way in which the muon gains momentum from the boson. The right hand side of Eq. (21.52) describes the result of this momentum change, i.e. describes the free muon-neutrino after the collision has taken place. The final momentum

$$p_3\left(\nu_\mu\right) = p_1\left(\mu^-\right) + k\left(W^-\right) \tag{21.53}$$

that emerges from the collision is therefore that of the muon-neutrino. This is therefore the objective way of describing a transmutation process in unified field theory.

By reference to diagram (21.4) and Eq. (21.35) the electron e^- that emerges form the collision has a final momentum p_4, which is defined by the initial momentum p_2 of the electron-neutrino minus the boson momentum k. The appropriate Evans equation for this process are accordingly:

$$\left(i\hbar\gamma^a\left(\partial_a - igW_a^-\right) - m_{\nu_e}c\right)\nu^b{}_e = \left(i\hbar\gamma^a\partial_a - m_e c\right)e^b = 0 \tag{21.54a}$$

$$\left(i\hbar\gamma^a\left(\partial_a + igW_a^-\right) - m_W c\right)W^b = 0. \tag{21.54b}$$

Therefore the complete process in diagram ((21.4)) can be described either by solving the two Equations (21.52a) and (21.52b) simultaneously or by solving the two equations (21.54a) and (21.54b) simultaneously. By adding Eqs (21.52a), (21.52b), (21.54a) and (21.54b) we obtain the conservation of energy/momentum equation for the complete process:

$$\begin{aligned}&\left(i\hbar\gamma^a\left(\partial_a + igW_a^-\right) - m_\mu c\right)\mu^b + \left(i\hbar\gamma^a\left(\partial_a - igW_a^-\right) - m_{\nu_e}c\right)\nu^b{}_e \\ &= \left(i\hbar\gamma^a\partial_a - m_{\nu_\mu}c\right)\nu^b{}_\mu + \left(i\hbar\gamma^a\partial_a - m_e c\right)e^b = 0.\end{aligned} \tag{21.55}$$

Finally eqn. (21.54) is expressed as the two SU(2) symmetry equations:

$$(i\hbar\gamma^a \begin{bmatrix}1\,0\\0\,1\end{bmatrix}\partial_a + i\hbar\gamma^a\begin{bmatrix}0 & igW_a\\ -igW_a & 0\end{bmatrix} - \begin{bmatrix}m_{\nu_e}c & 0\\ 0 & m_\mu c\end{bmatrix})\begin{bmatrix}\nu_e\\ \mu\end{bmatrix} = 0 \tag{21.56}$$

$$(i\hbar\gamma^a\begin{bmatrix}1\,0\\0\,1\end{bmatrix} - \begin{bmatrix}m_{\nu_e}c & 0\\ 0 & m_e c\end{bmatrix})\begin{bmatrix}\nu_{\mu_b}\\ e_b\end{bmatrix} = 0 \tag{21.57}$$

familiar from that part of the GWS/Higgs theory that is conventionally used to describe the weak charged current process of diagram (21.4).

21.3 Discussion

Glashow, Weinberg and Salaam independently arrived at some aspects of electro-weak (or GWS) theory based directly on the Higgs mechanism. We have shown in this paper that nearly all the assumptions of the GWS/Higgs theory contradict the basic tenet of physics, that of objectivity or general covariance. Therefore GWS/Higgs theory simply does not stand up to scholarly scrutiny. One of the basic ideas of GWS/Higgs is that the neutrino is massless, but recently the neutrino has been shown experimentally to have mass. Therefore it is an experimental fact that there are no massless fermions in nature. Indeed, it is now thought in some quarters that relic neutrinos are responsible for dark matter, and therefore for about 80 of the mass of the universe. Neutrino oscillations are observed experimentally in the muon-neutrino, which appears and disappears as it travels hundreds of kilometres through the earth (super Kamiokande collaboration). The electron-neutrino was first inferred theoretically by Fermi in about 1930 (from the observed energy deficit in beta particle decay) and the electron-neutrino was first observed experimentally in 1956. The muon-neutrino was discovered in 1961 and the tau-neutrino in 1974. All three types of neutrino have finite mass. This fact is inexplicable in the standard model but is explained in the Evans field theory as discussed in Section 21.2. In the equations of that section the neutrino mass always appears as non-zero. We exemplified the Evans field theory by considering the muon-neutrino in a charged weak current process, but the theory is generally applicable to all three types of neutrino and to all fermions and bosons in nature. Neutrino oscillation is explained in the Evans theory as a particular type of mass energy transmutation. Neutrino oscillation has no explanation in GWS/Higgs because the latter theory assumes that neutrino mass is always zero, so no neutrino oscillation (changing or transmutation of finite neutrino mass/energy) is possible in GWS/Higgs. It follows that the Higgs mechanism has been falsified experimentally and that there is no Higgs boson in nature, all that remains to us after forty years of speculation is an adjustable parameter in an experimentally falsified theory. Therefore there is no point in looking experimentally for a Higgs boson as in the heavy hadron collider experiments planned at CERN. It would be more logical to interpret the new heavy hadron collider data with the Evans theory and developments thereof.

Another basic idea of GWS/Higgs is that the neutrino (which is parity violating, or left handed only) forms a physically meaningful isospinor with a left handed fermion, but not with a right handed fermion. The arguments of Section 21.2 shows, however, that the isospinor has no particular physical meaning over and above that already present in conservation of energy/momentum. Eq. (21.56) or Eq. (21.57) shows that there is, rather, an ordinary column vector with two entries in a convenient mathematical representation of simultaneous Evans equations. The latter are generally covariant and are the fundamental and objective equations of electro-weak theory (radio-activity). It may be true experimentally that the neutrino is left handed (and so violates parity experimentally), but it does not follow that it must form an isospinor with a left handed fermion such as an electron or muon. So another fundamental tenet of GWS/Higgs has been shown to be false. It follows that if there is no isospinor in nature there can be no modified hypercharge as introduced by Weinberg and uncritically accepted in the standard model. It also follows that the SU(2) symmetry internal gauge space of GWS/Higgs theory has no particular physical significance. (Eq. (21.56) shows that this symmetry is conservation of momentum in the Evans equations.) In the standard model this SU(2) mathematical space is superimposed on a theory of special relativity which is only Lorentz covariant and so cannot be objective to all observers. Being a superimposed abstract space, its parameters are in the last analysis adjustable parameters which must be found from experimental data. They can never be used to predict data foundationally. All gauge theories of the standard model (the electromagnetic, weak and strong sectors) have this weakness inbuilt, so internally inconsistent gauge theory should be replaced by self-consistent general relativity, and the abstract fiber bundle of gauge theory replaced by the physically meaningful tangent bundle in differential geometry and general relativity as originally intended in Einsteins work. This is what has been done in Section 21.2 for the theory of radio activity. There is no purpose in accepting Einsteins work on the one hand, and rejecting it on the other. Yet this is what the standard model does all the time, it accepts Einsteinian general relativity in its gravitational sector and rejects it entirely in its other three sectors.

These elementary (i.e. foundational) considerations put the standard model in ever more serious difficulties, because the much vaunted quark model of the strong sector is built on an assumed SU(3) internal gauge space. It is certain that this gauge symmetry has no physical meaning in general relativity. It becomes ever clearer that all elementary particle physics should be interpreted with the generally covariant Evans theory, which is firmly based on the tetrad postulate of differential geometry and is a rigorously objective theory of Einsteinian natural philosophy as required. It is absurd to propose a theory of physics which is not objective to all observers, yet this is precisely what occurs in the standard model. The quark model can be criticised in several ways, the data for quarks are based on low angle scattering, and are equivocal. They could be interpreted as inhomogeneities due to the spatial characteristics of a given type of nuclear wavefunction, for example a proton wavefunction described by the inhomogeneities of the spherical harmonics (akin to the electronic s, p, d, ... orbitals). No one would claim that the spatially inhomogeneous electronic s, p, d,.... orbitals indicate the existence of a particle more fundamental than the electron, so why should the spatially inhomogeneous proton be made up of quarks? There is no reason in other words why an elementary particle such as a proton should be perfectly homogeneous, its internal density may vary from quantum mechanics. In other words there is no reason why the vague inhomogeneities of low angle scattering data should be interpreted in terms of other, more fundamental particles in nature (the quarks). It is true that elementary particle data appear to display an equally vague gauge symmetry akin to SU(3) but this is openly referred to as an approximate gauge symmetry, a term which should have no meaning whatsoever in objective physics or mathematics. So vague bumps in low angle scattering theory are all we really have on quarks, the rest is surmise in abstract gauge theory with loose parameters from an experimentally falsified Higgs mechanism. What the standard model really tells us is that quarks must exist only approximately, they are the only manifestations in nature of an approximate reality but must be confined so as to be unobservable. This is a great absurdity for which Nobel Prizes are habitually given. Add to this the many gross absurdities of renormalization and string theory then we must conclude that there is no physics at all in the standard model. Compared with this contemporary and hugely expensive contrivance, epicycles were models of foundational clarity.

In the Evans theory scattering data for all types of elementary particles can always be interpreted straightforwardly with equations of the type developed in Section 21.2, and no effort is made to look for the physically non-existent but mathematically convenient gauge symmetries of the standard model, symmetries which do not exist in general relativity and therefore do not exist in objective physics. Only in this way will it ever be possible to analyze the effect of gravity on nuclear processes or chemical reactions.

In addition to these severe foundational failings the GWS/Higgs theory cannot predict data from particle accelerators, as if often claimed. The theory can only fit data using adjustable parameters. The charged and neutral weak boson masses in GWS/Higgs are parameterized as follows:

$$m_W = g\eta/\sqrt{2} = m_Z \cos\theta_W \tag{21.58}$$

$$\eta^2 = \frac{1}{2\sqrt{2}G}, \tag{21.59}$$

therefore:

$$m_W = f\left(g, g^{'}, \eta\right) = f\left(g, g^{'}, G\right). \tag{21.60}$$

Here θ_W is the Weinberg angle, g and $g^{'}$ are coupling parameters, η is the Higgs parameter and G is the Fermi coefficient. Eq (21.59) shows that G is just replaced by η. In other words the straightforward G of Fermi is surmised to have something to do with a vacuum symmetry breaking which gives mass to mysterious, initially massless fermions but not to others. Since all fermions have mass experimentally this surmise is false and is erroneous in both special and general relativity. The origin of mass is now known from the Evans Lemma [3]–[27] to be least curvature. A massless particle corresponds to nothing at all. No experimental evidence for vacuum symmetry breaking is ever given in GWS/Higgs, what really happens is that two scattering peaks (for the neutral and charged boson) are fitted with three adjustable parameters (g, $g^{'}$ and η

or G). This could just as well be done with a curve fitting program without recourse to any physics at all.

The formal structure of the GWS/Higgs theory is a combination of

$$\left(i\hbar\gamma^a\left(\begin{bmatrix}1&0\\0&1\end{bmatrix}\partial_a+\frac{ig'}{2}\begin{bmatrix}1&0\\0&1\end{bmatrix}X_\mu-\frac{ig}{2}\begin{bmatrix}W^3{}_\mu & W^1{}_\mu-igW^2{}_\mu\\ W^1{}_\mu+igW^2{}_\mu & -W^3{}_\mu\end{bmatrix}\right)\right.$$
$$\left.-\begin{bmatrix}m_\nu & 0\\ 0 & m_e c\end{bmatrix}^c\right)\begin{bmatrix}\nu\\ e_L\end{bmatrix}=i\gamma^a\left(\partial_a+\frac{i}{2}\left(g'X_a-g\boldsymbol{\sigma}\cdot\boldsymbol{W}_\mu\right)\right)L \tag{21.61}$$

and

$$\left(i\hbar\gamma^a\left(\partial_a+ig'X_a\right)-m_e\right)\begin{bmatrix}1&0\\0&1\end{bmatrix}e_R$$
$$=\left(i\hbar\gamma^a\left(\partial_a+ig'X_a\right)-m_e\right)R \tag{21.62}$$

so the complete formal structure consists of two simultaneous equations:

$$i\hbar\gamma^a\left(\partial_a+\frac{i}{2}\left(g'X_a-gW^3{}_a\right)-m_\nu c\right)\nu+i\hbar\gamma^a\left(-\frac{i}{2}W_a\right)e_L$$
$$+\left(i\hbar\gamma^a\left(\partial_a+ig'X_a\right)-m_e c\right)e_R=0 \tag{21.63}$$

$$i\hbar\gamma^a\left(-\frac{ig}{2}W^*_a\right)\nu+i\hbar\gamma^a\left(\partial_a+\frac{i}{2}\left(g'X_a+gW^3{}_a\right)-m_e c\right)e_L$$
$$+\left(i\hbar\gamma^a\left(\partial_a+ig'X_a\right)-m_e c\right)e_R=0. \tag{21.64}$$

However, in the original GWS/Higgs theory the mass term is missing from these equations. The correct way of expressing the theory is the combination of Evans equations in Eqs. (21.63) and (21.64), a combination in which the mass terms appear correctly as a result of general relativity. It is then possible to define the electromagnetic field as:

$$A_\mu=\left(g'W^3{}_\mu+gX_\mu\right)/\left(g^2+g'^2\right)^{1/2}=W^3{}_\mu\sin\theta_W+X_\mu\cos\theta_W \tag{21.65}$$

and the weak neutral field by:

$$Z_\mu=\left(gW^3{}_\mu-g'X_\mu\right)/\left(g^2+g'^2\right)^{1/2}=W^3{}_\mu\cos\theta_W-X_\mu\sin\theta_W \tag{21.66}$$

However, no particular physical significance is attached to the Weinberg angle:

$$\theta_W=sin^{-1}g'/\left(g^2+g'^2\right)^{1/2} \tag{21.67}$$

Finally, the way in which particle scattering data is explained in the Evans theory is illustrated by taking two conjugate Evans equations such as:

$$\left(\gamma^a\left(i\hbar\partial_a - eA_a\right) - mc\right)q^b = 0 \tag{21.68}$$

$$\left(\gamma^a\left(-i\hbar\partial_a - eA_a^*\right) - mc\right)q^b = 0 \tag{21.69}$$

and from these equations forming the wave equation:

$$\left(\Box + \frac{emc}{\hbar^2}\gamma^a\left(A_a + A_a^*\right) + g^2 A^a A_a^* + \left(\frac{mc}{\hbar}\right)^2\right)q^b = 0. \tag{21.70}$$

The two equations (21.68) and (21.69) or the wave equation (21.70) illustrate the interaction of a an electron with a photon, and the interaction energy is defined by:

$$En_{int} = mc^2 = \hbar gc\left(A^a A_a^*\right)^{1/2} = ecA^{(0)} \tag{21.71}$$

in terms of the mean square amplitude $A^{(0)2}$ of the electromagnetic field. Similarly the interaction of the ${Z^3}_a$ boson with the neutrino (weak neutral current) is described from Eq. (21.28) of Section 21.2 by the interaction energy: It can be seen that there are no adjustable parameters in the Evans field theory. The mean square amplitudes $A^{(0)2}$ and $Z^{3(0)2}$ are foundational properties of the boson itself, as is the intrinsic boson mass. In the GWS/Higgs theory both the boson masses and the interaction energy must be described through the lose parameters g, g' and G or η. The boson masses are:

$$m_W = g\eta/\sqrt{2}, \quad m_3 = m_W/\cos\theta_W, \tag{21.72}$$

and the interaction energies are defined through:

$$\begin{aligned} m_W^2 &= \frac{e^2}{4\sqrt{2}\,G\sin^2\theta_W} = (78.6GeV/c^2)^2 \\ &= f\left(g, g', G\right). \end{aligned} \tag{21.73}$$

Acknowledgments

The Ted Annis Foundation and Craddock Inc. are thanked for funding and the AIAS Fellows and environment for many interesting discussions.

References

1. L. D. Ryder, *Quantum Field Theory*, (Cambridge, 1996, 2nd ed.).
2. Super Kamiokande Collaboration (1998).
3. M. W. Evans, A Unified Field Theory for Gravitation and Electromagnetism, Found. Phys. Lett., **16**, 367 (2003).
4. M. W. Evans, A Generally Covariant Wave Equation for Grand Unified Field Theory, *Found. Phys. Lett.*, **16**, 507 (2003).
5. M. W. Evans, The Equations of Grand Unified Field Theory in terms of the Maurer Cartan Structure Relations of Differential Geometry, *Found. Phys. Lett.*, **17**, 25 (2004).
6. M. W. Evans, Derivation of Diracs Equation from the Evans Wave Equation, *Found. Phys. Lett.*, **17**, 149 (2004).
7. M. W. Evans, Unification of the Gravitational and Strong Nuclear Fields, *Found. Phys. Lett.*, **17**, 267 (2004).
8. M. W. Evans, The Evans Lemma of Differential Geometry, *Found. Phys. Lett.*, **17**, 433 (2004).
9. M. W. Evans, Derivation of the Evans Wave Equation form the Lagrangian and Action, Origin of the Planck Constant in General Relativity, *Found. Phys. Lett.*, **17**, 535 (2004).
10. M. W. Evans and the AIAS Author Group, Development of the Evans Wave Equation in the Weak Field Limit, the Electrogravitic Equation, *Found. Phys. Lett.*, **17**, 497 (2004).
11. M. W. Evans, Physical Optics, the Sagnac Effect and the Aharonov Bohm Effect in the Evans Unified Field Theory, *Found. Phys. Lett.*, **17**, 301 (2004).
12. M. W. Evans, Derivation of the Geometrical Phase from the Evans Phase Law of Generally Covariant Unified Field Theory, Found. Phys., Lett., **17**, 393 (2004).
13. M. W. Evans, Derivation of the Lorentz Boost from the Evans Wave Equation, *Found. Phys. Lett.*, **17**, 663 (2004).
14. M. W. Evans, The Electromagnetic Sector of the Evans Field Theory, *Found. Phys. Lett.*, in press, preprint on www.aias.us.
15. M. W. Evans, New Concepts from the Evans Unified Field Theory, Part One: The Evolution of Curvature, Oscillatory Universe without Singularity and General Force and Field Equations, *Found. Phys. Lett.*, in press, preprint on www.aias.us.

16. M. W. Evans, New Concepts from the Evans Unified Field Theory, Part Two: Derivation of the Heisenberg Equation and Replacement of the Heisenberg Uncertainty Principle, *Found. Phys. Lett.*, in press, preprint on www.aias.us.
17. M. W. Evans, Derivation of O(3) Electrodynamics from Generally Covariant Unified Field Theory, *Found. Phys. Lett.*, in press, preprint on www.aias.us.
18. M. W. Evans, The Spinning and Curving of Spacetime, The Electromagnetic and Gravitational Fields in the Evans Unified Field Theory, *Found. Phys. Lett.*, in press, preprint on www.aias.us.
19. M. W. Evans, The Derivation of O(3) Electrodynamics from the Evans Unified Field Theory, *Found. Phys. Lett.*, in press, preprint on www.aias.us.
20. M. W. Evans, Calculation of the Anomalous Magnetic Moment of the Electron from the Evans Field Theory, *Found. Phys. Lett.*, in press, preprint on www.aias.us.
21. M. W. Evans, Generally Covariant Unified Field Theory, the Geometrization of Physics, (in press, 2005).
22. L. Felker, *The Evans Equations* (World Scientific, in prep, 2004 / 2005).
23. M. W. Evans and L. B. Crowell, *Classical and Quantum Electrodynamics and the* $\boldsymbol{B}^{(3)}$ *Field*, (World Scientific, Singapore, 2001).
24. M. W. Evans, (ed.), *Modern Non-Linear Optics*, a special topical issue of I. Prigogine and S. A. Rice (eds.), *Advances in Chemical Physics* (Wiley Interscience, New York, 2001, 2nd and e book eds.), vol. 119(1), 119(2) and 119(3).
25. M. W. Evans, J.-P, Vigier et alii, *The Enigmatic Photon* (Kluwer, Dordrecht, 1994 to 2002, paperback and hardback), in five volumes.
26. M. W. Evans and A. A. Hasanein, *The Photomagneton in Quantum Field Theory* (World Scientific, Singapore, 1994).
27. M. W. Evans, (ed.), first edition of ref. (24) (Wiley Interscience, New York, 1992, 1993, 1997 (softback)), vols. 85(1), 85(2) and 85(3).
28. A. Einstein, *The Meaning of Relativity* (Princeton Univ Press, 1921 and subsequent editions).
29. S. P. Carroll, *Lecture Notes in General Relativity*, (a graduate course at Harvard, UC Santa Barbara and Chicago, arXiv: gr-gq/9713019 v1 3 Dec 1997).
30. P. W. Atkins, *Molecular Quantum Mechanics* (Oxford Univ. Press, 1983, 2nd ed.),
31. P. Pinter, private communication and several source papers in cosmology, see www.aias.us.

22

Evans Field Theory of Neutrino Oscillations

Summary. Neutrino oscillations are described with the generally covariant Evans field theory, showing that gravitation is able to influence the transmutation of one type of neutrino into another. The neutrino mass is considered to originate in general relativity as eigenvalues of the Evans Lemma. The wavefunction of the neutrino is considered to be a tetrad made up of a complex sum of two types of neutrino multiplied by the Evans phase $e^{i\theta}$. The latter is also the result of general relativity, so the Evans field theory of neutrino oscillations is causal and generally covariant as required by Einsteinian natural philosophy. Neutrino oscillation is shown to be a confirmation of general relativity and causal quantum mechanics, (concepts that are unified in the Evans field theory). The Heisenberg Bohr quantum mechanics is therefore refuted experimentally by neutrino oscillation, and causal general relativity is preferred over the acausal assertions of the Copenhagen School.

Key words: Evans field theory; neutrino oscillations; Evans phase.

22.1 Introduction

Neutrino oscillation is the term give to the disappearance [1] of atmospheric muon- neutrinos and solar electron-neutrinos. It is thought to be a transmutation of mass/energy, a neutrino of one type changes into a neutrino of another type, with a different mass. This process is thought to be possible only if the neutrino mass is non-zero and only if the masses of the transmuting neutrinos are different [2]. For example, a muon-neutrino from the atmosphere is observed experimentally to disappear, and theoretically it is thought that this is due to the fact that it transmutes into a tau-neutrino with a different mass. The latter is thought to be essentially undetectable, so the overall effect is the apparent disappearance of the muon-neutrino without violation of the Noether Theorem. Similarly an electron-neutrino from the sun changes into a muon-neutrino or tau-neutrino [2], disappears, and leads to the experimentally reproducible and repeatable solar deficit of neutrinos. Given initially an electron-neutrino, oscillation means that it will be a mixture of different types of neutrino after propagating a certain distance as a matter wave. An electron-neutrino is therefore thought to be initially in a quantum state which is a mixture of different mass/energies. In the Evans field theory [3]–[28] this is a tetrad governed by the Evans Lemma, the eigenvalues of which are composite scalar curvatures. The neutrino masses in the Evans field theory are considered to originate in Einsteinian general relativity (causal physics or natural philosophy that is necessarily and always objective to an observer in any frame of reference moving arbitrarily with respect to any other frame of reference).

In the simplest case of oscillation between two neutrinos, Section 2 shows that the wavefunction or tetrad is a complex sum of components multiplied by the Evans phase [3]–[28], which in this paper is denoted $e^{i\theta}$. The Evans phase is the origin in general relativity of the Berry phase and various topological phases, and is considered in Section 2 to be responsible for the reproducible and repeatable phenomenon of neutrino oscillation. The real and imaginary parts of the wavefunction correspond to the two neutrino types, which transmute into each other as the result of the fundamental Evans phase. The propagation of the wavefunction is governed by differential geometry - the tetrad postulate [29] on which the Evans Lemma is based directly [3]–[28].

Section 22.3 is a brief discussion of the origin of the Evans phase that is inter alia responsible for neutrino oscillations and observed experimentally thereby.

22.2 The Mixing of Neutrino Wavefunctions Due to the Evans Phase

The experimental phenomenon of neutrino mass and neutrino oscillation is new to physics [2] and invalidates the standard model, in which the neutrino mass is zero. The available explanations are still essentially empirical. For example [2], it is thought that:

$$\begin{bmatrix} \nu_\mu \\ \nu_\tau \end{bmatrix} = \begin{bmatrix} \cos\theta & -\sin\theta \\ \sin\theta & \cos\theta \end{bmatrix} \begin{bmatrix} \nu_1 \\ \nu_2 \end{bmatrix} \tag{22.1}$$

where v_μ is the muon-neutrino eigenfunction and v_τ that of the tau-neutrino, and where v_1 and v_2 are neutrino flavors. The angle θ is used to mix the neutrino eigenfunctions through the flavors. This is little more than an empirical exercise. In the theory proposed here the physical origin of the angle θ is given through the Evans phase $e^{i\theta}$ of general relativity. The present empirical theory of neutrino oscillations [2] is based on the Schrödinger equation, which is a non-relativistic quantum limit of the Evans wave equation [3]–[28]. The latter is both generally covariant and an equation of causal wave mechanics (deterministic quantum mechanics).

In order to derive a physically more incisive explanation of neutrino oscillations we re-express Eq. (22.1) in this Section as a complex valued tetrad multiplied by the Evans phase $e^{i\theta}$. The latter is therefore part of the wavefunction of the Evans Lemma, the subsidiary geometrical proposition (lemma) leading to the Evans wave equation [3]–[28]. From Eq. (22.1):

$$\nu_\mu = \nu_1 \cos\theta - \nu_2 \sin\theta \tag{22.2}$$

$$\nu_\tau = \nu_1 \sin\theta + \nu_2 \cos\theta. \tag{22.3}$$

When $\theta = 0$:

$$\nu_\mu = \nu_1 \tag{22.4}$$

$$\nu_\tau = \nu_2 \tag{22.5}$$

and when $\theta = \frac{\pi}{2}$:

$$\nu_\mu = -\nu_2 \tag{22.6}$$

$$\nu_\tau = \nu_1. \tag{22.7}$$

It can be seen from Eqs. (22.4) and (22.7) that the muon-neutrino has become the tau-neutrino. The origin of this transmutation is the fundamental Evans phase $e^{i\theta}$ of general relativity. Therefore the empirical angle θ of the available theory [2] is identified as the Evans phase angle, θ, a fundamental spin invariant of non-Minkowski spacetime [3]–[28]. If this is accepted then neutrino oscillation becomes experimental confirmation of causal general relativity and thus of causal quantum mechanics. There is no explanation for the Evans phase in acausal Heisenberg Bohr quantum mechanics. This is because $e^{i\theta}$ (being part of the neutrino wavefunction) would be "unknowable" in the Heisenberg Bohr theory and therefore neutrinos would appear and disappear, without cause and at random. Such is not observed experimentally [2] and in the Copenhagen interpretation of wave mechanics neutrino oscillations would not be repeatable and reproducible as observed experimentally. They would be events that appeared without any cause, and so would appear in an "unknowable" way. This subjective assertion means that the oscillations would be caused by nothing knowable. Evidently this is not objective natural philosophy. Statistical averaging of wavefunctions occurs in the Evans field theory, and thus in general relativity, but this is entirely different from what is meant by the wavefunction of the Copenhagen School.

Using the well known formulae:

$$\cos\theta = \frac{1}{2}\left(e^{i\theta} + e^{-i\theta}\right) \tag{22.8}$$

$$\sin\theta = \frac{1}{2i}\left(e^{i\theta} - e^{-i\theta}\right) \tag{22.9}$$

it is seen that Eq. (22.1) is equivalent to:

$$\nu = \nu_\mu + i\nu_\tau = (\nu_1 + i\nu_2)\, e^{i\theta} \tag{22.10}$$

The origin of neutrino flavors in general relativity is given by Eq (22.10) - the neutrino flavors v_1 and v_2 are the pre-multipliers of the Evans phase of the complete neutrino wavefunction. Other types of elementary particle transmutation (for example that observed in the kaon [2]) can be explained similarly.

The Evans phase is a fundamental property in physics and has been used recently [3]–[28] to explain various electromagnetic effects such as the Aharonov Bohm and Sagnac effects, and also used to explain the origin in general relativity and unified field theory of everyday optical phenomena such reflection and interferometry. The Evans phase and Evans spin field $\boldsymbol{B}^{(3)}$ are spin invariants of the non-Minkowski spacetime of Evans field theory, a generally covariant unified field theory based directly on differential geometry. The theory is essentially the geometrization of all natural philosophy [3]–[28], and not just gravitation.

The eigenfunction (22.10) is governed by the Evans Lemma [3]–[28] and so must be possessed of a tangent bundle index a for all base manifolds. The complete description of the neutrino eigenfunction v in general relativity is therefore in terms of a tetrad ${\nu^a}_\nu$.

The latter is defined [29] in differential geometry as the invertible matrix linking vectors or basis elements in the tangent bundle (indexed a) and base manifold (indexed μ). The Evans Lemma then states that:

$$\Box {\nu^a}_\mu = R {\nu^a}_\mu \tag{22.11}$$

and can be written for all μ of the base manifold as

$$\Box \nu^a = R \nu^a . \tag{22.12}$$

Eq. (22.12) is then an equation, for all μ, in the flat or Minkowski spacetime of the tangent bundle. The effect of gravitation on this equation (and indeed on all equations of physics) is measured through elements of the tetrad [3]–[28]. Therefore one way of distinguishing between the Evans field theory of neutrino oscillations and the empirical theory [2] would be to look for the effect of gravitation on neutrino oscillations. In the available empirical theory of the type summarized in ref. [2], there is no effect of gravitation. Similarly, in the Evans field theory, gravitation affects the electro-weak field and therefore affects radio-activity [30], and gravitation evidently affects the electromagnetic field, weak field and strong field. These powerful results apply to all the physical and life sciences [30] and are expressions of general relativity itself, expressions of the objective nature of physics. The objective nature of physics invalidates the subjective assertions of the Copenhagen School. Therefore to invalidate Heisenberg Bohr quantum mechanics, effects of gravitation should be looked for experimentally as described in this paper.

The physical part of the tetrad is the real part of the mathematical expression:

$$\begin{aligned} \nu^a &= \left({\nu^a}_1 + i{\nu^a}_2 \right) e^{i\theta} \\ &= \left(\nu_\mu + i\nu_\tau \right). \end{aligned} \tag{22.13}$$

Initially, $\theta = 0$, and so:

$$\nu^a = \left({\nu^a}_\mu + i{\nu^a}_\tau \right), \tag{22.14}$$

which means that the real part of the neutrino wavefunction is initially that of a pure muon-neutrino ${\nu^a}_\nu$ with mass m_μ. The tau-neutrino is described initially (zero Evans angle θ and unit Evans phase $e^{i\theta}$) by the pure imaginary and therefore initially unphysical component of the complete neutrino wave function. If the masses of the muon-neutrino and the tau-neutrino were the same, the two neutrinos would be the same elementary particle, and their wave-functions would be the same. This means that mixing according to Eq. (22.13) could never occur and the Evans phase angle would always be zero. In this case the relevant equation in the Evans field theory would be:

$$\Box {\nu^a}_\mu = R {\nu^a}_\mu, \tag{22.15}$$

and would be a wave equation for a muon-neutrino which could never change into a tau-neutrino.. The observation of neutrino oscillations experimentally means however that the needed Evans equation must be Eq. (22.13). If the various neutrino masses were each zero there would be no Evans phase of the right type present to explain the observed neutrino oscillations. Zero mass is evidently a special case of neutrinos with the same mass.

22.3 Origin of the Evans Phase in General Relativity

The Evans angle is a spin invariant of general relativity defined [3]–[28] by the generally covariant Stokes Theorem:

$$\theta = \kappa \oint_{DS} q^a = \kappa \int_S D \wedge q^a = \kappa \int_S T^a \tag{22.16}$$

where q^a is the tetrad one-form and T^a is the torsion two-form. The Evans phase factor is then the complex exponential:

$$\Phi = \exp(i\theta) \tag{22.17}$$

and has been shown to be the origin of the Berry phase and the various topological phases, and to be the generally covariant electromagnetic phase needed for the correct description of optics and interferometry and so forth. Therefore the fundamental geometrical origin of the Evans angle is the definition of the torsion form, one of the two fundamental forms of differential geometry, the other [29] being the curvature or Riemann form. The torsion form is defined by the second Maurer-Cartan structure relation, and the curvature form by the first Maurer-Cartan structure relation. The Evans phase law is therefore the generally covariant form of all phase laws in physics, for example the Dirac and Berry phase laws.

In order to describe neutrino oscillations the Evans phase law has been used to mix the original wavefunctions. In the Evans field theory this means that there is a mixing of spacetime properties, easily understood geometrically through the fact that one type of geometry may be deformed into another in a causal manner.

Acknowledgement

The Ted Annis Foundation is thanked for funding and the Fellows of AIAS for many interesting discussions.

References

1. Super Kamiokande Collaboration, 1998 to present.
2. Univ California Irvine Website, ps.uci.edu/ superk/oscillation.html.
3. M. W. Evans, A Unified Field Theory for Gravitation and Electromagnetism, *Found. Phys. Lett.*, **16**, 367 (2003).
4. M. W. Evans, A Generally Covariant Wave Equation for Grand Unified Field Theory, *Found. Phys. Lett.*, **16**, 507 (2003).
5. M. W. Evans, The Equations of Grand Unified Field Theory in terms of the Maurer-Cartan Structure Relations of Grand Unified Field Theory, *Found. Phys. Lett.*, **17**, 25 (2004).
6. M. W. Evans, Derivation of Dirac's Equation from the Evans Wave Equation, *Found. Phys. Lett.*, **17**, 149 (2004).
7. M. W. Evans, Unification of the Gravitational and Strong Nuclear Fields, *Found. Phys. Lett.*, **17**, 267 (2004).
8. M. W. Evans, The Evans Lemma of Differential Geometry, *Found. Phys. Lett.*, **17**, 433 (2004).
9. M. W. Evans, Derivation of the Evans Wave Equation from the Lagrangian and Action: origin of the Planck Constant in General Relativity, *Found. Phys. Lett.*, **17**, 535 (2004).
10. M. W. Evans et alii, Development of the Evans Wave Equation in the Weak Field Limit: the Electrogravitic Equation, *Found. Phys. Lett.*, **17**, 497 (2004).
11. M. W. Evans, Physical Optics, the Sagnac Effect and the Aharonov Bohm Effect in the Evans Unified Field Theory, *Found. Phys. Lett.*, **17**, 301 (2004).
12. M. W. Evans, Derivation of the Geometrical Phase from the Evans Phase Law of Generally Covariant Unified Field Theory, *Found. Phys. Lett.*, **17**, 393 (2004).
13. M. W. Evans, Derivation of the Lorentz Boost from the Evans Wave Equation, *Found. Phys. Lett.*, **17**, 663 (2004).
14. M. W. Evans, The Electromagnetic Sector of the Evans Field Theory, *Found. Phys. Lett.*, in press, preprint on www.aias.us.
15. M. W. Evans, New Concepts from the Evans Field Theory: Part One, the Evolution of Curvature, Oscillatory Universe without Singularity, and General Force and Field Equations, *Found. Phys. Lett.*, in press, preprint on www.aias.us.
16. M. W. Evans, New Concepts from the Evans Field Theory: Part Two, Derivation of the Heisenberg Equation and Replacement of the Heisenberg Uncertainty Principle, *Found. Phys. Lett.*, in press, preprint on www.aias.us.

17. M. W. Evans, Derivation of O(3) Electrodynamics from Generally Covariant Unified Field Theory, *Found. Phys. Lett.*, submitted, preprint on www.aias.us.
18. M. W. Evans, The Spinning and Curving of Spacetime, the Electromagnetic and Gravitational Fields in the Evans Unified Field Theory, *Found. Phys. Lett.*, submitted, preprint on www.aias.us.
19. M. W. Evans, The Derivation of O(3) Electrodynamics from the Evans Unified Field Theory, *Found. Phys. Lett.*, submitted, preprint on www.aias.us.
20. M. W. Evans, Calculation of the Anomalous Magnetic Moment of the Electron from the Evans Field Theory, *Found. Phys. Lett.*, submitted, preprint on www.aias.us.
21. M. W. Evans, Generally Covariant Electro-weak Theory, *Found. Phys. Lett.*, submitted, preprint on www.aias.us.
22. M. W. Evans, *Generally Covariant Unified Field Theory: the Geometrization of Physics*, (in press, 2005) preprint on www.aias.us.
23. L. Felker, *The Evans Equations*, in prep (freshman level book).
24. M. W. Evans and L. B. Crowell, *Classical and Quantum Electrodynamics and the* $\boldsymbol{B}^{(3)}$ *Field*, (World Scientific, Singapore, 2001).
25. M. W. Evans (ed.), *Modern Non-Linear Optics*, a special topical issue of I. Prigogine and S. A. Rice (series eds.), *Advances in Chemical Physics*, (Wiley-Interscience, New York, 2001, second ed., hardback and e book), vols. 119(1) to 119(3).
26. M. W. Evans, J.-P. Vigier et alii, *The Enigmatic Photon*, (Kluwer, Dordrecht, 1994 to 2002, hardback and softback), in five volumes.
27. M. W. Evans and A. A. Hasanein, *The Photomagneton in Quantum Field Theory*, (World Scientific, Singapore, 1994).
28. M. W. Evans and S. Kielich (eds), first edition of ref. (25), volumes 85(1) to 85(3).
29. S. P. Carroll, *Lecture Notes in General Relativity* (a graduate course at Harvard, Univ California Santa Barbara and Univ Chicago, arXiv: gr-gq/9713019 v1 Dec 1997, public domain).
30. P. Pinter, personal communications on the Pinter Hypothesis on the causal origin of life .

23

The Interaction of Gravitation and Electromagnetism

Summary. The interaction of gravitation and electromagnetism in Evans field theory is governed by the first Bianchi identity of differential geometry. The Christoffel symbol of the unified field is in general asymmetric in its lower two indices. The Einstein field theory of gravitation is recovered when the Christoffel symbol becomes symmetric and the torsion tensor vanishes. The Maxwell Heaviside field theory of electromagnetism is recovered when the Christoffel symbol becomes antisymmetric. The theory of O(3) electrodynamics is recovered when the Christoffel symbol is antisymmetric and when the tangent bundle index is developed in a complex circular basis. The Evans unified field is described in general by an asymmetric Christoffel symbol. In this case gravitation and electromagnetism are mutually influential. The details of this interaction are found by solving the first Bianchi identity of differential geometry, which is the homogeneous Evans field equation within a C negative scalar. These details are important for the design of major new technologies which take electromagnetic energy from Evans spacetime defined by the asymmetric Christoffel symbol, and for the design of new counter-gravitational aerospace devices.

Key words: Evans field theory; gravitation; electromagnetism; electromagnetic energy from Evans spacetime; counter-gravitational technology.

23.1 Introduction

The Evans field theory [1]–[29] is the first generally covariant unified field theory, and goes beyond the standard model in several ways. An important consequence of the theory is that it is able to describe the interaction of gravitation and electromagnetism, leading to major new technologies. In this note the details of the interaction are defined. These details are important for the acquisition of electromagnetic energy in theoretically unlimited quantities from Evans spacetime, the term given to spacetime with an asymmetric Christoffel symbol. This appellation is a convenient way of distinguishing such a spacetime from the Riemann spacetime with symmetric Christoffel symbol. The latter defines Einstein's theory of gravitation [30] uninfluenced by electromagnetism. It is shown in Section 23.2 that the Einstein and Maxwell-Heaviside field theories are well defined limiting forms of the Evans unified field. However, neither the Einstein nor the Maxwell-Heaviside theory is capable of describing the mutual interaction of gravitation and electromagnetism, for this we must progress beyond the standard model and use the Evans unified field..

In Section 23.2 the Bianchi identity of differential geometry [31] is developed into the homogeneous Evans field equation. This is an identity of differential geometry, and may be developed in tensor notation. The familiar Bianchi identity of Einstein's gravitational field theory is recovered from the Evans homogeneous field equation when the Christoffel symbol becomes symmetric in its lower two indices and when the torsion tensor vanishes in consequence. The homogeneous field equation of the Maxwell Heaviside theory is recovered when the Christoffel symbol becomes antisymmetric. This is because there is no contribution to electromagnetism from a symmetric Christoffel symbol, electromagnetism is spacetime torsion [1]–[29], and the torsion tensor is defined as the difference of two Christoffel symbols:

$$T^{\kappa}{}_{\mu\nu} = \Gamma^{\kappa}{}_{\mu\nu} - \Gamma^{\kappa}{}_{\nu\mu} \tag{23.1}$$

This vanishes if the Christoffel symbol is symmetric, i.e. when:

$$\Gamma^{\kappa}{}_{\mu\nu} = \Gamma^{\kappa}{}_{\nu\mu} \tag{23.2}$$

A powerful new understanding therefore emerges in section two from Evans field theory: gravitation and electromagnetism can be mutually influential if and only if the Christoffel symbol is asymmetric in its lower two indices. Finally in section three a short discussion is given of the implications of this principle for urgently needed new technologies.

23.2 The First Bianchi Identity of Differential Geometry and the Homogeneous Evans Field Equation

The first Bianchi identity of differential geometry is written most succinctly as:

$$D \wedge T = R \wedge q \tag{23.3}$$

where indices have been suppressed to reveal the basic structure of the equation. Here T is shorthand for the torsion form, R shorthand for the curvature or Riemann form, q shorthand for the tetrad form and $D\wedge$ represents the covariant exterior derivative [1]–[29], [31]. Eq. (23.3) is valid in Evans spacetime with asymmetric Christoffel symbol and is the geometrical equation that can be developed as follows into the homogeneous Evans field equation. Multiply both sides of Eq. (23.3) by a C negative scalar-valued potential magnitude $A^{(0)}$ with the units of tesla metres or weber per metre:

$$A^{(0)} D \wedge T = A^{(0)} R \wedge q \tag{23.4}$$

and define:

$$F = A^{(0)} T \tag{23.5}$$

$$A = A^{(0)} q \tag{23.6}$$

to obtain the basic structure of the homogeneous Evans field equation:

$$D \wedge F = R \wedge A \tag{23.7}$$

it is seen that Eq. (23.7) is a restatement of Eq. (23.3), and so Eq. (23.7) is differential geometry. All of physics is causal and objective and is defined by differential geometry, a major advance from the contemporary standard model.

Now start to use the received terminology of electrodynamics to identify F as the electromagnetic field and A as the electromagnetic potential. These terms are used only in deference to the history of physics, because Eq (23.7) governs a new concept: the generally covariant unified Evans field. The homogeneous field equation is developed now as an equation of the tangent spacetime of differential geometry:

$$D \wedge F^a = R^a{}_b \wedge A^b \tag{23.8}$$

for all types of base manifold. The tangent spacetime with Latin indices is a Minkowski spacetime and the base manifold with Greek indices an Evans spacetime. The electromagnetic field F^a is a vector valued two-form of differential geometry [1]–[29], [31], and the electromagnetic potential $A^{(0)}$ is a vector valued one-form. The Riemann tensor$R^a{}_b$ is a tensor valued two-form.

In tensor notation Eq. (23.8) becomes [1]–[29], [31]:

$$D_\mu F^a{}_{\nu\rho} + D_\rho F^a{}_{\mu\nu} + D_\nu F^a{}_{\rho\mu} = -A^b{}_\mu R^a{}_{b\nu\rho} - A^b{}_\nu R^a{}_{b\rho\mu} - A^b{}_\rho R^a{}_{b\mu\nu} \tag{23.9}$$

where:

$$D_\mu F^a{}_{\nu\rho} = \partial_\mu F^a{}_{\nu\rho} + \omega^a{}_{\mu b} F^b{}_{\nu\rho} \quad \text{etc.} \tag{23.10}$$

Here $\omega^a{}_{\mu b}$ is the well-known spin connection of differential geometry, related to the Christoffel connection through the tetrad postulate:

$$D_\mu q^a{}_\nu = 0. \tag{23.11}$$

The Einstein theory of gravitation 30 is the limit:

$$R^\sigma{}_{\mu\nu\rho} + R^\sigma{}_{\nu\rho\mu} + R^\sigma{}_{\rho\mu\nu} = 0. \tag{23.12}$$

In this limit the torsion tensor vanishes because Eq (23.12) implies [31] that the Christoffel symbol is symmetric in its lower two indices. Self consistently therefore, there is no electromagnetism present under condition (23.12), only gravitation. It follows that the Einstein theory cannot be used to describe the mutual influence of gravitation and electromagnetism. For this we need Eq. (23.9) of the unified Evans field. The mutual interaction of gravitation and electromagnetism is however of paramount importance to the urgently needed question of energy acquisition, because only by using Eq. (23.9) can we understand how to obtain electromagnetic energy from Evans spacetime. The existence of this spacetime is the key to clean energy in unlimited quantities.

The Maxwell-Heaviside theory of electromagnetism [32],[33] is the limit of Eq. (23.9) defined by:

$$\partial_\mu F_{\nu\rho} + \partial_\nu F_{\rho\mu} + \partial_\rho F_{\mu\nu} = 0. \tag{23.13}$$

This limit is reached when:

$$d \wedge F^a = 0 \tag{23.14}$$

$$\begin{aligned} j^a &= \frac{1}{\mu_0} \left(R^a{}_b \wedge A^b - \omega^a{}_b \wedge F^b \right) \\ &= 0 \end{aligned} \tag{23.15}$$

$$A^{(0)} \left(R^a{}_b \wedge q^b - \omega^a{}_b \wedge T^b \right) = 0 \tag{23.16}$$

Eqs. (23.14) to (23.16) define a particular type of Evans spacetime and Eq. (23.16) may be rewritten as:

$$\left(\partial_\mu F_{\nu\rho} + \partial_\nu F_{\rho\mu} + \partial_\rho F_{\mu\nu} \right)^a = 0. \tag{23.17}$$

Eq. (23.17) is an equation of the base manifold (i.e. the Evans spacetime) for all indices a of the tangent bundle (Minkowski spacetime). It is inferred that Eq. (23.13) is a special case of Eq. (23.17) when the quantity inside the brackets of Eq. (23.17) vanishes for all a. In the Maxwell Heaviside theory, Eq. (23.13), the latter index is not present. This is therefore the meaning of the Maxwell Heaviside theory - in this limit the tangent spacetime of Evans spacetime has not been defined or recognized to exist (there is no index a in Eq. (23.13)) and in consequence the base manifold has not been distinguished from the tangent spacetime. The two spacetimes have merged conceptually into a single flat spacetime upon which is superimposed a separate nineteenth century concept - the Maxwell Heaviside electromagnetic field. A properly covariant description of electromagnetism always requires the presence of two indices, a and μ. The first realization of this requirement was O(3) electrodynamics, in which the experimentally observable Evans spin field was defined self consistently [1]–[29] for the first time. In O(3) electrodynamics the indices of the tangent spacetime are defined with a complex circular basis whose space components are:

$$a = (1), (2), (3). \tag{23.18}$$

In contrast Eq. (23.9) is generally covariant, i.e. is an equation of general relativity and thus of objective and causal physics. The Maxwell Heaviside field theory was inferred many years before the development of relativity, in an era when the electromagnetic field was considered to be an entity superimposed on separated space and time. Apart from the fusion of space and time into a four dimensional Minkowski ("flat") spacetime, this description and concept are still the ones adhered to in the contemporary standard model. This is not, however, a generally covariant description as required by general relativity (causal and objective physics). In the standard model the electromagnetic field is the archetypical field of special relativity. In the Evans field theory it is part of a generally covariant unified field, and due to the spinning and curving of Evans spacetime. In consequence the standard model loses a great deal of key information. It can be seen that Eq. (23.13) is a drastic simplification of Eq. (23.9) and so both the Einstein and Maxwell-Heaviside field theories are incomplete. Similarly, special relativity (developed about 1887 to1905) showed that Newtonian physics is incomplete.

23.3 Discussion

The unified Evans field theory is generally covariant in all its sectors [1]–[29] and shows that there are many hitherto unknown effects in nature which can be harnessed for the good of humankind. For example:

1. Gravitation has effects on electromagnetism and the latter may be generated from Evans spacetime. Prototype devices based on this inference are already available and have been shown to be reproducible and repeatable [34]. Hopefully they will lead to the replacement of fossil fuel and the elimination of harmful emissions therefrom.
2. Electromagnetism has effects on gravitation, leading in principle to various new aerospace technologies based on the counter-gravitational effect of an on-board electromagnetic field.

Eq (23.14) may be re-expressed as:

$$(D \wedge \omega) \wedge q = \omega \wedge (D \wedge q) \tag{23.19}$$

so Maxwell Heaviside field theory may be more fully identified as:

$$\omega^a{}_b = -\kappa \epsilon^a{}_{bc} q^c \tag{23.20}$$

In Maxwell Heaviside theory in free space, appropriate to the homogeneous field equation, the d'Alembert wave equation becomes:

$$\Box A_\sigma = 0 \tag{23.21}$$

and so:

$$\Box A^a{}_\sigma = \mu_0 \tilde{j}^a{}_\sigma \tag{23.22}$$

Eq. (23.22) can now be identified as a special case of the Evans Lemma 1-29:

$$\Box A^a{}_\mu = RA^a{}_\mu \tag{23.23}$$

where R is scalar curvature (not to be confused with the shorthand R in Eq. (23.3)).

This example shows that the Evans field theory is capable of giving a good deal of new insight to the meaning both of the Einstein and the Maxwell Heaviside field theories. This is the hallmark or characteristic of a paradigm shift: a new theory gives extra meaning to older theories to which it reduces in well defined limits.

Non-linear optics, for example [1]–[29], has shown in many ways that the Maxwell Heaviside theory is incomplete. One non-linear optical effect led to the inference of the Evans spin field $\boldsymbol{B}^{(3)}$ and subsequently O(3) electrodynamics. This effect is magnetization by a circularly polarized electromagnetic field, the inverse Faraday effect, whose magnetization is due to the Evans spin field:

$$\boldsymbol{B}^{(3)*} = -ig\boldsymbol{A}^{(1)} \times \boldsymbol{A}^{(2)} \tag{23.24}$$

It is seen that the well known conjugate product of potentials $\boldsymbol{A}^{(1)} \times \boldsymbol{A}^{(2)}$ has been defined in terms of the tangent bundle indices (1) and (2). Thus, O(3) electrodynamics is recognized as another well defined limiting form of Eq. (23.9):

$$j^a \to 0 \tag{23.25}$$

in which the tangent bundle index a appears and is well defined.

It may be deduced with great confidence that Eq. (23.3) (or equivalently Eqs. (23.7) to (23.9)) is the ONLY way to develop a unified field theory based on the principle of objectivity in physics, the principle of general relativity. Without objectivity, physics to one observer would be different from physics to another observer. Objectivity in physics was recognized by Einstein to be a manifestation of geometry, and differential geometry is the only type of geometry that is self-consistently capable of describing both torsion and curvature [31], the spinning and curving of Evans spacetime. The other governing principle of physics is causality, any event has a cause. Causal and objective physics therefore inexorably leads us to differential geometry. Conversely, differential geometry gives us physics. There are two main governing equations in physics, Eq. (23.9) and the Evans wave equation:

$$(\Box + kT)\, q^a{}_\mu = 0. \tag{23.26}$$

The subsidiary proposition leading to Eq. (23.26) is the Evans Lemma [1]–[29]:

$$\Box q^a{}_\mu = Rq^a{}_\mu . \tag{23.27}$$

Eq. (23.27) is an identity of differential geometry. It states that:

$$D^\mu \left(D_\mu q^a{}_\nu\right) = 0 \tag{23.28}$$

and thus originates [1]–[29] in the well-known tetrad postulate (23.11).

$$D^{\mu} D_{\mu} = \Box - R. \tag{23.29}$$

The wave equation is obtained from the Lemma using a generalization of Einstein's field equation of gravitation to all radiated and matter fields. In index contracted form:

$$R = -kT \tag{23.30}$$

where R is scalar curvature, k is Einstein's constant and T is the index contracted canonical energy momentum tensor (not to be confused with the shorthand torsion symbol T in Eq. (23.3)).

So these are the concepts and equations upon which to build the urgently needed new technologies mentioned already. Heisenberg uncertainty and Bohr complementarity have recently been refuted experimentally [35],[36] using for example advanced microscopy and careful Young interference experiments. In contrast Einsteinian objectivity (general relativity) and de Broglie wave particle dualism have withstood the test of experiment. The copious experimental evidence for the Evans spin field and O(3) electrodynamics is summarized in the literature [1]–[29].

Finally the multi-disciplinary Pinter hypothesis [37] argues that life evolved in a rigorously causal manner from the Evans unified field, in other words there would be no life on earth without the existence of a generally covariant unified field and without the interaction of gravitation and electromagnetism.

Acknolegements

The Fellows and Emeriti of AIAS are thanked for many interesting discussions and the Ted Annis Foundation, Craddock Inc and Applied Science Associates for funding.

References

1. M. W. Evans, A Unified Field Theory for Gravitation and Electromagnetism, *Found. Phys. Lett.*, **16**, 367 (2003).
2. M. W. Evans, A Generally Covariant Wave Equation for Grand Unified Field Theory, *Found. Phys. Lett.*, **16**, 507 (2003).
3. M. W. Evans, The Equations of Grand Unified Field Theory in terms of the Maurer Cartan Structure Relations of Differential Geometry, *Found. Phys. Lett.*, **17**, 25 (2004).
4. M. W. Evans, Derivation of Dirac's Equation from the Evans Wave Equation, *Found. Phys. Lett.*, **17**, 149 (2004).
5. M. W. Evans, Unification of the Gravitational and Strong Nuclear Fields, *Found. Phys. Lett.*, **17**, 267 (2004).
6. M. W. Evans, The Evans Lemma of Differential Geometry, *Found. Phys. Lett.*, **17**, 433 (2004).
7. M. W. Evans, Derivation of the Evans Wave Equation from the Lagrangian and Action: Origin of the Planck Constant in General Relativity, *Found. Phys. Lett.*, **17**, 535 (2004).
8. M. W. Evans et alii, Development of the Evans Wave Equation in the Weak Field Limit: the Electrogravitic Equation, *Found. Phys. Lett.*, **17**, 497 (2004).
9. M. W. Evans, Physical Optics, the Sagnac Effect and the Aharonov Bohm Effect in the Evans Unified Field Theory, *Found. Phys. Lett.*, **17**, 301 (2004).
10. M. W. Evans, Derivation of the Geometrical Phase from the Evans Phase Law of Generally Covariant Unified Field Theory, *Found. Phys. Lett.*, **17**, 393 (2004).
11. M. W. Evans, Derivation of the Lorentz Boost from the Evans Wave Equation, *Found. Phys. Lett.*, **17**, 663 (2004).
12. M. W. Evans, The Electromagnetic Sector of the Evans Field Theory, *Found. Phys. Lett.*, in press, preprint on www.aias.us.
13. M. W. Evans, New Concepts form the Evans Field Theory, Part One: The Evolution of Curvature, Oscillatory Universe without Singularity, and General Force and Field Equations, *Found. Phys. Lett.*, in press, preprint on www.aias.us.
14. M. W. Evans, New Concepts from the Evans Field Theory, Part Two: Derivation of the Heisenberg Equation and Reinterpretation of the Heisenberg Uncertainty Principle, *Found. Phys. Lett.*, in press, preprint on www.aias.us.
15. M. W. Evans, Derivation of O(3) Electrodynamics from Generally Covariant Unified Field Theory, *Found. Phys. Lett.*, submitted, preprint on www.aias.us.

16. M. W. Evans, The Spinning and Curving of Spacetime, the Electromagnetic and Gravitational Fields in the Evans Unified Field Theory, *Found. Phys. Lett.*, submitted, preprint on www.aias.us.
17. M. W. Evans, The Derivation of O(3) Electrodynamics form the Evans Unified Field Theory, *Found. Phys. Lett.*, submitted, preprint on www.aias.us.
18. M. W. Evans, Calculation of the Anomalous Magnetic Moment of the Electron from the Evans Field Theory, *Found. Phys. Lett.*, submitted, preprint on www.aias.us.
19. M. W. Evans, Generally Covariant Electro-weak Theory, *Found. Phys. Lett.*, submitted, preprint on www.aias.us.
20. M. W. Evans, Evans Field Theory of Neutrino Oscillations, *Found. Phys. Lett.*, submitted, preprint on www.aias.us.
21. M. W. Evans, *Generally Covariant Unified Field Theory: the Geometrization of Physics*, (in press, 2005) .
22. L. Felker, *The Evans Equations* (in prep, freshman level volume).
23. M. W. Evans and L. B. Crowell, *Classical and Quantum Electrodynamics and the $\boldsymbol{B}^{(3)}$ Field* (World Scientific , Singapore, 2001).
24. M. W. Evans (ed.), *Modern Non-linear Optics*, a special topical issue in three parts of I. Prigogine and S. A. Rice (series eds.), *Advances in Chemical Physics* (Wiley-Interscience, New York, 2001, hardback and e book, second edition), vol. 119(1), 119(2) and 119(3).
25. M. W. Evans, J.-P. Vigier et alii, *The Enigmatic Photon* (Kluwer, Dordrecht, 1994 to 2002, hardback and softback), vols. 1-5.
26. M. W. Evans and A. A. Hasanein, *The Photomagneton in Quantum Field Theory*, (World Scientific, Singapore, 1994),
27. M. W. Evans and S. Kielich (eds.), first edition of ref. (24), (Wiley-Interscience, New York, 1992, 1993, 1997, hardback and softback) vols. 85(1), 85(2), and 85(3).
28. M. W. Evans, *The Photon's Magnetic Field, Optical NMR Spectroscopy*, (World Scientific, Singapore, 1992).
29. M. W. Evans, Physica B, **182**, 227, 237 (1992) (first paper on the Evans spin field).
30. A. Einstein, *The Meaning of Relativity*, (Princeton Univ Press, 1921-1953 editions).
31. S. P. Carroll, *Lecture Notes in General Relativity* (a graduate course at Harvard, Univ California Santa Barbara and Univ. Chicago, arXiv: gr-gq / 973019 v1 Dec 1997, public domain).
32. P. W. Atkins, *Molecular Quantum Mechanics*, (Oxford Univ. Press, 1983, 2nd ed.).
33. L. H. Ryder, *Quantum Field Theory*, (Cambridge Univ Press, 1996, 2nd ed.).
34. AIAS group, confidential information.
35. J. R. Croca, *Towards a Nonlinear Quantum Physics*, (World Scientific, Singapore, 2003).
36. M. Chown, New Scientist, **183**, 30 (2004).
37. P. H. Pinter, see correspondence and details on www.aias.us.

24

The Fundamental Invariants of the Evans Field Theory

Summary. The fundamental invariants of the Evans field theory are constructed by using the Stokes formula in the structure equations and Bianchi identities of differential geometry. The structure invariants are independent of the base manifold, and can be used to define the rotation and translation generators in general relativity. The identity invariants are defined by integrating the Bianchi identities. In so doing a new theorem of differential geometry is obtained and given the appellation "inverse structure theorem". The latter defines the integrated tetrad and integrated spin connection. These methods of differential geometry are used to explain the class of all Aharonov Bohm effects and to provide a rigorous geometrical basis for the Heisenberg equation of motion and uncertainty principle. The origin of the Planck constant is discussed in terms of differential geometry.

Key words: Evans field theory, fundamental invariants, inverse structure theorem, Aharonov Bohm effects; origin of the Planck constant in differential geometry.

24.1 Introduction

It was shown by Wigner [1] in 1939 that there are two fundamental invariants that characterise any particle in special relativity. These are the mass and spin invariants, the first and second Casimir invariants of the Poincaré group. Thus, any particle is characterized by its mass and spin. The Poincare group of special relativity has ten generators. In order to extend Wigner's analysis to general relativity the sixteen generators of the Einstein group need to be considered in order to define the fundamental invariants of general relativity. In Section 24.2 the structure equations and Bianchi identities of differential geometry are integrated with the Stokes formula of differential geometry in order to identify two types of invariants, the structure and identity invariants. The latter convey new meaning to familiar quantities such as the displacement four-vector x^a and the rotation generator $\theta^a{}_b$.

In the original work by Wigner [1] the first Casimir invariant is:

$$C_1 = \boldsymbol{P}^\mu \boldsymbol{P}_\mu \tag{24.1}$$

and the second Casimir invariant is:

$$C_2 = W^\mu W_\mu . \tag{24.2}$$

Here:

$$\boldsymbol{P}_\mu = i\frac{\partial}{\partial x^\mu} \tag{24.3}$$

is the spacetime translation generator, which is the energy-momentum in special relativity within a factor $\hbar$ [2]. Spacetime translation is defined by:

$$x^\mu \rightarrow x^\mu + a^\mu . \tag{24.4}$$

The second Casimir invariant C_2 is defined by the Pauli Lubanski vector [2]:

$$W_\mu = -\frac{1}{2}\epsilon_{\mu\nu\rho\sigma} J^{\nu\rho} \boldsymbol{P}^\sigma \tag{24.5}$$

It is seen that C_1 and C_2 are the two Casimir invariants that characterize the Poincaré group of special relativity. They are the mass and spin invariants respectively and are invariant under Lorentz transform. For example:

$$p^\mu p_\mu = m^2 c^2 = E_0^2/c^2 \tag{24.6}$$

and the rest energy E_0 is the same in all frames of reference, being a fundamental scalar. Similarly it may be shown [2] that:

$$C_2 = m^2 S(S+1) \tag{24.7}$$

in reduced units, and this is also a scalar invariant. In Section 24.2 Wigner's analysis is extended to general relativity using methods based on differential geometry rather than group theory, and the fundamental structure and identity invariants defined. In Section 24.3 the inverse structure theorem is proven from integration of the Bianchi identities of differential geometry. In Section 4 the class of Aharonov Bohm effects is explained with the theory of Sections 24.2 and 24.3. Finally in Section 24.5 the Heisenberg equation of motion is deduced from the invariants and the origin of the fundamental constants such as the Planck constant discussed.

24.2 The Structure and Identity Invariants of Differential Geometry

The Maurer Cartan structure equations of differential geometry are [3]:

$$T^a = D \wedge q^a = d \wedge q^a + \omega^a{}_b \wedge q^b \tag{24.8}$$

$$R^a{}_b = D \wedge \omega^a{}_b = d \wedge \omega^a{}_b + \omega^a{}_c \wedge \omega^c{}_b \tag{24.9}$$

Here T^a is the torsion form, q^a is the tetrad form, $R^a{}_b$ is the curvature or Riemann form, $\omega^a{}_b$ is the spin connection one-form and $D\wedge$ denotes the covariant exterior derivative:

$$D\wedge = d\wedge + \omega^a{}_b\wedge \tag{24.10}$$

where $d\wedge$ is the exterior derivative.

The Bianchi identities of differential geometry are:

$$D\wedge T^a = d\wedge T^a + \omega^a{}_b \wedge T^b = R^a{}_b \wedge q^b \tag{24.11}$$

$$D\wedge R^a{}_b = d\wedge R^a{}_b + \omega^a{}_c \wedge R^c{}_b - R^a{}_c \wedge \omega^c{}_b = 0. \tag{24.12}$$

These equations [3] have been used recently to develop the Evans field theory [4]–[35], generally accepted [36] as the first objective unified field theory of physics. The structure equations and Bianchi identities are augmented in all manifolds and tangent spacetimes by the tetrad postulate:

$$D_\mu q^a{}_\nu = 0 \tag{24.13}$$

and the Evans Lemma:

$$\Box q^a = R q^a. \tag{24.14}$$

Here R is the scalar curvature. The Einstein postulate is:

$$R = -kT \tag{24.15}$$

where T is the index contracted canonical energy momentum tensor and k is the Einstein constant. In the Evans field theory the Einstein postulate is extended to all radiated and matter fields, and not restricted to the gravitational field.

All these equations retain their form under general coordinate transformation and are therefore rigorously objective equations of general relativity.

The rest curvature is defined by:

$$|R_0| = \left(\frac{mc}{\hbar}\right)^2 = p^\mu p_\mu/\hbar^2 = \boldsymbol{P}^\mu \boldsymbol{P}_\mu \tag{24.16}$$

where

$$p_\mu = \hbar \boldsymbol{P}_\mu \tag{24.17}$$

is the energy momentum vector in special relativity. So it is seen that R_0 is a Casimir invariant of differential geometry and of the Evans field theory. The important conclusion is reached that any particle in unified field theory is characterized by the eigenvalues R of the Evans Lemma. We therefore achieve a quantization of general relativity, an objective and causal quantization of unified field theory. The eigenfunction or wavefunction is the tetrad. Thus eigenvalues of the tetrad are fundamental invariants and observables of unified field theory.

In special relativity $\boldsymbol{p}_\mu$ is the generator of spacetime translations. In special relativity and Minkowski spacetime:

$$x^{\mu}x_{\mu} = c^2t^2 - X^2 - Y^2 - Z^2 \tag{24.18}$$

is the square of the line element, and is invariant under Lorentz transformation. The corresponding Casimir invariant is the mass invariant, C_1, of the Poincaré group. Therefore we look for the objective generalization of Eq. (24.18) using differential geometry, having already found that:

$$|R_0| = C_1. \tag{24.19}$$

To proceed use the Stokes formula of differential geometry:

$$\int_S d \wedge f^a = \oint f^a \tag{24.20}$$

where f^a is any differential form [3]. The Stokes formula is true for any manifold enclosed by any surface. On the left hand side of Eq. (24.20) appears a surface integral over the surface S. On the right hand side appears a contour integral over the boundary enclosing the surface. Therefore we find results such as:

$$\int_S d \wedge q^a = \oint q^a \tag{24.21}$$

$$\int_S d \wedge \omega^a{}_b = \oint \omega^a{}_b \tag{24.22}$$

$$\int_S d \wedge T^a = \oint T^a \tag{24.23}$$

$$\int_S d \wedge R^a{}_b = \oint R^a{}_b. \tag{24.24}$$

There are two fundamental structure invariants, which are found by surface integration of the structure equations (24.8) and (24.9):

$$x^a = \int_S T^a \tag{24.25}$$

$$\theta^a{}_b = \int_S R^a{}_b. \tag{24.26}$$

It is seen that x^a has the units of metres and that $\theta^a{}_b$ is unitless and antisymmetric in its indices a and b. The latter are indices of the tangent spacetime, a Minkowski spacetime [3]. The details of the base manifold have been integrated out in Eqs. (24.25) and (24.26). Therefore x^a and $\theta^a{}_b$ are invariants in the sense that they are independent of the base manifold by surface integration.

It is now possible to define the scalar invariants:

$$E_1 = x^a x_a \tag{24.27}$$

$$E_2 = \theta^a{}_b \theta^b{}_a \tag{24.28}$$

by index contraction. These play a similar role to the Casimir invariants of special relativity. and E_1 and E_2 are the fundamental structure invariants of differential geometry and thus of the Evans field theory. They are invariant under the general coordinate transformation and thus the same to any observer.

From Eqs. (24.8) and (24.21)

$$x^a = \oint q^a + \int_s \omega^a{}_b \wedge q^b \tag{24.29}$$

and from Eqs. (24.9) and (24.22)

$$\theta^a{}_b = \oint \omega^a{}_b + \int_s \omega^a{}_c \wedge \omega^c{}_b. \tag{24.30}$$

Eqs. (24.29) and (24.30) give considerable insight to the fundamental meaning of translation and rotation in general relativity and can be regarded as the integral form of the fundamental Maurer Cartan structure equations. Thus, they provide considerable insight to unified field theory from differential geometry [4]–[36].

The two identity invariants originate in integration of the two Bianchi identities , Eqs. (24.11) and (24.12). Using the Stokes formula:

$$\theta^a = \int_s D \wedge T^a = \oint T^a + \int_s \omega^a{}_b \wedge T^b = \int_s R^a{}_b \wedge q^b \tag{24.31}$$

$$\kappa^a{}_b = \int_s D \wedge R^a{}_b = \oint R^a{}_b + \int_s \omega^a{}_c \wedge R^c{}_b - \int_s R^a{}_c \wedge \omega^c{}_b \tag{24.32}$$

The first identity invariant θ^a is unitless an the second identity invariant $\kappa^a{}_b$ has the units of inverse metres. The identity invariants, unlike the structure invariants, depend on an index of the base manifold, i.e. :

$$Q^a{}_\mu = \theta^a{}_\mu = \left(\int_s D \wedge T^a \right)_\mu \tag{24.33}$$

$$\Omega^a{}_{b\mu} = \kappa^a{}_{b\mu} = \left(\int_s D \wedge R^a{}_b \right)_\mu . \tag{24.34}$$

The first identity invariant has the same index structure as the tetrad, and is also unitless, so it is given the appellation "integrated tetrad" $Q^a{}_\mu$. Similarly the second identity invariant may be identified as the "integrated spin connection" $\Omega^a{}_\mu$.

24.3 The Inverse Structure Theorem

The inverse structure theorem is true for all manifolds and tangent spacetimes, and states that if

$$T^a = D \wedge q^a \tag{24.35}$$

$$R^a{}_b = D \wedge \omega^a{}_b \tag{24.36}$$

then

$$Q^a = \int_s D \wedge T^a \tag{24.37}$$

$$\Omega^a{}_b = \int_s D \wedge R^a{}_b. \tag{24.38}$$

Using the second Bianchi identity it is seen that:

$$\Omega^a{}_b = 0 \tag{24.39}$$

i.e. the integrated spin connection vanishes for all manifolds and tangent spacetimes. In a differential geometry where:

$$D \wedge T^a = R^a{}_b \wedge q^b \neq 0 \tag{24.40}$$

i.e. the geometry of a unified field theory, then the integrated tetrad is non-zero:

$$Q^a = \int_s D \wedge T^a \neq 0. \tag{24.41}$$

An objective unified field theory is therefore always defined by:

$$\begin{aligned} Q^a &\neq 0, \\ \Omega^a{}_b &= 0. \end{aligned} \tag{24.42}$$

In a differential geometry where:

$$D \wedge T^a = R^a{}_b \wedge q^b = 0 \tag{24.43}$$

then:

$$\begin{aligned} Q^a &= 0, \\ \Omega^a{}_b &= 0. \end{aligned} \tag{24.44}$$

Eq. (24.44) defines the geometry when electromagnetism and gravitation have become independent.

24.4 The Aharonov Bohm Effects

These fundamental results may now be applied to give a straightforward explanation of the class of Aharonov Bohm (AB) effects [2]. Proceed by multiplying both sides of Eq. (24.29) by the fundamental C negative potential $A^{(0)}$ in volts [4]–[36]:

$$A^{(0)} x^a = A^{(0)} \oint q^a + A^{(0)} \int_s \omega^a{}_b \wedge q^b \tag{24.45}$$

and rewrite Eq. (24.45) as:

$$\Phi^a = \oint A^a + \int_s \omega^a{}_b \wedge A^b. \tag{24.46}$$

In Eq. (24.46) Φ^a has the units of magnetic flux (weber = volt meter). We therefore obtain:

$$\Phi^a = \int_s F^a = \oint A^a + \int_s \omega^a{}_b \wedge A^b. \tag{24.47}$$

This is the equation of the Evans field theory that explains the class of all AB effects in terms of differential geometry. The AB effects are observed when it is arranged experimentally that:

$$\oint A^a = 0 \tag{24.48}$$

For example, in the region enclosed by the electron beams in the Chambers experiment [2] condition (24.48) is true outside the iron whisker. In the Chambers experiment (and all other AB effects), the magnetic flux being observed is therefore:

$$\Phi^a = A^{(0)} \int_s \omega^a{}_b \wedge q^b. \tag{24.49}$$

It is seen that the class of AB effects is due to a structure invariant of differential geometry. The presence of the covariant exterior derivative means that if a magnetic flux is excluded from a region of the base manifold it is nonetheless observed in another region. The AB effects depend only on $A^{(0)}$, (which is a scalar voltage), and on differential geometry. They are therefore due to "spacetime geodynamics". The AB effects can be explained in causal general relativity and are effects of the Evans unified field theory. Conversely the AB effects all provide quantitative experimental evidence for the Evans unified field theory.

The integrated potential for all manifolds and surfaces is:

$$\alpha^a = \int_s D \wedge F^a = \oint F^a + \int_s \omega^a{}_b \wedge F^b = \int_s R^a{}_b \wedge A^b \tag{24.50}$$

where

$$\int_s F^a = \oint A^a + \int_s \omega^a{}_b \wedge A^b \tag{24.51}$$

In general therefore, the class of AB effects is described by Eqs. (24.50) and (24.51) for all manifolds and surfaces. When:

$$d \wedge F^a = R^a{}_b \wedge A^b - \omega^a{}_b \wedge F^b = 0 \tag{24.52}$$

then:

$$\oint F^a + \int_s \omega^a{}_b \wedge F^b = \int_s R^a{}_b \wedge A^b \tag{24.53}$$

Eqs. (24.51) and (24.53) are the integrated counterparts of:

$$F^a = D \wedge A^a \tag{24.54}$$

$$D \wedge F^a = R^a{}_b \wedge A^b \tag{24.55}$$

i.e. of:

$$F^a = d \wedge A^a + \omega^a{}_b \wedge A^b \tag{24.56}$$

$$d \wedge F^a + \omega^a{}_b \wedge F^b = R^a{}_b \wedge A^b. \tag{24.57}$$

In regions where it is arranged experimentally that:

$$\oint A^a = \int_s d \wedge A^a = 0 \tag{24.58}$$

then Eqs. (24.56) and (24.57) reduce to:

$$F^a = \omega^a{}_b \wedge A^b \tag{24.59}$$

$$d \wedge F^a = 0. \tag{24.60}$$

The integrated counterparts of Eqs. (24.59) and (24.60) are:

$$\Phi^a = \int_s F^a = \int_s \omega^a{}_b \wedge A^b. \tag{24.61}$$

$$\oint F^a = 0 \tag{24.62}$$

Eqs. (24.59) to (24.62) describe the class of all Aharonov Bohm effects, and more generally the class of all "non-local" effects in general relativity and quantum mechanics.

All these effects are due to spacetime geodynamics.

Objective physics, or general relativity, is needed to describe the AB effects, which are therefore evidence for the Evans field theory. In a non-objective theory of special relativity such as the Maxwell-Heaviside field theory:

$$F = d \wedge A \tag{24.63}$$

$$d \wedge F = 0 \tag{24.64}$$

or:

$$\int_s F = \oint A \tag{24.65}$$

$$\oint F = 0. \tag{24.66}$$

So if $\oint A$ is zero experimentally the magnetic flux disappears:

$$\Phi = \int_s F = 0 \tag{24.67}$$

and there are no Aharonov Bohm effects in the Maxwell Heaviside field theory. This is conclusive evidence in favor of the Evans field theory and general relativity over Maxwell Heaviside field theory and special relativity. The fundamental reason for the failure of the Maxwell Heaviside field theory is that it is a theory of Minkowski spacetime, (flat spacetime), and uses the exterior derivative (e.g. as in Eq. (24.63)). The correct description of electrodynamics uses the covariant exterior derivative in the Evans field theory. The use of the covariant derivative introduces the spin connection and provides a description of the AB effects. In the last analysis the AB effects indicate that spacetime is not the Minkowski spacetime. This inference had of course been arrived at by Einstein and Hilbert long before the Chambers experiment, but the Maxwell Heaviside theory was adhered to in trying to explain the experiment. This caused confusion for over forty years. It has been shown lately [4]–[35] that textbook explanations [2] of the AB effect in terms of gauge transformation are incorrect mathematically as well as conceptually.

The only quantity apart from differential geometry that enters into the Evans field theory is the fundamental potential $A^{(0)}$ (units of volts = $JsC^{-1}m^{-1}$). The unified theory of matter fields, gravitation, electrodynamics, and the weak and strong nuclear fields is then built up entirely from the structure equations and Bianchi identties, together with the Evans Lemma and Einstein postulate. The Evans Lemma is based directly on the tetrad postulate of differential geometry and leads to objective quantization in physics: and to all the fundamental wave and quantum equations of physics, the wavefunction being the tetrad. This is a consequence of the fact that physics must be objective and causal both on philosophical and on experimental grounds. Using the units of $A^{(0)}$ this fundamental quantity can be written as:

$$A^{(0)} = \frac{\hbar}{er_0} \tag{24.68}$$

where $\hbar$ is Planck's constant, e the proton charge and r_0 a fundamental length in metres. If this fundamental length is identified with the Compton wavelength:

$$r_0 = \lambda_0 = \frac{\hbar}{mc} \tag{24.69}$$

it follows that $A^{(0)}$ may be expressed as:

$$A^{(0)} = \frac{mc}{e} = \frac{En_0}{ec}, \tag{24.70}$$

$$En_0 = mc^2, \tag{24.71}$$

and thus:

$$A^{(0)} = \frac{mc}{e} = \frac{\hbar}{e\lambda_0}. \tag{24.72}$$

If we now identify the fundamental momentum

$$p_0 = eA^{(0)} = \hbar\kappa_0 = \hbar\omega_0 c \tag{24.73}$$

Eq. (24.72) becomes the de Broglie postulate [4]–[36]:

$$En_0 = \hbar\omega_0 = mc^2 \tag{24.74}$$

In differential geometry, Eq. (24.74) can be expressed as

$$\kappa_0 = \frac{1}{\lambda_0} = \frac{m_0 c}{\hbar} = \frac{e}{\hbar} A^{(0)} = gA^{(0)}, \tag{24.75}$$

the factor $gA^{(0)}$ entering into the analysis [4]–[36] through the fact that the covariant derivative has entered into the theory as:

$$D\wedge = d\wedge + gA\wedge. \tag{24.76}$$

In Eq. (24.76) a "barebones" notation has been used which suppresses all indices for clarity of presentation and to reveal the fundamental structure of the theory. Therefore the factor $eA^{(0)}/\hbar$ originates in the spin connection of spacetime geodynamics. Fundamentally:

$$mc = eA^{(0)} \tag{24.77}$$

and so the rest energy divided by c has been factorized into the product of two C negative quantities, e (the fundamental charge) and $A^{(0)}$ (the fundamental voltage). This procedure means that we may always write:

$$mc = eA^{(0)} = e(\frac{\hbar\kappa_0}{e}) \tag{24.78}$$

and for two signs of charge there is only one sign of mass. This inference is observed experimentally in particles and antiparticles. The latter are predicted by the Evans wave equation, which in a well defined limit reduces to the Dirac equation of special relativity. The origin of charge and voltage is the same as that of mass (i.e. rest energy), and this origin is spacetime geodynamics.

24.5 Derivation of the Heisenberg Equation of Motion and Origin of the Planck Constant

Due to the antisymmetry of the unitless structure invariant $\theta^a{}_b$ it can always be written for all base manifolds as a commutator in the tangent spacetime:

$$\theta^a{}_b = [x^a, \kappa_b] \tag{24.79}$$

where κ_b has the units of inverse metres or wavenumber. The commutator is the basis for the Heisenberg equation of motion and therefore this equation has its origin in spacetime geodynamics, in common with all equations of physics. Using the de Broglie postulate in the form:

$$p_b = \hbar\kappa_b \tag{24.80}$$

we obtain the Heisenberg equation directly and in an objective or generally covariant format:

$$[x^a, p_b] = \hbar\theta^a{}_b. \tag{24.81}$$

It is seen that the structure invariants are fundamental to the Heisenberg equation. The de Broglie postulate or wave particle duality in Eq. (24.80) expresses momentum in terms of a geometric quantity κ_b. The latter is the result of the existence of the two structure invariants of differential geometry, invariants which are always related through the Bianchi identity:

$$D \wedge T^a = R^a{}_b \wedge q^b. \tag{24.82}$$

In the particular case:

$$D \wedge T^a = R^a{}_b \wedge q^b = 0 \tag{24.83}$$

the curvature with symmetric connection:

$$R^a{}_b \wedge q^b = 0 \tag{24.84}$$

can always be expressed as the torsion with anti-symmetric connection:

$$D \wedge T^a = 0. \tag{24.85}$$

In other words Eq. (24.84) is true if and only if the connection is symmetric, and Eq. (24.85) is true if and only if the connection is anti-symmetric. More generally (Eq. (24.82)) the connection must therefore be asymmetric. The Bianchi identity is therefore the geometrical reason why the Heisenberg equation is a relation between the structure invariants. In general differential geometry is at the root of everything in physics, and this inference is an obvious development to unified field theory of the fact that Riemann geometry is at the root of everything in gravitation. The particular type of Riemann geometry used in the Einstein theory of gravitation is one with a Riemann tensor and symmetric Christoffel connection. The relation between x^a and $\theta^a{}_b$ can also be made clear by integrating both sides of the Bianchi identity:

$$\int_s D \wedge T^a = \int_s R^a{}_b \wedge q^b. \tag{24.86}$$

In the special cases:

$$\int_s D \wedge T^a = D \wedge \int_s T^a \tag{24.87}$$

$$\int_s R^a{}_b \wedge q^b = \left(\int_s R^a{}_b \right) \wedge q^b \tag{24.88}$$

we obtain the following tangent spacetime equation for all base manifolds:

$$D \wedge x^a = \theta^a{}_b \wedge q^b. \tag{24.89}$$

Eq. (24.89) is similar in structure to the Heisenberg equation because both sides are commutators, and because $D\wedge$ also has the units of inverse metres.

The origin and fundamental meaning of the Planck constant may be elucidated by comparing the de Broglie duality equation (24.80) with Einstein postulate (24.15). Thereby we obtain a relation between $\hbar$ and k using:

$$R = \kappa_b \kappa^b = \frac{1}{\hbar} p_b \kappa^b = -kT, \tag{24.90}$$

a relation which implies:

$$\hbar k = -\frac{1}{T} p_b \kappa^b. \tag{24.91}$$

The origin of $\hbar$ and k is therefore essentially the same. We may write:

$$|T| = \frac{1}{k} |R| \tag{24.92}$$

and since $|R|$ is quantized by the Evans Lemma, Eq. (24.92) is a Planck relation. From the Lemma we are able to define [4]–[36] the fundamental volume:

$$V_0 = \frac{k\hbar^2}{mc^2} = \frac{k\hbar^2}{En_0} \tag{24.93}$$

and therefrom the fundamental frequency and wavenumber:

$$\omega_0 = \frac{En_0}{\hbar}. \tag{24.94}$$

It therefore follows that the Planck constant and the Einstein constant are inversely related:

$$\hbar k = V_0 \omega_0 \tag{24.95}$$

through the fundamental volume V_0 and the fundamental frequency ω_0. Expressing Eqs. (24.95) as

$$\frac{\hbar k}{c} = V_0 \kappa_0 \tag{24.96}$$

suggests the existence of a fundamental curvature (in inverse square metres):

$$R_u = \frac{c}{\hbar k} \left(= 1.52348 \times 10^{52} m^{-2}\right) \tag{24.97}$$

Using the following values of the universal constants:

$$\left.\begin{aligned} c &= 2.997925 \times 10^{8} m s^{-1} \\ k &= 1.86595 \times 10^{-26} N s^{2} kg^{-2} \\ \hbar &= 1.05459 \times 10^{-34} J s \end{aligned}\right\} \tag{24.98}$$

it is found that:

$$R_u = 1.52348 \times 10^{52} m^{-2} \tag{24.99}$$

is a universal constant. Using the Lemma:

$$\Box q^a{}_\mu = R_u q^a{}_\mu \tag{24.100}$$

it is suggested that R_u be interpreted as the initial curvature of the universe, its maximum possible curvature, associated with its minimum possible volume. This curvature and volume define the dimensions of the universe at an initial event in spacetime. This conclusion is similar to the Big Bang model, except insofar as the initial event is not a mathematical singularity. A mathematical singularity is incompatible with objective physics.

The minimum curvature of the universe is a universal constant which defines and is defined by the values of $c, \hbar$ and k observed experimentally. It should not be confused with the rest curvature of a particle [4]–[36]:

$$|R_0| = \frac{m^2 c^2}{\hbar^2} = \frac{m}{\hbar^2} E n_0. \tag{24.101}$$

The latter is the eigenvalue of the Evans Lemma in the limit when there is no interaction between particles and so when the gravitational field is infinitesimally different from zero. In this sense R_0 is a "least curvature", i.e. the curvature attained by a particle when the gravitational field is minimized. The minimum possible unit of action in this limit (the limit of special relativity) is a universal constant, the Planck constant. Action is quantized into multiples of the minimum possible action $\hbar$ because of the existence of the Evans Lemma in differential geometry, more accurately described as spacetime geodynamics. In this same limit the rest volume of the particle is defined by Eq. (24.93). Thus $\hbar$ is defined by spacetime geodynamics.

Arguing by analogy, the rest curvature of the universe is defined by that initial event in spacetime when the net gravitational field in the universe is zero: there is no net gravitational attraction or repulsion in the universe. The latter has contracted to this initial volume and curvature because of gravitational attraction, but the equilibrium thus attained is an unstable equilibrium, so expansion occurs once more as gravitational repulsion builds up. The eigenvalue of the oscillatory universe is the tetrad and the process is a dynamical one, it is governed by the Evans Lemma and wave equation of the universe.

The following terms are therefore introduced to distinguish these two types of fundamental curvature, the universal curvature, Eq. (24.97), and the rest curvature, defined in terms of the particle rest energy by:

$$R_0 = -\frac{m}{\hbar^2} En_0 . \tag{24.102}$$

The standard form of the Heisenberg equation is obtained if the following definition is used:

$$\theta^a{}_b = \frac{i}{\hbar} J^a{}_b , \tag{24.103}$$

a definition which implies:

$$[x^a, \kappa_b] = \frac{i}{\hbar} J^a{}_b . \tag{24.104}$$

Using the de Broglie wave particle duality:

$$[x^a, \hbar\kappa_b] = [x^a, p_b] = \frac{i}{\hbar} J^a{}_b , \tag{24.105}$$

we obtain the well known position / momentum form of the Heisenberg equation:

$$[x^a, p_b] = \frac{i}{\hbar} J^a{}_b . \tag{24.106}$$

The units of $J^a{}_b$ are those of action or angular momentum. The least action principle of Hamilton asserts that the classical action is minimized in the universe. The last possible action in the universe is $\hbar$, so:

$$[x^a, p_b]_{\text{min}} = i\hbar\epsilon^a{}_b \tag{24.107}$$

where $\epsilon^a{}_b$ is the two dimensional anti-symmetric unit tensor of the tangent spacetime for all base manifolds. The fundamental reason why $\hbar$ occurs in quanta is the equation governing the evolution of the tetrad [4]–[36]:

$$q^a{}_\mu(t, \boldsymbol{r}) = \exp\left(i\frac{S}{\hbar}\right) q^a{}_\mu(0, \boldsymbol{0}) \tag{24.108}$$

The tetrad is quantized, so x^a, p_b and $\theta^a{}_b$ are also quantized, being always related to the tetrad through an equation such as (24.89) of spacetime geodynamics.

Finally, the origin and meaning of mass may be elucidated using the proportionality:

$$|R_0| = \xi |R_u|, \tag{24.109}$$

to obtain an identity of spacetime geodynamics:

$$m^2 = \frac{\hbar}{ck} \frac{|R_0|}{|R_u|}. \tag{24.110}$$

It is concluded that the square of the mass of any particle is the ratio of two fundamental curvatures within the universal constant $\hbar/(ck)$. The fundamental reason for the occurrence of particles is the Evans least curvature principle [4]–[36], one of whose ramifications is Eq. (24.108). Charge e is then the factorization of mass as discussed in Eq. (24.78), which shows that there are two signs of charge for one sign of mass.

Acknowledgements

The Ted Annis Foundation and Craddock Inc. are thanked for funding and the Fellows and Emeriti of AIAS for many interesting discussions.

References

1. E. P. Wigner, Ann. Math., **40**, 149 (1939).
2. L. H. Ryder, *Quantum Field Theory*, (Cambridge Univ Press, 1996, 2nd ed.).
3. S. P. Carroll, *Lecture Notes in General Relativity*, (a graduate course at Harvard, UC Santa Barbara and U Chicago, arXiv: gr-gq / 973019 v1 Dec 1997, public domain).
4. M. W. Evans, A Unified Field Theory for Gravitation and Electromagnetism. *Found. Phys. Lett.*, **16**, 367 (2003).
5. M. W. Evans, A Generally Covariant Wave Equation for Grand Unified Field Theory, Found. Phys. Lett, **16**, 507 (2003).
6. M. W. Evans, The Equations of Grand Unified Field Theory in terms of the Maurer Cartan Structure Relations of Differential Geometry, *Found. Phys. Lett.*, **17**, 25 (2003).
7. M. W. Evans, Derivation of Dirac's Equation from the Evans Wave Equation, *Found. Phys. Lett.*, **17**, 149 (2004).
8. M. W. Evans, Unification of the Gravitational and Strong Nuclear Fields, *Found. Phys. Lett.*, **17**, 267 (2004).
9. M. W. Evans, The Evans Lemma of Differential Geometry, *Found. Phys. Lett.*, **17**, 433 (2004).
10. M. W. Evans, Derivation of the Evans Wave Equation from the Lagrangian and Action, Origin of the Planck Constant in General Relativity, *Found. Phys. Lett.*, **17**, 535 (2004).
11. M. W. Evans et alii (AIAS Author Group), Development of the Evans Wave Equation in the Weak Field Limit: the Electrogravitic Equation, *Found. Phys. Lett.*, **17**, 497 (2004).
12. M. W. Evans, Physical Optics, the Sagnac Effect and the Aharonov Bohm Effect in the Evans Unified Field Theory, *Found. Phys. Lett.*, **17**, 301 (2004).
13. M. W. Evans, Derivation of the Geometrical Phase from the Evans Phase Law of Generally Covariant Unified Field Theory, *Found. Phys. Lett.*, **17**, 393 (2004).
14. M. W. Evans, Derivation of the Lorentz Boost from the Evans Wave Equation, *Found. Phys. Lett.*, **17**, 663 (2004).
15. M. W. Evans, The Electromagnetic Sector of the Evans Field Theory, *Found. Phys. Lett.*, **18**, 37 (2005), preprint on www.aias.us.
16. M. W. Evans, New Concepts from the Evans Field Theory Part One: The Evolution of Curvature, Oscillatory Universe without Singularity, and Gen-

eral force and Field Equations, *Found. Phys. Lett.*, in press (2005); preprint on www.aias.us.

17. M. W. Evans, New Concepts from the Evans Field Theory Part Two: Derivation of the Heisenberg Equation and Reinterpretation of the Heisenberg Uncertainty Principle, *Found. Phys. Lett.*, in press (2005), preprint on www.aias.us.
18. M. W. Evans, Derivation of O(3) Electrodynamics from Generally Covariant Unified Field Theory, *Found. Phys. Lett.*, submitted, preprint on www.aias.us.
19. M. W. Evans, The Spinning and Curving of Spacetime, The Electromagnetic and Gravitational Fields in the Evans Unified Field Theory, *Found. Phys. Lett.*, submitted, preprint on www.aias.us.
20. M. W. Evans, Derivation of O(3) Electrodynamics from the Evans Unified Field Theory, *Found. Phys. Lett.*, submitted, preprint on www.aias.us.
21. M. W. Evans, Calculation of the Anomalous Magnetic Moment of the Electron, *Found. Phys. Lett.*, submitted, preprint on www.aias.us.
22. M. W. Evans, Generally Covariant Electro-weak Theory, *Found. Phys. Lett.*, submitted, preprint on www.aias.us.
23. M. W. Evans, Evans field Theory of Neutrino Oscillations, *Found. Phys. Lett.*, submitted, preprint on www.aias.us.
24. M. W. Evans, The Interaction of Gravitation and Electromagnetism, *Found. Phys. Lett.*, submitted, preprint on www.aias.us.
25. M. W. Evans, Electromagnetic Energy from Gravitation, *Found. Phys. Lett.*, submitted, preprint on www.aias.us.
26. M. W. Evans, *Generally Covariant Unified Field Theory, the Geometrization of Physics*, (in press, 2005).
27. L. Felker, *The Evans Equations* (freshman text, in prep).
28. M. W. Evans and L. B. Crowell, *Classical and Quantum Electrodynamics and the $B^{(3)}$ Field* (World Scientific, Singapore, 2001).
29. M. W. Evans (ed.), *Modern Non-Linear Optics*, a special topical issue in three parts of I Prigogine and S. A. Rice, (series eds.), *Advances in Chemical Physics* (Wiley-Interscience, New York, 2001, 2nd ed., hardback and e book) vols. 119(1), 119(2) and 119(3).
30. M. W. Evans and S. Kielich (eds.), *Modern Non-Linear Optics*, a special topical issue in three parts of I. Prigogine and S. A. Rice (series eds.), *Advances in Chemical Physics* (Wiley-Interscience, New York, 1992, 1993, 1997, hardback and softback) vols. 85(1), 85(2) and 85(3).
31. M. W. Evans, J.-P. Vigier et alii, *The Enigmatic Photon* (van der Merwe Series, Kluwer, Dordecht, 1994 to 2002, hardback and softback) in five volumes.
32. M. W. Evans and A. A. Hasanein, *The Photomagneton in Quantum Field Theory*, (World Scientific, Singapore, 1994).
33. M. W. Evans, *The Photon's Magnetic Field: Optical NMR Spectroscopy* (World Scientific, Singapore, 1992).
34. M. W. Evans, Physica B, **182**, 227 and 237 (first scientific papers on the $B^{(3)}$ field).
35. M. W. Evans, Collected Scientific Papers website, Artspeed, California, in prep., a collection of over six hundred and fifty scientific papers.
36. over one million two hundred thousand page views of www.aias.us and aias.rfsafe.com, Jan-Sept 2004.

25

Electromagnetic Energy from Gravitation

Summary. The geometric interaction of gravitation and electromagnetism in the Evans field theory is shown to lead to a new cross-current term which does not appear in the standard model, but which is, in theory, an important new source of electromagnetic energy. The structure of the new current term is developed with the Hodge dual of the homogeneous Evans field equation. Such a development shows that the new current term causes a tiny imbalance in the Faraday law of induction and Gauss law of magnetism. This gravitational effect (which does not exist in the standard model) changes the polarization of electromagnetic radiation in free space (or vacuum) and so can be detected in theory by looking for tiny changes in polarization of light grazing the sun in a total eclipse (Eddington type experiment), or in radiation grazing an intensely gravitating object such as a quasar or pulsar. This tests for the relic effect of gravitation on electromagnetism several billion years after the initial expansion event known as "Big Bang". This effect is expected to be very tiny, but very important to new forms of energy. So it must be looked for with the highest possible contemporary precision and the greatest experimental care. It is already possible, however, to amplify this current to useful levels, and reliable devices are available which achieve this aim.

Key words: Evans field theory, homogeneous Evans field equation, Hodge duality, electromagnetic current and energy due to gravitation, Faraday law of induction.

25.1 Introduction

Recently [1]–[30] the first objective (i.e. generally covariant) unified field theory has been developed based on differential geometry. The theory is known as the Evans unified field theory [31] (or simply as Evans field theory), an appellation intended to distinguish it from the standard model. It is already generally accepted [31], and for good reason, that the Evans field theory is a major advance from the standard model, bringing with it some important new technologies. In this note the homogeneous equation of the Evans field theory is developed into its Hodge dual [32, 33] in order to analyse a new current term which does not appear in the standard model, but which is in theory an important new source of electromagnetic energy from spacetime. In Section 25.2 the Hodge dual equation is deduced in tensor notation and written out in vector notation, whereupon it becomes clear that the new current causes a tiny imbalance of gravitational origin in the Faraday law of induction and the Gauss law of magnetism. In Section 25.3 suggestions are given for an experimental test of the theory by looking for tiny changes in polarization in electromagnetic radiation reaching a space or earthbound telescope/polarimeter after grazing the sun in a total eclipse (Eddington type of experiment) or after grazing an intensely gravitating object such as a quasar or pulsar in deep space.

25.2 Development of the Hodge Dual of the Homogeneous Equation of Evans Field Theory

In developing the Hodge dual [32, 33] some important mathematical details must be noted, because the development is taking place in a general four dimensional manifold (non-Minkowski spacetime). The starting point must therefore be the rigorous definition [32] of the Hodge dual in general relativity for the n dimensional manifold. The Hodge dual is obtained from the homogeneous Evans field equation [1, 30] in differential form notation:

$$d \wedge F^a = \mu_0 j^a \tag{25.1}$$

where $D\wedge$ is the covariant exterior derivative, F^a is the vector valued electromagnetic field two-form, j^a is the vector valued charge-current density three-form, and μ_0 is the S.I. permeability in vacuo. The current in Eq. (25.1) is defined by the wedge product [32]

$$j^a = \frac{1}{\mu_0}\left(R^a{}_b \wedge A^b - \omega^a{}_b \wedge F^b\right) \tag{25.2}$$

where $R^a{}_b$ is the tensor-valued Riemann or curvature two-form [32] and where A^b is the vector valued potential one-form. The latter is defined by the vector-valued tetrad one-form as follows [1, 30]:

$$A^b = A^{(0)} q^b \tag{25.3}$$

where $A^{(0)}$ is a C negative scalar with the units of the electromagnetic potential (tesla meters or webers per metre). As per standard practice [32] in differential geometry only the Latin indices of the tangent spacetime appear in Eqs. (25.1) - (25.3), the Greek indices of the base manifold are always the same on both sides of the equation and so can be omitted for ease of notation [32]. The homogeneous electromagnetic field equation of the standard model is the homogeneous Maxwell-Heaviside equation:

$$d \wedge F = 0. \tag{25.4}$$

In Eq. (25.4) $d\wedge$ is the ordinary (or flat spacetime) exterior derivative, and F is the scalar valued electromagnetic field two-form [32, 33].

It can be seen that there is a new vector-valued current three-form, j^a, present in the Evans field theory, which is a rigorously objective theory of physics. Conversely, in order that physics be a rigorously objective subject, one we can all agree upon, this current term has to be recognised to exist in physics. By inspection of the structure of Eq. (25.1) the new current three-form is non-zero if and only if the Christoffel connection [32] is asymmetric in its lower two indices. If the Christoffel connection is symmetric in its lower two indices the torsion form vanishes, and the Evans field reduces to the Einstein gravitational field. The latter therefore becomes independent of electromagnetism when the Christoffel symbol becomes symmetric. In this limit the Christoffel symbol of the decoupled electromagnetic field can only be anti-symmetric: the electromagnetic field has become self-consistently independent of the gravitational field. It follows that electromagnetism and gravitation can be mutually influential if and only if the Christoffel symbol is asymmetric. This is of basic importance and can be regarded as a new and rigorously objective law of physics. When the two fields are independent the only possibility is:

$$d \wedge F^a = 0, \tag{25.5}$$

$$j^a = \frac{1}{\mu_0}\left(R^a{}_b \wedge A^b - \omega^a{}_b \wedge F^b\right) = 0, \tag{25.6}$$

and the current term in Eq. (25.1) vanishes.

The Evans field theory indicates therefore that this current MAY be non-zero. The Evans field theory does not imply that the current MUST be non-zero, only experiment can distinguish between these two theoretical possibilities. However the Evans field theory implies objectively and geometrically that the electromagnetic field must be a vector-valued two-form, F^a, not a scalar valued two-form,F, as in the standard model. This means that we must reject the standard model as being unobjective and incomplete. An early sign of this incompleteness was the inference [34] of the fundamental Evans spin field, denoted $\boldsymbol{B}^{(3)}$, whose origin is now understood [1, 30] to be the complex circular index $a = (3)$ of the tangent spacetime of Evans field theory. This index does not exist in the standard model, which therefore cannot be used to analyse the Evans spin field, a well known observable [1, 30] of the inverse Faraday effect. Reliable experimental devices are already available which indicate that the current, j^a, is non-zero [35]. These devices produce potentially very important energy savings which cannot be explained qualitatively by Maxwell Heaviside theory, Eq. (25.4). As yet, these devices provide only qualitative indications, but these are important indications for several good reasons, both fundamental and technological. At electronic scales in a circuit taking electromagnetic energy from spacetime the curvature of spacetime becomes very large. Indeed, in the limit of a point electron the scalar curvature is infinite (infinite spacetime compression). So it is not surprising that the new current is significant at electronic scales because it is directly proportional to the Riemann curvature for a given tetrad. These important concepts do not exist in the standard model, and there is no indication in the standard model that electromagnetic energy can be obtained from spacetime.

The general definition of the Hodge dual is given by Carroll [32]:

$$A_{\mu_1 \ldots \mu_{n-p}} = \frac{1}{p!} \epsilon^{\nu_1 \ldots \nu_p}{}_{\mu_1 \ldots \mu_{n-p}} A_{\nu_1 \ldots \nu_p} \tag{25.7}$$

Eq. (25.7) maps from a p-form of differential geometry to an (n-p)-form in a general n dimensional manifold. The general Levi-Civita symbol is defined [32] in any manifold to be:

$$\widetilde{\epsilon}_{\mu_1 \mu_2 \ldots \mu_n} = \begin{cases} 1 \text{ if } \mu_1 \mu_2 \ldots \mu_n \text{ is an even permutation} \\ -1 \text{ if } \mu_1 \mu_2 \ldots \mu_n \text{ is an odd permutation} \\ 0 \qquad\qquad \text{otherwise} \end{cases} \tag{25.8}$$

The Levi-Civita tensor used in the definition of the Hodge dual is however [32]:

$$\epsilon_{\mu_1 \mu_2 \ldots \mu_n} = (|g|)^{1/2} \, \widetilde{\epsilon}_{\mu_1 \mu_2 \ldots \mu_n} \tag{25.9}$$

where $|g|$ is the numerical magnitude of the determinant of the metric. This procedure is necessary to define a valid tensor, Eq. (25.9), whose indices can be raised or lowered using the metric tensor for the general manifold. These mathematical details are important for numerical computation of the Evans equation (25.1). In the general manifold the metric and inverse metric tensors are defined by the Kronecker delta:

$$g^{\mu\nu} g_{\nu\sigma} = \delta^{\mu}_{\sigma} \tag{25.10}$$

Tensor operations such as contraction, symmetrization, and so on are unchanged in the general manifold from their equivalents in Minkowski spacetime, but in the general manifold the tensor $g^{\mu\nu}$ is not the same as the tensor $g_{\nu\sigma}$, and as we have just argued, the Levi-Civita symbol is not the same. These details can however be easily coded into a form that can be used routinely to evaluate the new current by computation. In this way the amount of electromagnetic energy available from spacetime can be estimated in a given circuit design.

In order to clarify the meaning of Eq. (25.7) some examples are given as follows.

1. In a three dimensional manifold (space) a one-form is the Hodge dual of a two-form. This is the well known result that an axial vector is dual to an

antisymmetric tensor. However, it seems not to be well known that this result is true for the general three dimensional space as well as Euclidean or flat three dimensional space. From Eq. (25.7) we obtain:

$$^{*}A_{\mu_1} = \frac{1}{2!}\epsilon^{\nu_1\nu_2}{}_{\mu_1}A_{\nu_1\nu_2} \tag{25.11}$$

when $p = 2, n = 3$. Re-labelling indices we obtain:

$$^{*}A_{\rho} = \frac{1}{2}\epsilon^{\nu\mu}{}_{\rho}A_{\nu\mu}. \tag{25.12}$$

Eq. (25.12) is the generalization to non-Euclidean space of the familiar Euclidean:

$$^{*}A_{k} = \frac{1}{2}\epsilon^{ij}{}_{k}A_{ij}. \tag{25.13}$$

2. In a four-dimensional manifold a one-form is dual to a three-form:

$$^{*}A_{\mu_1} = \frac{1}{3!}\epsilon^{\nu_1\nu_2\nu_3}{}_{\mu_1}A_{\nu_1\nu_2\nu_3} \tag{25.14}$$

or:

$$^{*}A_{\mu} = \frac{1}{6}\epsilon^{\mu\rho\sigma}{}_{\mu}A_{\nu\rho\sigma} \tag{25.15}$$

with $n = 4, p = 3$.

3. In a four-dimensional manifold a three-form is dual to a one-form:

$$^{*}A_{\mu_1\mu_2\mu_3} = \epsilon^{\nu_1}{}_{\mu_1\mu_2\mu_3}A_{\nu_1} \tag{25.16}$$

or

$$^{*}A_{\mu\nu\rho} = \epsilon^{\sigma}{}_{\mu\nu\rho}A_{\sigma} \tag{25.17}$$

with $n = 4, p = 1$.

4. In a four-dimensional manifold a two-form is dual to a two-form:

$$^{*}A_{\mu_1\mu_2} = \frac{1}{2}\epsilon^{\nu_1\nu_2}{}_{\mu_1\mu_2}A_{\nu_1\nu_2} \tag{25.18}$$

or

$$^{*}A_{\mu\nu} = \frac{1}{2}\epsilon^{\rho\sigma}{}_{\mu\nu}A_{\rho\sigma} \tag{25.19}$$

with $n = 4, p = 2$.

The indices in the Levi-Civita tensor defined by Eq,. (25.9) are raised or lowered with the appropriate metric tensor of the general manifold, for example:

$$\epsilon_{\sigma\mu\nu\rho} = g_{\sigma\kappa}\epsilon^{\kappa}{}_{\mu\nu\rho} \tag{25.20}$$

This is an important difference from the Levi-Civita tensor of Minkowski spacetime, but here again computer code is easily written provided that the mathematical fundamentals are precisely defined. Some important inferences can be made without resorting to computation.

Eq. (25.1) in tensor notation is [1, 30, 32]:

$$\partial_\mu F^a{}_{\nu\rho} + \partial_\nu F^a{}_{\rho\mu} + \partial_\rho F^a{}_{\mu\nu} = \mu_0 \left(j^a{}_{\mu\nu\rho} + j^a{}_{\nu\rho\mu} + j^a{}_{\rho\mu\nu} \right). \tag{25.21}$$

This is ideal for computation, but the equivalent Hodge dual equation, which we develop as follows, is ideal for physical inference. Using Eq. (25.15):

$$\begin{aligned} {}^*j^a{}_\sigma &= \frac{1}{6}\epsilon^{\mu\nu\rho}{}_\sigma j^a{}_{\mu\nu\rho} \\ &= \frac{1}{6}\epsilon^{\nu\rho\mu}{}_\sigma j^a{}_{\nu\rho\mu} \\ &= \frac{1}{6}\epsilon^{\rho\mu\nu}{}_\sigma j^a{}_{\rho\mu\nu} \end{aligned} \tag{25.22}$$

so the Hodge dual of the right hand side of Eq. (25.21) is $3_{\mu_0} {}^*j^a{}_\rho$. This Hodge dual current is a vector valued one-form. Similarly the Hodge dual of the left hand side of Eq. (25.21) is found using Eq. (25.19):

$$\begin{aligned} {}^*F^a{}_{\mu\sigma} &= \frac{1}{2}\epsilon^{\nu\rho}{}_{\mu\sigma} F^a{}_{\nu\rho}, \\ {}^*F^a{}_{\nu\sigma} &= \frac{1}{2}\epsilon^{\rho\mu}{}_{\nu\sigma} F^a{}_{\rho\mu}, \\ {}^*F^a{}_{\rho\sigma} &= \frac{1}{2}\epsilon^{\mu\nu}{}_{\rho\sigma} F^a{}_{\mu\nu}. \end{aligned} \tag{25.23}$$

It follows that

$$\begin{aligned} \partial^\mu \ {}^*F^a{}_{\mu\sigma} &= \frac{1}{2}\partial^\mu \left(\epsilon^{\nu\rho}{}_{\mu\sigma} F^a{}_{\nu\rho} \right), \\ \partial^\nu \ {}^*F^a{}_{\nu\sigma} &= \frac{1}{2}\partial^\nu \left(\epsilon^{\rho\mu}{}_{\nu\sigma} F^a{}_{\rho\mu} \right), \\ \partial^\rho \ {}^*F^a{}_{\rho\sigma} &= \frac{1}{2}\partial^\rho \left(\epsilon^{\mu\nu}{}_{\rho\sigma} F^a{}_{\mu\nu} \right). \end{aligned} \tag{25.24}$$

and re-arranging dummy indices it follows that the Hodge dual of the left hand side of Eq. (25.21) is $3\partial^\mu \ {}^*F^a{}_{\mu\sigma}$. Therefore the complete Hodge dual equation is found to be:

$$\partial^\mu \ {}^*F^a{}_{\mu\nu} = \mu_0 \ {}^*j^a{}_\nu. \tag{25.25}$$

In field theory the Hodge dual is written with a tilde instead of an asterisk so we obtain:

$$\partial^\mu \widetilde{F}^a{}_{\mu\nu} = \mu_0 \widetilde{j}^a{}_\nu. \tag{25.26}$$

The equivalent equation in the standard model is well known [1],[30, 32], [33] to be:

$$\partial^\mu \widetilde{F}_{\mu\nu} = 0 \tag{25.27}$$

Eq. (25.27) is a combination of the Gauss law of magnetism and the Faraday law of induction:

$$\nabla \cdot \boldsymbol{B} = \boldsymbol{0}, \tag{25.28}$$

$$\nabla \times \boldsymbol{E} + \frac{\partial \boldsymbol{B}}{\partial t} = \boldsymbol{0} \tag{25.29}$$

The Evans field theory is more richly structured, notably, the presence of the tangent spacetime index a means that there are more states of polarization and more field components than in the standard model. One of these polarization states or fields is the Evans spin field observed in the inverse Faraday effect [1, 30]. The vector structure of Eq. (25.26) is as follows:

$$\nabla \cdot \boldsymbol{B}^a = \mu_0 \widetilde{j_0^a}, \tag{25.30}$$

$$\nabla \times \boldsymbol{E}^a + \frac{\partial \boldsymbol{B}^a}{\partial t} = c\mu_0 \widetilde{\boldsymbol{j}}^a, \tag{25.31}$$

so there is an additional charge density due to gravitation in the Gauss law of magnetism and an additional current density due to gravitation in the Faraday law of induction. It may be deduced that in the presence of gravitation, the circular polarization of an electromagnetic wave is changed because the right hand sides of Eqs. (25.30) and (25.31) are no longer zero. In other words circularly polarized solutions of Eqs (25.28) and (25.29) are no longer true for Eqs. (25.30) and (25.31). This deduction assumes, of course, that j of Eq. (25.1) is non-zero, and this can only be determined experimentally. The existence of reliable devices as described already strongly suggests, however, that j^a is non-zero.

25.3 Eddington Type Experiment

In order to observe the expected changes in circular polarization predicted theoretically in section 25.2 a source of intense gravitation is needed to maximize the cross-current j^a. Then changes in polarization are observed in theory in electromagnetic radiation grazing this source of gravitation. One such possibility is to look for polarization changes in light grazing the sun during a total eclipse. The instrument needed for this observation is a high accuracy polarimeter mounted on a telescope: either a space telescope or an earthbound telescope. The polarization of light grazing the sun would be compared with the polarization of light from the same source but in the absence of the gravitating object, in this case the sun. The experiment could be repeated for electromagnetic radiation reaching the earth from a source with an intervening intense gravitational field, such as that of a quasar or pulsar. Several variations on this experiment are possible in contemporary cosmology, in each case the objective of the experiment would be to evaluate the effect of gravitation on the polarization of electromagnetic radiation.

Other types of experiments have been suggested recently [36] in order to look for the mutual effects of gravitation and electromagnetism. The governing principle of each experiment is to initially balance a given design and to look for changes in the balance with high precision apparatus. For example to look for changes in a high precision gravimeter due to intense pulses of electromagnetic radiation. On an electronic scale in a given circuit design, we expect to see significant effects, as argued already in Section 25.2.

Acknowledgements

The Fellows and Emeriti of AIAS are thanked for many interesting discussions. The Ted Annis Foundation, Craddock Inc., Applied Science Associates and several individual scholars are thanked for funding this work .

References

1. M. W. Evans, A Unified Field Theory for and Gravitation and Electromagnetism, *Found. Phys. Lett.*, **16**, 367 (2003).
2. M. W. Evans, A Generally Covariant Wave Equation for Grand Unified Field Theory, *Found. Phys. Lett.*, **16**, 507 (2003).
3. M. W. Evans, The Equations of Grand Unified Field Theory in terms of the Maurer Cartan Structure Relations of Differential Geometry, *Found Phys. Lett.*, **17**, 25 (2004).
4. M. W. Evans, Derivation of Dirac's Equation from the Evans Wave Equation, *Found. Phys. Lett.*, **17**, 149 (2004).
5. M. W. Evans, Unification of the Gravitational and Strong Nuclear Fields, *Found. Phys. Lett.*,**17**, 267 (2004).
6. M. W. Evans, The Evans Lemma of Differential Geometry, *Found. Phys. Lett.*, **17**, 433 (2004).
7. M. W. Evans, Derivation of the Evans Wave Equation from the Lagrangian and Action, Origin of the Planck Constant in General Relativity, *Found. Phys. Lett.*, **17**, 535 (2004).
8. M. W. Evans et alii, (AIAS Author Group), Development of the Evans Wave Equation in the Weak Field Limit: the Electrogravitic Equation, *Found. Phys. Lett.*, **17**, 497 (2004).
9. M. W. Evans, Physical Optics, the Sagnac Effect and the Aharonov Bohm Effect in the Evans Unified Field Theory, *Found. Phys. Lett.*, **17**, 301 (2004).
10. M. W. Evans, Derivation of the Geometrical Phase from the Evans Phase Law of Generally Covariant Unified Field Theory, *Found. Phys. Lett.*, **17**, 393 (2004).
11. M. W. Evans, Derivation of the Lorentz Boost from the Evans Wave Equation, *Found. Phys. Lett.*, **17**, 663 (2004).
12. M. W. Evans, The Electromagnetic Sector of the Evans Field Theory, *Found. Phys. Lett.*, in press, preprint on www.aias.us.
13. M. W. Evans, New Concepts from the Evans Field Theory, Part One: The Evolution of Curvature, Oscillatory Universe without Singularity, and General Force and Field Equations, *Found. Phys. Lett.*, in press, preprint on www.aias.us.
14. M. W. Evans, New Concepts from the Evans Field Theory, Part Two: Derivation of the Heisenberg Equation and Reinterpretation of the Heisenberg Uncertainty Principle, *Found. Phys. Lett.*, in press, preprint on www.aias.us.

15. M. W. Evans, Derivation of O(3) Electrodynamics from Generally Covariant Unified Field Theory, *Found. Phys. Lett.*, submitted, preprint on www.aias.us.
16. M. W. Evans, The Spinning and Curving of Spacetime, the Electromagnetic and Gravitational Fields in the Evans Unified Field Theory, *Found. Phys. Lett.*, submitted, preprint on www.aias.us.
17. M. W. Evans, Derivation of O(3) Electrodynamics from the Evans Unified Field Theory, *Found Phys. Lett.*, submitted, preprint on www.aias.us.
18. M. W. Evans, Calculation of the Anomalous Magnetic Moment of the Electron from the Evans Field Theory, *Found. Phys. Lett.*, submitted, preprint on www.aias.us.
19. M. W. Evans, Generally Covariant Electro-weak Theory, *Found. Phys. Lett.*, submitted, preprint on www.aias.us.
20. M. W. Evans, Evans Field Theory of Neutrino Oscillations, *Found. Phys. Lett.*, submitted, preprint on www.aias.us.
21. M. W. Evans, The Interaction of Gravitation and Electromagnetism, *Found. Phys. Lett.*, submitted, preprint on www.aias.us.
22. M. W. Evans, *Generally Covariant Unified Field Theory, the Geometrization of Physics*, (in press, 2005).
23. L. Felker, *The Evans Equations*, (freshman level text in prep).
24. M. W. Evans and L. B. Crowell, *Classical and Quantum Electrodynamics and the* $\boldsymbol{B}^{(3)}$ *Field* (World Scientific, Singapore, 2001).
25. M. W. Evans (ed.), *Modern Non-Linear Optics*, a special topical issue in three parts of I. Prigogine and S. A. Rice, *Advances in Chemical Physics*, (Wiley-Interscience, New York, 2001, second and e book editions), vols. 119(1), 119(2) and 119(3)
26. M. W. Evans, J.-P. Vigier et alii, *The Enigmatic Photon* (van der Merwe series, Kluwer, 1994 to 2002 , hardback and softback), in five volumes.
27. M. W. Evans and A. A. Hasanein, *The Photomagneton in Quantum Field Theory* (World Scientific, Singapore, 1994).
28. M. W. Evans and S. Kielich (eds.), *Modern Non-Linear Optics*, a special topical issue in three parts of I. Prigogine and S A Rice, (series Eds.), *Advances in Chemical Physics*, (Wiley-Interscience, New York, 1992, 1993 and 1997, hardback and softback), vols. 85(1), 85(2) and 85(3).
29. M. W. Evans, *The Photon's Magnetic Field, Optical NMR Spectrocopy* (World Scientific, Singapore, 1992).
30. M. W. Evans, Collected Papaers Website of circa 650 scientific papers, Artspeed, California, in prep., webmaster M. Anderson.
31. Over one million page views for www.aias.us and aias.rfsafe.com in 2004 to date, from all leading physics institutions worldwide.
32. S. P. Carroll, *Lecture Notes in General Relativity*, (a graduate course at Harvard, Univ California Santa Barbara and Univ Chicago, arXiv: gr-gq / 973019 v1 Dec 1997, public domain).
33. L. H. Ryder, *Quantum Field Theory* (Cambridge Univ Press, 1996, 2nd ed., softback).
34. M. W. Evans, Physica B, **182**, 227, 237 (1992)
35. AIAS Confidential Internal Memorandum.
36. Two papers accepted for STAIF2004, preprints on www.aias.us.

26

The Homogeneous and Inhomogeneous Evans Field Equations

Summary. The homogeneous (HE) and inhomogeneous (IE) Evans unified field equations are deduced from differential geometry and the Hodge dual in the general four dimensional manifold (Evans spacetime) of unified field theory. The HE is the first Bianchi identity of differential geometry multiplied on both sides by the fundamental voltage $A^{(0)}$, a scalar valued electromagnetic potential magnitude. The IE is deduced by evaluating the Hodge dual of the Riemann form and the Hodge dual of the torsion form in the first Bianchi identity, then multiplying both sides of the resultant equation by $A^{(0)}$. This procedure generalizes the well known Hodge dual relation between the anti-symmetric electromagnetic field tensors of the homogeneous and inhomogeneous Maxwell Heaviside field equations of the standard model. The resulting HE and IE equations are correctly objective, or generally covariant, whereas the MH equations are valid only in the Minkowski spacetime of special relativity, and so are not generally covariant or objective equations of physics. For this reason the HE and IE field equations are able to analyze the mutual effects of gravitation on electromagnetism and vice versa, whereas the MH equations fail qualitatively in this objective.

Key words: Homogeneous and inhomogeneous Evans field equations; Hodge dual; Evans unified field theory.

26.1 Introduction

It is well known that the homogeneous and inhomogeneous Maxwell Heaviside (MH) field equations of the standard model are respectively [1, 2]:

$$d \wedge F = 0 \tag{26.1}$$

$$d \wedge \widetilde{F} = \mu_0 J \tag{26.2}$$

in differential geometry. Here $d\wedge$ is the exterior derivative, F is the scalar valued electromagnetic field two-form; $\widetilde{F}$ is the Hodge dual [1] of F in Minkowski spacetime and so is another scalar valued two-form, and J is the scalar valued charge-current density three-form. Eqs. (26.1) and (26.2) are written in S.I. units and μ_0 is the S.I. permeability in vacuo. In Section 2 the MH equations of the standard model are made objective or generally covariant equations of the Evans unified field theory [3, 32]. Eq. (26.1) is developed into the homogeneous Evans field equation (HE) and Eq. (26.2) is developed into the inhomogeneous Evans field equation (IE). The resulting equations are written in the general four dimensional manifold known as Evans spacetime and are:

$$d \wedge F^a = R^a{}_b \wedge A^b - \omega^a{}_b \wedge F^b = \mu_0 j^a \tag{26.3}$$

$$d \wedge \widetilde{F}^a = \widetilde{R}^a{}_b \wedge A^b - \omega^a{}_b \wedge \widetilde{F}^b = \mu_0 J^a. \tag{26.4}$$

Here

$$D \wedge F^a = d \wedge F^a + \omega^a{}_b \wedge F^b \tag{26.5}$$

is the covariant exterior derivative where $\omega^a{}_b$ is the spin connection [1] of differential geometry. In Eqs (26.3) and (26.4) F^a is the vector valued electromagnetic field two-form; $\widetilde{F}^a$ is its Hodge dual [1] in Evans spacetime, $R^a{}_b$ is the tensor valued Riemann or curvature two-form; $\widetilde{R}^a{}_b$ is the Hodge dual of $\widetilde{R}^a{}_b$ in Evans spacetime; A^a is the vector-valued electromagnetic potential one-form. Eq. (26.3) is the first Bianchi identity of differential geometry [1] mutiplied on both sides by the fundamental voltage $A^{(0)}$ [3, 32]; a scalar valued electromagnetic potential magnitude in volts. Thus the HE equation is:

$$A^{(0)}(D \wedge T^a) = A^{(0)}(R^a{}_b \wedge q^b). \tag{26.6}$$

Similarly the IE equation is:

$$A^{(0)}(D \wedge \widetilde{T}^a) = A^{(0)}(\widetilde{R}^a{}_b \wedge q^b). \tag{26.7}$$

In Eqs. (26.6) and (26.7) T is the vector-valued torsion two-form of differential geometry and q^b is the vector valued tetrad one-form of differential geometry. Thus [3, 32]:

$$F^a = A^{(0)} T^a \tag{26.8}$$

$$A^a = A^{(0)} q^a \tag{26.9}$$

Eqs. (26.8) and (26.9) convert differential geometry to the unified field theory.

In Section 26.2 the current terms j^a and J^a are developed in terms of the fundamental differential forms of geometry. In section 26.3 the mutual effects of gravitation and electromagnetism are discussed in some detail and suggestions made for numerical solutions of the HE and IE field equations. Essentially this work shows that all of physics originates in spacetime geodynamics [?].

26.2 Development Of The Current Terms j and J

It is convenient to summarize the notation used in the unified field theory as follows, first for the HE and then for the IE field equations.

The HE equation in "barebones notation", with all indices suppressed is:

$$D \wedge F = R \wedge A. \tag{26.10}$$

The tangent bundle indices are first restored to give:

$$D \wedge F^a = R^a{}_b \wedge A^b. \tag{26.11}$$

Secondly the indices of the base manifold are restored to give:

$$(D \wedge F^a)_{\mu\nu\rho} = \left(R^a{}_b \wedge A^b\right)_{\mu\nu\rho}. \tag{26.12}$$

In tensor notation, Eq (26.12) is developed as follows:

$$\begin{aligned} \partial_\mu F^a{}_{\nu\rho} + \partial_\nu F^a{}_{\rho\mu} + \partial_\rho F^a{}_{\mu\nu} + \omega^a{}_{\mu b} F^b{}_{\nu\rho} + \omega^a{}_{\nu b} F^b{}_{\rho\mu} + \omega^a{}_{\rho b} F^b{}_{\mu\nu} \\ = R^a{}_{b\mu\nu} A^b{}_\rho + R^a{}_{b\nu\rho} A^b{}_\mu + R^a{}_{b\rho\mu} A^b{}_\nu. \end{aligned} \tag{26.13}$$

It can therefore be seen that the condensed notation of Eq (26.10) is equivalent to solving simultaneous partial differential equations for given initial and boundary conditions. This problem can be addressed numerically and the result will be an estimate of the effect of gravitation on electromagnetism for a given experimental situation. Analytical methods of approximation and experimental data where available should always be used as guidelines to the numerical solution.

The homogeneous MH equation of the standard model loses a lot of information in comparison with the HE equation. This can be seen as follows. In barebones notation the equivalent of Eq. (26.10) in the standard model is:

$$d \wedge F = 0. \tag{26.14}$$

There are no tangent bundle indices in equation (26.14) because it is an equation of Minkowski spacetime in which the electromagnetic field is thought of as a separate entity, an entity superimposed on the frame (the Minkowski spacetime). The Evans field theory on the other hand is a unified field theory of general relativity in which the unified field is the Evans spacetime, or general four dimensional manifold. Tangent bundles [1] are defined on this base manifold, and so tangent bundle indices appear in Eq. (26.11). These indices define states of polarization of the electromagnetic field. The indices of the base manifold can be restored in Eq. (26.14) to give:

$$(d \wedge F)_{\mu\nu} = 0 \tag{26.15}$$

an equation which when written out in full is:

$$\partial_\mu F_{\nu\rho} + \partial_\nu F_{\rho\mu} + \partial_\rho F_{\mu\nu} = 0. \tag{26.16}$$

In comparison with the HE equation it is seen that there is no spin connection and no Riemann form in Eq. (26.15). The latter cannot therefore describe the effect of gravitation on electromagnetism and vice versa. This is an obvious and well known failure of the standard model, meaning that the Evans field theory should be preferred to the standard model. The latter is not an objective theory of nature, while the Evans field theory is objective and thus the first generally covariant theory of ALL natural philosophy - the first workable unified field theory.

The charge-current density of the HE equation is the vector valued three-form:

$$j = \frac{1}{\mu_0}(R \wedge A - \omega \wedge F) \tag{26.17}$$

in barebones notation. It is therefore a balance of terms. Eq. (26.3) describes the Gauss Law of magnetism and the Faraday Law of induction in the unified field theory. Experimentally it is known that these laws are obeyed to high precision, for example in standards laboratories, implying:

$$d \wedge F^a = 0. \tag{26.18}$$

Therefore:

$$j^a = \frac{1}{\mu_0}\left(R^a{}_b \wedge A^b - \omega^a{}_b \wedge F^b\right) \sim 0, \tag{26.19}$$

i.e. j is very small experimentally in laboratory experiments. In a cosmological context however it may become possible to detect j experimentally, for example in electromagnetic radiation grazing an intensely gravitating object (Eddington type experiment) or in anomalous redshift data or similar. If j vanishes identically it follows from Eqs. (26.6) and (26.19) that:

$$R^a{}_b \wedge q^b = \omega^a{}_b \wedge T^b. \tag{26.20}$$

Using the structure equations of differential geometry:

$$T^b = D \wedge q^b \tag{26.21}$$

$$R^a{}_b = D \wedge \omega^a{}_b \tag{26.22}$$

Eq. (26.20) implies:

$$\left(D \wedge \omega^a{}_b\right) \wedge q^b = \omega^a{}_b \wedge \left(D \wedge q^b\right) \tag{26.23}$$

One possible solution of Eq. (26.23) is:

$$\omega^a{}_b = -\kappa \epsilon^a{}_{bc} q^c \tag{26.24}$$

and this is true in circular polarization [3]-[32]. The first Bianchi identity of differential geometry, Eq. (26.6), allows for:

$$R^a{}_b \wedge q^b \neq \omega^a{}_b \wedge T^b, \tag{26.25}$$

and so j can be non-zero mathematically. For j to be non-zero the gamma connection of Riemann geometry [1] must be asymmetric in its lower two indices, and when this condition is true gravitation and electromagnetism are mutually influential [3, 32]. Initially circularly polarized electromagnetic radiation is changed by intense gravitation into elliptical polarization when j is non-zero. In the laboratory this is expected to be a very small effect, because Eq. (26.18) is true experimentally to high precision in the laboratory, but in a cosmological context j may become observable as discussed already. In the standard model j is always zero. So this type of experiment may be used to test the difference between Evans field theory and the standard model. Numerical methods would be needed to simulate and design such an experiment and to estimate j for a given R and given spin connection in the presence of gravitation. "The presence of gravitation" means that there is a contribution to the complete (in general asymmetric) gamma connection from a component of the gamma connection which is symmetric in its lower two indices (the well known Christoffel, Levi-Civita or Riemannian connection [1]). "The absence of gravitation" means that the gamma connection is anti-symmetric in its lower two indices. This is the gamma connection of electromagnetism uninfluenced by gravitation. "The absence of electromagnetism" means that the gamma connection is symmetric in its lower two indices, and that the torsion form vanishes by definition [1]. In all three cases note carefully that the Riemann tensor or Riemann form is non-zero. The Riemann form of electromagnetism is therefore non-zero in the absence of gravitation but is undefined in the standard model. On the other hand the torsion form vanishes by definition when the gamma connection is symmetric in its lower two indices. If the Riemann form and connection form are both zero, the spacetime is Minkowski spacetime, and gravitation and electromagnetism are not defined because the unified Evans field is not defined in a Minkowski spacetime. In the self-inconsistent and incomplete standard model electromagnetism is a separate philosophical entity superimposed on the frame of Minkowski spacetime, and gravitation is in essence the Christoffel connection.

The charge-current density J of the IE equation can become much larger than j and is of much greater practical importance in the acquisition of electric power from Evans spacetime. The practically important current J is defined as:

$$J = \frac{1}{\mu_0}\left(\widetilde{R} \wedge A - \omega \wedge \widetilde{F}\right) \tag{26.26}$$

in barebones notation. Reinstating indices:

$$J^a{}_{\mu\nu\rho} = \frac{1}{\mu_0}\left(\widetilde{R}^a{}_b \wedge A^b - \omega^a{}_b \wedge \widetilde{F}^b\right)_{\mu\nu\rho} \tag{26.27}$$

and when written out in tensor notation we obtain:

$$J^a{}_{\mu\nu\rho} + J^a{}_{\nu\rho\mu} + J^a{}_{\rho\mu\nu} = \frac{1}{\mu_0}\Big(\widetilde{R}^a{}_{b\mu\nu}A^b{}_\rho + \widetilde{R}^a{}_{b\nu\rho}A^b{}_\mu + \widetilde{R}^a{}_{b\rho\mu}A^b{}_\nu \\ -\omega^a{}_{b\mu}\widetilde{F}^b{}_{\nu\rho} - \omega^a{}_{b\nu}\widetilde{F}^b{}_{\rho\mu} - \omega^a{}_{b\rho}\widetilde{F}^b{}_{\mu\nu}\Big). \tag{26.28}$$

The IE equation is obtained in analogy with the inhomogeneous MH equation (26.2), in which the scalar valued [1] electromagnetic field two-form F is replaced by its Hodge dual [1] $\widetilde{F}$ in Minkowski spacetime. Therefore in order to obtain the IE equation from the HE equation the Hodge duals are defined of F^a and $R^a{}_b$ in Evans spacetime. Here F^a is a vector valued [1] two-form of the Evans (i.e. unified) field, and $R^a{}_b$ is a tensor valued two-form of the Evans field. A differential two-form is antisymmetric in the indices of the base manifold. Thus:

$$F^a{}_{\mu\nu} = -F^a{}_{\nu\mu} \tag{26.29}$$

$$R^a{}_{b\mu\nu} = -R^a{}_{b\nu\mu}. \tag{26.30}$$

The Hodge dual [1] of a vector valued two-form in the general four-dimensional manifold is another vector valued two-form, another antisymmetric tensor with respect to the base manifold. Similarly the Hodge dual of a tensor valued two-form is another tensor valued two-form. Thus by anti-symmetry (Eqs. (26.29) and (26.30)) we obtain the IE from the HE.

In so doing care must be taken to define the Hodge dual correctly. The general definition of the Hodge dual for any differential form in any manifold is [1]:

$$\widetilde{\mathcal{X}}_{\mu_1\ldots\mu_{\mathbf{n-p}}} = \frac{1}{p!}\epsilon^{\nu_1\ldots\nu_{\mathbf{p}}}{}_{\mu_1\ldots\mu_{n-\mathbf{p}}}\mathcal{X}_{\nu_1\ldots\nu_{\mathbf{p}}}. \tag{26.31}$$

Eq. (26.31) maps from a p-form to an (n - p)-form the general n dimensional manifold. The Levi-Civita symbol in the general manifold is defined [1] as:

$$\epsilon_{\mu_1\mu_2\ldots\mu_n} = \begin{cases} 1 \text{ if } \mu_1\mu_2\ldots\mu_n \text{ is an even permutation} \\ -1 \text{ if } \mu_1\mu_2\ldots\mu_n \text{ is an odd permutation} \\ 0 \qquad \text{otherwise}. \end{cases} \tag{26.32}$$

The Levi-Civita symbol used in the definition of the Hodge dual is [1]:

$$\epsilon_{\mu_1\mu_2\ldots\mu_n} = (|g|)^{1/2}\,\widetilde{\epsilon}_{\mu_1\mu_2\ldots\mu_n} \tag{26.33}$$

where $|g|$ is the numerical magnitude of the determinant of the metric. In the general four-dimensional manifold (Evans spacetime), a two-form is dual to a two-form:

$$\widetilde{\mathcal{X}}_{\mu_1\mu_2} = \frac{1}{2}\epsilon^{\nu_1\nu_2}{}_{\mu_1\mu_2}\mathcal{X}_{\nu_1\nu_2}. \tag{26.34}$$

Indices are raised and lowered in the Levi-Civita tensor by use of the metric tensor. The latter is normalized by:

$$g^{\mu\nu}g_{\mu\nu} = 4. \tag{26.35}$$

Therefore:

$$\widetilde{\mathcal{X}}_{\mu\nu} = \frac{1}{2}\epsilon^{\rho\sigma}{}_{\mu\nu}\mathcal{X}_{\rho\sigma}. \tag{26.36}$$

This is the correct definition of the Hodge dual of a differential two-form in Evans spacetime. For numerical solutions this definition must be coded correctly in order to define the IE equation and the current $J^a{}_{\mu\nu\rho}$. The definition (26.6) is true both for a vector valued and tensor valued two-form of Evans spacetime.

None of these fundamentally important concepts of differential geometry appear in the standard model but they are of basic importance for the engineering of electric power from Evans spacetime through the current $J^a{}_{\mu\nu\rho}$.

26.3 Discussion Of The Interaction Of Electromagnetism and Gravitation

It is important to realize that in the Evans field theory the currents j and J are a logical consequence of differential geometry, and also govern the way in which electromagnetism influences gravitation and vice versa. In the older MH theory the current j is not recognized to exist, and J is introduced empirically without reference to geometry as required in general relativity. The Evans field theory therefore has fundamental advantages over the MH field theory. It is proved as follows that if j is zero experimentally to high precision, then J is non-zero as also observed experimentally. It is required to prove that if:

$$R^a{}_b \wedge q^b = \omega^a{}_b \wedge T^b \tag{26.37}$$

then:

$$\widetilde{R}^a{}_b \wedge q^b \neq \omega^a{}_b \wedge \widetilde{T}^b. \tag{26.38}$$

In tensor notation Eq. (26.37) is:

$$R^a{}_{b\mu\nu}q^b{}_\rho + R^a{}_{b\nu\rho}q^b{}_\mu + R^a{}_{b\rho\mu}q^a{}_\nu = \omega^a{}_{b\mu}T^b{}_{\nu\rho} + \omega^a{}_{b\nu}T^b{}_{\rho\mu} + \omega^a{}_{b\rho}T^b{}_{\mu\nu} \tag{26.39}$$

and this is equivalent [3, 32] to:

$$\widetilde{R}^a{}_b{}^{\mu\nu}q^b{}_\mu = \omega^a{}_{\mu b}\widetilde{T}^{\mu\nu b}. \tag{26.40}$$

It follows from Eq. (26.40) that in general:

$$\widetilde{R}^a{}_{b\mu\nu}q^b{}_\rho + \widetilde{R}^a{}_{b\nu\rho}q^b{}_\mu + \widetilde{R}^a{}_{b\rho\mu}q^a{}_\nu \neq \omega^a{}_{b\mu}\widetilde{T}^b{}_{\nu\rho} + \omega^a{}_{b\nu}\widetilde{T}^b{}_{\rho\mu} + \omega^a{}_{b\rho}\widetilde{T}^b{}_{\mu\nu} \tag{26.41}$$

So the proof is complete (q.e.d.) and rigorously geometrical in nature. In the Evans field theory it has been proven rigorously that if $j \sim 0$ experimentally then J is not zero. This proof then shows the origin of charge current density (J) in general relativity or objective physics.

The Einsteinian theory of gravitation uses the Christoffel symbol, as discussed already, and in this theory there is no influence of gravitation on electromagnetism. The use of a Christoffel symbol implies:

$$R \wedge q = 0 \tag{26.42}$$

which is the familiar Bianchi identity:

$$R_{\sigma\mu\nu\rho} + R_{\sigma\nu\rho\mu} + R_{\sigma\rho\mu\nu} = 0 \tag{26.43}$$

of the Einstein theory. Eq. (26.42) implies that electromagnetism is described in objective physics with the use of an anti-symmetric gamma connection in the HE:

$$\begin{aligned} d \wedge F^a &= R^a{}_b \wedge A^b - \omega^a{}_b \wedge F^b \\ &= \mu_0 j^a \sim 0. \end{aligned} \tag{26.44}$$

There is no contribution from gravitation, so the latter, self consistently, does not influence electromagnetism in the Einstein limit.

When there is interaction between gravitation and electromagnetism, the gamma connection is in general asymmetric in the HE and IE equations. This means that the gamma connection is a sum of antisymmetric and symmetric components, a conclusion which follows from the well known theorem that an asymmetric matrix is always the well defined sum of a symmetric matrix and an anti-symmetric matrix. This INTERACTION between gravitation and electromagnetism is the critically important factor in extracting electric power from Evans spacetime in situations where the standard model fails qualitatively. The standard model is not capable of describing this interaction, so is not capacble of describing the critically important extra source of electric power that appears in general in both j and J.

Acknowledgements

Craddock Inc., the Ted Annis Foundation, and Prof. John B. Hart and others are gratefully acknowledged for funding this work, and the staff of AIAS for many interesting discussions.

References

1. S. P. Carroll, *Lecture Notes in General Relativity*, (a graduate course at Harvard, UC Santa Barbara and U. Chicago, public domain, arXiv : gr-gq 973019 v1 Dec 1991).
2. L. H. Ryder, *Quantum Field Theory* (Cambridge, 1996, 2nd ed.).
3. M. W. Evans, A Unified Field Theory for Gravitation and Electromagnetism, *Found. Phys. Lett.*, **16**, 367 (2003).
4. M. W. Evans, A Generally Covariant Wave Equation for Grand Unified Field Theory, *Found. Phys. Lett.*, **16**, 507 (2003).
5. M. W. Evans, The Equations of Grand Unified Field Theory in Terms of the Maurer Cartan Structure Relations of Differential Geometry, *Found. Phys. Lett.*, **17**, 25 (2004).
6. M. W. Evans, Derivation of Dirac's Equation from the Evans Wave Equation, *Found. Phys. Lett.*, **17**, 149 (2004).
7. M. W. Evans, Unification of the Gravitational and Strong Nuclear Fields, *Found. Phys. Lett.*, **17**, 267 (2004).
8. M. W. Evans, The Evans Lemma of Differential Geometry, *Found. Phys. Lett.*, **17**, 433 (2004).
9. M. W. Evans, Derivation of the Evans Wave Equation from the Lagrangian and Action, Origin of the Planck Constant in General Relativity, *Found. Phys. Lett.*, **17**, 535 (2004).
10. M. W. Evans et alii (AIAS Author Group), Development of the Evans Wave Equation in the Weak Field Limit: the Electrogravitic Equation, *Found. Phys. Lett.*, **17**, 497 (2004).
11. M. W. Evans, Physical Optics, the Sagnac Effect and the Aharonov Bohm Effect in the Evans Unified Field Theory, *Found. Phys. Lett.*, **17**, 301 (2004).
12. M. W. Evans, Derivation of the Geometrical Phase from the Evans Phase Law of Generally Covariant Unified Field Theory, *Found. Phys. Lett.*, **17**, 393 (2004).
13. M. W. Evans, Derivation of the Lorentz Boost from the Evans Wave Equation, *Found. Phys. Lett.*, **17**, 663 (2004).
14. M. W. Evans, The Electromagnetic Sector of the Evans Field Theory, *Found. Phys. Lett.*, in press, (2005), preprint on www.aias.us.
15. M. W. Evans, New Concepts from the Evans Field Theory, Part One: The Evolution of Curvature, Oscillatory Universe without Singularity and General

force and Field Equations, *Found. Phys. Lett.*, in press (2005), preprint on www.aias.us.

16. M. W. Evans, New Concepts form the Evans field Theory Part Two: Derivation of the Heisenberg Equation and Reinterpretation of the Heisenberg Uncertainty Principle, *Found. Phys. Lett.*, in press (2005), preprint on www.aias.us.
17. M. W. Evans, The Spinning and Curving of Spacetime, the Electromagnetic and Gravitational Fields in the Evans Unified Field Theory, *Found. Phys. Lett.*, in press (2005), preprint on www.aias.us.
18. M. W. Evans, Derivation of O(3) Electrodynamics from Generally Covariant Unified Field Theory, *Found. Phys. Lett.*, submitted, preprint on www.aias.us.
19. M. W. Evans, Derivation of O(3) Electrodynamics from the Evans Unified Field Theory, *Found. Phys. Lett.*, submitted, preprint, on www.aias.us.
20. M. W. Evans, Calculation of the Anomalous Magnetic Moment of the Electron, *Found. Phys. Lett.*, submitted, preprint on www.aias.us.
21. M. W. Evans, Generally Covariant Electro-weak Theory, *Found. Phys. Lett.*, submitted, preprint on www.aias.us.
22. M. W. Evans, Evans Field Theory of Neutrino Oscillations, *Found. Phys. Lett.*, submitted, preprint on www.aias.us.
23. M. W. Evans, The Interaction of Gravitation and Electromagnetism, *Found. Phys. Lett.*, submitted, preprint on www.aias.us.
24. M. W. Evans, Electromagnetic Energy from Gravitation, *Found. Phys. Lett.*, submiteed, preprint on www.aias.us.
25. M. W. Evans, The Fundamental Invariants of the Evans Field Theory; *Found. Phys. Lett.*, submitted, preprint on www.aias.us.
26. M. W. Evans, *Generally Covariant Unified Field Theory, the Geometrization of Physics*, (in press, 2005).
27. L. Felker, *The Evans Equations* (freshman text, in prep).
28. M.Anderson, Artspeed Website for the collected works of Myron Evans, www.myronevanscollectedworks.com
29. M. W. Evans and L. B. Crowell, *Classical and Quantum Electrodynamics and the* $\boldsymbol{B}^{(3)}$ *Field*, (World Scientific, Singapore, 2001).
30. M. W. Evans (ed.), *Modern Nonlinear Optics*, a special topical issue in three parts of I. Prigogine and S. A. Rice (series eds.), *Advances in Chemical Physics*, (Wiley Interscience, New York, 2001, 2nd ed., and 1992, 1993, 1997, 1st ed.), vols. 119 and 85.
31. M. W. Evans, J. P. Vigier et alii, *The Enigmatic Photon* (Kluwer, Dordrecht, 1994 to 2002, hardback and softback), in five volumes.
32. M. W. Evans and A. A. Hasanein, *The Photomagneton in Quantum Field Theory*, (World Scientific, Singapore, 1994).

27

Derivation of the Gauss Law of Magnetism, The Faraday Law of Induction, and O(3) Electrodynamics from The Evans Field Theory

Summary. The Gauss Laws of magnetism and the Faraday law of induction are derived from the Evans unified field theory. The geometrical constraints imposed on the general field theory by these well known laws lead self consistently to O(3) electrodynamics.

Key words: Evans field theory, Gauss law applied to magnetism; Faraday law of induction; O(3) electrodynamics.

27.1 Introduction

In this paper the Gauss law applied to magnetism [1] and the Faraday Law of induction [1, 2] are derived from the Evans field theory [3]–[33] by the imposition of well defined constraints in differential geometry. Therefore the origin of these well known laws is traced to differential geometry and the properties of the general four dimensional manifold known as Evans spacetime. Such inferences are not possible in the Maxwell-Heaviside (MH) theory of the standard model [1, 2] because MH is not an objective theory of physics, rather it is a theory of special relativity covariant only under the Lorentz transformation. An objective theory of physics must be covariant under any coordinate transformation[1] and this is a fundamental philosophical requirement for all physics, as first realized by Einstein. This fundamental requirement is known as general relativity and the general coordinate transformation leads to general covariance in contrast to the Lorentz covariance of special relativity. The fundamental lack of objectivity in the Lorentz covariant MH theory means that it is not able to describe the important mutual effects of gravitation and electromagnetism. In contrast the Evans unified field theory is generally covariant and is a direct logical consequence of Einstein's general relativity, which is essentially the geometrization of physics. The unified field theory is able by definition to analyze the effects of gravitation on electromagnetism and vice-versa.

In Section 2 a fundamental geometrical constraint on the general field theory is derived from a consideration of the first Bianchi identity of differential geometry. It is then shown that this constraint leads to O(3) electrodynamics [3]–[33] directly from the Bianchi identity. These inferences trace the origin of the Gauss law applied to magnetism and the Faraday law of induction to differential geometry and general relativity, as required by Einsteinian natural philosophy. Section 3 is a discussion of the numerical methods needed to solve the general and restricted Evans field equations.

27.2 Geometrical Condition Needed for the Gauss Law of Magnetism and the Faraday Law of Induction and Derivation of O(3) Electrodynamics

The geometrical origin of these laws in the Evans field theory is the first Bianchi identity of differential geometry [1]:

$$D \wedge T^a = R^a{}_b \wedge q^b \tag{27.1}$$

which can be rewritten as:

$$d \wedge T^a = R^a{}_b \wedge q^b - \omega^a{}_b \wedge T^b. \tag{27.2}$$

Here T^a is the vector valued torsion two-form, $R^a{}_b$ is the tensor valued curvature or Riemann two-form; q^b is the vector valued tetrad one-form;$\omega^a{}_b$ is the spin connection, which can be regarded as a one-form [1]. The symbol $D\wedge$ denotes the covariant exterior derivative and $d\wedge$ denotes the ordinary exterior derivative.

The Bianchi identity becomes the homogeneous Evans field equation (HE) using:

$$A^a = A^{(0)} q^a, \tag{27.3}$$

$$F^a = A^{(0)} T^a. \tag{27.4}$$

Here $A^{(0)}$ is a scalar valued electromagnetic potential magnitude (whose S.I. unit is volt s / m). Thus A^a is the vector valued electromagnetic potential one-form and F^a is the vector valued electromagnetic field two-form. The HE is therefore:

$$\begin{aligned} d \wedge F^a &= R^a{}_b \wedge A^b - \omega^a{}_b \wedge F^b \\ &= \mu_0 j^a \end{aligned} \tag{27.5}$$

where:

$$j^a = \frac{1}{\mu_0} \left(R^a{}_b \wedge A^b - \omega^a{}_b \wedge F^b \right) \tag{27.6}$$

is the homogeneous current, a vector valued three-form. Here μ_0 is the S. I. vacuum permeability.

The homogeneous current is theoretically non-zero. However it is known experimentally to great precision that:

$$d \wedge F^a \sim 0. \tag{27.7}$$

Eqn. (27.7) encapsulates the two laws which are to be derived here from Evans field theory. These are usually written in vector notation as follows. The Gauss law applied to magnetism is:

$$\nabla \cdot \boldsymbol{B}^a \sim 0, \tag{27.8}$$

where $\boldsymbol{B}^a$ is magnetic flux density. The Faraday law of induction is:

$$\nabla \times \boldsymbol{E}^a + \frac{\partial \boldsymbol{B}^a}{\partial t} \sim 0 \tag{27.9}$$

where $\boldsymbol{E}^a$ is electric field strength. The index a appearing in Eqns. (27.8) and (27.9) comes from equation (27.5), i.e. from general relativity as required by Einstein. The physical meaning of a is that it indicates a basis set of the tangent bundle spacetime, a Minkowski or flat spacetime. Any basis elements (e.g. unit vectors or Pauli matrices) can be used [1]in the tangent spacetime of differential geometry, and the basis elements can be used to describe states of polarization [3]–[33], for example circular polarization first discovered experimentally by Arago in 1811. Arago was the first to observe what is now known as the two transverse states of circular polarization. It is convenient [3]–[33] to describe these states of circular polarization by the well known [34] complex circular basis:

$$a = (1)\,, (2)\ \text{and}\ (3) \tag{27.10}$$

whose unit vectors are:

$$\boldsymbol{e}^{(1)} = \frac{1}{\sqrt{2}}\left(\boldsymbol{i} - i\boldsymbol{j}\right) = \boldsymbol{e}^{(2)*}, \tag{27.11}$$

$$\boldsymbol{e}^{(3)} = \boldsymbol{k} \tag{27.12}$$

where * denotes complex conjugation. Each state of circular polarization can be described by two complex conjugates. One sense of circularly polarized radiation is described by the complex conjugates:

$$\boldsymbol{A}_1^{(1)} = \frac{A^{(0)}}{\sqrt{2}}\left(\boldsymbol{i} - i\boldsymbol{j}\right)e^{i\phi}, \tag{27.13}$$

$$\boldsymbol{A}_1^{(2)} = \frac{A^{(0)}}{\sqrt{2}}\left(\boldsymbol{i} + i\boldsymbol{j}\right)e^{-i\phi}. \tag{27.14}$$

The other sense of circularly polarized radiation is described by the complex conjugates:

$$\boldsymbol{A}_2^{(1)} = \frac{A^{(0)}}{\sqrt{2}} (\boldsymbol{i} + i\boldsymbol{j}) \, e^{i\phi}, \tag{27.15}$$

$$\boldsymbol{A}_2^{(2)} = \frac{A^{(0)}}{\sqrt{2}} (\boldsymbol{i} - i\boldsymbol{j}) \, e^{-i\phi}. \tag{27.16}$$

Here ϕ is the electromagnetic phase and Eqns. (27.13) and (27.16) are solutions of Eq. (27.7).

The experimentally observable Evans spin field $\boldsymbol{B}^{(3)}$ [3]–[33] is defined by the vector cross product of one conjugate with the other. In non-linear optics [3]–[33] this is known as the conjugate product, and is observed experimentally in the inverse Faraday effect, (IFE), the magnetization of any material matter by circularly polarized electromagnetic radiation. In the sense of circular polarization defined by Eqns. (27.13) and (27.14):

$$\boldsymbol{B}_1^{(3)} = -ig\boldsymbol{A}_1^{(1)} \times \boldsymbol{A}_1^{(2)} = B^{(0)}\boldsymbol{k} \tag{27.17}$$

where

$$g = \frac{\kappa}{A^{(0)}} \tag{27.18}$$

and where κ is the wavenumber. In the sense of circular polarization defined by Eqns. (27.15) and (27.16) $\boldsymbol{B}^{(3)}$ reverses sign:

$$\boldsymbol{B}_2^{(3)} = -ig\boldsymbol{A}_2^{(1)} \times \boldsymbol{A}_2^{(2)} = -B^{(0)}\boldsymbol{k} \tag{27.19}$$

and this is observed experimentally [3]–[33] because the observable magnetization changes sign when the handedness or sense of circular polarization is reversed. Linear polarization is the sum of 50% left and 50% right circular polarization and in this state the IFE is observed to disappear. Thus $\boldsymbol{B}^{(3)}$ in a linearly polarized beam vanishes because half the beam has positive $\boldsymbol{B}^{(3)}$ and the other half negative $\boldsymbol{B}^{(3)}$. The $\boldsymbol{B}^{(3)}$ field was first inferred by Evans in 1992 [34] and it was recognized for the first time that the phase free magnetization of the IFE is due to a third (spin) state of polarization now recognized as $a = (3)$ in the unified field theory. The Gauss law and the Faraday law of induction hold for $a = (3)$, but it is observed experimentally that $\boldsymbol{A}^{(3)} = 0$ [3]–[33]. Therefore:

$$\nabla \cdot \boldsymbol{B}^{(3)} \sim 0 \tag{27.20}$$

$$\frac{\partial \boldsymbol{B}^{(3)}}{\partial t} \sim 0. \tag{27.21}$$

The fundamental reason for this is that the spin of the electromagnetic field produces an angular momentum which is observed experimentally in the Beth effect [3]–[33]. The electromagnetic field is negative under charge conjugation symmetry (C), so the Beth angular momentum produces $\boldsymbol{B}^{(3)}$ directly, angular momentum and magnetic field being both axial vectors. The putative radiated $\boldsymbol{E}^{(3)}$ would be a polar vector if it existed, and would not be produced by spin. There is however no electric analogue of the inverse Faraday effect, a circularly polarized electromagnetic field does not produce an electric polarization experimentally, only a magnetization. Similarly, in the original Faraday effect, a static magnetic field rotates the plane of linearly polarized radiation, but a static electric field does not. The Faraday effect and the IFE are explained using the same hyperpolarizability tensor in the standard model, and in the Evans field theory by a term in the well defined Maclaurin expansion of the spin connection in terms of the tetrad, producing the IFE magnetization:

$$\boldsymbol{M}^{(3)} = -\frac{i}{\mu_0} g' \boldsymbol{A}^{(1)} \times \boldsymbol{A}^{(2)} \tag{27.22}$$

where $\boldsymbol{A}^{(1)}$ and $\boldsymbol{A}^{(2)}$ are complex conjugate tetrad elements combined into vectors [3]–[33]. Similarly all non-linear optical effects in the Evans field theory are, self consistently, properties of the Evans spacetime or general four-dimensional base manifold. The unified field theory therefore allows non-linear optics to be built up from spacetime, as required in general relativity. In the MH theory the spacetime is flat and cannot be changed, so non-linear optics must be described using constitutive relations extraneous to the original linear theory.

In the unified field theory the Evans spin field and the conjugate product are deduced self consistently from the experimental observation:

$$j^a \sim 0. \tag{27.23}$$

Eqs.(27.23) and (27.6) imply:

$$R^a{}_b \wedge q^b = \omega^a{}_b \wedge T^b \tag{27.24}$$

to high precision.In other words the Gauss law and Faraday law of induction appear to be true within contemporary experimental precision.The reason for this in general relativity (i.e. objective physics) is Eq.(27.24), a constraint of differential geometry. Using the Maurer-Cartan structure equations of differential geometry [1]:

$$T^a = D \wedge q^a \tag{27.25}$$

$$R^a{}_b = D \wedge \omega^a{}_b. \tag{27.26}$$

Eq.(27.24) becomes the following experimentally implied constraint on the general unified field theory:

$$\left(D \wedge \omega^a{}_b\right) \wedge q^b = \omega^a{}_b \wedge \left(D \wedge q^b\right). \tag{27.27}$$

A particular solution of Eq.(27.27) is:

$$\omega^a{}_b = -\kappa \epsilon^a{}_{bc} q^c \tag{27.28}$$

where $\epsilon^{a}{}_{bc}$ is the Levi-Civita tensor in the flat tangent bundle spacetime.Being a flat spacetime, Latin indices can be raised and lowered in contravariant covariant notation and so we may rewrite Eq.(27.28) as:

$$\omega_{ab} = \kappa \epsilon_{abc} q^{c}. \tag{27.29}$$

Eqn.(27.29) states that the spin connection is an antisymmetric tensor dual to the axial vector within a scalar valued factor with the dimensions of inverse metres. Thus Eq.(27.29) defines the wave-number magnitude, κ, in the unified field theory. It follows from Eqn.(27.29) that the covariant derivative defining the torsion form in the first Maurer-Cartan structure equation (27.25) can be written as:

$$T^{a} = d \wedge q^{a} + \omega^{a}{}_{b} \wedge q^{b} \tag{27.30}$$

$$= d \wedge q^{a} + \kappa q^{b} \wedge q^{c} \tag{27.31}$$

□

from which it follows,using Eqs.(27.3) and (27.4), that:

$$F^{a} = d \wedge A^{a} + g A^{b} \wedge A^{c} \tag{27.32}$$

In the complex circular basis,Eq.(27.32) can be expanded as the cyclically symmetric set of three equations:

$$F^{(1)*} = d \wedge A^{(1)*} - ig A^{(2)} \wedge A^{(3)} \tag{27.33}$$

$$F^{(2)*} = d \wedge A^{(2)*} - ig A^{(3)} \wedge A^{(1)} \tag{27.34}$$

$$F^{(3)*} = d \wedge A^{(3)*} - ig A^{(1)} \wedge A^{(2)} \tag{27.35}$$

with O(3) symmetry [3]–[33]. These are the defining relations of O(3) electrodynamics, developed by Evans from 1992 to 2003 [3]–[33]

It has been shown that O(3) electrodynamics is a direct result of the unified field theory given the experimental constraints imposed by the Gauss law and the Faraday law of induction. It follows that O(3) electrodynamics automatically produces these laws, i.e.:

$$d \wedge F^{(a)} = 0, \quad a = 1, 2, 3 \tag{27.36}$$

as observed experimentally.The existence of the conjugate product has been DEDUCED in Eq.(27.32) from differential geometry,and it follows that the spin field and the inverse Faraday effect have also been deduced from differential geometry and the Evans unified field theory of general relativity or objective physics. This is a major advance from the standard model and the MH theory of special relativity.

27.3 Numerical Methods of Solutions

In general the homogeneous and inhomogeneous Evans field equations must be solved simultaneously for given initial and boundary conditions. In this section the two equations are written out in tensor notation and subsidiary information summarized. The homogeneous field equation in tensor notation is:

$$\begin{aligned} \partial_\mu F^a{}_{\nu\rho} + \partial_\nu F^a{}_{\rho\mu} + \partial_\rho F^a{}_{\mu\nu} &= R^a{}_{b\mu\nu} A^b{}_\rho + R^a{}_{b\nu\rho} A^b{}_\mu + R^a{}_{b\rho\mu} A^b{}_\nu \\ &\quad - \omega^a{}_{b\mu} F^b{}_{\nu\rho} - \omega^a{}_{b\nu} F^b{}_{\rho\mu} - \omega^a{}_{b\rho} F^b{}_{\mu\nu} \end{aligned} \tag{27.37}$$

and this is equivalent to the barebones or minimalist notation of differential geometry:

$$d \wedge F = R \wedge A - \omega \wedge F \tag{27.38}$$

where all indices have been suppressed. For the purposes of electrical engineering Eq.(27.37) is,to an excellent approximation:

$$\partial_\mu F^a{}_{\nu\rho} + \partial_\nu F^a{}_{\rho\mu} + \partial_\rho F^a{}_{\mu\nu} = 0 \tag{27.39}$$

Eq.(27.39) is equivalent to the Gauss Law applied to magnetism:

$$\nabla \cdot \boldsymbol{B}^a = 0 \tag{27.40}$$

and the Faraday Law of induction:

$$\nabla \times \boldsymbol{B}^a + \frac{\partial \boldsymbol{E}^a}{\partial t} = 0 \tag{27.41}$$

for all polarization states a.

The exceedingly important influence of gravitation on electromagnetism and vice versa must however be computed in general when the right hand side of Eq.(27.37) is non-zero. This tiny but in general non-zero influence leads to a violation of the well known laws (27.40) and (27.41), and this must be searched for with high precision instrumentation. Eq.(27.39) may be rewritten as the Hodge dual equation:

$$\partial^\mu \widetilde{F}^a{}_{\mu\nu} = 0 \tag{27.42}$$

and for each index a this is a homogeneous Maxwell-Heaviside field equation. In general $\widetilde{F}^a_{\mu\nu}$ is the Hodge dual of $F^a_{\mu\nu}$ in Evans spacetime and $\widetilde{j}^a_\nu$ is the Hodge dual of the charge-current density three-form defined by the right hand side of Eq.(27.37), i.e.by:

$$\partial_\mu F^a{}_{\nu\rho} + \partial_\nu F^a{}_{\rho\mu} + \partial_\rho F^a{}_{\mu\nu} = \mu_0 \left(j^a{}_{\mu\nu\rho} + j^a{}_{\nu\rho\mu} + j^a{}_{\rho\mu\nu} \right) \tag{27.43}$$

where:

$$j^a{}_{\mu\nu\rho} = \frac{1}{\mu_0}\left(R^a{}_{b\mu\nu}A^b{}_\rho - \omega^a{}_{\mu b}F^b{}_{\nu\rho}\right) \tag{27.44}$$

etc.

Thus:

$$\begin{aligned} \widetilde{j}^a{}_\sigma &= \frac{1}{6}\epsilon^{\mu\nu\rho}{}_\sigma j^a{}_{\mu\nu\rho} \\ &= \frac{1}{6}\epsilon^{\nu\rho\mu}{}_\sigma j^a{}_{\nu\rho\mu} \\ &= \frac{1}{6}\epsilon^{\rho\mu\nu}{}_\sigma j^a{}_{\rho\mu\nu} \end{aligned} \tag{27.45}$$

and

$$\begin{aligned} \widetilde{F}^a{}_{\mu\sigma} &= \frac{1}{2}\epsilon^{\nu\rho}{}_{\mu\sigma}F^a{}_{\nu\rho}, \\ \widetilde{F}^a{}_{\nu\sigma} &= \frac{1}{2}\epsilon^{\rho\mu}{}_{\nu\sigma}F^a{}_{\rho\mu}, \\ \widetilde{F}^a{}_{\rho\sigma} &= \frac{1}{2}\epsilon^{\mu\nu}{}_{\rho\sigma}F^a{}_{\mu\nu}. \end{aligned} \tag{27.46}$$

In computing these Hodge duals the correct general definition maps from a p-form of differential geometry to an (n-p)-form of differential geometry in the general n dimensional manifold. The general four dimensional manifold is the Evans spacetime, so named to distinguish it from the Riemannian spacetime used in Einstein's field theory of gravitation. The Hodge dual is in general [1]:

$$\widetilde{\mathcal{X}}_{\mu_1\ldots\mu_{n-p}} = \frac{1}{p!}\epsilon^{\nu_1\ldots\nu_p}{}_{\mu_1\ldots\mu_{n-p}}\mathcal{X}_{\nu_1\ldots\nu_p}. \tag{27.47}$$

The general Levi-Civita symbol is:

$$\epsilon'_{\mu_1\mu_2\ldots\mu_n} = \begin{cases} 1 \text{ if } \mu_1\mu_2\ldots\mu_n \text{ is an even permutation} \\ -1 \text{ if } \mu_1\mu_2\ldots\mu_n \text{ is an odd permutation} \\ 0 \qquad\qquad \text{otherwise} \end{cases} \tag{27.48}$$

and the Levi-Civita tensor used in Eq.(27.47) is:

$$\epsilon_{\mu_1\mu_2\ldots\mu_n} = (|g|)^{1/2}\,\epsilon'_{\mu_1\mu_2\ldots\mu_n} \tag{27.49}$$

where $|g|$ is the numerical value (i.e. a number) of the determinant of the metric tensor $g_{\mu\nu}$. The field tensor is defined by the torsion tensor:

$$F^a{}_{\mu\nu} = A^{(0)}T^a{}_{\mu\nu} \tag{27.50}$$

and the potential is defined by the tetrad:

$$A^a{}_\mu = A^{(0)} q^a{}_\mu . \tag{27.51}$$

The metric is factorized into a dot product of tetrads:

$$g_{\mu\nu} = q^a{}_\mu q^b{}_\nu \eta_{ab} \tag{27.52}$$

where:

$$\eta_{ab} = \begin{bmatrix} 1 & 0 & 0 & 0 \\ 0 & -1 & 0 & 0 \\ 0 & 0 & -1 & 0 \\ 0 & 0 & 0 & -1 \end{bmatrix} \tag{27.53}$$

is the diagonal metric tensor of the tangent bundle spacetime, a Minkowski or flat spacetime. The gamma and spin connections are related by the tetrad postulate:

$$D_\mu q^a{}_\nu = \partial_\mu q^a{}_\nu + \omega^a{}_{\mu b} q^b{}_\nu - \Gamma^\lambda{}_{\mu\nu} q^a{}_\lambda = 0. \tag{27.54}$$

The torsion and curvature (or Riemann) tensors are defined by the Maurer-Cartan structure relations of differential geometry. The torsion tensor is the covariant derivative of the tetrad and is:

$$T^\lambda{}_{\mu\nu} = q^\lambda{}_a T^a{}_{\mu\nu} = \Gamma^\lambda{}_{\mu\nu} - \Gamma^\lambda{}_{\nu\mu} . \tag{27.55}$$

It therefore vanishes for the Christoffel connection of Einstein's gravitational theory:

$$\Gamma^\lambda{}_{\mu\nu} = \Gamma^\lambda{}_{\nu\mu} . \tag{27.56}$$

The curvature tensor is the covariant derivative of the spin connection and is:

$$R^\sigma{}_{\lambda\nu\mu} = q^\sigma{}_a q^b{}_\lambda R^a{}_{b\nu\mu} . \tag{27.57}$$

It follows that:

$$R^a{}_{b\nu\mu} = \partial_\nu \omega^a{}_{\mu b} - \partial_\mu \omega^a{}_{\nu b} + \omega^a{}_{\nu c} \omega^c{}_{\mu b} - \omega^a{}_{\mu c} \omega^c{}_{\nu b} \tag{27.58}$$

and

$$R^\sigma{}_{\lambda\nu\mu} = \partial_\nu \Gamma^\sigma{}_{\mu\lambda} - \partial_\mu \Gamma^\sigma{}_{\nu\lambda} + \Gamma^\sigma{}_{\nu\rho} \Gamma^\rho{}_{\mu\lambda} - \Gamma^\sigma{}_{\mu\rho} \Gamma^\rho{}_{\nu\lambda} . \tag{27.59}$$

The Evans spacetime is therefore completely defined by the structure relations. The homogeneous field equation is the first Bianchi identity of differential geometry within the C negative factor $A^{(0)}$. The second Bianchi identity leads to the Noether Theorem of Evans field theory, and states that the covariant derivative of the curvature tensor vanishes identically.

Note that Eq.(27.39) is equivalent to the use of Riemann normal coordinates and a locally flat spacetime, because it is an equation in ordinary derivatives, not covariant derivatives. In general, $|g|$, the modulus of the determinant of the metric, in Eq.(27.49) is a function of x^μ, but at a point p in the Evans spacetime or manifold M it is always possible to define the Riemann normal coordinate system so the metric is in canonical form, and:

$$\partial_\mu g = 0. \tag{27.60}$$

This defines the locally flat spacetime. Note carefully that the general homogeneous equation (27.37) cannot be expressed as a Hodge dual equation of type (27.39), so the appropriate equation for numerical solution must always be Eq.(27.37). Eq.(27.39) is mentioned only because of the traditional method of expressing the homogeneous Maxwell-Heaviside equation of Minkowski spacetime (HME) as the Hodge dual equation. The correct form of the HME is the following Bianchi identity of Minkowski spacetime:

$$\partial_\lambda F_{\mu\nu} + \partial_\mu F_{\nu\lambda} + \partial_\nu F_{\lambda\mu} = 0 \tag{27.61}$$

a spacetime in which the Hodge dual is definable as:

$$\widetilde{F}^{\mu\nu} = \frac{1}{2}\epsilon^{\mu\nu\rho\sigma} F_{\rho\sigma}. \tag{27.62}$$

It may then be proven that Eq.(27.61) is the same equation as:

$$\partial_\mu \widetilde{F}^{\mu\nu} = 0. \tag{27.63}$$

The proof is as follows. From Eq.(27.63):

$$\partial_\lambda \widetilde{F}^{\lambda\rho} + \partial_\mu \widetilde{F}^{\mu\rho} + \partial_\nu \widetilde{F}^{\nu\rho} = 0 \tag{27.64}$$

and using Eq.(27.62):

$$\frac{1}{2}\left(\partial_\lambda\left(\epsilon^{\lambda\rho\mu\nu}F_{\mu\nu}\right) + \partial_\mu\left(\epsilon^{\mu\rho\nu\lambda}F_{\nu\lambda}\right) + \partial_\nu\left(\epsilon^{\nu\rho\lambda\mu}F_{\lambda\mu}\right)\right) = 0. \tag{27.65}$$

Using the Leibniz Theorem and the constancy of $\epsilon^{\mu\nu\rho\sigma}$ in Minkowski spacetime, Eq.(27.65) becomes:

$$\epsilon^{\lambda\rho\mu\nu}\partial_\lambda F_{\mu\nu} + \epsilon^{\mu\rho\nu\lambda}\partial_\mu F_{\nu\lambda} + \epsilon^{\nu\rho\lambda\mu}\partial_\nu F_{\lambda\mu} = 0. \tag{27.66}$$

We may add individual indices of Eq.(27.66), to give, for example:

$$\epsilon^{1\rho 23}\partial_1 F_{23} + \epsilon^{2\rho 31}\partial_2 F_{31} + \epsilon^{3\rho 12}\partial_3 F_{12} + \cdots = 0 \tag{27.67}$$

which is

$$\partial_1 F_{23} + \partial_2 F_{31} + \partial_3 F_{12} + \cdots = 0 \tag{27.68}$$

upon using:

$$\epsilon^{1023} = \epsilon^{2031} = \epsilon^{3012} = -1. \tag{27.69}$$

Proceeding in this way we see that Eq.(27.61) is the same as Eq.(27.63).Note carefully that this proof is not true in general for Evans spacetime, because in that spacetime we obtain results such as:

$$\partial^{\mu}\widetilde{F}^{a}{}_{\mu\sigma}=\frac{1}{2}\partial^{\mu}\left(\epsilon^{\nu\rho}{}_{\mu\sigma}F^{a}{}_{\nu\rho}\right)=\frac{1}{2}\left(\epsilon^{\nu\rho}{}_{\mu\sigma}\partial^{\mu}F^{a}{}_{\nu\rho}+\left(\partial^{\mu}\epsilon^{\nu\rho}{}_{\mu\sigma}\right)F^{a}{}_{\nu\rho}\right) \quad (27.70)$$

and since depends on, one obtains:

$$\partial^{\mu}\epsilon^{\nu\rho}{}_{\mu\sigma}\neq 0 \quad (27.71)$$

and there is an extra term which does not appear in Eq.(27.63) of Minkowski spacetime. Only in the special case of Eq.(27.62) do we obtain:

$$\partial^{\mu}\epsilon^{\nu\rho}{}_{\mu\sigma}=0. \quad (27.72)$$

Therefore in general, the homogeneous Evans equation (27.37) must be solved simultaneously with the inhomogeneous Evans equation.The latter is an expression for the covariant exterior derivative of the Hodge dual of $F^{a}_{\mu\nu}$. In the minimalist or barebones notation of differential geometry this expression is:

$$\begin{aligned} d\wedge\widetilde{F}&=\widetilde{R}\wedge A-\omega\wedge\widetilde{F}\\ &=\mu_0 J \end{aligned} \quad (27.73)$$

where $\widetilde{R}$ denotes the Hodge dual of the curvature form. Eq.(27.73) is the objective or generally covariant expression in unified field theory of the inhomogeneous Maxwell-Heaviside equation (IMH) of special relativity:

$$d\wedge\widetilde{F}=\mu_0 J. \quad (27.74)$$

It is seen by comparison of Eqs.(27.73) and (27.74) that in the unified field theory $d\wedge$ has been replaced by $D\wedge$ as required. Since both F and R in Eq.(27.38) are two-forms it follows by symmetry that if one takes the Hodge dual of F on the left hand side, one must take the Hodge dual of R on the right hand side. The reason is that the Hodge dual of any two-form in Evans spacetime is always another two-form. It is convenient to rewrite Eq.(27.73) as:

$$D\wedge\widetilde{F}=\widetilde{R}\wedge A \quad (27.75)$$

and in analogy with Eq.(27.37), the tensorial expression for Eq.(27.73) is:

$$\begin{aligned} \partial_{\mu}\widetilde{F}^{a}{}_{\nu\rho}+\partial_{\nu}\widetilde{F}^{a}{}_{\rho\mu}+\partial_{\rho}\widetilde{F}^{a}{}_{\mu\nu}&=\widetilde{R}^{a}{}_{b\mu\nu}A^{b}{}_{\rho}+\widetilde{R}^{a}{}_{b\nu\rho}A^{b}{}_{\mu}+\widetilde{R}^{a}{}_{b\rho\mu}A^{b}{}_{\nu}\\ &\quad-\omega^{a}{}_{b\mu}\widetilde{F}^{b}{}_{\nu\rho}-\omega^{a}{}_{b\nu}\widetilde{F}^{b}{}_{\rho\mu}-\omega^{a}{}_{b\rho}\widetilde{F}^{b}{}_{\mu\nu}. \end{aligned} \quad (27.76)$$

Therefore the general computational task is to solve Eqs.(27.37) and (27.76) simultaneously for given initial and boundary conditions. In the last analysis this is a problem in simultaneous partial differential equations. We may now define the inhomogeneous current

$$\begin{aligned} J_{\mu\nu\rho}&=\tfrac{1}{\mu_0}\left(\widetilde{R}^{a}{}_{b\mu\nu}A^{b}{}_{\rho}-\omega^{a}{}_{\mu b}\widetilde{F}^{b}{}_{\nu\rho}\right)\\ &\vdots \end{aligned} \quad (27.77)$$

and we may note that $J_{\mu\nu\rho}$ is in general much larger than the homogeneous current $j_{\mu\nu\rho}$. Therefore $J_{\mu\nu\rho}$ is of great practical importance for the acquisition of electric power from Evans spacetime and for counter gravitational technology in the aerospace industry.

Finally a convenient form of the inhomogeneous field equation may be obtained from the following considerations.We first construct the following Hodge dual:

$$\widetilde{F}^{\mu\nu} = \frac{1}{2}\epsilon^{\mu\nu\rho\sigma}F_{\rho\sigma} \tag{27.78}$$

in the MH limit:

$$d \wedge F = 0 \tag{27.79}$$

$$d \wedge \widetilde{F} = \mu_0 J \tag{27.80}$$

and note that:

$$\left(d \wedge \widetilde{F}\right)^{\mu\nu} \neq \frac{1}{2}\epsilon^{\mu\nu\rho\sigma}(d \wedge F)_{\rho\sigma}. \tag{27.81}$$

In convenient shorthand notation this result may be written as:

$$d \wedge \widetilde{F} \neq \frac{1}{2}\epsilon^{'} d \wedge F \tag{27.82}$$

a result which becomes:

$$d \wedge \widetilde{F} \neq \frac{1}{2}|g|^{\frac{1}{2}}\epsilon^{'} d \wedge F \tag{27.83}$$

in Evans spacetime. This is a key result because it shows that J^a is not zero if j^a is zero or almost zero, and it is J^a that is the important current term for the acquisition of electric power from Evans spacetime.

For the purposes of computation a systematic method of constructing the inhomogeneous Evans field equation (IE) is needed, where every term needed for coding is defined precisely. In order to do this begin with the fundamental definitions of differential geometry, the two Maurer-Cartan structure equations:

$$T^a = D \wedge q^a \tag{27.84}$$

$$R^a{}_b = D \wedge \omega^a{}_b \tag{27.85}$$

and the two Bianchi identities:

$$D \wedge T^a = R^a{}_b \wedge q^b \tag{27.86}$$

$$D \wedge R^a{}_b = 0. \tag{27.87}$$

In order to correctly construct the Hodge duals of T^a and R^a_b appearing in the IE the determinant of the metric must be defined correctly in each case (see Eq.(27.49)). In order to proceed consider the limit of the Evans unified field theory that gives Einstein's field theory of gravitation uninfluenced by electromagnetism. In the Einstein limit the metric tensor is symmetric and is defined by the inner or dot product of two tetrads:

$$g_{\mu\nu}{}^{(S)} = q^a{}_\mu{}^{(S)} q^b{}_\nu{}^{(S)} \eta_{ab}. \tag{27.88}$$

The differential geometry appropriate to the Einstein theory is then:

$$T^{a(S)} = 0 \tag{27.89}$$

$$d \wedge q^{a(S)} = -\omega^a{}_b{}^{(S)} \wedge q^{b(S)}, \tag{27.90}$$

$$R^a{}_b{}^{(S)} \wedge q^{b(S)} = 0. \tag{27.91}$$

The determinant of the metric is defined in this limit by:

$$g^{(S)} = |g_{\mu\nu}{}^{(S)}| \tag{27.92}$$

and so the Hodge dual of the Riemann form is defined in Einstein's theory of gravitation by:

$$\widetilde{R}^a{}_b{}^{(S)} = \frac{1}{2}|g_{\mu\nu}{}^{(S)}|^{\frac{1}{2}}\epsilon' R^a{}_b{}^{(S)}. \tag{27.93}$$

Consider next the limit of the Evans field theory that gives the free electromagnetic field when there is no field matter interaction, matter being defined by the presence of non-zero mass. The differential geometry that defines this limit is:

$$T^{a(A)} = D \wedge q^{a(A)}, \tag{27.94}$$

$$R^a{}_b{}^{(A)} = D \wedge \omega^a{}_b{}^{(A)}, \tag{27.95}$$

$$g^c{}_{\mu\nu}{}^{(A)} = q^a{}_\mu{}^{(A)} \wedge q^b{}_\nu{}^{(A)}, \tag{27.96}$$

and the determinant of the metric is:

$$g^{(A)} = |g^c{}_{\mu\nu}{}^{(A)}|. \tag{27.97}$$

The Hodge duals of the torsion and Riemann forms for the free electromagnetic field are therefore:

$$\widetilde{T}^{a(A)} = \frac{1}{2}|g^{(A)}|^{\frac{1}{2}}\epsilon' T^{a(A)}, \tag{27.98}$$

$$\widetilde{R}^a{}_b{}^{(A)} = \frac{1}{2}|g^{(A)}|^{\frac{1}{2}}\epsilon' R^a{}_b{}^{(A)}. \tag{27.99}$$

Thirdly, when, the electromagnetic field interacts with matter, as in the IE, the appropriate differential geometry is:

$$T^a = D \wedge q^a, \tag{27.100}$$

$$R^a{}_b = D \wedge \omega^a{}_b, \tag{27.101}$$

$$g^{ab}{}_{\mu\nu} = q^a{}_\mu q^b{}_\nu, \tag{27.102}$$

and the determinant of the metric tensor is now:

$$g = |g^{ab}{}_{\mu\nu}|. \tag{27.103}$$

The metric tensor itself is the sum of symmetric and anti-symmetric component metric tensors:

$$q^a{}_\mu q^b{}_\nu = \frac{1}{2}\left(\left(q^a{}_\mu q^b{}_\nu\right)^{(S)} + \left(q^a{}_\mu q^b{}_\nu\right)^{(A)}\right). \tag{27.104}$$

The Hodge dual of the Riemann form in the IE is therefore:

$$\widetilde{R}^a{}_b = \frac{1}{2}|g|^{\frac{1}{2}}\epsilon' R^a{}_b, \tag{27.105}$$

because in general both the symmetric and antisymmetric metrics contribute to the Riemann tensor or Riemann form when there is field matter interaction

However, the Hodge dual of the torsion form in the IE is:

$$\widetilde{T}^a = \frac{1}{2}|g^{(A)}|^{\frac{1}{2}}\epsilon' T^a, \tag{27.106}$$

because only the antisymmetric metric contributes to the torsion tensor or torsion form from Eq.(27.94).

Therefore the computational algorithm is fully defined, and the computational task in general is to solve the HE and IE SIMULTANEOUSLY for given initial and boundary conditions. The HE and IE contain more information than the equivalents in Maxwell-Heaviside field theory, and the engineering task is to CAD/CAM a circuit taking electric power from Evans spacetime, defined by the general four dimensional manifold in which the Riemann and torsion tensors are both non-zero. In the Minkowski spacetime of Maxwell-Heaviside field theory both tensors are zero. We must use the computer to define this extra source of power and to optimize circuits which utilize this extra source of power in practical devices.

Acknowledgments

The Ted Annis Foundation, Craddock Inc., John B.Hart and other scholars are thanked for funding this work, and the AIAS group for many interesting discussions.

References

1. S.P.Carroll., *Lecture Notes in General Relativity* (a graduate course at Harvard, UC Santa Barbara and U.Chicago, public domain arXic:gr-gq 973019 v1 Dec 1997, public domain).
2. L.H.Ryder, *Quantum Field Theory*, (Cambridge Univ Press,1996,2nd ed.).
3. M.W.Evans, A Unified Field Theory for Gravitation and Electromagnetism, *Found.Phys.Lett.*, **16**, 367 (2003).
4. M.W.Evans, A Generally Covariant Wave Equation for Grand Unified Field Theory, *Found.Phys.Lett.*, **16**, 507 (2003).
5. M.W.Evans, The Equations of Grand Unified Field Theory in Terms of the Maurer-Cartan Structure relations of Differential Geometry, *Found.Phys.Lett.*,**17**,25 (2004).
6. M.W.Evans, Derivation of Dirac's Equation from the Evans Wave Equation, *Found.Phys.Lett.*,**17**,149 (2004).
7. M.W.Evans, Unification of the Gravitational and Strong Nuclear Fields, *Found.Phys.Lett.*, **17**,267 (2004).
8. M.W.Evans, The Evans Lemma of Differential Geometry, *Found.Phys.Lett.*, **17**,433 (2004).
9. M.W.Evans, Derivation of the Evans Wave Equation form the Lagrangian and Action,Origin of the Planck Constant in General Relativity, *Found.Phys.Lett.*,**17**,535 (2004).
10. M.W.Evans et. al. (AIAS Author Group), Development of the Evans Wave Equation in the Weak Field Limit: the Electrogravitic Equation, *Found.Phys.Lett.*, **17**,497 (2004).
11. M.W.Evans, Physical Optics, the Sagnac Effect and the Aharonov Bohm Effect in the Evans Unified Field Theory, *Found.Phys.Lett.*, **17**, 301 (2004).
12. M.W.Evans, Derivation of the Geometrical Phase from the Evans Phase Law of Generally Covariant Unified Field Theory, *Found.Phys.Lett.*,**17**,393 (2004).
13. M.W.Evans, Derivation of the Lorentz Boost from the Evans Wave Equation, *Found.Phys.Lett.*,**17**,663 (2004).
14. M.W.Evans,The Electromagnetic Sector of the Evans Field Theory, *Found.Phys.Lett.*, in press (2005),preprint on www.aias.us.
15. M.W.Evans, New Concepts from the Evans Field Theory,Part One:The Evolution of Curvature, Oscillatory Universe without Singularity and general Force and Field Equations, *Found.Phys.Lett.*, in press (2005),preprint on www.aias.us

16. M.W.Evans, New Concepts from the Evans Field Theory,Part Two: Derivation of the Heisenberg Equation and Reinterpretation of the Heisenberg Uncertainty Principle, *Found.Phys.Lett.*, in press (2005), preprint on www.aias.us.
17. M.W.Evans, The Spinning and Curving of Spacetime, the Electromagnetic and Gravitational Fields in the Evans Unified Field Theory, *Found.Phys.Lett.*, in press,(2005), preprint on www.aias.us.
18. M.W.Evans, Derivation of O(3) Electrodynamics from Generally Covariant Unified Field Theory, *Found.Phys.Lett.*, in press, preprint on www.aias.us.
19. M.W.Evans, Derivation of O(3) Electrodynamics form the Evans Unified Field Theory, *Found.Phys.Lett.*, in press, preprint on www.aias.us.
20. M.W.Evans, Calculation of the Anomalous Magnetic Moment of the Electron, *Found.Phys.Lett.*, in press, preprint on www.aias.us.
21. M.W.Evans, Generally Covariant Electro-weak Theory, *Found.Phys.Lett.*, in press, preprint on www.aias.us.
22. M.W.Evans, Evans Field Theory of Neutrino Oscillations, *Found.Phys.Lett.*, in press, preprint on www.aias.us.
23. M.W.Evans, The Interaction of Gravitation and Electromagnetism, *Found.Phys.Lett.*, preprint on www.aias.us.
24. M.W.Evans, Electromagnetic Energy from Gravitation, *Found.Phys.Lett.*, in press, preprint on www.aias.us.
25. M.W.Evans, The Fundamental Invariants of the Evans Field Theory, *Found.Phys.Lett.*, in press, preprint on www.aias.us.
26. M.W.Evans, The Homogeneous and Inhomogeneous Evans Field Equations, *Found.Phys.Lett.*, in press, preprint on www.aias.us.
27. M.W.Evans, *Generally Covariant Unified Field Theory: the Geometrization of Physics*, (in press, 2005), preprint on www.aias.us.
28. L.Felker, *The Evans Equations*, (freshman text in prep.).
29. M.Anderson, Artspeed Website for the Collected Works of Myron Evans, www.myronevanscollectedworks.com.
30. M.W.Evans and L.B.Crowell, *Classical and Quantum Electrodynamics and the* $\boldsymbol{B}^{(3)}$ *Field* (World Scientific,Singapore,2001).
31. M.W.Evans (ed.), *Modern Nonlinear Optics*, a special topical issue in three parts of I.Prigogine and S.A.Rice (series eds.), *Advances in Chemical Physics*, (Wile Interscience, New York, 2001, 2nd ed., and 1992,1993,1997,1st ed.), vols 119 and 85.
32. M.W.Evans, J.-P.Vigier et al, *The Enigmatic Photon*, (Kluwer,Dordrecht,1994 to 2002,hardback and softback), in five volumes.
33. M.W.Evans and A.A.Hasanein, *The Photomagneton in Quantum Field Theory* (World Scientific,Singapore,1994)
34. M.W.Evans, Physica B, **182**,227,237 (1992).

Part III

Technical References

A

Evans Reference Charts

A.1 Evans Unified Field Theory

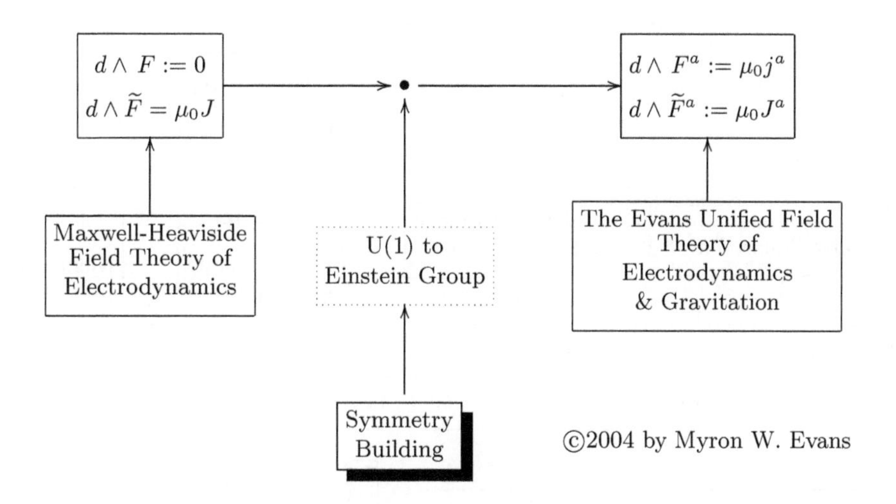

Fig. A.1. Evans Unified Field Theory

A.2 Evans Unified Field Theory

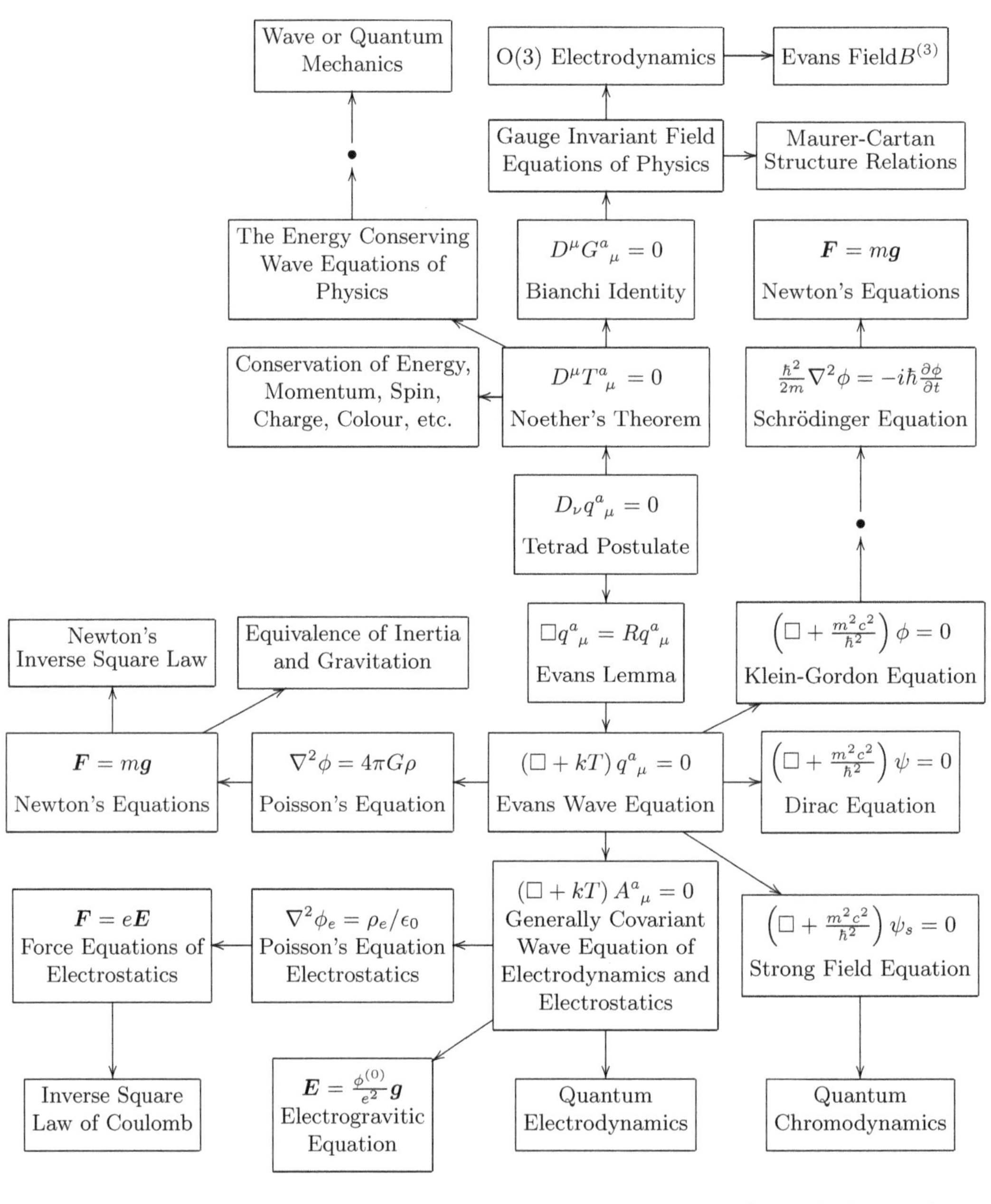

Fig. A.2. Evans Unified Field Theory

©2004 by Myron Evans

A.3 Evans Unified Field Theory

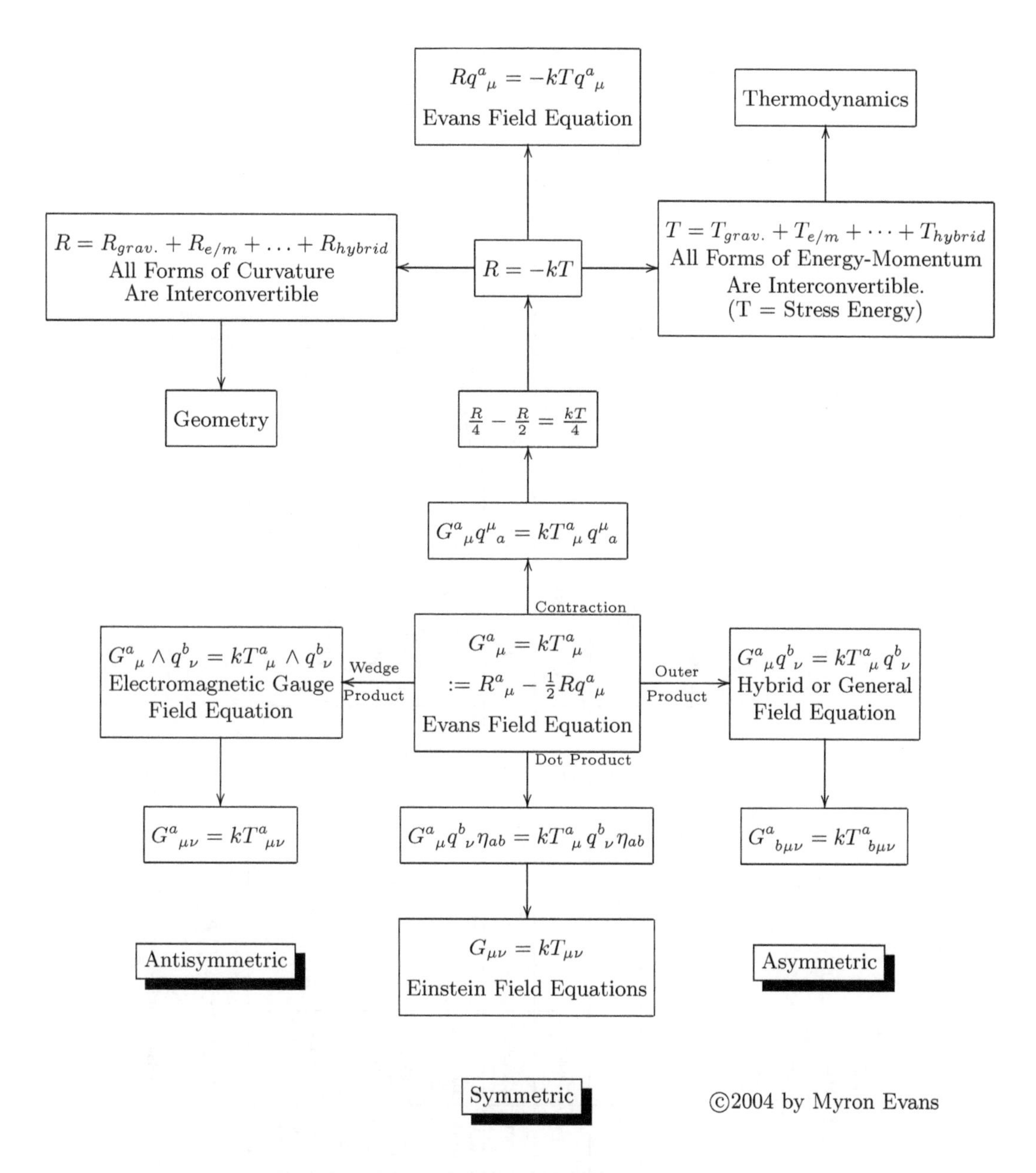

Fig. A.3. Evans Unified Field Theory

A.4 Table: Field Theories of General Relativity

TABLE: FIELD THEORIES OF GENERAL RELATIVITY

ORIGIN	TYPE	POTENTIAL FIELD	GAUGE FIELD	FIELD EQUATION (CLASSICAL MECHANICS)	WAVE EQUATION (QUANTUM MECHAINS)	CONTRACTED ENERGY / MOMENTUM	SCALAR CURVATURE
Einstein / Hilbert (1915)	Central Gravitational	$q_\mu^{a\,(S)}$	$R_{b\mu\nu}^{a\ \ \ (A)}$	$R_\mu^{a\,(S)} - \frac{R}{2} q_\mu^{a\,(S)} = k\, T_\mu^{a\,(S)}$	$(\square + kT) q_\mu^{a\,(S)} = 0$ Evans (2003)	T_{grav} (gravitation)	R_{grav} (gravitation)
Evans (2003)	Unified	q_μ^a	$R_{b\mu\nu}^a$	$R_\mu^a - \frac{R}{2} q_\mu^a = k\, T_\mu^a$	$(\square + kT) q_\mu^a = 0$	T_{unified} (hybrid energy)	R_{unified} (hybrid energy)
Evans (2004)	Dark Matter	$q_\mu^{a\,(A)}$	$\tau_{\mu\nu}^c$	$R_\mu^{a\,(A)} - \frac{R}{2} q_\mu^{a\,(A)} = k\, T_\mu^{a\,(A)}$	$(\square + kT) q_\mu^{a\,(A)} = 0$	T_{dark} (dark energy)	R_{dark} (dark energy)
Evans (2003) Evans (2004)	Electro-dynamics	$A_\mu^{a\,(A)} = A^{(0)}\, q_\mu^{a\,(A)}$	$A^{(0)}\, \tau_{\mu\nu}^c$	$G_\mu^{a\,(A)} = A^{(0)} k\, T_\mu^{a\,(A)}$ $= A^{(0)} \left(R_\mu^{a\,(A)} - \frac{R}{2} q_\mu^{a\,(A)} \right)$	$(\square + kT) A_\mu^{a\,(A)} = 0$	$T_{\text{e/m}}$ (electro-dynamic)	$R_{\text{e/m}}$ (electro-dynamic)
Evans (2003) Evans (2004)	Electro-statics	$A_\mu^{a\,(S)} = A^{(0)}\, q_\mu^{a\,(S)}$	$A^{(0)}\, R_{b\mu\nu}^{a\ \ \ (A)}$	$G_\mu^{a\,(S)} = A^{(0)} k\, T_\mu^{a\,(S)}$ $= A^{(0)} \left(R_\mu^{a\,(S)} - \frac{R}{2} q_\mu^{a\,(S)} \right)$	$(\square + kT) A_\mu^{a\,(S)} = 0$	$T_{\text{e/s}}$ (electrostatic)	$R_{\text{e/s}}$ (electrostatic)

1) Duality: $\tau^c = \varepsilon_a^{cb}\, R_{\ b}^{a\,(A)}$

2) Basic Matrix property: $q_\mu^a = q_\mu^{a\,(S)} + q_\mu^{a\,(A)}$

Fig. A.4. Field Theories of General Relativity

A.5 Development of Maxwell-Heaviside Field Theory Into Generally Covariant Electrodynamics

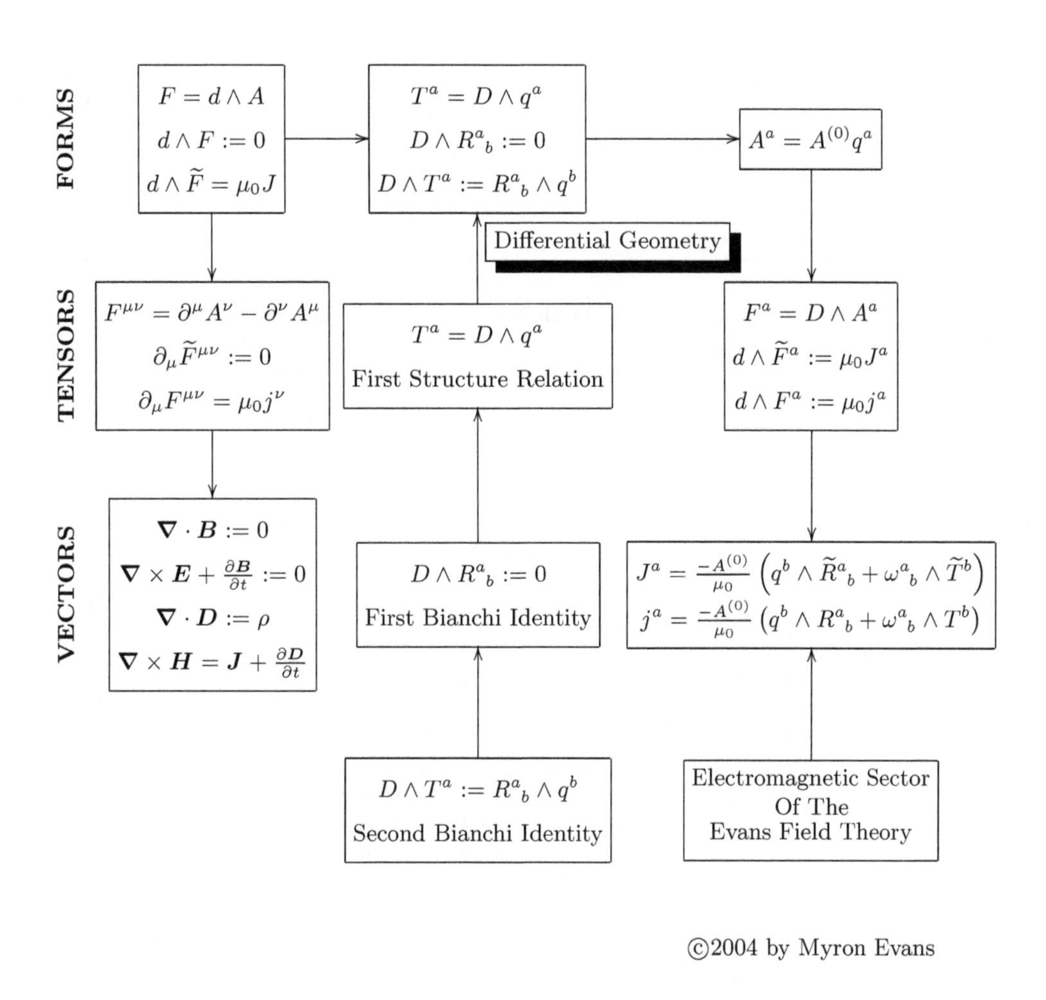

Fig. A.5. MH Theory into Generally Covariant Electrodynamics

A.6 Evans Unified Field Theory

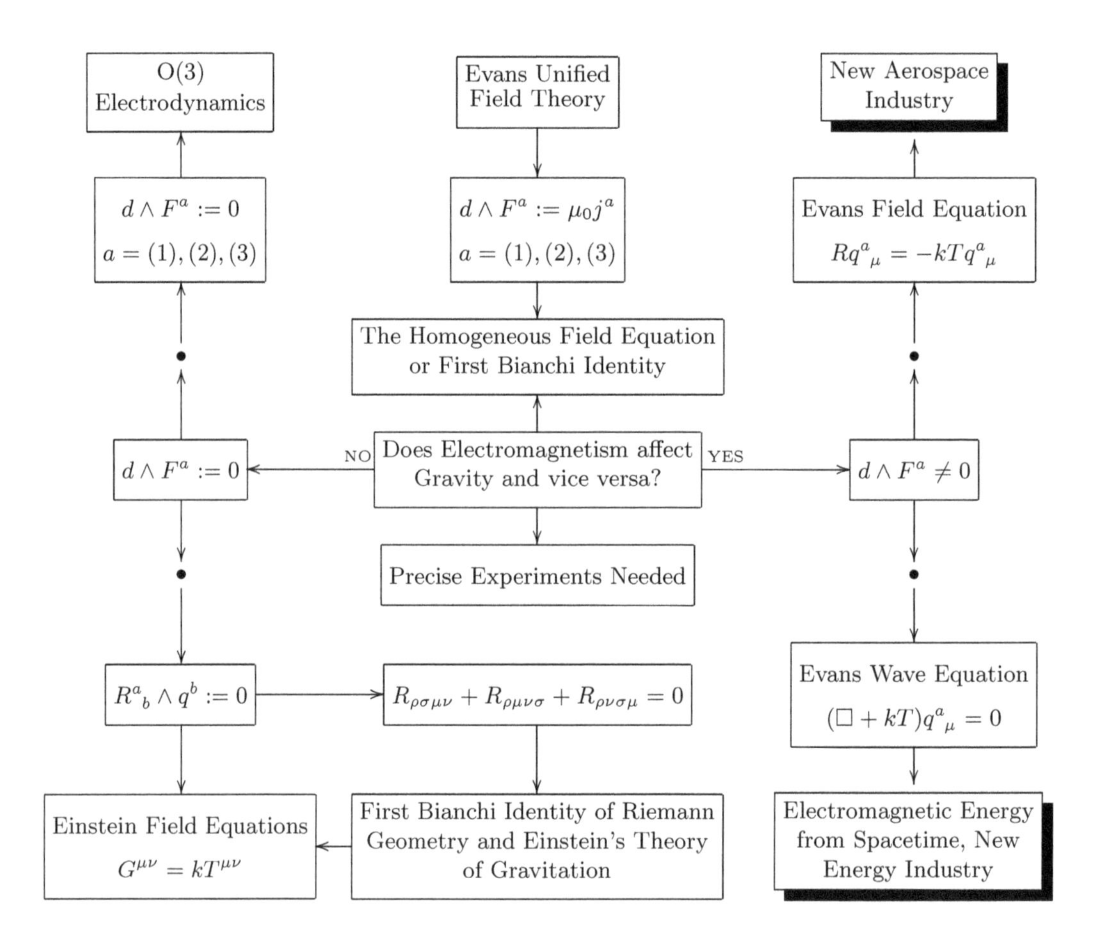

Fig. A.6. Evans Unified Field Theory

B

Standard Tensor Formulation of the Inhomogeneous and Homogeneous Maxwell-Heaviside Field Equations

B.1 Inhomogeneous Field Equation

$$\partial_\mu F^{\mu\nu} = \mu_0 c J^\nu \tag{B.1}$$

where

$$F^{\mu\nu} = \begin{bmatrix} 0 & -E^1 & -E^2 & -E^3 \\ E^1 & 0 & -cB^3 & cB^2 \\ E^2 & cB^3 & 0 & -cB^1 \\ E^3 & -cB^2 & cB^1 & 0 \end{bmatrix}. \tag{B.2}$$

B.1.1 Coulomb's Law($\nu = 0$)

$$\partial_1 F^{10} + \partial_2 F^{20} + \partial_3 F^{30} = \mu_0 c J^0. \tag{B.3}$$

B.1.2 Ampère-Maxwell Law ($\nu = 1, 2, 3$)

$$\partial_0 F^{01} + \partial_2 F^{21} + \partial_3 F^{31} = \mu_0 c J^1 \tag{B.4}$$

$$\partial_0 F^{02} + \partial_1 F^{12} + \partial_3 F^{32} = \mu_0 c J^2 \tag{B.5}$$

$$\partial_0 F^{03} + \partial_1 F^{13} + \partial_2 F^{23} = \mu_0 c J^3. \tag{B.6}$$

B.1.3 Coulomb Law

From eqn.(B.3)

$$\boxed{\nabla \cdot \boldsymbol{E} = \mu_0 c J^0 = \mu_0 c^2 \rho = \frac{\rho}{\epsilon_0}} \tag{B.7}$$

where

$$J^\mu = (c\rho, \boldsymbol{J}) \tag{B.8}$$

$$\mu_0 \epsilon_0 = \frac{1}{c^2}. \tag{B.9}$$

B.1.4 Ampère-Maxwell Law

$$-\partial_0 E^1 + c(\partial_2 B^3 - \partial_3 B^2) = \mu_0 c J^1 \tag{B.10}$$

$$-\partial_0 E^2 - c(\partial_1 B^3 - \partial_3 B^1) = \mu_0 c J^2 \tag{B.11}$$

$$-\partial_0 E^3 + c(\partial_1 B^2 - \partial_2 B^1) = \mu_0 c J^3 \tag{B.12}$$

e.g. from eqn.(B.12)

$$-\frac{1}{c^2}\frac{\partial E_z}{\partial t} + \frac{\partial B_y}{\partial x} - \frac{\partial B_x}{\partial y} = \mu_0 J_z. \tag{B.13}$$

Therefore

$$\boxed{\nabla \times \boldsymbol{B} = \frac{1}{c^2}\frac{\partial \boldsymbol{E}}{\partial t} + \mu_0 \boldsymbol{J}}. \tag{B.14}$$

This is the Ampère-Maxwell Law.

B.2 Homogeneous Field Equation

The dual tensor is

$$\widetilde{F}^{\mu\nu} = \frac{1}{2}\epsilon^{\mu\nu\rho\sigma}F_{\rho\sigma} \tag{B.15}$$

where

$$F^{\mu\nu} = g^{\mu\rho}g^{\nu\sigma}F_{\rho\sigma} \tag{B.16}$$

$$g^{\mu\nu} = g_{\mu\nu} = \begin{bmatrix} 1 & 0 & 0 & 0 \\ 0 & -1 & 0 & 0 \\ 0 & 0 & -1 & 0 \\ 0 & 0 & 0 & -1 \end{bmatrix}. \tag{B.17}$$

Therefore, for example

$$F^{23} = g^{22}g^{33}F_{23} = F_{23}, \tag{B.18}$$

$$F^{23} = -B^1 \tag{B.19}$$

$$F_{23} = B_1 \tag{B.20}$$

$$B^1 = -B_1. \tag{B.21}$$

So

$$\widetilde{F}^{01} = \epsilon^{0123}F_{23} = \epsilon^{0123}F^{23} = cB_1 = -cB^1 \tag{B.22}$$

$$\widetilde{F}^{12} = \epsilon^{1230}F_{30} = \epsilon^{1230}F^{30} = E^3 \tag{B.23}$$

and so on.

In eqn.(B.23) we have used

$$\epsilon^{1230} = -\epsilon^{1203} = \epsilon^{1023} = -\epsilon^{0123} = -1 \tag{B.24}$$

$$F^{30} = g^{33} g^{00} F_{30} = -F_{30}. \tag{B.25}$$

The homogeneous Maxwell-Heaviside field equation is therefore

$$\boxed{\partial_\mu \widetilde{F}^{\mu\nu} = 0} \tag{B.26}$$

$$\widetilde{F}^{\mu\nu} = \begin{bmatrix} 0 & -cB^1 & -cB^2 & -cB^3 \\ cB^1 & 0 & E^3 & -E^2 \\ cB^2 & -E^3 & 0 & E^1 \\ cB^3 & E^2 & -E^1 & 0 \end{bmatrix}. \tag{B.27}$$

B.2.1 The Gauss Law ($\nu = 0$)

$$\partial_1 \widetilde{F}^{10} + \partial_2 \widetilde{F}^{20} + \partial_3 \widetilde{F}^{30} = 0. \tag{B.28}$$

Therefore

$$\boxed{\boldsymbol{\nabla} \cdot \boldsymbol{B} = 0}. \tag{B.29}$$

B.2.2 The Faraday Law of Induction ($\nu = 1, 2, 3$)

$$\partial_0 \widetilde{F}^{01} + \partial_2 \widetilde{F}^{21} + \partial_3 \widetilde{F}^{31} = 0 \tag{B.30}$$

$$\partial_0 \widetilde{F}^{02} + \partial_1 \widetilde{F}^{12} + \partial_3 \widetilde{F}^{32} = 0 \tag{B.31}$$

$$\partial_0 \widetilde{F}^{03} + \partial_1 \widetilde{F}^{13} + \partial_2 \widetilde{F}^{23} = 0. \tag{B.32}$$

From eqns (B.30) to (B.32)

$$\boxed{\frac{\partial \boldsymbol{B}}{\partial t} + \boldsymbol{\nabla} \times \boldsymbol{E} = \boldsymbol{0}}. \tag{B.33}$$

B.3 Overall Result

$$\boxed{\begin{aligned} d \wedge F &= 0 \\ d \wedge \widetilde{F} &= \mu_0 c J \end{aligned}} \longrightarrow \boxed{\begin{aligned} \partial_\mu \widetilde{F}^{\mu\nu} &= 0 \\ \partial_\mu F^{\mu\nu} &= \mu_0 c J^\nu \end{aligned}} \longrightarrow \boxed{\begin{aligned} \boldsymbol{\nabla} \cdot \boldsymbol{B} &= 0 \\ \frac{\partial \boldsymbol{B}}{\partial t} + \boldsymbol{\nabla} \times \boldsymbol{E} &= \boldsymbol{0} \\ \boldsymbol{\nabla} \cdot \boldsymbol{E} &= \frac{\rho}{\epsilon_0} \\ \boldsymbol{\nabla} \times \boldsymbol{B} &= \frac{1}{c^2} \frac{\partial \boldsymbol{E}}{\partial t} + \mu_0 \boldsymbol{J} \end{aligned}}$$

B.3.1 Differential Form Notation

$$d \wedge F = 0 \tag{B.34}$$

$$d \wedge \widetilde{F} = \mu_0 c J \tag{B.35}$$

F = field two-form, $\widetilde{F}$ = dual of F, J = charge-current density three-form.

B.3.2 Evans Unified Field Theory

$$\boxed{\begin{aligned} d \wedge F^a &= \mu_0 j^a \\ d \wedge \widetilde{F}^a &= \mu_0 J^a \end{aligned}} \tag{B.36}$$

where

$$j^a = \frac{A^{(0)}}{\mu_0}(R^a{}_b \wedge q^b - \omega^a{}_b \wedge T^b) \sim 0 \tag{B.37}$$

$$J^a = \frac{A^{(0)}}{\mu_0}(\widetilde{R}^a{}_b \wedge q^b - \omega^a{}_b \wedge \widetilde{T}^b) > 0. \tag{B.38}$$

B.4 Units and Constants

$$\begin{aligned}
\epsilon_0 &= 8.854188 \times 10^{-12} J^{-1} C^2 m^{-1} \\
\mu_0 &= 4\pi \times 10^{-7} J s^2 C^{-2} m^{-1} \\
\boldsymbol{E} &= V m^{-1} = J C^{-1} m^{-1} \\
\rho &= C m^{-3} \\
\boldsymbol{B} &= T = Wb m^{-2} = V s m^{-2} = J s C^{-1} m^{-2} \\
\boldsymbol{J} &= C s^{-1} m^{-2}
\end{aligned}$$

These are the S.I. system of units

B.5 Notes

If we include polarization and magnetization then

$$\boldsymbol{D} = \epsilon_0 \boldsymbol{E} + \boldsymbol{P}$$

$$\boldsymbol{B} = \mu_0(\boldsymbol{H} + \boldsymbol{M})$$

and

$$\nabla \cdot \boldsymbol{D} = \rho$$

$$\nabla \times \boldsymbol{H} = \boldsymbol{J} + \frac{\partial \boldsymbol{D}}{\partial t}$$

$\boldsymbol{D} = Cm^{-2}$ = electric displacement
$\boldsymbol{H} = Am^{-1}$ = magnetic field strength

C

The Complex Circular Basis

The complex circular basis is well known, and is an O(3) symmetry basis for 3-D Euclidean space. First consider the Cartesian basis:

$$\boldsymbol{i} \times \boldsymbol{j} = \boldsymbol{k} \tag{C.1}$$

$$\boldsymbol{k} \times \boldsymbol{i} = \boldsymbol{j} \tag{C.2}$$

$$\boldsymbol{j} \times \boldsymbol{k} = \boldsymbol{i}. \tag{C.3}$$

It can be seen that this has a **cyclic** symmetry.

Now define the complex circular basis using the following unit vectors:

$$\boldsymbol{e}^{(1)} = \frac{1}{\sqrt{2}} (\boldsymbol{i} - i\boldsymbol{j}) \tag{C.4}$$

$$\boldsymbol{e}^{(2)} = \frac{1}{\sqrt{2}} (\boldsymbol{i} + i\boldsymbol{j}) \tag{C.5}$$

$$\boldsymbol{e}^{(3)} = \boldsymbol{k}. \tag{C.6}$$

It can be seen that:

$$\boldsymbol{e}^{(1)} = \boldsymbol{e}^{(2)*} \tag{C.7}$$

where * denotes complex conjugation.

Next form the vector cross product of $\boldsymbol{e}^{(1)}$ and $\boldsymbol{e}^{(2)}$ as follows:

$$\boldsymbol{e}^{(1)} \times \boldsymbol{e}^{(2)} = \frac{1}{2} \begin{vmatrix} \boldsymbol{i} & \boldsymbol{j} & \boldsymbol{k} \\ 1 & -i & 0 \\ 1 & i & 0 \end{vmatrix} = i\boldsymbol{k}. \tag{C.8}$$

Therefore:

$$\boldsymbol{e}^{(1)} \times \boldsymbol{e}^{(2)} = i\boldsymbol{e}^{(3)}. \tag{C.9}$$

It is convenient to write this as:

$$e^{(1)} \times e^{(2)} = ie^{(3)*}. \tag{C.10}$$

Now form the vector cross product of $e^{(2)}$ and $e^{(3)}$:

$$e^{(2)} \times e^{(3)} = \frac{1}{\sqrt{2}} \begin{vmatrix} \boldsymbol{i} & \boldsymbol{j} & \boldsymbol{k} \\ 1 & i & 0 \\ 0 & 0 & 1 \end{vmatrix} = \frac{i}{\sqrt{2}} (\boldsymbol{i} + i\boldsymbol{j}) \tag{C.11}$$

$$= ie^{(1)*}.$$

Finally form the vector cross product of $e^{(3)}$ and $e^{(1)}$:

$$e^{(3)} \times e^{(1)} = \frac{1}{\sqrt{2}} \begin{vmatrix} \boldsymbol{i} & \boldsymbol{j} & \boldsymbol{k} \\ 0 & 0 & 1 \\ 1 & -i & 0 \end{vmatrix} = \frac{i}{\sqrt{2}} (\boldsymbol{i} - i\boldsymbol{j}) \tag{C.12}$$

$$= ie^{(2)*}.$$

We therefore obtain the result that:

$$e^{(1)} \times e^{(2)} = ie^{(3)*} \tag{C.13}$$

$$e^{(2)} \times e^{(3)} = ie^{(1)*} \tag{C.14}$$

$$e^{(3)} \times e^{(1)} = ie^{(2)*} \tag{C.15}$$

□

Eqn.(C.13) to (C.15) are those of the complex circular basis. This has the same type of O(3) symmetry as the Cartesian basis $\boldsymbol{i}, \boldsymbol{j}, \boldsymbol{k}$ of eqn. (C.1) to (C.3).

C.1 O(3) Electrodynamics

The transverse plane waves of the radiated magnetic field are defined as follows:

$$\boldsymbol{B}^{(1)} = B^{(0)} e^{(1)} e^{i\phi} \tag{C.16}$$

$$\boldsymbol{B}^{(2)} = B^{(0)} e^{(2)} e^{-i\phi}. \tag{C.17}$$

The Evans spin field is:

$$\boldsymbol{B}^{(3)} = B^{(0)}\boldsymbol{e}^{(3)}. \tag{C.18}$$

The B Cyclic Theorem is therefore:

$$\boldsymbol{B}^{(1)} \times \boldsymbol{B}^{(2)} = iB^{(0)}\boldsymbol{B}^{(3)*} \tag{C.19}$$

$$\boldsymbol{B}^{(2)} \times \boldsymbol{B}^{(3)} = iB^{(0)}\boldsymbol{B}^{(1)*} \tag{C.20}$$

$$\boldsymbol{B}^{(3)} \times \boldsymbol{B}^{(1)} = iB^{(0)}\boldsymbol{B}^{(2)*}. \tag{C.21}$$

Here ϕ is the phase of the wave. Multiply both sides of eqn.(C.13) to (C.15) by $\boldsymbol{B}^{(0)2}$ to give eqns. (C.19) - (C.21).

C.2 Notes and References

- The complex circular basis is well known and is described for example in B.L.Silver, "Irreducible Tensorial Sets" (Academic, New York, 1976)
- The complex circular basis becomes the B Cyclic theorem through equations (C.16) to (C.18), and so the complex circular basis describes circular polarization, as is well known.
- For considerable development of these notes see: M.W.Evans, J.-P. Vigier et al., "The Enigmatic Photon" (Kluwer, Dordrecht, 1994-2002, hardback and softback)

D

The AntiSymmetric Metric

It is well known in differential geometry that the tetrad is defined by:

$$\boxed{V^a = q^a{}_\mu V^\mu}. \tag{D.1}$$

Here V^a is a four-vector defined in the Minkowski spacetime of the tangent bundle at point P to the base manifold. The latter is the general 4-D spacetime in which the vector is defined by V^μ.
The metric tensor used by Einstein in his field theory of gravitation (1915) is (Carroll):

$$g_{\mu\nu}{}^{(S)} = q^a{}_\mu q^b{}_\nu \eta_{ab}. \tag{D.2}$$

In eqn. (D.2) η_{ab} is the metric of the tangent bundle. Eqn. (D.2) defines a symmetric metric $g_{\mu\nu}{}^{(S)}$, through an inner or dot product of two tetrads.

It is seen in eqn. (D.1) that there is summation over repeated indices. This is the Einstein convention. One index μ is a subscript (covariant) on the right hand side of eqn. (D.1).

Thus, written out in full eqn. (D.1) is:

$$V^a = q^a{}_0 V^0 + q^a{}_1 V^1 + q^a{}_2 V^2 + q^a{}_3 V^3. \tag{D.3}$$

Similarly, eqn. (D.2) is:

$$g_{\mu\nu}{}^{(S)} = q^0{}_\mu q^0{}_\nu \eta_{00} + \cdots + q^3{}_\mu q^3{}_\nu \eta_{33}. \tag{D.4}$$

In eqn. (D.4) it is seen that all possible combinations of a, b are summed. Another example is given by Einstein in his famous book “The Meaning of Relativity” (Princeton, 1921-1954):

$$g_{\mu\nu}{}^{(S)} g^{\mu\nu(S)} = 4. \tag{D.5}$$

It is seen that the double summation over μ and ν in eqn. (D.5) produces a scalar (the number 4). In differential geometry a scalar is a zero-form.

It is seen from the basic and well known definition (D.2) that is possible to define the wedge product of two tetrads:

$$q^c{}_{\mu\nu} = q^a{}_\mu \wedge q^b{}_\nu . \tag{D.6}$$

The wedge product is a generalization to any dimension of the vector cross product in 3-D. In eqn. (D.6) $q^c{}_{\mu\nu}$ is a two-form of differential geometry, i.e. a tensor antisymmetric in μ and ν. It is a vector-valued two-form due to the presence of the index c. This is the antisymmetric metric:

$$q^c{}_{\mu\nu}{}^{(A)} = q^c{}_{\mu\nu} . \tag{D.7}$$

The antisymmetric metric is part of the more general tensor metric formed by the outer product of two tetrads:

$$q^{ab}{}_{\mu\nu} = q^a{}_\mu q^b{}_\nu . \tag{D.8}$$

It is seen that the indices μ and ν are always the same on both sides, so can be left out for clarity of presentation (see Carroll). Thus we obtain:

$$q^{ab} = q^a q^b \tag{D.9}$$

$$g^{c(A)} = q^a \wedge q^b \tag{D.10}$$

$$g^{(S)} = q^a q^b \eta_{ab} \tag{D.11}$$

This notation shows clearly that q^{ab} is a tensor; $g^{c(A)}$ is a vector; $g^{(S)}$ is a scalar. It is well known that any tensor is the sum of a symmetric and antisymmetric component:

$$q^{ab} = q^{\mathrm{ab}(S)} + q^{\mathrm{ab}(A)} . \tag{D.12}$$

Furthermore, $q^{\mathrm{ab}(S)}$ is the sum of an off-diagonal symmetric tensor and a diagonal tensor. The sum of the elements of the diagonal tensor is known as the trace.

Thus, $g^{c(A)}$ is the antisymmetric part of q^{ab}:

$$\boxed{g^{c(A)} = \frac{1}{2}\epsilon^{abc} q^{\mathrm{ab}(A)}} \tag{D.13}$$

In eqn. (D.11):

$$\eta_{ab} = \begin{vmatrix} 1 & 0 & 0 & 0 \\ 0 & -1 & 0 & 0 \\ 0 & 0 & -1 & 0 \\ 0 & 0 & 0 & -1 \end{vmatrix} \tag{D.14}$$

thus:

$$\begin{aligned} g^{(S)} &= q^0 q^0 \eta_{00} + q^1 q^1 \eta_{11} + q^2 q^2 \eta_{22} + q^3 q^3 \eta_{33} \\ &= q^0 q^0 - q^1 q^1 - q^2 q^2 - q^3 q^3 \end{aligned} \tag{D.15}$$

and so:

$$\boxed{g^{(S)} = \mathrm{Trace}\, q^{ab}.} \tag{D.16}$$

From eqn. (D.9), (D.13) and (D.16) it is seen that the existence of the anti-symmetric metric is implied by the existence of the symmetric metric.

□

In the notation of eqn. (2.33) of Evans, Chapter 2:

$$\omega_2 = -\frac{1}{2} q^{\mu\nu(A)} du_\mu \wedge du_\nu \tag{D.17}$$

From the definition of the wedge product, eqn. (D.6), eqn. (D.17) is:

$$\omega_2 = -\frac{1}{2} q^{\mu\nu(A)} q_{\mu\nu}{}^{(A)}, \tag{D.18}$$

and by comparison with Einstein's eqn. (D.5), it is seen that ω_2 is a scalar.

□

E

Tensorial Structure of the Inhomogeneous Field Equations (IE)

E.1 Introduction (HE & IE)

The tensorial structure of the IE is derived from the Bianchi identity of differential geometry

$$D \wedge T^a = R^a{}_b \wedge q^b \tag{E.1}$$

i.e.

$$d \wedge T^a = R^a{}_b \wedge q^b - \omega^a{}_b \wedge T^b \tag{E.2}$$

or

$$d \wedge T^a = -q^b \wedge R^a{}_b - \omega^a{}_b \wedge T^b \tag{E.3}$$

Restoring the indices of the base manifold in eqn. (E.3)

$$\boxed{d \wedge T^a{}_{\mu\nu} = -\left(q^b \wedge R^a{}_{b\mu\nu} + \omega^a{}_b \wedge T^b{}_{\mu\nu}\right)} \tag{E.4}$$

Eqn. (E.4) becomes the homogeneous field equations (HE) using the rules:

$$A^a{}_\mu = A^{(0)} q^a{}_\mu \tag{E.5}$$

$$F^a{}_{\mu\nu} = A^{(0)} T^a{}_{\mu\nu} \tag{E.6}$$

Therefore the HE is:

$$\boxed{\begin{aligned} d \wedge F^a{}_{\mu\nu} &= -A^{(0)}(q^b \wedge R^a{}_{b\mu\nu} + \omega^a{}_b \wedge T^a{}_{\mu\nu}) \\ &\sim 0\,(\text{experimentally}) \end{aligned}} \tag{E.7}$$

The IE is obtained from eqn (E.7) by taking the appropriate Hodge duals:

$$\tilde{F}^a{}_{\rho\sigma} = \frac{1}{2}|g|^{\frac{1}{2}} \epsilon_{\rho\sigma}{}^{\mu\nu} F^a{}_{\mu\nu} \tag{E.8}$$

$$\tilde{R}^a{}_{b\rho\sigma} = \frac{1}{2}|g|^{\frac{1}{2}}\epsilon_{\rho\sigma}{}^{\mu\nu} R^a{}_{b\mu\nu} \tag{E.9}$$

in the general 4-D manifold.
We therefore multiply both sides of eqn (E.7) by:

$$\frac{1}{2}|g|\epsilon_{\rho\sigma}{}^{\mu\nu}$$

to obtain the tensorial representation of the IE:

$$d \wedge \tilde{F}^a{}_{\mu\nu} = -A^{(0)}(q^b \wedge \tilde{R}^a{}_{b\mu\nu} + \omega^a{}_b \wedge \tilde{T}^b{}_{\mu\nu}) \tag{E.10}$$

The charge-current density of field theory is therefore:

$$J^a = -\frac{A^{(0)}}{\mu_0}\left(q^b \wedge \tilde{R}^a{}_b + \omega^a{}_b \wedge \tilde{T}^b\right) \tag{E.11}$$

and is a vector valued three-form of differential geometry, defined by:

$$\boxed{d \wedge \tilde{F}^a = \mu_0 J^a} \tag{E.12}$$

Eqn. (E.12) is the same equation as:

$$\boxed{\begin{aligned} \partial_\mu F^{a\mu\nu} &= -A^{(0)}(q^b{}_\mu R^a{}_b{}^{\mu\nu} + \omega^a{}_{b\mu} T^{b\mu\nu}) \\ &= \mu_0 J^{a\nu} \end{aligned}} \tag{E.13}$$

E.2 The Coulomb Law ($\nu = 0$)

The Coulomb Law in the unified field theory is given by $\nu = 0$, and is:

$$\partial_1 F^{a10} + \partial_2 F^{a20} + \partial_3 F^{a30} = \mu_0 J^{a0} \tag{E.14}$$

where:

$$\begin{aligned} J^{a0} = -\frac{A^{(0)}}{\mu_0} & (q^b{}_1 R^a{}_b{}^{10} + q^b{}_2 R^a{}_b{}^{20} + q^b{}_3 R^a{}_b{}^{30} \\ & +\omega^a{}_{b1} T^{b10} + \omega^a{}_{b2} T^{b20} + \omega^a{}_{b3} T^{b30}) \end{aligned} \tag{E.15}$$

i.e.

$$\boxed{\begin{aligned} \nabla \cdot \boldsymbol{E}^a = -\,\phi^{(0)} & (q^b{}_1 R^a{}_b{}^{10} + q^b{}_2 R^a{}_b{}^{20} + q^b{}_3 R^a{}_b{}^{30} \\ & +\omega^a{}_{b1} T^{b10} + \omega^a{}_{b2} T^{b20} + \omega^a{}_{b3} T^{b30}) \end{aligned}} \tag{E.16}$$

E.2.1 Notes on the Coulomb Law

It can be seen from the familar vector notation in eqn. (E.16) that the origin of the Coulomb Law is spacetime. Using the constitutive equations:

$$T^a = d \wedge q^a + \omega^a{}_b \wedge q^b \tag{E.17}$$

$$R^a{}_b = d \wedge \omega^a{}_b + \omega^a{}_c \wedge \omega^c{}_b \tag{E.18}$$

the right hand side of eqn (E.16) becomes a function only of the tetrad and the spin connection. The Riemann form must always obey the second Bianchi identity.

$$D \wedge R^a{}_b = 0 \tag{E.19}$$

Therefore eqn (E.17) to (E.19) are constraints on the variables on the right hand side of eqn. (E.16). Einstein field theory of gravitation is given by the limit:

$$T^a = 0 \tag{E.20}$$

$$D \wedge T^a = R^a{}_b \wedge q^b = 0 \tag{E.21}$$

and Newtonian gravitation is the weak field limit of eqn (E.20) and (E.21). In the Einstein and Newton theories of pure gravitation therefore:

$$q^b{}_1 R^a{}_b{}^{10} + q^b{}_2 R^a{}_b{}^{20} + q^b{}_3 R^a{}_b{}^{30} \neq 0 \tag{E.22}$$

The electromagnetic field is torsion of spacetime (eqn.(E.6)) so from eqn. (E.20) there is no electro-magnetism in the Einstein or Newton theories of gravitation. This is of course a self-consistent result. However, it is seen further from eqns (E.16) and (E.22) that Einstein or Newtonian gravitation does not affect the Coulumb Law.

In order for gravitational forces to change the Coulomb Law the condition is:

$$\boxed{q^b{}_1 R^a{}_b{}^{10} + q^b{}_2 R^a{}_b{}^{20} + q^b{}_3 R^a{}_b{}^{30} \neq 0} \tag{E.23}$$

this condition is compatible with:

$$R^a{}_b \wedge q^b = 0 \tag{E.24}$$

and to:

$$R_{\mu\nu\rho\sigma} + R_{\rho\mu\nu\sigma} + R_{\nu\rho\mu\sigma} = 0 \tag{E.25}$$

Eqn. (E.25) means that:

$$\Gamma^\kappa{}_{\mu\nu} = \Gamma^\kappa{}_{\nu\mu} \tag{E.26}$$

where $\Gamma^\kappa_{\mu\nu}$ is the Christoffel symbol.

Eqn (E.26) means that the gravitational torsion tensor vanishes:

$$T^\kappa{}_{\mu\nu} = q^\kappa{}_\mu T^a{}_{\mu\nu} = \Gamma^\kappa{}_{\mu\nu} - \Gamma^\kappa{}_{\nu\mu} = 0. \tag{E.27}$$

E.3 The Ampère Maxwell Law (ν = 1,2,3)

The Ampère Maxwell Law is:

$$\partial_0 F^{a01} + \partial_2 F^{a21} + \partial_3 F^{a31} = \mu_0 J^{a1} \; (\nu = 1) \tag{E.28}$$

$$\partial_0 F^{a02} + \partial_1 F^{a12} + \partial_3 F^{a32} = \mu_0 J^{a2} \; (\nu = 2) \tag{E.29}$$

$$\partial_0 F^{a03} + \partial_1 F^{a13} + \partial_2 F^{a23} = \mu_0 J^{a3} \; (\nu = 3) \tag{E.30}$$

where:

$$\begin{aligned} J^{a1} = &- \frac{A^{(0)}}{\mu_0} \left(q^b{}_0 R^a{}_b{}^{01} + q^b{}_2 R^a{}_b{}^{21} + q^b{}_3 R^a{}_b{}^{31} \right. \\ &\left. + \omega^a{}_{b0} T^{b01} + \omega^a{}_{b2} T^{b21} + \omega^a{}_{b3} T^{b31} \right) \end{aligned} \tag{E.31}$$

$$\begin{aligned} J^{a2} = &- \frac{A^{(0)}}{\mu_0} \left(q^b{}_0 R^a{}_b{}^{02} + q^b{}_1 R^a{}_b{}^{12} + q^b{}_3 R^a{}_b{}^{32} \right. \\ &\left. + \omega^a{}_{b0} T^{b02} + \omega^a{}_{b1} T^{b12} + \omega^a{}_{b3} T^{b32} \right) \end{aligned} \tag{E.32}$$

$$\begin{aligned} J^{a3} = &- \frac{A^{(0)}}{\mu_0} \left(q^b{}_0 R^a{}_b{}^{03} + q^b{}_1 R^a{}_b{}^{13} + q^b{}_2 R^a{}_b{}^{23} \right. \\ &\left. + \omega^a{}_{b0} T^{b03} + \omega^a{}_{b1} T^{b13} + \omega^a{}_{b2} T^{b23} \right) \end{aligned} \tag{E.33}$$

E.3.1 Notes on the Ampère Maxwell Law

It is seen from eqns (E.31) to (E.33) that electric current is spacetime. The same consideration apply as to the Coulomb Law, because both laws are part of the IE. Therefore gravitational torsion is needed to generate electric current and electric power.

F

Some Notes On The I.E. (Inhomogeneous Evans Field Equation)

F.1 Introduction

The IE is deduced from the Bianchi identity of differential geometry:

$$d \wedge T^a = R^a{}_b \wedge q^b - \omega^a{}_b \wedge T^b \tag{F.1}$$

in the general 4-D manifold or Evans spacetime. Here T^a is the torsion form, $R^a{}_b$ the curvature or Riemann form, $\omega^a{}_b$ the spin connection and $d\wedge$ the exterior derivative.

The geometry is converted to the field theory using:

$$A^a = A^{(0)} q^a \tag{F.2}$$

$$F^a = A^{(0)} T^a. \tag{F.3}$$

Here $A^{(0)}$ is the fundamental potential magnitude, with units of volt s/m, A^a is the potential form and F^a the field form. From eqn. (F.1) to (F.3) we obtain **the homogeneous Evans field equations (HE)**:

$$\boxed{\begin{aligned} d \wedge F^a &= R^a{}_b \wedge A^b - \omega^a{}_b \wedge F^b \\ &= -A^{(0)} \left(q^b \wedge R^a{}_b + \omega^a{}_b \wedge T^b \right) \end{aligned}}. \tag{F.4}$$

With the geometrical constraint:

$$R^a{}_b \wedge q^b = \omega^a{}_b \wedge T^b \tag{F.5}$$

eqn. (F.4) reduces to the homogeneous Maxwell-Heaviside field equations:

$$d \wedge F^a = 0 \tag{F.6}$$

for each index a. The latter is the tangent bundle index of differential geometry, and physically indicates states of polarization. Thus, for each a:

$$d \wedge F = 0. \tag{F.7}$$

Eqn. (F.7) is the way in which the homogeneous MH eqn is usually written in differential geometry. It is seen that the homogeneous MH eqn. (F.4) is **a particular case**. More generally, gravitation affects the homogeneous MH eqn when the eqn. (F.5) is not obeyed. **These effects should be investigated experimentally.** Using the structure equations:

$$\begin{aligned} T^a &= D \wedge q^a = d \wedge q^a + \omega^a{}_b \wedge q^b \\ R^a{}_b &= D \wedge \omega^a{}_b = d \wedge \omega^a{}_b + \omega^a{}_c \wedge \omega^c{}_b \end{aligned} \tag{F.8}$$

of differential geometry, eqn (F.5) becomes:

$$\left(D \wedge \omega^a{}_b\right) \wedge q^b = \omega^a{}_b \wedge \left(D \wedge q^b\right) \tag{F.9}$$

one possible solution of eqn. (F.9) is:

$$\omega^a{}_b = -\kappa \epsilon^a{}_{bc} q^c \tag{F.10}$$

$$R^a{}_b = \kappa \epsilon^a{}_{bc} T^c \tag{F.11}$$

where κ is the wavenumber.

Eqns (F.10) and (F.11) mean that in free space, $\omega^a{}_b$ is the antisymmetric tensor dual to the vector q^c, and $R^a{}_b$ is the antisymmetric tensor dual to T^c. These equations define free space electromagnetism free of gravitation influence (mass). In this case the spin connection and tetrad are duals and the curvature and torsion forms are duals.
In tensor notation, eqn. (F.7) is:

$$\partial_\mu F_{\nu\rho} + \partial_\nu F_{\rho\mu} + \partial_\rho F_{\mu\nu} = 0 \tag{F.12}$$

or

$$\partial_\mu \widetilde{F}^{\mu\nu} = 0 \tag{F.13}$$

where

$$\widetilde{F}^{\mu\nu} = \frac{1}{2} \epsilon^{\mu\nu\rho\sigma} F_{\rho\sigma}. \tag{F.14}$$

Eqn. (F.13a) is a combination of two fundamental laws:

$$\nabla \cdot \boldsymbol{B} = 0 \tag{F.15}$$

$$\nabla \times \boldsymbol{E} + \frac{\partial \boldsymbol{B}}{\partial t} = \boldsymbol{0}. \tag{F.16}$$

Eqn. (F.15) is the Gauss Law applied to magnetism, and eqn. (F.16) is the Faraday Law of induction.

These laws are therefore special cases of the unified field theory, eqn. (F.4).

Physically, these laws describe electromagnetism assuming that it is not influenced by gravitation. In the laboratory this is an excellent approximation but in cosmology, a beam of light near a black hole will obey eqn. (F.4), and small changes in the laws (F.15) and (F.16) are expected.

This is a test of Einstein general relativity, upon which the Evans field theory is based directly. Einstein's field theory of gravitation assumes that gravitation is described by Riemann spacetime and a Christoffel connection. Evans' unified field theory of all radiated and matter fields assumes that the fields are governed by the more general Evans spacetime, the 4-D manifold in which torsion and curvature are defined by the structure relation of differential geometry. Evans spacetime reduces to the Riemann spacetime used by Einstein when:

$$T^a = 0 \tag{F.17}$$

$$R^a{}_b \wedge q^b = 0. \tag{F.18}$$

Eqn. (F.17) means that there is no electromagnetic field present:

$$F^a = 0. \tag{F.19}$$

Eqn. (F.18) is the well known Bianchi identity used by Einstein:

$$R_{\rho\mu\nu\sigma} + R_{\rho\sigma\mu\nu} + R_{\rho\nu\sigma\mu} = 0. \tag{F.20}$$

Eqn. (F.20) is true if and only if:

$$\Gamma^\kappa{}_{\mu\nu} = \Gamma^\kappa{}_{\nu\mu} \tag{F.21}$$

and

$$T^\kappa{}_{\mu\nu} = \Gamma^\kappa{}_{\mu\nu} - \Gamma^\kappa{}_{\nu\mu} = 0. \tag{F.22}$$

Here $R_{\rho\mu\nu\sigma}$ is the Riemann tensor, $\Gamma^\kappa{}_{\mu\nu}$ is the Christoffel connection, $T^\kappa{}_{\mu\nu}$ is the torsion tensor. Eqn. (F.22) in tensor notation is equivalent to eqn. (F.17) in differential form notation. Eqn. (F.1) is a generalization of eqn. (F.20) to Evans spacetime, the 4-D manifold of differential geometry.

Before proceeding to a discussion of the IE it is intructive to give the following proofs. Firstly, we prove eqn. (F.13) from eqn. (F.12).

Proof of Eqn. (F.13)

Consider the case:

$$\partial_1 F_{23} + \partial_3 F_{12} + \partial_2 F_{31} = 0 \tag{F.23}$$

Note that:

$$\epsilon^{2301} \partial_1 F_{23} + \epsilon^{1203} \partial_3 F_{12} + \epsilon^{3102} \partial_2 F_{31} = 0 \tag{F.24}$$

because:

$$\epsilon^{2301} = \epsilon^{1203} = \epsilon^{3102} = 1 \tag{F.25}$$

Using the Leibniz theorem:

$$\begin{aligned} \partial_1 \left(\epsilon^{2301} F_{23} \right) &= \left(\partial_1 \epsilon^{2301} \right) F_{23} + \epsilon^{2301} \partial_1 F_{23} \\ &= \epsilon^{2301} \partial_1 F_{23} \end{aligned} \tag{F.26}$$

and

$$\begin{aligned} \epsilon^{2301} \partial_1 F_{23} &= \tfrac{1}{2} \left(\partial_1 \left(\epsilon^{2301} F_{23} + \epsilon^{3201} F_{32} \right) \right) \\ &= \partial_1 \widetilde{F}^{01} \end{aligned} \tag{F.27}$$

Therefore eqn. (F.23) is the same as:

$$\partial_1 \widetilde{F}^{01} + \partial_2 \widetilde{F}^{02} + \partial_3 \widetilde{F}^{03} = 0 \tag{F.28}$$

or:

$$\nabla \cdot \boldsymbol{B} = 0 \tag{F.29}$$

□

Similarly the HE, eqn. (F.4) is the same as:

$$\partial_\mu \widetilde{F}^{\mu\nu,a} = A^b{}_\mu \widetilde{R}^a{}_b{}^{\mu\nu} - \omega^a{}_{b\mu} \widetilde{F}^{\mu\nu,b} \tag{F.30}$$

The geometrical condition for free space electromagnetism is therefore:

$$q^b{}_\mu \widetilde{R}^a{}_b{}^{\mu\nu} = \omega^a{}_{b\mu} \widetilde{T}^{\mu\nu,b} \tag{F.31}$$

i.e. **the 4-D dot products on the left and right must be the same.** Eqn. (F.31) defines free space electromagnetism free of any gravitational influence. More generally eqn. (F.31) does not hold, and electromagnetism and gravitation are mutually influential.

This case is very important for new technologies, and is the case where the dot products in eqn. (F.31) are not the same, i.e.:

$$\begin{aligned} \omega^a{}_b &\neq -\kappa \epsilon^a{}_{bc} q^c \\ R^a{}_b &\neq -\kappa \epsilon^a{}_{bc} T^c \end{aligned} \tag{F.32}$$

and:

$$\begin{aligned} \nabla \cdot \boldsymbol{B} &\neq 0 \\ \nabla \times \boldsymbol{E} + \frac{\partial \boldsymbol{B}}{\partial t} &\neq \boldsymbol{0}. \end{aligned} \tag{F.33}$$

Eqns. (F.32) and (F.33) mean, for example, that the polarization of a beam of light grazing an intensely gravitating object will be changed.
Secondly we prove that if:

$$d \wedge F = 0 \tag{F.34}$$

then:

$$d \wedge \widetilde{F} \neq 0 \tag{F.35}$$

in general.

Proof of Eqn. (F.35)

It must be proven that in general:

$$\partial_\mu \widetilde{F}_{\nu\rho} + \partial_\rho \widetilde{F}_{\mu\nu} + \partial_\nu \widetilde{F}_{\rho\mu} \neq 0 \tag{F.36}$$

if

$$\partial_\mu F_{\nu\rho} + \partial_\rho F_{\mu\nu} + \partial_\nu F_{\rho\mu} = 0 \tag{F.37}$$

Consider the example:

$$\begin{aligned} \partial_1 \widetilde{F}_{23} + \partial_3 \widetilde{F}_{12} + \partial_2 \widetilde{F}_{31} &= \frac{1}{2} \left(\epsilon_{\rho\sigma 12} \partial_3 F^{\rho\sigma} + \cdots \right) \\ &= \frac{1}{2} \left(\epsilon_{0123} \partial_1 F^{01} + \epsilon_{1023} \partial_1 F^{10} + \cdots \right) \\ &= \partial_1 F^{01} + \partial_2 F^{02} + \partial_3 F^{03} \\ &= \nabla \cdot \boldsymbol{E} = \frac{\rho}{\epsilon_0} \end{aligned} \tag{F.38}$$

Eqn. (F.38) is the Coulomb law of electromagnetism. Eqn. (F.36) is therefore the same as:

$$\partial_\mu F^{\mu\nu} = \mu_0 J^\nu \tag{F.39}$$

where J^ν is the charge -current density four vector. Eqn. (F.39) is the inhomogeneous Maxwell Heaviside field equation. In differential form notation it is:

$$d \wedge \widetilde{F} = \mu_0 J \tag{F.40}$$

where J is the charge-current density three-form.

□

It is to be expected that eqn.(F.40) is a special case of the more general **inhomogeneous Evans field equation (IE):**

$$\boxed{\begin{aligned} d\wedge\widetilde{F}^a &= \widetilde{R}^a{}_b\wedge A^b-\omega^a{}_b\wedge\widetilde{F}^b \\ &= A^{(0)}(\widetilde{R}^a{}_b\wedge q^b-\omega^a{}_b\wedge\widetilde{T}^b)\end{aligned}} \tag{F.41}$$

where:

$$\widetilde{R}^a{}_b\wedge q^b\neq\omega^a{}_b\wedge\widetilde{T}^b \tag{F.42}$$

Proof of Eqn. (F.41)

Eqn. (F.41) is derived from eqn. (F.4) by considering the duals of F^a and $R^a{}_b$ in Evans spacetime.

$$\widetilde{F}^a{}_{\mu\nu}=\frac{1}{2}|g|^{\frac{1}{2}}\epsilon^{\rho\sigma}{}_{\mu\nu}F^a{}_{\rho\sigma} \tag{F.43}$$

$$\widetilde{R}^a{}_{b\mu\nu}=\frac{1}{2}|g|^{\frac{1}{2}}\epsilon^{\rho\sigma}{}_{\mu\nu}\widetilde{R}^a{}_{b\rho\sigma} \tag{F.44}$$

where $|g|^{\frac{1}{2}}$ is the square root of the metric determinant and $\epsilon^{\rho\sigma}{}_{\mu\nu}$ the 4-D Levi-Civita symbol. In 4-D the Hodge dual, $\widetilde{F}^a$, of F^a is another two-form. Similarly, the Hodge dual, $\widetilde{R}^a{}_b$, of $R^a{}_b$ is also another two-form. Eqns. (F.43) and (F.44) are defined by the general definition of a Hodge dual in the n-dimensional manifold.

Using eqns. (F.43) and (F.44) the IE, eqn. (F.41) follows from eqn. (F.4) by correctly defining the Hodge duals on both sides of eqn. (F.4). This procedure generates the IE from the HE and the fundamental Bianchi identity (F.1). **We therefore arrive at a generalization of the Coulomb Law and the Ampère-Maxwell law.**

To check and illustrate this important result consider a particular term of eqn (F.4) such as:

$$\partial_1 F^a{}_{23}=R^a{}_{b12}A^b{}_3-\omega^a{}_{b1}F^b{}_{23} \tag{F.45}$$

Integrating:

$$F^a{}_{23}=\int\left(R^a{}_{b12}A^b{}_3-\omega^a{}_{b1}F^b{}_{23}\right)dx^1 \tag{F.46}$$

Therefore:

$$\widetilde{F}^a{}_{23}=\int\left(\widetilde{R}^a{}_{b12}A^b{}_3-\omega^a{}_{b1}\widetilde{F}^b{}_{23}\right)dx^1 \tag{F.47}$$

and:

$$\partial_1\widetilde{F}^a{}_{23}=\widetilde{R}^a{}_{b12}A^b{}_3-\omega^a{}_{b1}\widetilde{F}^b{}_{23} \tag{F.48}$$

Eqn. (F.47) is the only possible way of defining Hodge duals so that indices match on both sides. The Hodge dual of the product $R^a{}_{b12}A^b{}_3$ must be a one-form in 4-D and this cannot be equated to the two-form on the LHS.

The Hodge dual of $A^b{}_3$ must be a three-form, and this again does not give the right answer.

Therefore the correct Hodge dual structure of eqn.(F.4) is eqn. (F.41).

□

The tensorial structure of the IE is:

$$\partial_\mu \widetilde{F}^a{}_{\nu\rho} + \partial_\rho \widetilde{F}^a{}_{\mu\nu} + \partial_\nu \widetilde{F}^a{}_{\rho\mu} = A^{(0)}(\widetilde{R}^a{}_{b\mu\nu} q^b{}_\rho + \widetilde{R}^a{}_{b\rho\mu} q^b{}_\nu + \widetilde{R}^a{}_{b\nu\rho} q^b{}_\mu - \omega^a{}_{b\mu} \widetilde{T}^b{}_{\nu\rho} - \omega^a{}_{b\rho} \widetilde{T}^b{}_{\mu\nu} - \omega^a{}_{b\nu} \widetilde{T}^b{}_{\rho\mu}) \quad (F.49)$$

which is the same as:

$$\boxed{\partial_\mu F^{\mu\nu,a} = -A^{(0)}(q^b{}_\mu R^a{}_{b\mu\nu} + \omega^a{}_{b\mu} T^{\mu\nu,b})} \quad (F.50)$$

F.2 The Coulomb Law in the Evans field Theory

The Coulomb Law is defined by:

$$\begin{aligned}\partial_1 \widetilde{F}^a{}_{23} + \partial_3 \widetilde{F}^a{}_{12} + \partial_2 \widetilde{F}^a{}_{31} &= \partial_1 F^{01,a} + \partial_2 F^{02,a} + \partial_3 F^{03,a} \\ &= \nabla \cdot \boldsymbol{E}\end{aligned}$$

$$\boxed{\begin{aligned}\nabla \cdot \boldsymbol{E}^a = -\phi^{(0)}(&q^b{}_1 R^a{}_b{}^{01} + q^b{}_2 R^a{}_b{}^{02} + q^b{}_3 R^a{}_b{}^{03} \\ &+ \omega^a{}_{b1} T^{01b} + \omega^a{}_{b2} T^{02b} + \omega^a{}_{b3} T^{03b}).\end{aligned}} \quad (F.51)$$

In eqn. (F.51), $\phi^{(0)}$ is in volts.
Using the result:

$$R^a{}_1{}^{01} = R^a{}_b{}^{01} q^b{}_1 \quad (F.52)$$

eqn. (F.51) simplifies to:

$$\boxed{\begin{aligned}\nabla \cdot \boldsymbol{E}^a &= -\phi^{(0)}(R^a{}_1{}^{01} + R^a{}_1{}^{02} + R^a{}_1{}^{03} + \omega^a{}_{b1} T^{01b} + \omega^a{}_{b2} T^{02b} + \omega^a{}_{b3} T^{03b}) \\ &= \frac{\rho}{\epsilon_0}\end{aligned}} \quad (F.53)$$

Eqn. (F.53) is the Coulomb Law in the Evans field theory.

F.3 Discussion

Eqn. (F.53) is the direct result of the Bianchi identity (F.1) of differential geometry, and reveals the origin of charge density, ρ, in general relativity. In the limit of Einstein's field theory of gravitation

$$\boldsymbol{E}^a = 0 \text{ and } T^{a,\mu\nu} = 0 \tag{F.54}$$

but in this limit.

$$R^a{}_1{}^{01} + R^a{}_1{}^{02} + R^a{}_1{}^{03} \neq 0 \tag{F.55}$$

This is the limit where electromagnetism is absent, and eqn. (F.55) is the same as:

$$\widetilde{R}^a{}_{123} + \widetilde{R}^a{}_{312} + \widetilde{R}^a{}_{231} \neq 0 \tag{F.56}$$

In the limit of free space electromagnetism:

$$\nabla \cdot \boldsymbol{E}^a = 0 \tag{F.57}$$

$$\boldsymbol{E}^a \neq \boldsymbol{0} \tag{F.58}$$

and so:

$$R^a{}_1{}^{01} + R^a{}_2{}^{02} + R^a{}_3{}^{03} = \omega^a{}_{b1}T^{01b} + \omega^a{}_{b2}T^{02b} + \omega^a{}_{b3}T^{03b} \tag{F.59}$$

The Coulomb Law therefore indicates that the electromagnetic and gravitational fields are influencing each other, and this defines field-matter interaction.

These are important insights of the Evans field theory which cannot be obtained from the Maxwell-Heaviside field theory.

These insights mean that gravitation can be used to influence electromagnetism and vice-versa. The simplest example is the Coulomb Law, one of the oldest laws of physics.

G

O(3) Electrodynamics From General Relativity and Unified Field Theory

G.1 Introduction

In generally covariant or objective unified field theory the field tensor is defined by the first Maurer-Cartan structure relation of differential geometry:

$$F^a = d \wedge A^a + \omega^a{}_b \wedge A^b \tag{G.1}$$

(where the base manifold indices have been suppressed for clarity of presentation). In eqn. (G.1) F^a is the field two-form, A^a is the potential one form, and $\omega^a{}_b$ is the spin connection.

The homogeneous and inhomogeneous field equations (HE and IE respectively) are found from the first Bianchi identity of differential geometry:

G.1.1 Homogeneous Field Equation (HE)

$$\boxed{d \wedge F^a = \mu_0 j^a = -A^{(0)} \left(q^b \wedge R^a{}_b + \omega^a{}_b \wedge T^b \right)} \tag{G.2}$$

G.1.2 Inhomogeneous Field Equation (IE)

$$\boxed{d \wedge \widetilde{F}^a = \mu_0 J^a = -A^{(0)} \left(q^b \wedge \widetilde{R}^a{}_b + \omega^a{}_b \wedge \widetilde{T}^b \right)} \tag{G.3}$$

Eqn. (G.1) to (G.3) are the correctly objective equations of electrodynamics. They are the direct logical consequence of the basic principle of objectivity in physics. This is the principle of **general relativity**, all equations of physics must retain their form under any type of coordinate transformation - the equations must be **generally covariant**.
The Maxwell-Heaviside field theory (MH) does not obey this basic principle of objectivity because it is a theory of special relativity. It is well known that

special relativity is **Lorentz covariant** but not generally covariant. The Maxwell Heaviside equations corresponding to eqn. (G.1) to (G.3) are well known to be:

$$F = d \wedge A \tag{G.4}$$

$$d \wedge F = 0 \tag{G.5}$$

$$d \wedge \widetilde{F} = \mu_0 J \tag{G.6}$$

In eqns. (G.3) and (G.6) the tilde denotes the Hodge dual. In eqn. (G.2) T^b is the torsion form of differential geometry, $R^a{}_b$ is the curvature or Riemann form, q^b is the tetrad form. The number of independent variable on the right hand side of eqn. (G.2) is reduced by the first and second Maurer-Cartan equations of differential geometry:

$$T^a = d \wedge q^a + \omega^a{}_b \wedge q^b \tag{G.7}$$

$$R^a{}_b = d \wedge \omega^a{}_b + \omega^a{}_c \wedge \omega^c{}_b \tag{G.8}$$

Therefore the independent variables are q^a and $\omega^a{}_b$. Eqn. (G.7) is transformed into eqn. (G.1) using the fundamental rules:

$$A^a = A^{(0)} q^a \tag{G.9}$$

$$F^a = A^{(0)} T^a \tag{G.10}$$

The Hodge duals in eqn. (G.3) are defined by the rules of general relativity (see Carroll):

$$\widetilde{T}^b{}_{\mu\nu} = \frac{1}{2} |g|^{\frac{1}{2}} \epsilon^{\rho\sigma}{}_{\mu\nu} T^b{}_{\rho\sigma} \tag{G.11}$$

$$\widetilde{R}^a{}_{b\mu\nu} = \frac{1}{2} |g|^{\frac{1}{2}} \epsilon^{\rho\sigma}{}_{\mu\nu} R^a{}_{b\rho\sigma} \tag{G.12}$$

Here

$$g = |g_{\mu\nu}| \tag{G.13}$$

is the determinant of the metric and $\epsilon^{\rho\sigma}{}_{\mu\nu}$ the Levi-Civita symbol in the general 4-D manifold (Evans spacetime).

It is seen that q^a and $\omega^a{}_b$ are the same in eqn. (G.2) and (G.3), but duals are used of the torsion and curvature forms.

Experimentally, we know that:

$$j^a \sim 0 \tag{G.14}$$

$$J^a > 0 \tag{G.15}$$

Therefore:

$$d \wedge F^a = 0 \tag{G.16}$$

within contemporary instrumental precision. For each index a eqn. (G.16) is eqn. (G.5). The latter is a combination of the **Gauss law of magnetism**:

$$\nabla \cdot \boldsymbol{B} = 0 \tag{G.17}$$

and the **Faraday law of induction**:

$$\nabla \times \boldsymbol{E} + \frac{\partial \boldsymbol{B}}{\partial t} = \boldsymbol{0} \tag{G.18}$$

Both laws are tested to high precision, so eqn. (G.14) follows experimentally. In general however j^a is not zero when gravitation influences electromagnetism. The latter influence is missing completely from MH theory because MH is a theory only of special relativity.

From eqn. (G.2) and (G.16) we obtain **the free space condition**:

$$\boxed{\omega^a{}_b \wedge T^b = R^a{}_b \wedge q^b} \tag{G.19}$$

which is a condition or constraint on eqns. (G.1) to (G.3) imposed by the laws (G.17) and (G.18). Eqn. (G.19) is therefore an experimental constraint. Using the Maurer-Cartan structure equations. (G.7) and (G.8):

$$T^a = D \wedge q^a \tag{G.20}$$

$$R^a{}_b = D \wedge \omega^a{}_b \tag{G.21}$$

where $D\wedge$ is the covariant exterior derivative, eqn. (G.19) becomes a relation between q^a and $\omega^a{}_b$:

$$\boxed{\omega^a{}_b \wedge (D \wedge q^b) = (D \wedge \omega^a{}_b) \wedge q^b} \tag{G.22}$$

Therefore there is only one independent variable on the right hand sides of eqn. (G.2) and (G.3).
A solution of eqn. (G.22) is:

$$\omega^a{}_b = -\frac{\kappa}{2} \epsilon^a{}_{bc} q^c \tag{G.23}$$

where κ is the wavenumber and $\epsilon^a{}_{bc}$ is the Levi-Civita symbol of the Minkowski spacetime of the tangent bundle. Given the experimental constraint (G.22), eqn. (G.23) is true both for eqn. (G.2) and (G.3), and also for eqn. (G.1). The Levi-Civita symbol is defined by:

$$\epsilon^a{}_{bc} = g^{ad} \epsilon_{dbc} \tag{G.24}$$

where:

$$g^{ad} = \text{diag}(1, -1, -1, -1)$$
$$= \begin{vmatrix} 1 & 0 & 0 & 0 \\ 0 & -1 & 0 & 0 \\ 0 & 0 & -1 & 0 \\ 0 & 0 & 0 & -1 \end{vmatrix} \tag{G.25}$$

Thus:

$$\epsilon^{1}{}_{23} = g^{1d}\epsilon_{d23} = g^{11}\epsilon_{123} = -\epsilon_{123} \tag{G.26}$$
$$\epsilon^{3}{}_{12} = g^{3d}\epsilon_{d12} = g^{33}\epsilon_{312} = -\epsilon_{312} \tag{G.27}$$
$$\epsilon^{2}{}_{31} = g^{2d}\epsilon_{d31} = g^{22}\epsilon_{231} = -\epsilon_{231} \tag{G.28}$$

and:

$$\begin{aligned} F^1 &= d \wedge A^1 + \frac{g}{2}\left(\epsilon^{1}{}_{23}A^3 \wedge A^2 + \epsilon^{1}{}_{32}A^2 \wedge A^3\right) \\ &= d \wedge A^1 - \frac{g}{2}\left(\epsilon_{123}A^3 \wedge A^2 + \epsilon_{132}A^2 \wedge A^3\right) \\ F^1 &= d \wedge A^1 + gA^2 \wedge A^3 \\ F^2 &= d \wedge A^2 + gA^3 \wedge A^1 \\ F^3 &= d \wedge A^3 + gA^1 \wedge A^2 \end{aligned} \tag{G.29}$$

Eqns. (G.29) are the fundamental definition of the field tensor of electromagnetism in objective physics.

In these equations:

$$g = \frac{\kappa}{A^{(0)}} \tag{G.30}$$

and should not be confused with the determinant of the metric. In the complex circular basis eqns. (G.29) define **O(3) electrodynamics**:

$$F^{(1)*} = d \wedge A^{(1)*} - igA^{(2)} \wedge A^{(3)} \tag{G.31}$$
$$F^{(2)*} = d \wedge A^{(2)*} - igA^{(3)} \wedge A^{(1)} \tag{G.32}$$
$$F^{(3)*} = d \wedge A^{(3)*} - igA^{(1)} \wedge A^{(2)} \tag{G.33}$$

and **the Evans spin field**:

$$\boxed{B^{(3)*} = -igA^{(1)} \wedge A^{(2)}} \tag{G.34}$$

observed experimentally in the inverse Faraday effect and the generally covariant phase of electromagnetism.

It is seen that the spin field and O(3) electrodynamics originate

in general relativity, in the fundamental requirement that physics be objective to any observer. The spin field originates in the spin connection of eqn. (G.1), and the spin connection originates in the realization that **electromagnetism** is **spinning spacetime**. Similarly gravitation is curving spacetime. Spinning is described by the torsion form T^a and curving described by the Riemann form R^a_b.

To translate from form notation to vector notation proceed as follows:

$$\boxed{\begin{aligned} F^a &= d \wedge A^a + gA^b \wedge A^c \\ F^a{}_{\mu\nu} &= (d \wedge A^a)_{\mu\nu} + gA^b{}_\mu \wedge A^c{}_\nu \\ F^3{}_{12}{}^* &= \left(d \wedge A^{(3)*}\right)_{12} - igA^{(1)}{}_1 \wedge A^{(2)}{}_2 \\ B^{(3)}{}_{12}{}^* &= -igA^{(1)}{}_1 \wedge A^{(2)}{}_2 \\ B^{(3)}{}_3{}^* &= \frac{1}{2}\left(\epsilon_{123} B^{(3)}{}_{12}{}^* + \epsilon_{213} B^{(3)}{}_{21}{}^*\right) \\ \boldsymbol{B}^{(3)*} &= -ig\boldsymbol{A}^{(1)} \wedge \boldsymbol{A}^{(2)} \end{aligned}} \tag{G.35}$$

There is no $\boldsymbol{E}^{(3)}$ field because:

$$\boxed{\begin{aligned} F^{(3)}{}_{03}{}^* &= \left(d \wedge A^{(3)}\right)^*_{03} - igA^{(1)}{}_0 \wedge A^{(2)}{}_3 \\ &= 0 \end{aligned}} \tag{G.36}$$

as observed experimentally, **there being no electric analogue of the inverse Faraday effect.**

H

Illustration That The Evans Field Theory is Completely Determined Mathematically

In the case of circular polarization, the free space condition applies (Appendix ??):

$$\omega^a{}_b \wedge \left(D \wedge q^b\right) = \left(D \wedge \omega^a{}_b\right) \wedge q^b \tag{H.1}$$

and the tetrad is defined by:

$$T^a = d \wedge q^a + \omega^a{}_b \wedge q^b \tag{H.2}$$

$$d \wedge T^a = 0 \tag{H.3}$$

$$d \wedge \widetilde{T}^a = 0 \tag{H.4}$$

The spin connection and the gamma connection are related through the tetrad postulate:

$$\partial_\mu q^a{}_\lambda + \omega^a{}_{\mu b} q^b{}_\lambda - \Gamma^\nu{}_{\mu\lambda} q^a{}_\nu = 0 \tag{H.5}$$

The Riemann Form for circular polarization is defined by:

$$\begin{aligned} R^a{}_b &= D \wedge \omega^a{}_b \\ &= d \wedge \omega^a{}_b + \omega^a{}_c \wedge \omega^c{}_b \end{aligned} \tag{H.6}$$

and the torsion form for circular polarization by eqn.(H.2). For circular polarization we have seen in Appendix G that

$$\omega^a{}_b = -\frac{1}{2} \epsilon^a{}_{bc} q^c \tag{H.7}$$

Eqns. (H.1), (H.3) and (H.7) mean that the mathematical problem is to determine the tetrad $q^a{}_\mu$. In circular polarization the elements of the tetrad are $q^{(1)}{}_x$, $q^{(1)}{}_y$, $q^{(2)}{}_x$, $q^{(2)}{}_y$, and $q^{(3)}{}_z$. There are therefore five scalar unknowns bu it is also known that:

$$q^{(1)}{}_x = q^{(2)*}{}_x \tag{H.8}$$

$$q^{(2)}{}_x = q^{(1)*}{}_x \tag{H.9}$$

$$q^{(1)}{}_y = q^{(2)*}{}_y \tag{H.10}$$

$$q^{(2)}{}_y = q^{(1)*}{}_y \tag{H.11}$$

It follows that the complex circular basis (Appendix C) determines the problem completely, because:

$$\mathbf{q}^{(1)} \times \mathbf{q}^{(2)} = i\mathbf{q}^{(3)*} \tag{H.12}$$

$$\mathbf{q}^{(2)} \times \mathbf{q}^{(3)} = i\mathbf{q}^{(1)*} \tag{H.13}$$

$$\mathbf{q}^{(3)} \times \mathbf{q}^{(1)} = i\mathbf{q}^{(2)*} \tag{H.14}$$

Therefore the tetrad elements are:

$$q^{(1)}{}_x = \frac{1}{\sqrt{2}}e^{i\phi}, q^{(1)}{}_y = -\frac{i}{\sqrt{2}}e^{i\phi}, \tag{H.15}$$

$$q^{(2)}{}_x = \frac{1}{\sqrt{2}}e^{-i\phi}, q^{(2)}{}_y = \frac{i}{\sqrt{2}}e^{-i\phi}, \tag{H.16}$$

$$q^{(3)}{}_z = 1 \tag{H.17}$$

From eqn. (H.15) to (H.17) it is possible to deduce all the scalar elements governing the differential geometry of circular polarization. The method is illustrated in the Appendix . For convenience denote:

$$q^1{}_x = q^{(1)}{}_x \,, q^1{}_y = q^{(2)}{}_y \quad etc. \tag{H.18}$$

It follows that the spin connection elements are given in general by:

$$\omega^1{}_2 = \frac{\kappa}{2}q^3 \tag{H.19}$$

$$\omega^2{}_3 = \frac{\kappa}{2}q^1 \tag{H.20}$$

$$\omega^3{}_1 = \frac{\kappa}{2}q^2 \tag{H.21}$$

from the free space condition.

The gamma connection elements are then obtained from the tetrad postulate. For example, for $\lambda = x, \mu = 0, a = 1$:

$$\partial_0 q^1{}_x + \omega^1{}_{0b} q^b{}_x - \Gamma^\nu{}_{0x}\, q^1{}_\nu = 0 \tag{H.22}$$

and since $\omega^1_{x1} = 0$, we have:

$$\partial_0 q^1{}_x + \omega^1{}_{02} q^2{}_x - \Gamma^x{}_{0x} q^1{}_x - \Gamma^y{}_{0x} q^1{}_y = 0 \tag{H.23}$$

However,

$$\omega^1{}_{02} = \frac{\kappa}{2} q^3{}_0 = 0 \tag{H.24}$$

and using

$$\phi = \omega t - \kappa z \tag{H.25}$$

we obtain:

$$\Gamma^y{}_{0x} + i\Gamma^x{}_{0x} = -\kappa \tag{H.26}$$

Similarly, for $\lambda = y, \mu = 0, a = 1$:

$$\Gamma^x{}_{0y} - i\Gamma^y{}_{0y} = \kappa, \tag{H.27}$$

for $\lambda = x, \mu = 0, a = 2$:

$$i\Gamma^x{}_{0x} - \Gamma^y{}_{0x} = \kappa, \tag{H.28}$$

and for $\lambda = y, \mu = 0, a = 2$:

$$\Gamma^x{}_{0y} + i\Gamma^y{}_{0y} = \kappa \tag{H.29}$$

Thus:

$$\Gamma^x{}_{0y} = -\Gamma^y{}_{0x} = \kappa, \tag{H.30}$$

$$\Gamma^x{}_{0x} = \Gamma^y{}_{0y} = 0 \tag{H.31}$$

similarly, for $\lambda = z, \mu = 0, a = 3$ we obtain:

$$\Gamma^z{}_{0z} = 0 \tag{H.32}$$

It is seen that all the scalar elements of the gamma connection are exactly determined.

Again for $\lambda = x, \mu = 3, a = 1$:

$$\partial_3 q^1{}_x + \omega^1{}_{3b} q^b{}_x - \Gamma^\nu{}_{3x} q^1{}_\nu = 0 \tag{H.33}$$

i.e.

$$\partial_z q^1{}_x + \omega^1{}_{z2} q^2{}_x - \Gamma^x{}_{zx} q^1{}_x - \Gamma^y{}_{zx} q^1{}_y = 0 \tag{H.34}$$

Now use:

$$\omega^1{}_{z2} = \frac{\kappa}{2} q^3{}_z = \frac{\kappa}{2} \tag{H.35}$$

to obtain:

$$\Gamma^y{}_{zx} + i\Gamma^x{}_{zx} = \kappa \left(1 + \frac{i}{2} e^{-i\phi}\right) \tag{H.36}$$

Similarly, for $\lambda = x, \mu = 3, a = 2$:

$$\Gamma^{y}{}_{zx} - i\Gamma^{x}{}_{zx} = \kappa \left(1 + \frac{i}{2} e^{i\phi} \right) \tag{H.37}$$

It follows that:

$$\begin{aligned} \Gamma^{y}{}_{zx} &= \kappa \left(1 + \frac{i}{4} \left(e^{i\phi} + e^{-i\phi} \right) \right) \\ &= \kappa \left(1 + \frac{i}{2} \cos\phi \right) \end{aligned} \tag{H.38}$$

and

$$\begin{aligned} \Gamma^{x}{}_{zx} &= -i\kappa \left(1 + \frac{i}{4} \left(e^{-i\phi} - e^{i\phi} \right) \right) \\ &= -i\kappa \left(1 - \frac{i}{2} \sin\phi \right) \end{aligned} \tag{H.39}$$

The torsion tensor elements may then be deduced from the gamma connection elements using:

$$T^{\kappa}{}_{\mu\nu} = \Gamma^{\kappa}{}_{\mu\nu} - \Gamma^{\kappa}{}_{\nu\mu} \tag{H.40}$$

and the Riemann tensor elements from:

$$R^{\rho}{}_{\sigma\mu\nu} = \partial_{\mu}\Gamma^{\rho}{}_{\nu\sigma} - \partial_{\nu}\Gamma^{\rho}{}_{\mu\sigma} + \Gamma^{\rho}{}_{\mu\lambda}\Gamma^{\lambda}{}_{\nu\sigma} - \Gamma^{\rho}{}_{\nu\lambda}\Gamma^{\lambda}{}_{\mu\sigma} \tag{H.41}$$

Therefore all scalar elements have been deduced of the differential geometry of circular polarization.

□

I

A Summary of the Evans Field Theory

I.1 The Homogeneous Field Equation

Barebones Notation

$$D \wedge F = R \wedge A \tag{I.1}$$

Tangent Bundle Notation

$$D \wedge F^a = R^a{}_b \wedge A^b \tag{I.2}$$

Complete Index Notation

$$(D \wedge F^a)_{\mu\nu\rho} = \left(R^a{}_b \wedge A^b\right)_{\mu\nu\rho} \tag{I.3}$$

Spin Connection Notation

$$\left(d \wedge F^a + \omega^a{}_b \wedge F^b\right)_{\mu\nu\rho} = \left(R^a{}_b \wedge A^b\right)_{\mu\nu\rho} \tag{I.4}$$

Tensor Notation

$$\begin{aligned}(d \wedge F)^a_{\mu\nu\rho} &= \partial_\mu F^a{}_{\nu\rho} + \partial_\nu F^a{}_{\rho\mu} + \partial_\rho F^a{}_{\mu\nu} \\ (\omega \wedge F)^a{}_{\mu\nu\rho} &= \omega^a{}_{\mu b} F^b{}_{\nu\rho} + \omega^a{}_{\nu b} F^b{}_{\rho\mu} + \omega^a{}_{\rho b} F^b{}_{\mu\nu} \\ (R \wedge A)^a{}_{\mu\nu\rho} &= R^a{}_{b\mu\nu} A^b{}_\rho + R^a{}_{b\nu\rho} A^b{}_\mu + R^a{}_{b\rho\mu} A^b{}_\nu\end{aligned} \tag{I.5}$$

Eq.(I.5) give the most complete expression of the homogeneous field equation, i.e.:

$$\begin{aligned}\partial_\mu F^a{}_{\nu\rho} + \partial_\nu F^a{}_{\rho\mu} + \partial_\rho F^a{}_{\mu\nu} + \omega^a{}_{\mu b} F^b{}_{\nu\rho} + \omega^a{}_{\nu b} F^b{}_{\rho\mu} + \omega^a{}_{\rho b} F^b{}_{\mu\nu} = \\ R^a{}_{b\mu\nu} A^b{}_\rho + R^a{}_{b\nu\rho} A^b{}_\mu + R^a{}_{b\rho\mu} A^b{}_\nu\end{aligned} \tag{I.6}$$

Maxwell-Heaviside Limit

$$\begin{aligned}R^a{}_b \wedge A^b &= \omega^a{}_b \wedge F^b \\ \partial_\mu F^a{}_{\nu\rho} + \partial_\nu F^a{}_{\rho\mu} + \partial_\rho F^a{}_{\mu\nu} &= 0\end{aligned} \tag{I.7}$$

The tangent bundle is not identified, so:

$$\partial_\mu F_{\nu\rho} + \partial_\nu F_{\rho\mu} + \partial_\rho F_{\mu\nu} = 0 \tag{I.8}$$

Eq.(I.8) is the homogeneous field equation of the MH theory. Eq. (I.8) cannot describe the effect of gravitation on electromagnetism and vice-versa, because it is a flat spacetime equation of special relativity.

I.2 Hodge Dual of the Homogeneous Field Equation

The Hodge dual of the homogenous field equation is a re-expression of the Bianchi identity and therefore contains the same information. In the general 4-D manifold the Hodge dual must be precisely defined as follows.

The general definition of the Hodge dual is given by Carroll.

$$\widetilde{\mathcal{X}}_{\mu_1 \ldots \mu_{n-p}} = \frac{1}{p!} \epsilon^{\nu_1 \ldots \nu_p}{}_{\mu_1 \ldots \mu_{n-p}} \mathcal{X}_{\nu_1 \ldots \nu_p} \tag{I.9}$$

In a general n-dimensional manifold Eq.(I.9) maps from a p-form of differential geometry to an (n-p)-form. The general Levi-Civita symbol is defined in any manifold to be:

$$\epsilon_{\mu_1 \mu_2 \ldots \mu_n} = \begin{cases} 1 \text{ if } \mu_1 \mu_2 \ldots \mu_n \text{ is an even permutation} \\ -1 \text{ if } \mu_1 \mu_2 \ldots \mu_n \text{ is an odd permutation} \\ 0 \qquad\qquad \text{otherwise} \end{cases} \tag{I.10}$$

The Levi-Civita tensor used in the definition of the Hodge dual, is:

$$\epsilon_{\mu_1 \mu_2 \ldots \mu_n} = \left(|g|\right)^{1/2} \widetilde{\epsilon}_{\mu_1 \mu_2 \ldots \mu_n} \tag{I.11}$$

where $|g|$ is the numerical magnitude of the determinant of the metric. In a four-dimensional manifold a two-form is dual to a two-form:

$$\widetilde{\mathcal{X}}_{\mu_1 \mu_2} = \frac{1}{2} \epsilon^{\nu_1 \nu_2}{}_{\mu_1 \mu_2} \mathcal{X}_{\nu_1 \nu_2} \tag{I.12}$$

Indices are raised and lowered on the Levi-Civita tensor by use of the metric tensor. The latter is normalized by:

$$g^{\mu\nu} g_{\mu\nu} = 4 \tag{I.13}$$

Therefore we have results such as:

$$\widetilde{\mathcal{X}}_{\mu\nu} = \frac{1}{2} \epsilon^{\rho\sigma}{}_{\mu\nu} \mathcal{X}_{\rho\sigma} \tag{I.14}$$

with, for example:

$$\epsilon_{\sigma\mu\nu\rho} = g_{\sigma\kappa}\epsilon^{\kappa}{}_{\mu\nu\rho} \tag{I.15}$$

We may then rewrite eqn.(I.4):

$$\boxed{\partial^{\mu}\widetilde{F}^{a}{}_{\mu\nu} = \mu_0 \widetilde{j}^{a}{}_{\nu}} \tag{I.16}$$

Note that eqn.(I.4) and (I.16) contain the same information, they are both expressions of the homogeneous field equation of objective or generally covariant physics.

I.3 The Inhomogeneous Field Equation

The homogeneous field equations eqn.(I.4) and (I.16) is the generally covariant form of Gauss Law applied to magnetism and the Faraday Law of induction. The inhomogeneous field equation is deduced from eqn. (I.1) as follows:

$$\boxed{D \wedge \widetilde{F} = \widetilde{R} \wedge A} \tag{I.17}$$

The complete description is therefore:

$$\boxed{\begin{aligned} d \wedge F &= R \wedge A - \omega \wedge F = \mu_0 j \\ d \wedge \widetilde{F} &= \widetilde{R} \wedge A - \omega \wedge \widetilde{F} = \mu_0 J \end{aligned}} \tag{I.18}$$

eqn.(I.18) are the generally covariant forms of the four fundamental laws of electromagnetics.

In order to CAD/CAM a circuit working from the general 4-D manifold known as "Evans Spacetime" eqn.(I.18) must be solved simultaneously. The mathematical problem is one of solving simultaneous partial differential tensor equations with given initial and boundary conditions.

I.3.1 The Standard Model

The equivalents of eqn.(I.18) is the standard model are:

$$\boxed{\begin{aligned} d \wedge F &= 0 \\ d \wedge \widetilde{F} &= \mu_0 J \end{aligned}} \tag{I.19}$$

Eqs (I.19) are equations of a flat or Minkowski spacetime. Eqs (I.19) are not objective equations of physics because they are not equations of general relativity. There is no indication in eqs (I.19) that J is derived from the wedge product of the Riemann and tetrad forms. The Evans field equations (I.18) show that

$$\boxed{J = \frac{A^{(0)}}{\mu_0}\left(\widetilde{R} \wedge q - \omega \wedge \widetilde{T}\right)} \tag{I.20}$$

Eqn. (I.20) shows that Evans spacetime is a source of electric charge/current density.

It is therefore very important to evaluate J numerically. For a given potential:

$$A = A^{(0)} q \tag{I.21}$$

and given curvature:

$$R = D \wedge \omega \tag{I.22}$$

we need to calculate J. In eqn. (I.22) ω is the spin connection, related to the Christoffel connection.

I.4 Summary of the Unified Field Theory

Any situation in field theory is described by:

$$D \wedge F^a = R^a{}_b \wedge A^b \tag{I.23}$$

$$D \wedge \widetilde{F}^a = \widetilde{R}^a{}_b \wedge A^b \tag{I.24}$$

where:

$$F^a = A^{(0)} T^a \tag{I.25}$$

$$A^a = A^{(0)} q^a \tag{I.26}$$

and:

$$D \wedge F^a = d \wedge F^a + \omega^a{}_b \wedge F^b \tag{I.27}$$

the fundamental charge-current three-forms are defined by:

$$j^a = \frac{1}{\mu_0}\left(R^a{}_b \wedge A^b - \omega^a{}_b \wedge F^b\right) \tag{I.28}$$

$$J^a = \frac{1}{\mu_0}\left(\widetilde{R}^a{}_b \wedge A^b - \omega^a{}_b \wedge \widetilde{F}^b\right) \tag{I.29}$$

In eqs (I.23–I.29):

$A^{(0)}$ =Fundamental potential mangnitude in volts
q^a =vector valued tetrad one-form
T^a =vector valued torsion two-form
$R^a{}_b$ =tensor valued curvature two-form
$\omega^a{}_b$ =spin-connection one-form
F^a =vector valued electromagnetic field two-form
A^a =vector valued electromagnetic potential one-form
j^a =homogeneous current three-form
J^a =inhomogeneous current three-form
$d\wedge$ =exterior derivative
$D\wedge$ =covariant exterior derivative
μ_0 =permeability is vacuo (SI)

The equations below are in SI units. $\widetilde{F^a}$ is the hodge dual of F^a in the general 4-d manifold (Evans spacetime) and $\widetilde{R^a{}_b}$ is the Hodge dual of $R^a{}_b$ in Evans spacetime.

The following are well-known limiting forms of the Evans field theory.

I.5 Einstein Field Theory of Gravitation

This limit is defined by:

$$F^a = 0 \tag{I.30}$$

$$R^a{}_b \wedge A^b = 0 \tag{I.31}$$

$$\widetilde{F^a} = 0 \tag{I.32}$$

$$\widetilde{R^a{}_b} \wedge A^b \neq 0 \tag{I.33}$$

Eqn.(I.31) implies that the Christoffel symbol is symmetric in its lower two indices. This self-consistantly implies:

$$T^a(\text{Einstein}) = 0 \tag{I.34}$$

Self-consistantly therefore, in the Einstein field theory of gravitation, there is no electromagnetic field present. In this theory the torsion tensor is zero because it is defined as the difference:

$$T^{\kappa}{}_{\mu\nu} = \Gamma^{\kappa}{}_{\mu\nu} - \Gamma^{\kappa}{}_{\nu\mu} = 0 \tag{I.35}$$

Metrics that obey conditions (I.35) cannot be used in a unified field theory, they can only be used to describe gravitation.

I.6 Maxwell-Heaviside Field Theory of Electromagnetism

This is the limit described by:

$$d \wedge F^a = 0 \tag{I.36}$$

$$d \wedge \widetilde{F}^a = \mu_0 J^a \tag{I.37}$$

Therefore:

$$R^a{}_b \wedge A^b = \omega^a{}_b \wedge F^b \tag{I.38}$$

$$j^a = 0 \tag{I.39}$$

$$\widetilde{R}^a{}_b \wedge A^b \neq \omega^a{}_b \wedge \widetilde{F}^b \tag{I.40}$$

$$J^a \neq 0 \tag{I.41}$$

The Evans spacetime reduces to Minkowski spacetime, so $D\wedge$ is replaced by $d\wedge$. In the original nineteenth century Maxwell Heaviside theory, the field is an entity superimposed on a flat Euclidean 3-D space and the concept of time is distinct from that of space. The existence of the tangent bundle index a is not recognized, neither is that of the spin connection $\omega^a{}_b$ and curvature $R^a{}_b$. The current J is essentially empirical in MH field Theory.

J

Mathematical Proofs

J.1 The Free Space Condition

$$\omega^a{}_b = -\kappa \epsilon^a{}_{bc} q^c$$

This fundamental condition is a solution of:

$$\boxed{R^a{}_b \wedge q^b = \omega^a{}_b \wedge T^b} \tag{J.1}$$

$$\left(D \wedge \omega^a{}_b\right) \wedge q^b = \omega^a{}_b \wedge \left(D \wedge q^b\right) \tag{J.2}$$

$$\left(d \wedge \omega^a{}_b\right) \wedge q^b + \left(\omega^a{}_c \wedge \omega^c{}_b\right) \wedge q^b = \omega^a{}_b \wedge \left(d \wedge q^b\right) + \omega^a{}_b \wedge \left(\omega^b{}_c \wedge q^c\right) \tag{J.3}$$

To Prove:

$$\left(d \wedge \omega^a{}_b\right) \wedge q^b = \omega^a{}_b \wedge \left(d \wedge q^b\right) \tag{J.4}$$

Proof for $a = 1$

$$\left(d \wedge \omega^1{}_2\right) \wedge q^2 + \left(d \wedge \omega^1{}_3\right) \wedge q^3 = \omega^1{}_2 \wedge \left(d \wedge q^2\right) + \omega^1{}_3 \wedge \left(d \wedge q^3\right) \tag{J.5}$$

Eqn. (J.5)

$$\omega^1{}_2 = -\kappa \epsilon^1{}_{23} q^3 = \kappa \epsilon_{123} q^3 = \kappa q^3 \tag{J.6}$$

$$\omega^1{}_3 = -\kappa \epsilon^1{}_{23} q^2 = \kappa \epsilon_{132} q^2 = -\kappa q^2 \tag{J.7}$$

i.e.

$$\left(d \wedge q^3\right) \wedge q^2 - \left(d \wedge q^2\right) \wedge q^3 = q^3 \wedge \left(d \wedge q^2\right) - q^2 \wedge \left(d \wedge q^3\right) \tag{J.8}$$

$$\begin{aligned} \Longrightarrow \left(d \wedge q^3\right) \wedge q^2 &= -q^2 \wedge \left(d \wedge q^3\right) \\ - \left(d \wedge q^2\right) \wedge q^3 &= q^3 \wedge \left(d \wedge q^2\right) \end{aligned} \tag{J.9}$$

To prove

$$\left(\omega^a{}_c \wedge \omega^c{}_b\right) \wedge q^b = \omega^a{}_b \wedge \left(\omega^b{}_c \wedge q^c\right) \tag{J.10}$$

Proof for a=1, b=2, c=3;

$$\left(\omega^1{}_3 \wedge \omega^3{}_2\right) \wedge q^2 = \omega^1{}_2 \wedge \left(\omega^2{}_3 \wedge q^3\right) \tag{J.11}$$

where

$$\omega^1{}_2 = \kappa q^3; \;\; \omega^1{}_3 = -\kappa q^2$$
$$\omega^3{}_2 = -\kappa q^1; \;\; \omega^2{}_3 = \kappa q^1$$

therefore

$$\left(q^2 \wedge q^1\right) \wedge q^2 = -q^3 \wedge \left(q^1 \wedge q^3\right)$$

i.e.

$$\begin{aligned} q^3 \wedge q^2 &= -q^3 \wedge \left(-q^2\right) \\ &= q^3 \wedge q^2 \end{aligned} \tag{J.12}$$

For O(3) electrodynamics we choose:

$$\omega^a{}_b = -\frac{1}{2}\kappa \epsilon^a{}_{bc} q^c \tag{J.13}$$

in the structure relation:

$$D \wedge q^a = d \wedge q^a + \omega^a{}_b \wedge q^b \tag{J.14}$$

Proof For a=1:

$$D \wedge q^1 = d \wedge q^1 + \frac{1}{2}\left(\epsilon^1{}_{23} q^3 \wedge q^2 + \epsilon^1{}_{32} q^2 \wedge q^3\right) \tag{J.15}$$

$$\boxed{D \wedge q^1 = d \wedge q^1 + \kappa q^2 \wedge q^3}$$

In the O(3) circular complex basis this gives O(3) electrodynamics.

This allows the tetrad of the free field to be identified as the potential, and also the spin connection. O(3) electrodynamics is therefore a fundamental theory of general relativity.

J.2 The Tetrad Postulate

The tetrad postulate follows from the fact that a tensor is independent of the way it is written. The postulate follows from a consideration of the covariant derivative of a vector in two different bases. We denote these by J.16 and J.17. thus:

$$(DX)_1 = (DX)_2 \tag{J.16}$$

It follows that:

$$D_\nu q^a{}_\mu = 0. \tag{J.17}$$

For those interested a detailed proof is given as follows but eqn. (J.16) is enough to know where the tetrad postulate comes from.

Detailed Proof

In the coordinate basis (see Carroll(3.129))

$$\begin{aligned} DX &= (D_\mu X^\nu)\, dx^\mu \otimes \partial_\nu \\ &= \left(\partial_\mu X^\nu + \Gamma^\nu_{\ \mu\lambda} X^\lambda\right) dx^\mu \otimes \partial_\nu \end{aligned} \tag{J.18}$$

In the mixed basis:

$$\begin{aligned} DX &= (D_\mu X^a)\, dx^\mu \otimes \hat{e}_{(a)} \\ &= \left(\partial_\mu X^a + \omega^a_{\ \mu b} X^b\right) dx^\mu \otimes \hat{e}_{(a)} \end{aligned} \tag{J.19}$$

$$\begin{aligned} &= \left(\partial_\mu \left(q^a_{\ \nu} X^\nu\right) + \omega^a_{\ \mu b} q^b_{\ \lambda} X^\lambda\right) dx^\mu \otimes \left(q^\sigma_{\ a} \partial_\sigma\right) \\ &= q^\sigma_{\ a} \left(q^a_{\ \nu} \partial_\mu X^\nu + X^\nu \partial_\mu q^a_{\ \nu} + \omega^a_{\ \mu b} q^b_{\ \lambda} X^\lambda\right) dx^\mu \otimes \partial_\sigma \end{aligned} \tag{J.20}$$

where we have used the commutator rule. Now switch σ to μ and use:

$$q^\nu_{\ a} q^a_{\ \nu} = 1 \tag{J.21}$$

to obtain:

$$DX = \left(\partial_\mu X^\nu + q^\nu_{\ a} \partial_\mu q^a_{\ \lambda} X^\lambda + q^\nu_{\ a} q^b_{\ \lambda} \omega^a_{\ \mu b} X^\lambda\right) dx^\mu \otimes \partial_\nu \tag{J.22}$$

Now compare eqn. (J.18) and (J.22) to give:

$$\Gamma^\nu_{\ \mu\lambda} = q^\nu_{\ a} \partial_\mu q^a_{\ \lambda} + q^\nu_{\ a} q^b_{\ \lambda} \omega^a_{\ \mu b} \tag{J.23}$$

multiply both sides of eqn.(J.23) by $q^a_{\ \nu}$:

$$q^a_{\ \nu} \Gamma^\nu_{\ \mu\lambda} = \partial_\mu q^a_{\ \lambda} + q^b_{\ \lambda} \omega^a_{\ \mu b} \tag{J.24}$$

i.e.

$$\boxed{D_\mu q^a_{\ \lambda} = \partial_\mu q^a_{\ \lambda} + \omega^a_{\ \mu b} q^b_{\ \lambda} - \Gamma^\nu_{\ \mu\lambda} q^a_{\ \nu} = 0} \tag{J.25}$$

□

Eqn. (J.25) is known as **the tetrad postulate**, and is true for all connections.

Meaning of the Tetrad Postulate

The tetrad postulate means that the basis chosen for DX does not affect the result. The tetrad postulate originates in the definition of the tetrad itself:

$$V^a = q^a_{\ \mu} V^\mu \tag{J.26}$$

where a refers to the tangent spacetime and μ to the base manifold.

J.3 The Evans Lemma

The Evans Lemma is the direct result of **the tetrad postulate** of differential geometry:

$$\boxed{D_\mu q^a{}_\lambda = \partial_\mu q^a{}_\lambda + \omega^a{}_{\mu b} q^b{}_\lambda - \Gamma^\nu{}_{\mu\lambda} q^a{}_\nu = 0} \tag{J.27}$$

using the notation of the text. It follows from eqn. (J.27) that:

$$D^\mu (D_\mu q^a{}_\lambda) = \partial^\mu \left(D_\mu q^a{}_\lambda\right) = 0, \tag{J.28}$$

i.e.

$$\partial^\mu \left(\partial_\mu q^a{}_\lambda + \omega^a{}_{\mu b} q^b{}_\lambda - \Gamma^\nu{}_{\mu\lambda} q^a{}_\nu\right) = 0, \tag{J.29}$$

or

$$\Box q^a{}_\lambda = \partial^\mu \left(\Gamma^\nu{}_{\mu\lambda} q^a{}_\nu\right) - \partial^\mu \left(\omega^a{}_{\mu b} q^b{}_\lambda\right). \tag{J.30}$$

Define:

$$R q^a{}_\lambda := \partial^\mu \left(\Gamma^\nu{}_{\mu\lambda} q^a{}_\nu\right) - \partial^\mu \left(\omega^a{}_{\mu b} q^b{}_\lambda\right) \tag{J.31}$$

to obtain **the Evans Lemma**:

$$\boxed{\Box q^a{}_\lambda = R q^a{}_\lambda} \tag{J.32}$$

K

The Four Fundamental Laws of Electrodynamics In The Unified Field Theory: Vector Notation

K.1 The Inhomogeneous Laws

Those are obtained from:

$$\partial_\mu F^{a\mu\nu} = \mu_0 c J^{a\nu} \tag{K.1}$$

where:

$$J^{a\nu} = -\frac{A^{(0)}}{\mu_0}\left(q^b{}_\mu R^a{}_b{}^{\mu\nu} + \omega^a{}_{\mu b} T^{b\mu\nu}\right) \tag{K.2}$$

K.2 Coulomb Law ($\nu = 0, \mu = 1, 2, 3$)

The charge density is:

$$\begin{aligned} J^{a0} = &-\frac{A^{(0)}}{\mu_0}(q^b{}_1 R^a{}_b{}^{10} + q^b{}_2 R^a{}_b{}^{20} + q^b{}_3 R^a{}_b{}^{30} \\ &+ \omega^a{}_{1b} T^{b10} + \omega^a{}_{2b} T^{b20} + \omega^a{}_{3b} T^{b30}) \end{aligned} \tag{K.3}$$

so

$$\boxed{\nabla \cdot \boldsymbol{E}^a = \frac{\rho^a}{\epsilon_0} = \mu_0 c J^{a0}} \tag{K.4}$$

It is seen that charge density originates in the geometry of spacetime. Therefore if gravitation changes spacetime it has an effect on the charge density. This is a direct result of differential geometry.

K.3 Ampère Maxwell law ($\nu = 1, 2, 3$)

For $\nu = 1, \mu = 0, 2, 3$

$$\begin{aligned} J^a{}_x = J^{a1} = & -\frac{A^{(0)}}{\mu_0}(q^b{}_0 R^a{}_b{}^{01} + q^b{}_2 R^a{}_b{}^{21} + q^b{}_3 R^a{}_b{}^{31} \\ & + \omega^a{}_{0b} T^{b02} + \omega^a{}_{2b} T^{b21} + \omega^a{}_{3b} T^{b31}) \end{aligned} \tag{K.5}$$

For $\nu = 2, \mu = 0, 1, 3$

$$\begin{aligned} J^a{}_y = J^{a2} = & -\frac{A^{(0)}}{\mu_0}(q^b{}_0 R^a{}_b{}^{02} + q^b{}_1 R^a{}_b{}^{12} + q^b{}_3 R^a{}_b{}^{32} \\ & + \omega^a{}_{0b} T^{b02} + \omega^a{}_{1b} T^{b12} + \omega^a{}_{3b} T^{b32}) \end{aligned} \tag{K.6}$$

For $\nu = 3, \mu = 0, 1, 2$

$$\begin{aligned} J^a{}_z = J^{a3} = & -\frac{A^{(0)}}{\mu_0}(q^b{}_0 R^a{}_b{}^{03} + q^b{}_1 R^a{}_b{}^{13} + q^b{}_2 R^a{}_b{}^{23} \\ & + \omega^a{}_{0b} T^{b03} + \omega^a{}_{1b} T^{b13} + \omega^a{}_{2b} T^{b23}) \end{aligned} \tag{K.7}$$

Thus:

$$\boxed{\nabla \times \boldsymbol{B}^a = \frac{1}{c^2}\frac{\partial \boldsymbol{E}^a}{\partial t} + \mu_0 \boldsymbol{J}^a} \tag{K.8}$$

in which the scalar elements of current are given in eqns.(K.5) to (K.7).

K.4 The Gauss Law of Magnetism

For all practical purposes:

$$\boxed{\nabla \cdot \boldsymbol{B}^a = 0} \tag{K.9}$$

K.5 The Faraday Law of Induction

For all practical purposes:

$$\boxed{\nabla \times \boldsymbol{E}^a + \frac{\partial \boldsymbol{B}^a}{\partial t} = \boldsymbol{0}} \tag{K.10}$$

If we take into consideration the very tiny homogeneous current then:

$$\begin{aligned} j^{a\nu} & = -\frac{A^{(0)}}{\mu_0}\left(q^b{}_\mu \widetilde{R}^a{}_b{}^{\mu\nu} + \omega^a{}_{\mu b} \widetilde{T}^{b\mu\nu}\right) \\ & \sim 0 \end{aligned} \tag{K.11}$$

and very tiny terms appear on the right hand sides of eqns. (K.9) and (K.10).

K.6 Simplification Of The IE

An important simplification of the structure of the IE is possible as follows.

In free space, it is known that:

$$d \wedge F^a = 0 \tag{K.12}$$

$$d \wedge \widetilde{F}^a = 0 \tag{K.13}$$

and so:

$$R^a{}_b \wedge q^b = \omega^a{}_b \wedge T^b \tag{K.14}$$

$$\widetilde{R}^a{}_b \wedge q^b = \omega^a{}_b \wedge \widetilde{T}^b \tag{K.15}$$

The free space condition means that:

$$\omega^a{}_b = -\kappa \epsilon^a{}_{bc} q^c \tag{K.16}$$

$$R^a{}_b = -\kappa \epsilon^a{}_{bc} T^c \tag{K.17}$$

The IE describes **field-matter interaction** as follows:

$$\begin{aligned} d \wedge \widetilde{F}^a &= A^{(0)} \left(\widetilde{R}^a{}_b \wedge q^b - \omega^a{}_b \wedge \widetilde{T}^b \right) \\ &\neq 0. \end{aligned} \tag{K.18}$$

Comparison of eqn. (K.15) and (K.18) means that the pressence of mass changes the equality (K.15). The reason for this is **gravitation**, i.e. the presence of mass.

For Einsteinian or Newtonian gravitation:

$$R^a{}_b \wedge q^b = 0 \tag{K.19}$$

but:

$$\left(\widetilde{R}^a{}_b \wedge q^b \right)_{grav.} \neq 0. \tag{K.20}$$

However, for electromagnetism:

$$\left(\widetilde{R}^a{}_b \wedge q^b - \omega^a{}_b \wedge \widetilde{T}^b \right)_{e/m} = 0. \tag{K.21}$$

Therefore the IE simplifies to:

$$\boxed{d \wedge \widetilde{F}^a = A^{(0)} \left(\widetilde{R}^a{}_b \wedge q^b \right)_{grav}} \tag{K.22}$$

and:

$$J^a = \frac{A^{(0)}}{\mu_0} \left(\widetilde{R}^a{}_b \wedge q^b \right)_{grav.} \tag{K.23}$$

In the weak field limit eqn. (K.23) means that he Coulomb and Newton inverse square laws have the same distance dependence, as known experimentally.

APART FROM THE $A^{(0)}$ FACTOR J^a ORIGINATES ENTIRELY IN CENTRALLY DIRECTED GRAVITATION.

K.7 Derivation of The Coulomb Law From The Evans Unified Field Theory

The Coulomb Law is derived from the inhomogeneous Evans field equation (barebones notation):

$$\begin{aligned} d \wedge \widetilde{F} &= A^{(0)}(\widetilde{R} \wedge q - \omega \wedge \widetilde{T}) \\ &= -A^{(0)}(q \wedge \widetilde{R} + \omega \wedge \widetilde{T}). \end{aligned} \tag{K.24}$$

For electromagnetic radiation in free space:

$$q \wedge \widetilde{R} + \omega \wedge \widetilde{T} = 0. \tag{K.25}$$

It is assumed that condition (K.25) continues to be true in field-matter interaction. (This is equivalent to the standard **minimal prescription** where p^μ is replaced by $p^\mu + eA^\mu$.)

For central gravitation (Einstein/Newton):

$$T = 0 \tag{K.26}$$

$$R \wedge q = 0 \tag{K.27}$$

$$\widetilde{R} \wedge q \neq 0. \tag{K.28}$$

Therefore from eqns. (K.25) to (K.28):

$$\boxed{d \wedge \widetilde{F} = -A^{(0)}(\widetilde{R} \wedge q)_{grav}} \tag{K.29}$$

This is the inhomogeneous field equation linking electromagnetism to gravitation. Any type of electromagnetic field matter interaction is described by eqn. (K.29) provided eqn. (K.25) remains true for the electromagnetic field when the latter interacts with matter.

K.7.1 Tensor Notation

Eqn. (K.29) is:

$$\partial_\mu \widetilde{F}^a{}_{\nu\rho} + \partial_\rho \widetilde{F}^a{}_{\mu\nu} + \partial_\nu \widetilde{F}^a{}_{\rho\mu} = -A^{(0)}(q^b{}_\mu \widetilde{R}^a{}_{b\nu\rho} + q^b{}_\rho \widetilde{R}^a{}_{b\mu\nu} + q^b{}_\nu \widetilde{R}^a{}_{b\rho\mu}) \tag{K.30}$$

which is the same equation as:

$$\partial_\mu F^{a\mu\nu} = -A^{(0)} q^b{}_\mu R^a{}_b{}^{\mu\nu} \tag{K.31}$$

$$\boxed{\partial_\mu F^{a\mu\nu} = -A^{(0)} R^a{}_\mu{}^{\mu\nu}} \tag{K.32}$$

using:

$$R^a{}_{\lambda\nu\mu} = q^b{}_\lambda R^a{}_{b\nu\mu}. \tag{K.33}$$

Eqn.(K.32) is the simplest tensor formulation of the inhomogeneous Evans field equation.

K.7.2 Vector Notation

In vector notation eqn (K.32) gives the Coulomb Law and the Ampère-Maxwell Law.

Coulomb Law (ν=0, μ=1,2,3)

$$\begin{aligned}\partial_1 F^{a10} + \partial_2 F^{a20} + \partial_3 F^{a30} &= -A^{(0)}(R^a{}_1{}^{10} + R^a{}_2{}^{20} + R^a{}_3{}^{30}) \\ &= -A^{(0)} R^a{}_i{}^{i0}\end{aligned} \tag{K.34}$$

where summation over repeated i is implied. Now denote **the fundamental voltage,** $\phi^{(0)}$ by:

$$\phi^{(0)} = cA^{(0)} \tag{K.35}$$

to obtain:

$$\nabla \cdot \boldsymbol{E}^a = -cA^{(0)} R^a{}_i{}^{i0} \tag{K.36}$$

i.e.

$$\boxed{\nabla \cdot \boldsymbol{E}^a = -\phi^{(0)} R^a{}_i{}^{i0}} \tag{K.37}$$

and

$$\boxed{\rho^a = -\epsilon_0 \phi^{(0)} R^a{}_i{}^{i0}}. \tag{K.38}$$

Eqn. (K.37) is the Coulomb Law unified with the Newton inverse square law.

Notes.

The units on both sides of eqn. (K.37) are volt/m squared and it is seen in eqn (K.38) that charge density originates in $R^a{}_i{}^{i0}$, **the sum of three Riemann curvature elements.** These elements describe gravitation in the Einstein theory of general relativity. The elements are therefore governed by theEinstein field equation. In the weak field limit this becomes **the Newton inverse square law.**

Given the existence of $\phi^{(0)}$ it is seen from eqn. (K.37) and (K.38) that **an electric field can be generated by gravitation.**

Ampère-Maxwell Law (ν=1,2,3)

This is given by:

$$\boxed{\nabla \times \boldsymbol{B}^a = \frac{1}{c^2}\frac{\partial \boldsymbol{E}^a}{\partial t} + \mu_0 \boldsymbol{J}^a} \tag{K.39}$$

where:

$$\boldsymbol{J}^a = J_x^a \boldsymbol{i} + J_y^a \boldsymbol{j} + J_z^a \boldsymbol{k} \tag{K.40}$$

and:

$$J^a{}_x = -\frac{A^{(0)}}{\mu_0}\left(R^a{}_0{}^{10} + R^a{}_2{}^{12} + R^a{}_3{}^{13}\right) \tag{K.41}$$

$$J^a{}_y = -\frac{A^{(0)}}{\mu_0}\left(R^a{}_0{}^{20} + R^a{}_1{}^{21} + R^a{}_3{}^{23}\right) \tag{K.42}$$

$$J^a{}_z = -\frac{A^{(0)}}{\mu_0}\left(R^a{}_0{}^{30} + R^a{}_1{}^{31} + R^a{}_2{}^{32}\right). \tag{K.43}$$

From eqns. (K.40) to (K.43) it is seen that **current density originates in sums over Riemann tensor elements.**

This finding has the important consequence that electric current can be generated by spacetime curvature. The relevant Riemann tensor elements are again calculated from the Einstein field equation for gravitation.

Index

A

F

S

Lightning Source UK Ltd.
Milton Keynes UK
UKHW030706280321
381086UK00004B/334